D0782489

Life of Fred

Geometry

Life of Fred

Geometry

Stanley F. Schmidt, Ph.D.

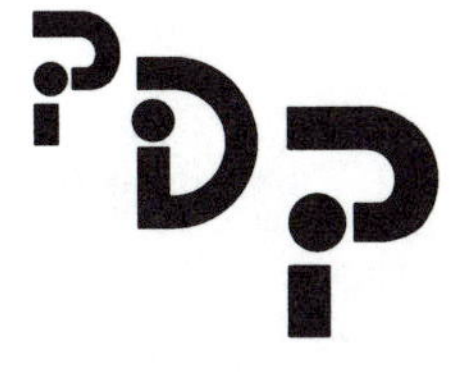

Polka Dot Publishing

ISBN: 0-9709995-4-2
Library of Congress Catalog Number: 2003095414
Printed and bound in the United States of America

Polka Dot Publishing P. O. Box 8458 Reno NV 89507-8458

For full information regarding price and availability of this book, e-mail Polka Dot Publishing at: lifeoffred@yahoo.com

for Goodness' sake

or as J.S. Bach—who was
never noted for his plain
English—often expressed it:

Ad Majorem Dei Gloriam

(to the greater glory of God)

If you happen to spot an error that the author, the publisher, and the printer missed, please let us know with an e-mail to: lifeoffred@yahoo.com

SPECIAL OFFER

As a reward, we'll e-mail back to you a list of all the corrections that readers have reported.

Note to Students

One day in the life of Fred. A Thursday just after his sixth birthday. In those few hours you will experience a full course in geometry. Everything's here: the formulas, the definitions, the theorems, the proofs, and the constructions.

WHAT YOU'LL NEED TO HAVE

You'll need to know a little algebra before you start this Thursday with Fred—not a whole lot. In geometry we will talk about some of the numbers on the real number line.

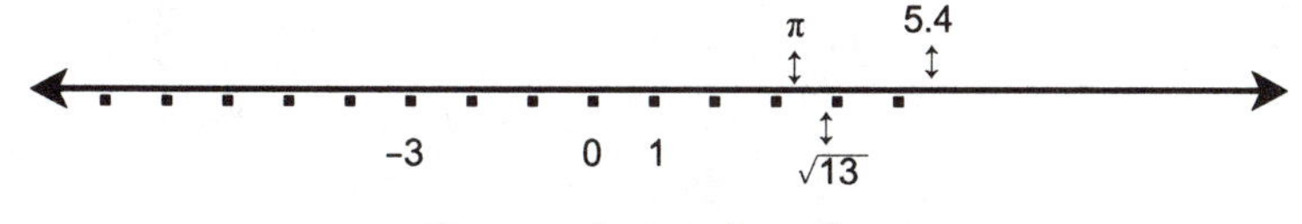

the real number line

On p. 26 we will go from $y + y = z$ to $y = \frac{1}{2} z$.

On p. 78 we will have three equations that look something like:

$$a = b$$
$$a + c = 180$$
$$b + d = 180$$

and then we arrive at $c = d$. If you can figure out on scratch paper how those first three equations can yield $c = d$, your algebra background is probably just fine.

Besides a bit of algebra, you'll need the usual supplies for geometry:

✓ paper and a pencil
✓ a ruler
✓ a compass  (for use in chapter 11 when we do constructions)

✓ maybe a protractor 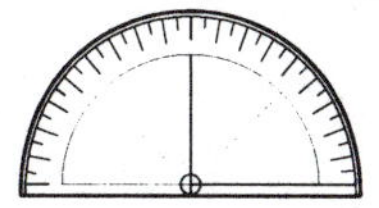 (to measure angles)

✓ maybe your old hand-held calculator with +, –, ×, ÷, and √‾ keys on it.

The paper and pencil and ruler you probably already have. A compass may cost you a dollar or two. If you're superrich and you want to blow another buck for a protractor, that's your choice. You won't really need one for this course.

WHAT YOU'LL NEED TO DO

Throughout this book are sections called *Your Turn to Play*, which are opportunities for you to interact with the geometry. Complete solutions are given for all the problems in the *Your Turn to Play* sections. If you really want a solid grounding in geometry, just reading the problems and the solutions without working them out for yourself really won't work.

At the end of each chapter are six sets of questions, each set named after a city in the United States. You may not have heard of some of the cities such as Elmira or Parkdale, but they all exist. Answers are supplied to half of the questions in the Cities at the end of each chapter.

If you would like to learn geometry, the general rule is easy: Personally work out everything for which a solution is supplied. That's all of the *Your Turn to Play* and half of the Cities questions.

WHAT GEOMETRY OFFERS YOU

In the usual sequence of the study of mathematics—

arithmetic
beginning algebra
Geometry
advanced algebra
trig
calculus—

there is one course that is different from all the rest. In five out of these six courses, the emphasis is on calculating, manipulating and computing answers.

In arithmetic, you find what 6% of $1200 is.

In beginning algebra, you solve $\frac{x^2}{x+2} = \frac{8}{3}$

In advanced algebra, you use logs to find the answer to: *If my waistline grows by 2% each year, how long will it be before my waist is one-third larger than it is now?* [Taken from a Cities problem at the end of chapter three in *Life of Fred: Advanced Algebra*.]

In trigonometry, you find the five different answers to $x^5 = 1$.

In calculus, you find the arc length of the curve $y^2 = 4x^3$ from $x = 0$ to $x = 2$. [From a Cities problem at the end of chapter 17 in *LOF: Calculus*. The answer, in case you're wondering, is $(2/27)(19^{3/2} - 1)$.]

In contrast, in geometry there are proofs to be created. It is much more like solving puzzles than grinding out numerical answers. For example, if you start out with a triangle that has two sides of equal length, you are asked to show that it has two angles that have the same size.

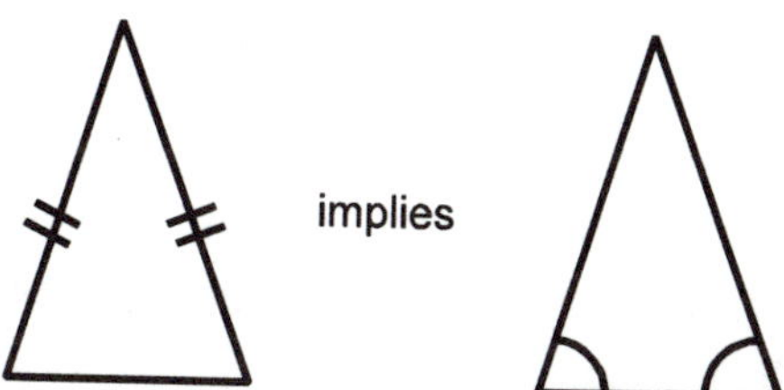

There are at least four different ways to prove that this is true. The proof that you create may be different than someone else's. Things in geometry are much more creative than in the other five courses. It was because of the fun I experienced in geometry that I decided to become a mathematician.

The surprising (and delightful) truth is that geometry is much more representative of mathematics than are arithmetic, beginning algebra, advanced algebra, trig, or calculus. Once you get beyond all the "number stuff" of those five courses, math becomes a playground like geometry. On p. 209 I describe eight (of the many) math courses which follow calculus. All of them have the can-you-show-that spirit that geometry has.

I wish you the best of luck in your adventures in geometry.

A Note to Teachers

This book is perhaps the best teaching assistant you have ever had. You and *Life of Fred: Geometry* can teach the course together. For your students this book will be much more than just a source of homework problems. Open this book at random and you will see why many of your students will actually read this textbook. (Gasp! Shock!) That will make your job significantly easier.

This book has *all* of geometry. You will not have to cover material that isn't in the book. If students miss a lecture they can find what they missed in the textbook. Look over the Table of Contents (and the index) and you will find that it's all there.

In fact, there is more than enough material for most classroom courses. The six "Other Worlds" chapters (which are chapters 5½, 7½, 8½, 11½, 12½ and 13½) may be included or excluded at your discretion. That will give you plenty of flexibility in dealing with the length of the academic calendar over which you have little control.

Some Ideas on Teaching with *LOF: Geometry*

1. Each chapter has several *Your Turn to Play* sections with representative problems and their complete solutions. Many teachers find this the ideal place to begin their discussions of the material.

2. At the end of each chapter are six sets of problems (called Cities). Each City may take your students 20–30 minutes to work through. The first two Cities have all the answers given. The second two have the odd answers supplied in the text. The last pair can be used for tests or quizzes or as lecture material. (Answers are in booklets supplied without cost to instructors.)

Why the word "Cities"? This makes it easier on the instructor who only has to say, "Do San Francisco for homework" rather than the old, "Do every third problem on page 231."

3. It is expected that each student will work though every problem that has an answer supplied—all the *Your Turn to Play* sections and half of the Cities problems.
4. Ask your students to read the material the night before you cover it in class. *The nature of this book makes that kind of assignment possible.* You will benefit since it will make teaching the material much more pleasant.

5. On p. 21 I ask the students to take out a sheet of paper and start keeping a list of the definitions as we encounter them. In #4 on p. 42, I ask them to also keep a page on which they list the theorems as they encounter them, and to keep another page on which they list the symbols and abbreviations. They should also have a fourth page on which they list the postulates. There are two reasons for these requests. First, students will often remember the material better if they write it down. Second, having these lists will make it easier for them to create proofs of the new theorems. If you insist that they keep these four lists, it will make their lives easier.

As a secret between you and me, on p. 516, is the **A.R.T.** section (**A**ll **R**eorganized **T**ogether) in which I list all the theorems, definitions, and postulates in the order they are encountered in this book. Some teachers use this list along with the *Your Turn to Play* and the Cities problems as their lecture notes.

Contents

Fred's two little eyes popped open. It was several minutes before dawn on an early spring morning. He awoke with a smile. There were so many things for which to be grateful. He ticked them off in his mind one by one.

✔ He was home, safe and sound.

✔ It was Thursday—one of his seven favorite days of the week.

✔ He had received a wonderful pet llama at his sixth birthday party last night.

✔ His math teaching position at KITTENS University.

What a wonderful life! was his morning prayer. He stood up and stretched to his full height of thirty-six inches. Fred's home for the last five years or so was his office on the third floor of the math building at the university. By most standards he had been quite young when he first came to KITTENS.

Years ago, when he had arrived at the school, they had assigned him his office. He had never had a room of his own before. He was so tuckered out from all the newness in his life (a new job, a new state to live in, a new home) that he had just closed the door to his office and found a nice cozy spot (under his desk) and had taken a nap.

And every night since then, that is where he slept.

He had the world's shortest commute to work. And no expenses for an apartment or a car.

"Good morning Lambda!" he said to his llama. He had named her Lambda in honor of the Greek letter lambda (λ). He enjoyed the alliteration of "Lambda the llama."

She was busy chewing on the wooden fence that he had erected in his office.

"I just thought you'd like to get out and get some exercise with me," he continued. Fred hopped into his jogging clothes. The pair headed down the two flights of stairs and out into an icy Kansas morning.

λ

Fred was worried that his new pet wouldn't be able to keep up with him as he jogged. He had been jogging for years and a ten-mile run was nothing for him. Fred's fears, however, were unfounded. His six-foot-tall llama had no trouble matching the pace of Fred's little legs.

In fact, after a few minutes, Lambda spotted the new bocci ball lawn and raced ahead to enjoy some breakfast. By the time Fred caught up to her, she had mowed a straight line right across the lawn. Fred's little snacks that he had given her last night had left her hungry.

Oh no! Fred thought. Larry is gonna be mad. He put a lot of effort into that lawn. The international bocci ball tournament is scheduled to be here next week.

"Lambda, please come here. You're not supposed to be on Mr. Wistrom's lawn."

She, being a good llama, obeyed. On her way back to Fred she munched a second line in the grass. Fred loved his pet and didn't know that he was supposed to discipline her. Besides, he thought to himself, those are such nice parallel lines.

Parallel lines

Actually, to be perfectly accurate, those are line segments, he corrected himself. A line segment is just part of a line. It has two endpoints. A line is infinitely long in both directions. Anybody who's studied geometry knows that. Fred enjoyed the precision that mathematical language afforded him.

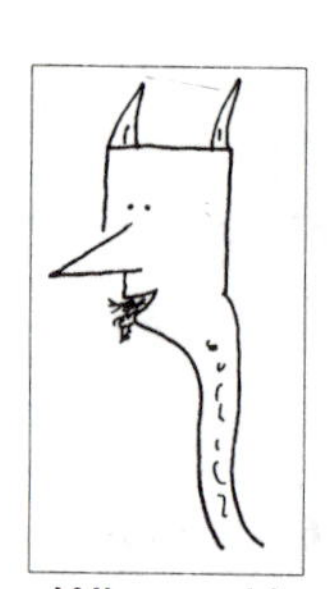

Who could discipline such a lovely creature?

When Fred drew lines on the blackboard in his geometry class, he'd put arrows on ends to indicate that the lines went on forever. He labeled lines with lower case letters.

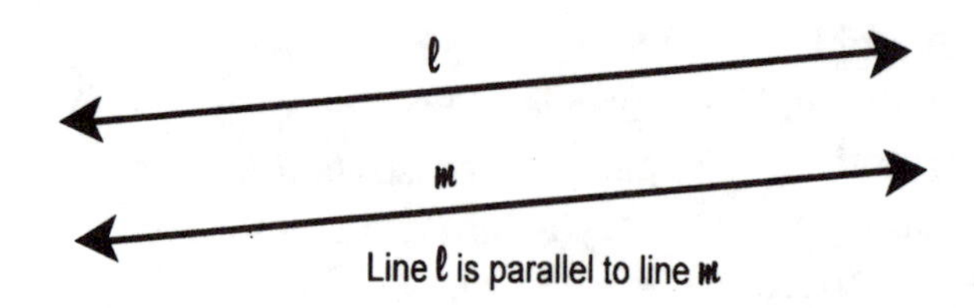

Line ℓ is parallel to line m

To draw a line segment was easy. You didn't need any arrows.

Line segment $\overline{AB}$ with endpoints A and B

Points like A and B are written with capital letters, and lines like ℓ and m are written with lower case letters.

$\overline{AB}$ is the notation for the line segment with endpoints A and B.

$\overleftrightarrow{AB}$ is the notation for the line which contains A and B.

Line $\overleftrightarrow{AB}$ which passes through A and B

And just to make things complete: AB is the *distance* between points A and B. That makes AB a number (like six feet) whereas $\overline{AB}$ and $\overleftrightarrow{AB}$ are geometrical objects (a segment and a line).

Your Turn to Play

1. Line segments (sometimes called segments for short) can come in lots of different lengths. Name a number that *couldn't* be a length of a line segment. (Please try to figure out the answer before you look at the solution furnished below.)

2. If AB = 4 and AC = 4, does that mean that A, B and C all lie on the same line?

3. If points D, E and F are collinear with E between D and F, it would *not* be right to say that $\overline{DE} + \overline{EF} = \overline{DF}$. Why not?

D E F

4. (A tougher question) If GH + HI = GI, must it be true that:

   a) points G, H and I are collinear?

   b) H is between G and I?

The way to figure out the answer to this question is to get out a piece of paper and draw three points so that the distance from G to H plus the distance from H to I is equal to the distance from G to I. Be convinced in your own mind what the answers to questions a) and b) are before you look at the solutions below. You will learn very little by just reading the questions and then glancing at the answers.

COMPLETE SOLUTIONS

1. If you phone a travel agency and ask how far it is between San Francisco and Yosemite National Park, they might say something like 209 miles. If you phone them the next day and ask them how far it is from Yosemite National Park to San Francisco, they would again say 209 miles. The distance between two points is never negative. So the answer to question 1 might be something like -5 or -978267 or $-\pi$ or $-\sqrt{7}$. It could be zero since the distance from San Francisco to San Francisco is zero. In symbols, it's always true that AA = 0.

2. On this map of Kansas, the distance from Garden City to Colby is about the same as the distance from Garden City to Hays. Those three cities are not **collinear**. (Collinear = lie on the same line.)

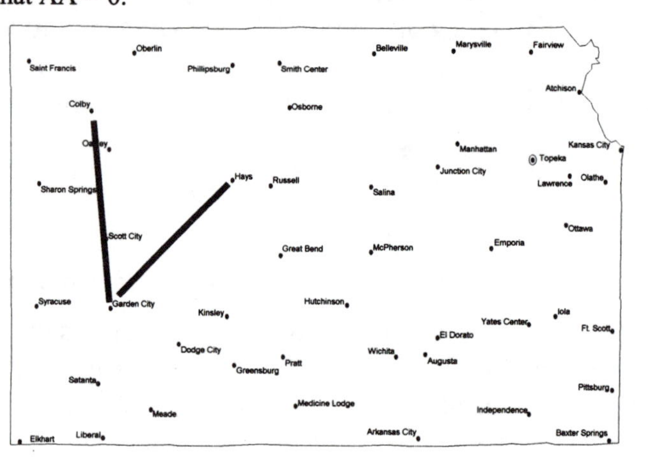

3. It is true that DE + EF = DF (which states that the distance from D to E plus the distance from E to F is equal to the distance from D to F). We can add numbers together. But $\overline{DE}$ isn't a number. It's a line segment which is a geometrical object. The only thing we know how to add are numbers. Could you add Martin Luther and pizza? Could you divide the Red Cross by a stop sign?

Similarly, you can't draw AB since AB is a number (the distance between points A and B).

4. Suppose for a moment that H *weren't* on the line that passes through G and I.

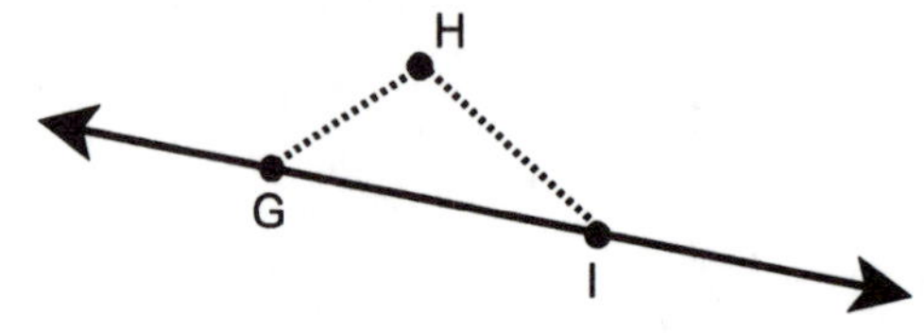

To say that GH + HI = GI is to say that the journey from G to H and then to I is the same distance as the journey from G straight to I. That's nonsense unless H is on the road from G to I. Namely, H is between G and I.

If H isn't between G and I, then the side trip to H always takes longer. If H isn't between G and I, then GH + HI > GI. (> is the symbol from algebra for *greater than.*)

We have arrived at:

<u>Definition 1</u>: H is **between** G and I if and only if GH + HI = GI.

Some notes about definition 1:

♪#1: The word we're defining is in **boldface type**.

♪#2: Every definition in mathematics has the phrase "if and only if" in it. That means two things:

I) If H is between G and I then it's true that GH + HI = GI, and

II) If GH + HI = GI, then H is between G and I.

Either part of the definition implies the other part.

♪#3: Later on in geometry when we're doing a proof and we know, for example, that AB + BC = AC, then we can say, "B is between A and C" and give as a reason, "By the definition of between."

♪#4: Definitions are entirely optional. We could do all of geometry without using a single definition. However, we use definitions to *make our lives easier*. We could always talk about "the geometrical object consisting of three non-collinear points A, B and C and the line segments $\overline{AB}$, $\overline{BC}$ and $\overline{CA}$," but don't you think it is a lot easier if we just say "triangle ABC"?

♪#5: Oops! We just made our second definition in geometry (in the previous note). We will call it <u>Definition 2</u>. It is the definition of triangle ABC (which we can write in symbols as $\triangle ABC$).

♪#6: Since we are introducing symbols, we will write "H is between G and I" as G-H-I.

♪#7: Do you want to learn geometry? To make the whole thing a happy experience, here's a suggestion which has worked for many students over the years. Take out a sheet of paper and head it, "Definitions." There will be lots of definitions in geometry and it will be a lot easier if you copy them down as you encounter them. Then you will have them all in one place for easy reference. So far, your definitions page and symbols page would look like:

Definitions

<u>Definition 1</u>: H is between G and I if and only if GH + HI = GI.

<u>Definition 2</u>: <u>Triangle ABC</u> is defined as non-collinear points A, B and C and the line segments $\overline{AB}$, $\overline{BC}$ and $\overline{CA}$.

List of Symbols

$\overline{AB}$ is the notation for the line segment with endpoints A and B.

Line segment $\overline{AB}$ with endpoints A and B

$\overleftrightarrow{AB}$ is the notation for the line which contains A and B.

Line $\overleftrightarrow{AB}$ which passes through A and B

AB is the *distance between points* A and B.

A-B-C means that B is between A and C.

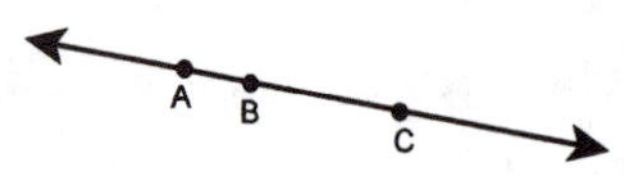

△ABC is the triangle whose vertices are A, B and C.

♪#8: Definitions just keep popping up. Definition 3 is that the **vertices** of △ABC are A, B and C. They are the "corners" of the triangle. The singular of *vertices* is *vertex*. We can say that B is a vertex of △ABC.

Fred and Lambda jogged toward the north end of the campus and away from the bocci ball lawn. Fred carefully selected their route so that his pet wouldn't get into any more trouble. They passed the student parking lot, the tennis courts, the university chapel, the outdoor ice rink (which is only open during the ten months of the year when temperatures are freezing) and finally they got to the two-lane road which borders the campus. Some students, using a bit of irony,* call that the North Freeway.

Fred and Lambda headed east on the Tangent Road (as he calls it).

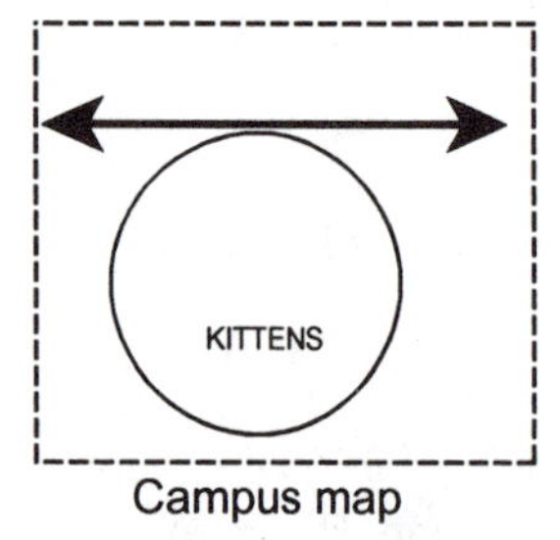

Campus map

As they ran their miles, Fred thought about Larry Wistrom's lawn. Larry is the university gardener and Fred knew that a lot of work went into making that lawn perfect for lawn bowling. Even if no one saw his llama do the damage, Fred knew that it was only right to take responsibility. Larry was often hard to find since he was often away at Landscapers' Conventions. When they got back from their run, he would see Larry's brother, Sam, and would tell him that he would pay for fixing the lawn. Sam Wistrom was the janitor for the math building and every Thursday morning he empties the office wastebaskets. Fred was sure to see him as he made his rounds.

With his conscience clear, Fred looked out at the bright cold scenery. *Today seems like a geometrical day,* Fred thought to himself: *the*

---

* irony = you say one thing, but you mean the opposite of your literal words. If you call Fred a "dumb bunny," everyone knows you're making an ironic statement (and actually complimenting him). It would be like calling Jack LaLanne physically unfit or Jeanette MacDonald a poor singer or Mother Teresa uncaring.

parallel lines that Lambda ate in the grass, the tangent road, the funny square head of my pet, polygons were everywhere (triangles, quadrilaterals, pentagons, hexagons, heptagons, octagons, . . .), the circles within circles of a pepperoni pizza (there was the chance that Fred would go out tonight for pizza with his favorite students, Betty and Alexander), the shape of a slice of pizza (◁), which is called a sector.

STOP

An octagon

He stopped to comb Lambda's hair which had gotten mussed up in the wind. More parallel lines.

Fred noticed that his precious pet was trembling a bit. Maybe it was the c-c-c-cold. Maybe it was the nine miles that they had run. Maybe it was that she was hungry. (In truth, it was all three.)

Fred turned around and headed west on Tangent Road. In another nine miles we'll be back in my office all safe and sound. At least that's what Fred thought.

Five miles later at the point where Tangent Road, Archimedes Lane and Newton Street are concurrent, Lambda slowed to a walk and her trembling increased.

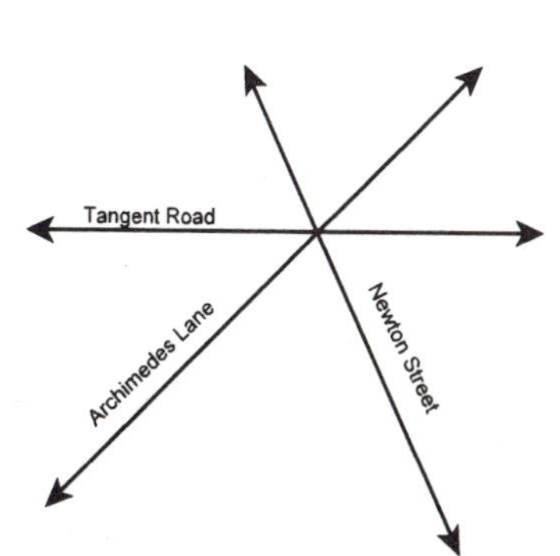

Definition: Two or more lines are **concurrent** if and only if they all share a common point.

What Fred didn't know was that his birthday present pet was in mighty poor physical condition. His friends had obtained the llama through an ad in the local newspaper which read, "Llama (used). Former community pet at Kansas Two-Star Nursing Home. 35¢ or best offer."

Lambda had spent most of her life indoors as a shared pet of all the old folks at Two-Star. There was not much for the residents to do at Two-Star (that's two stars out of a possible five). The television set was broken and no one ever came to visit. The joke among the residents was that the facility should be renamed, "The Finishing School." The only thing to do was play with the llama. A subtle competition broke out among the residents as each one tried to win the special affections of Lambda. When she wandered from room to room, each resident would try to entice her with some special food treat. Most of the time it was chocolate. By the time that Fred received her, she was very overweight and underexercised.

Fred paused at the intersection and pulled a copy of the KITTENS University newspaper from the kiosk. Awfully early for the paper to be out. There must be some special news, Fred muttered to himself. He was shocked to read:

THE KITTEN Caboodle

The Official Campus Newspaper of KITTENS University   Thursday 10¢

Vandals Hit the Campus President Declares, "The Hunt Is On"

Our President

"I'm in great sorrow," he declared in an exclusive Caboodle interview.
(continued on page 16)

KANSAS : For the first time in the seven-year history of this university, vandals have made a pre-dawn attack upon our beloved campus.

Lawrence L. Wistrom, head gardener for the southeast part of the campus, discovered the crime as he was heading to his car to attend the biweekly Landscapers' Convention in Nebraska.

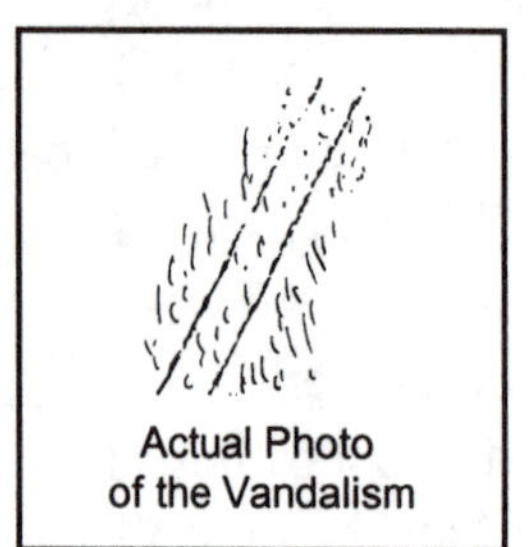

Actual Photo of the Vandalism

Wistrom reported to the university president that the perpetrators had apparently run a four-wheel-drive truck over the new bocci ball lawn.

The entire campus is in great sorrow over this evil act even though most students/staff/faculty/administrators have not yet heard the news because they are still asleep. (But Caboodle has heard from an inside source that they will be sad.)
(continued on page 10)

"Look Lambda!" Fred exclaimed. "Some crooks wrecked the university bocci ball lawn. It must have just happened, since when we saw the lawn this morning is was just perfect." Then Fred realized that maybe it wasn't a truck that cut those parallel lines in the lawn. Fred turned white as a sheet of paper.

Fred looked down the road. He knew: The police are coming! And they're using bloodhounds and Dobermans and pit bulls to track us down! I'm an escaped felon and Lambda is my accomplice. Is this a capital offense? Do they hang people in Kansas or just electrocute them? Fred put up his hands and shouted, "Don't shoot! I give up."

Actually, the group that was coming toward Fred was quite a ways away. If we mark the place where Tangent Road touches the campus as point T, and place Fred at point F and the police at point P, the diagram would look something like:

where the distance from Fred to the police was 1.25 miles and the distance from the police to where the road touches the campus was also 1.25 miles. FP = 1.25 and PT = 1.25.

Your Turn to Play

1. What's wrong with, "P is the midpoint of TF"? (As usual, please make the effort to figure out the answer before you look at the solution given below. Thinking is truly hard work. Why do you imagine that most people avoid it? But it is also very well paid in comparison to being a "work beast" as Jack London* thought of himself in the early part of his life.)

2. What's wrong with the following definition? "P is the midpoint of $\overline{TF}$ if and only if TP = PF."

3. Prove that "If P is the midpoint of $\overline{TF}$, then PF = ½ TF." This may be the first proof you've ever done in your life. After you have seen half a dozen proofs done in geometry, you will start to get the hang of it.

---

* Jack London was the highest-paid American author of about a century ago. He wrote *Call of the Wild* and a bunch of other books.

It is a little like learning to eat spaghetti. It takes a little practice so that you don't get it in your hair or all over the floor. But after you learn, it can be a very pleasurable experience.

1. You can't have a midpoint of a number, and TF is a number. It's the distance between points T and F. What is true is that P is the midpoint of $\overline{TF}$.

2. If you have a line segment $\overline{TF}$, there are lots of points P where TP = PF. Tons of them. And only one of them is a midpoint.

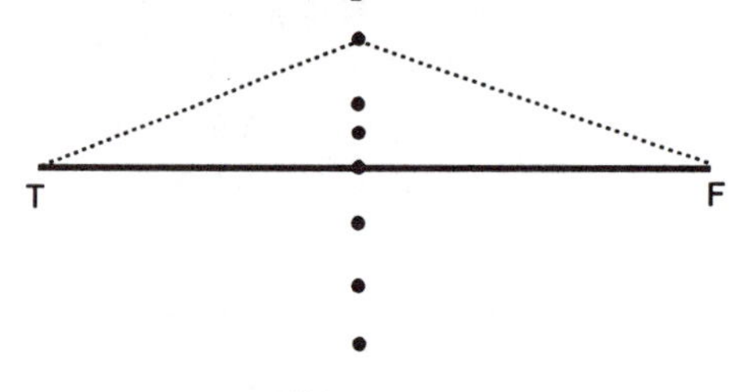

We needed to also say that P is on $\overline{TF}$. The easiest way to do that is to say that P is between T and F (since we've already defined "between"). Here is the official definition: Point M is the **midpoint** of $\overline{AB}$ if and only if AM = MB and A–M–B.

3. To prove: "If P is the midpoint of $\overline{TF}$, then PF = ½ TF."

Proof: If we know that P is the midpoint of $\overline{TF}$, then (by definition of midpoint) we know that TP = PF and we also know that T–P–F. Then (by definition of T–P–F) we know that TP + PF = TF.

These are two algebra equations that we now know are true:

TP = PF and
TP + PF = TF.

If that's hard to look at, then replace TP by x
PF by y and
TF by z

and you'd get $x = y$
and $x + y = z$.

Substitute the first equation into the second and you have $y + y = z$ or PF + PF = TF

and using a little more algebra $y = ½ z$ or PF = ½ TF

which finishes the proof.

This was a proof of the theorem in paragraph form. A **theorem** is a statement that can be/is being/was proved. In contrast, you can't prove a definition. A definition is just an agreement to shorten down some big long phrase into a compact notation. When we write "ΔABC," it's just shorthand for, "the geometrical object consisting of three non-collinear points A, B and C and the line segments $\overline{AB}$, $\overline{BC}$ and $\overline{CA}$."

There is another way to write proofs in addition to the paragraph form. They can be written in Statement–Reason form:

Theorem 1: If P is the midpoint of $\overline{TF}$, then PF = ½ TF.

Proof:

Statement	*Reason*
1. P is the midpoint of $\overline{TF}$	1. Given
2. TP = PF and TP + PF = TF	2. Definition of midpoint
3. PF = ½ TF	3. Algebra

Now I know that this is awfully silly—proving that a midpoint cuts a segment into two pieces which are half the length of the whole segment. You knew that already. But I wanted to start out easy. How would you have felt if your first geometry proof were a bit more complex?

Take, for example, G. A. Wentworth's exercise number 130 in his book *Plane Geometry,* copyright 1899.

Ex. 130. If two straight lines are drawn through any point in a diagonal of a square parallel to the sides of the square, the points where these lines meet the sides lie on the circumference of a circle whose centre is the point of intersection of the diagonals.

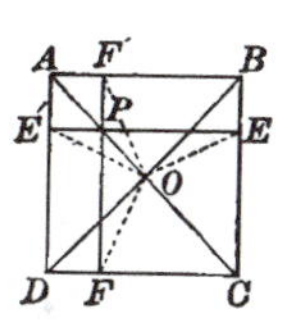

On second thought, don't take it. Let's get back to Fred and his adventures.

Fred's vision at 1.25 miles with his little pointy eyeballs and a brain filled with terror didn't give him a clear picture of what was coming toward him. It wasn't the police with bloodhounds, Dobermans and pit bulls. It was just a fellow taking his puppy out for a walk.

When they got closer, the puppy came up and gave Lambda a lick. She moaned and winced. She was very tired and wasn't used to being around *animals*.

Fred was in a quandary. He didn't know what to do next. Should he head back to his office and put Lambda to bed so that she could rest and recover her strength or should he head directly to the police station and turn himself in?

When Lambda started coughing and spit up some grass, Fred made up his mind. He'd head to his office first.

Newton Street was the most direct way to the math building.

Tangent Road was a tangent to the circle since it intersected the circle in only one point.

Newton Street was a **secant line** since it intersected the circle in two points.

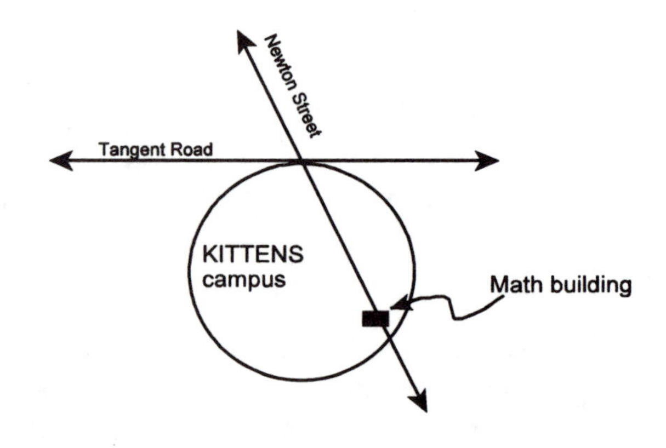

He needed to get from point T (where Tangent Road, Newton Street and the border of the KITTENS campus intersected) to the Math building, which we will call point B.

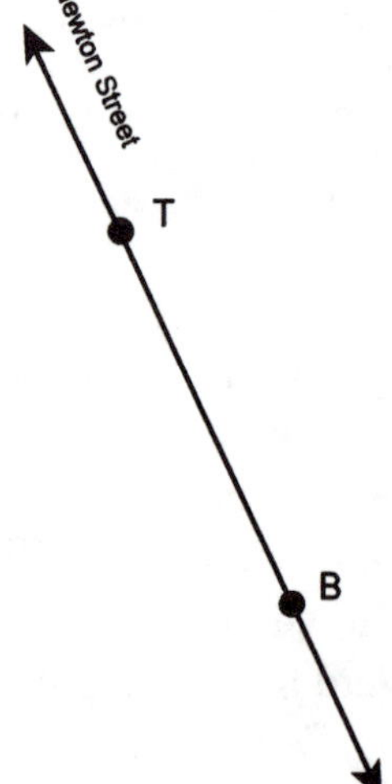

Fred got out his handkerchief and wiped Lambda's mouth. Then they faced southeast and slowly started off toward point B (the Math building). In order to help her walk, Fred put his shoulder under part of the llama's belly as he walked along beside her. He was surprised at how fat she was.

After about a dozen steps, Lambda fell over. There are two ways for a llama to fall over: to the left or to the right. Fred was on the left side of her, and she did not fall to the right.

Fred was trapped underneath her.

The man with the puppy rushed over to them and extricated the 37 pounds of mashed Fred. Fred's rib, the one that had been broken last Sunday night (in *Life of Fred: Beginning Algebra*), was starting to hurt again.

Fred thanked the man. (He had read several of the *Miss Manners* books.)

Fred was grateful that Newton Street was a line. (In geometry when we say *line*, we always mean a *straight* line.) A line would be the shortest way from point T to point B. There were some walking paths on the KITTENS campus that weren't lines. On summer evenings Darlene and Joe (whom you will hear about later) liked to stroll on those winding curves through the campus. It was more romantic.

The path that Darlene and Joe liked to take

As Fred and Lambda inched their way along, Fred tried to take his mind off his worries and pains by thinking about math. Mathematics, especially geometry, was his best anodyne. (ANN-eh-dine. That which relieves distress or discomfort.)

And the best part of geometry was finding ways of showing that things were true. Demonstrations. Proofs. Deductive reasoning. Proving theorems. These were fun because they were creative acts. He really didn't much like memorizing formulas. That always seemed to him like using his

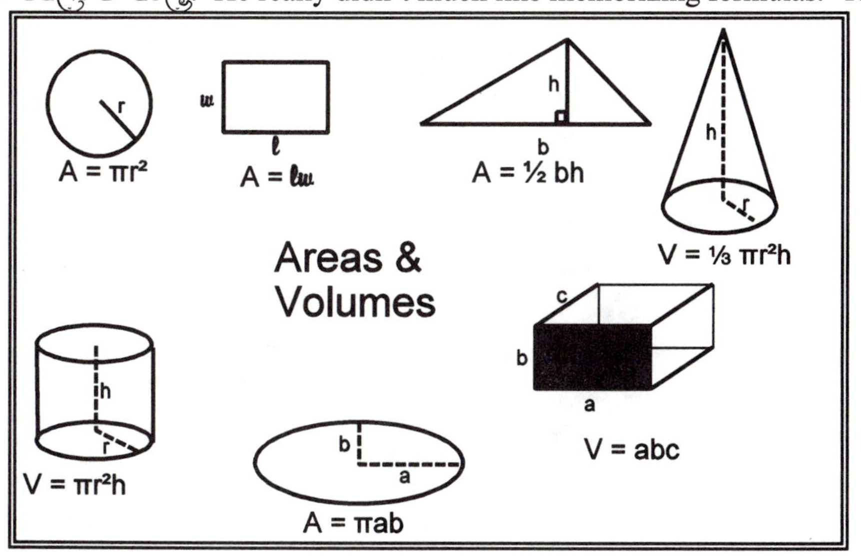

mind as a tape recorder. What a waste. Fred had never sat down and memorized that the volume of a cone is $\frac{1}{3}\pi r^2h$. He knew the formula because he had used it so often. That was the natural (and painless) way to memorize things.

Now as Fred walked on the right side of his llama (hoping that if she fell over again, she would continue to fall over on her left side), he thought of how he might prove that the line through points T and B was straight.

He couldn't think of a way.

What about using a definition? he thought to himself. *What if I define a line as that which is straight? But wait! Then I gotta define what straight means. I could define "straight" as "like a line." That won't work. That's a circular definition.*

A couple of years ago Fred was reading a Jane Austen novel and ran across a passage where Mr. X was *unreticent*. Not knowing what *unreticent* meant, he looked it up in the dictionary that he always kept nearby when he was reading. The dictionary defined *unreticent* as *brusque*. That wasn't a whole lot of help since he didn't know what *brusque* meant. So he looked up *brusque* and found that it meant *forthright*. He didn't know what that meant. All he could imagine was that *forthright* might have something to do with being the 4th one on the right hand side (but that's not correct). So he looked up *forthright* and you know what he found. *Forthright* was defined as *unreticent*.

He was trapped in a circular definition. (Later he found out that those three words have to do with being outspoken, unreserved, and bold.)

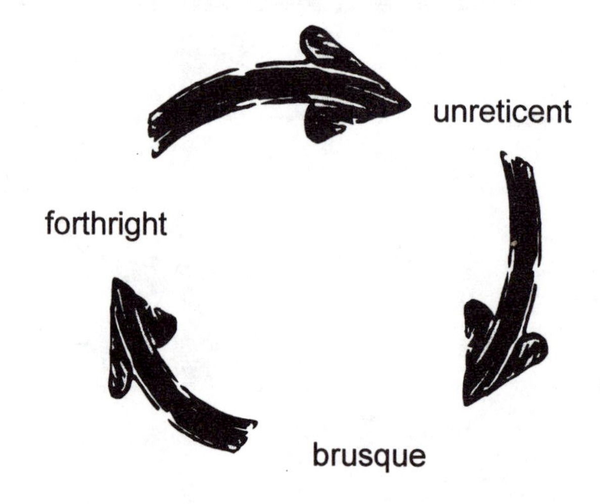

In order to avoid circular definitions, in every mathematical structure that we build (such as geometry, algebra, logic, or set theory), we have to start with some words that we never define.

The **undefined terms** for geometry are point, line, and plane. We can informally chat about what we are thinking about when we mention a point or line or plane, but we will never define them.

If we were to define a plane as something with two dimensions and a line as that which has only one dimension and a point as something with zero dimensions, then *dimension* would have to be an undefined term and that's a lot harder to think about than points, lines and planes.

When Fred thinks about a plane, he imagines a flat sheet of very thin paper that stretches out to infinity. (Virtually everything in this book will happen on a single plane. Except for chapter 12, this book could have been entitled *Life of Fred: Plane Geometry.**)

For Fred, lines are straight and very, very thin and go on forever in both directions.

Thinking about points on a line that always seems to give students the most trouble. They are *not* like little beads on a wire. Points are not like bowling balls in a gutter.

Points are finer than talcum powder. You can always find another point between any two points on a line. No two points are "right next to each other."

That is also true with numbers. You will never find two different numbers that are right next to each other. For example, between 2.6 and 2.7 is the number 2.65. All you need to do to find a number between two given numbers is average them—add them and divide by two.

---

* In chapter 12 we will deal with some solid geometry and talk about cylinders, spheres, pyramids, and cones.

A student once challenged this in Fred's classroom. He said he could name two different numbers that no one could find a number between. That sounded exciting. He came to the blackboard and wrote down his two numbers:

2.99999999999999999. . . and
3.

"Those numbers are pretty close to each other," Fred admitted. He smiled. "In fact," Fred continued, "they are equal to each other.

"Let x equal 2.999999999999999999999999. . . .
If x were also equal to 3, that would prove they were the same number.

"If x = 2.9999999999999999999999. . . then
10x = 29.99999999999999999999999999. . .
since to multiply by ten, all you need to do is move the decimal place over one to the right."

He rewrote the two equations, putting the 10x equation on top: 10x =29.9999999999999999999. . .
x = 2.9999999999999999999. . .
and subtracted one equation from the other:
9x = 27.000000000000000000000. . .

Then divided both sides by 9 and obtained x = 3.

Even if you imagine points as atoms, or as dots on atoms, those points would be too "fat." Here's a real picture of a geometrical point magnified one hundred zillion times:

Genuine picture of a real point

If you can't see it, that's okay. Nobody else can either. No amount of magnification will make a point (or a line) visible.

"Aha!" one of his students exclaimed. "Then they really don't exist. There's no such thing as a point or a line. I'll never go out in the street and find a point just sitting there."

"You're right that you'll never see a point or a line," Fred answered. "I've never seen one either. But, come to think of it, I've never seen the North Pole or my liver."

"I'm not just talking about eyeballing something," his student said. "Points are just plain not tangible. You can't touch them or weigh them."

"That's true," Fred admitted. "They aren't *out there.* The only place you'll ever find a point is in your thoughts." He drew a picture of himself on the blackboard:

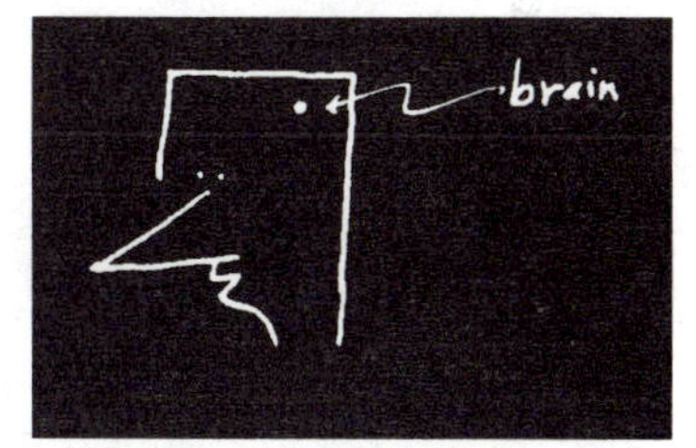

Where a point can be located

Another student joined in the discussion. "So this whole geometry thing is just in my head?"

"Yes," Fred said. "But that's the most important place in our galaxy. It's the only place that you can find love . . . or anything else that is truly important in life." The classroom became very quiet for a moment.

Meanwhile, Fred was trying to think of a way of expressing the "straightness" of a line through points T and B. No definition would work. He thought to himself, *If lines are going to be straight, then there can only be one line that passes through T and B.* He wrote in his notebook:

Postulate 1: One and only one line can be drawn through any two points.

A **postulate** is a statement that is assumed (rather than proven). Postulates are the last item that we need in order to build a mathematical structure for geometry (or algebra or any other part of mathematics). The postulates will be our beginning assumptions which we need to get started in building the structure (which is called a **theory**.) If we all agree that, "One and only one line can be drawn through any two points," then together we can explore what must be true because of that assumption. Our theorems (which are statements that are proven instead of assumed) will follow from our postulates, undefined terms, and definitions.

Here's what a theory looks like:

Sometimes when mathematicians illustrate what a theory looks like they draw buildings (where the foundation is postulates and the bricks are definitions, etc.)

Sometimes they think of creating a mathematical structure as a pizza where the crust is the undefined terms and the toppings are the theorems.

In any event, our lines are now straight since Postulate 1 declares that only one line will pass through any two points.

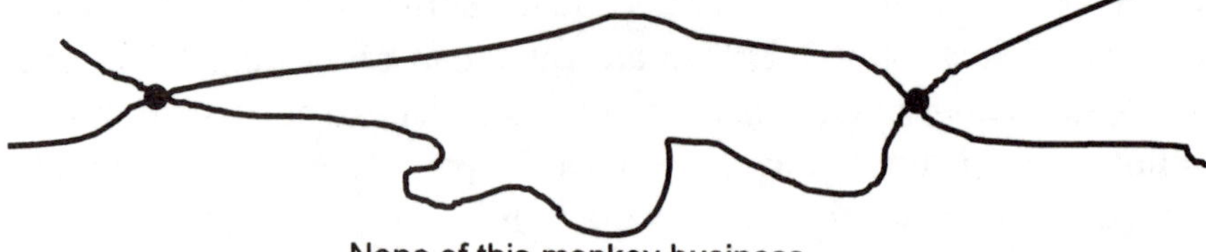

None of this monkey business

Sometimes Postulate 1 is expressed as, "Exactly one line passes through any two points." Perhaps the most traditional way it is expressed is, "Two points determine a line."

Now, if we are given two points, A and B, we can say that $\overleftrightarrow{AB}$ is the unique line containing A and B. The reason will be Postulate 1.

Your Turn to Play

1. A student once claimed that he could draw two straight lines through two points:

Where did he go wrong?

2. Suppose you were trying to prove the statement, "If ΔABC, then it is not the case that A–B–C." What would you put in place of the "?" in the following proof?

Statement	***Reason***
1. ?	1. Given

3. What will be the last line in the proof for question 2?

4. What does the symbolism "A–B–C" stand for?

5. If ΔABC is the first statement of the proof, then the second statement of the proof will be, "A, B and C are not collinear." What reason we will cite?

6. Find a number between 0.333 and 1/3.

7. (For English majors) Express Postulate 1 in *If* ••• *then* ••• form.

8. What is the hypothesis of Postulate 1?

9. A) Does Postulate 1 say that two points exist?

   B) Does Postulate 1 say that if a line exists, then it must have two points on it?

   C) What if, by some magic, you knew that no lines exist. There aren't any at all. What could you say about the existence of points on the basis of Postulate 1?

10. A) Give an *If* ••• *then* ••• statement which is true, but whose converse is false.

    B) Give an *If* ••• *then* ••• statement which is true, and whose converse is also true.

COMPLETE SOLUTIONS

1. When we draw points and lines on paper, we are not drawing real points and lines. The tiniest pencil point that you could draw still has zillions of carbon atoms in it. Our drawings are just to help us imagine the ideal geometrical objects which only exist in our imagination.

2. Whenever we have a theorem to prove which is in the form: *If ••• then •••*, the given is what follows the "*If.*" This is called the **hypothesis** of the statement. The part that follows the "*then*" is called the **conclusion**. That's the thing we are trying to prove. In this proof, the hypothesis is "$\triangle$ABC " which is statement #1.

3. "It is not the case that A–B–C." This is the conclusion of the theorem.

4. It means that B is between A and C.

5. There are only three possible reasons that we have used so far in doing proofs. They are:

① Given, or ② Definition, or ③ Algebra.

In the present situation, if statement 1 reads $\triangle$ABC, the only thing we could do with that is invoke the definition of triangle which states that $\triangle$ABC consists of non-collinear points A, B and C together with line segments $\overline{AB}$, $\overline{BC}$ and $\overline{CA}$. So the reason for statement 2 will be "Definition of a triangle."

6. 1/3 is equal to 0.3333333333333333333333333333. . . so some possible numbers between 1/3 and 0.333 are 0.333333337 or 0.33333 or 0.3333333339202.

7. If we are given two points, then there exists exactly one line that contains them.

8. The hypothesis is, "We are given two points."

9. A) Postulate 1 states that *if* we have two points, then there has to be a line that contains them. It doesn't say that two points exist. It just says, "If." On a date once, many years ago, a woman said to me, "If you had a million bucks in your checking account, then you should buy me a new car.". That didn't mean that I did have $1,000,000 in my checking account. (All it did mean is that I might have been dating the wrong person.)

B) No. That's the converse of Postulate 1. (A **converse** of an *If ••• then •••* statement interchanges the *If* part with the *then* part.) A converse of a statement is not necessarily true.

This statement is true: "If you weigh more than 350 pounds, then you will be dead within 135 years."

It's converse is false: "If you will be dead within 135 years, then you weigh more than 350 pounds." Lots of skinny people die. (In fact, 100% of them.)

The converse of *If* P, *then* Q is *If* Q, *then* P. When you study logic, this will be written with symbols:

The converse of P ➜ Q is Q ➜ P. ("➜" may be read "implies.")

C) Postulate 1 states: Two points ➔ there exists a unique line containing them.

If there were no lines at all, then there couldn't be two points. That's true since the minute there are two points, then there has to be a line. There might be just one point, or no points at all, but no more than that.

What we're dealing with here is a statement and its contrapositive. (The **contrapositive** of an *If* ••• *then* ••• statement interchanges the *If* part with the *then* part AND negates both parts.)

The contrapositive of *If* we have two points, *then* there is a line that contains them, is *If* there is no line, *then* we don't have two points.

In logic this is symbolized: the contrapositive of P ➔ Q is not-Q ➔ not-P.

A statement and its contrapositive are logically equivalent: they are both true or they are both false.

The statement: *If the country of Freedonia abolishes income taxes, then the citizens will prosper* is logically equivalent to the contrapositive: *If the citizens of Freedonia don't prosper, then income taxes were not abolished.*

If an *If* ••• *then* ••• statement is true, its converse need not be true, but its contrapositive must be true.

10. A) If you've ever gone jogging for more than ten miles with an overweight llama, then you know what trouble is. Its converse isn't true. If you know what trouble is, that doesn't mean you've gone jogging for more than ten miles with an overweight llama. There are other ways to experience trouble.

B) If a triangle has two sides which are equal in length, then it has two angles that have the same size. This is true.

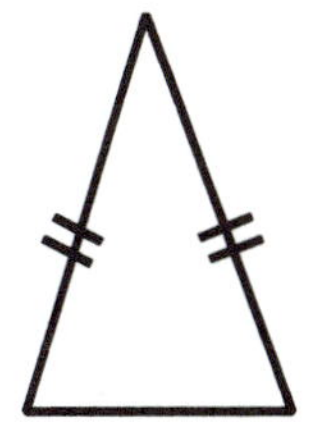

implies

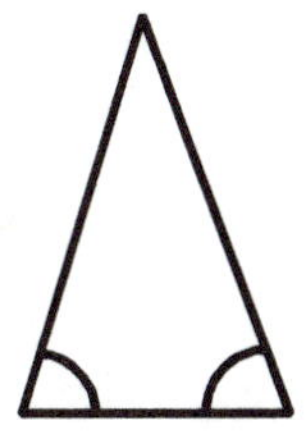

Its converse (If a triangle has two angles that have the same size, then it has two sides that are equal in length) is also true:

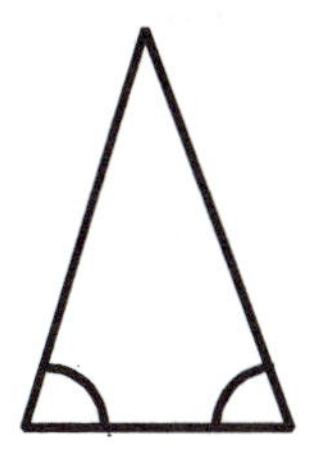

implies

All these thoughts that Fred had—

Of how much more fun proofs are than memorizing formulas;

Of circular definitions and how to avoid them using undefined terms;

Of how two points are never right next to each other;

Of the converse of P → Q being Q → P;

Of the contrapositive of P → Q being not-Q → not-P;

Of how to make lines straight using *Two points determine a line*;

Of how theories are built from undefined terms, postulates, definitions and theorems

—spun around in Fred's head as he pushed, pulled and dragged Lambda back to the Math building. They took the freight elevator rather than tackle the two flights of stairs. When the doors opened on the third floor, they found themselves right in front of the vending machines in the hallway.

"Are you hungry?" he asked his pet.

She nodded and Fred pumped a lot of quarters into the machines. As fast as the machines could vend their products, Lambda scarfed up:

and then washed it all down with a large (64-oz.) container of extra-sugar Sluice. In short, she was enjoying your basic American breakfast.

After a short trip down the hall to their office/home in room 314, Lambda collapsed in her pen. Fred knew it was time to get to the police station and make restitution for the damage that his llama had done to the university bocci ball lawn. There was nothing more he could do for his darling right now, so he gave her a little pat on the head, grabbed his checkbook, and headed out the door.

He closed the door quietly since she already seemed to be asleep.

Sam, the janitor, was in the hallway with a shovel and a garbage can working in front of the vending machines. He didn't look very happy. He asked Fred, "Would you look at this! What kind of practical joke is this? Did you see any students up here with a horse or a really big dog?"

"Can't say that I have," Fred answered. He knew that it wasn't Lambda that had made the mess since he saw her eat every last scrap of all the snack foods that he had purchased for her.

Alas, Fred was woefully ignorant of—how shall we say it?—the facts of life when it came to having a pet. All he knew about was feeding his pet, combing her hair, and saying nice things to her.

"And please don't make too much noise," Fred added. "She's sleeping in my office and we don't want to wake her."

As Fred headed down the stairs, Sam scratched his head and wondered to himself, "If that don't beat all. He's got a girl in his office and since it's only 6 A.M., she's probably been there overnight. But I know that Fred isn't that kind of guy. I see him every Sunday in church and know he's a believer. Heck, since he's only six years old, he probably isn't foolin' around with no female anyway."

There were lots of things that Fred might have been thinking about on his six-minute walk to the campus police station. He could have been thinking about what he was going to say to the police. But he wasn't, since he knew that he would just have to pay to fix up the lawn.

He could have been thinking about his darling little pet. But he wasn't, since he knew that he had done every nice thing that he could think of for her. He had exercised her, combed her hair, fed her breakfast, and tucked her in her pen.

Instead, he was worried about his geometry. *There are no points in my geometry!* he thought to himself. *This has got to be remedied. I want points, lots of points. As many points as there are real numbers.* He thought about when he had taught beginning algebra and had drawn the real number line on the blackboard.

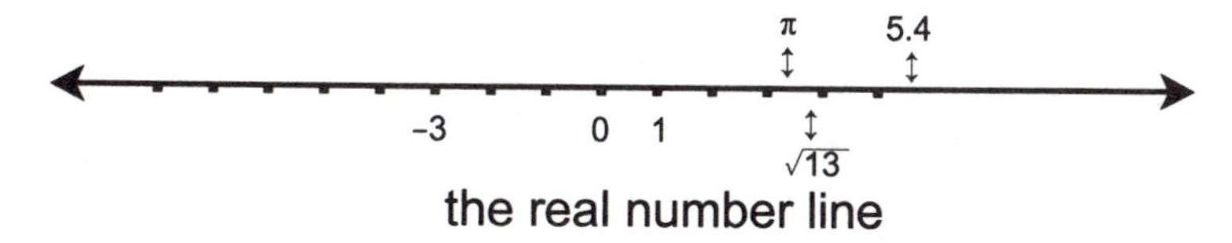

the real number line

Fred took out a notebook, which he carried in his shirt pocket, and wrote out: Postulate 2: You can match up every real number with a point on the line.

He rephrased it: Postulate 2: There is a one-to-one correspondence between the points on a line and the real numbers so that every point matches up with a single real number and every real number matches up with a single point.

He tried a third time with: Postulate 2: The points on a line are like men and the real numbers are like women and everyone's married. (And no polygamy.)

He was so engrossed in trying to get the wording right for his second postulate that he walked right by the sign that was posted everywhere on campus:

Police artist rendering of eyewitness description

A reward of $4.39 is being offered for the capture of these felons who viciously tore up the new bocci ball lawn on campus.

It is amazing to see how eyewitnesses can be mistaken. The witness had reported that he had seen two of them on the lawn. He testified: One of them looked more like an animal than a human, but since animals aren't allowed on campus, it must have been some kind of ugly person. The hairy one was called "Darling" by the short one. The hair must have been on the guy and not the gal named Darling. There must have been a car since there were tire tracks, but I can't say what kind since it was pretty dark outside at that time of morning. He had shouted to her something like, "Hey! Get your bod over here right now," so he must have been rough and tough with muscles and some tattoos. That's all I can remember.

In the police spotlight

Police cars were patrolling all over the campus. As they passed Fred they shined their spotlight on him and then one occupant of the car said to the other, "Nah. That ain't our perp. He's only a kid with a stupid bow tie."

He still couldn't get the wording the way he wanted it for his second postulate. He tried using the language of functions and wrote, Postulate 2: There exists a function whose domain is the points on a line and whose codomain is the real numbers such that no two points on the line are mapped to the same real number and every element in the codomain is the image of some point in the domain.

But this was far too wordy. He decided to stick with calling Postulate 2 the **Ruler Postulate** and staying with the idea of matching up the points on a line with the real numbers. The only thing Fred wished was that rulers had both positive and negative numbers. But even with that shortcoming, he thought that calling it the Ruler Postulate was a lot more elegant than calling it the Points-And-Numbers-Match-Up Postulate.

As he walked to the campus police station, he now had a little bounce in his step. There were now genuine distinct points in his geometry. Lots of them.

He wrote down a little theorem in his notebook:

Theorem 2: On any line ℓ, there are five different points on ℓ.

"Five" was purely an arbitrary number. Fred could have chosen six or 98620955. He was just proving this theorem for fun to make sure that the Ruler Postulate was working the way he hoped it would.

Here's the proof he wrote down:

Statement	*Reason*
1. We have a line ℓ.	1. Given
2. 4, 5, π, $\sqrt{7}$, and –0.391 are distinct numbers.	2. Algebra
3. There are distinct points A, B, C, D and E on line ℓ that correspond to 4, 5, π, $\sqrt{7}$, and –0.391.	3. The Ruler Postulate ("There is a one-to-one correspondence between the points on a line and the real numbers so that different numbers match up with different points.")

Here are some notes about Fred's proof:

♪#1: If Theorem 2 were written in *If ••• then •••* form, it would read: "If we have a line ℓ, then there are five different points on ℓ. The hypothesis of the theorem is, "If we have a line ℓ."

♪#2: In order to avoid the long expression, "the number 4 is the number which corresponds to the point A by the Ruler Postulate," we'll make the following definition:

Definition: The **coordinate** of a point is the number which corresponds to that point according to the Ruler Postulate.

♪#3: We'll keep numbering the postulates in this book, but we won't continue to number the theorems and the definitions. When we refer to a theorem or a definition, it is common practice to not write out the entire statement, "The **coordinate** of a point is the number which corresponds to that point according to the Ruler Postulate," but rather to shorten it to something like, "Def of coordinate."

♪#4: If this book had been written in the 1920s, you might now encounter the sentence, "Here's a *swell* idea." If it were the 1950s, "Here's a *nifty* thought." In the 1960s: "Let me lay this cool cogitation on your brainpan." In some parts of the world in the 1930s, it would have been expressed as: "You will now take out a sheet of paper and head it Theorems and you will keep track of each theorem that is presented in this book by writing it down on your theorem page. Or else!" Nowadays: "Pretty please. It will make your life easier if you keep a notebook with the definitions written out in one section, the symbols written out in another section and the theorems in a third section."

To help you get started, here's how your theorems page might look:

Theorems

Thm 1: If P is the midpoint of $\overline{TF}$, then PF = ½ TF.

Thm 2: On any line ℓ, there are five different points on ℓ.

♪#5: Of course, you wrote down the definition of *coordinate* on your definitions page. (I hope. I hope.) Besides keeping all the definitions in one place so they are easy to refer to, the actual act of writing them down helps many people remember them.

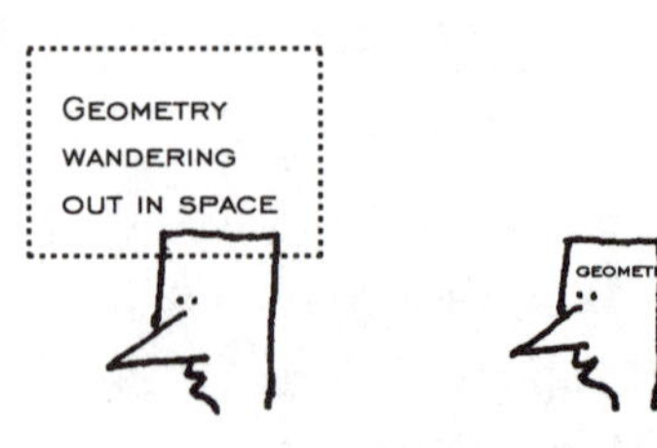

Note how writing down the definitions and theorems inserts geometry into one's mind

♪#6: When we use "Algebra" as a reason in a proof, this can be used to justify anything that you learned in algebra.

Many other geometry books make the whole process a lot harder by demanding that the writer of a proof give the super-specific reason for any piece of algebra that's performed. Here are some examples of some of the agony that they put their readers through:

Statement	*Reason*
1. $6x + 2y = 2y + 6x$	1. commutative law of addition

Statement	*Reason*
1. $(3 + x) + 7p = 3 + (x + 7p)$	1. associative law of addition

Statement	*Reason*
1. $w = 888r$	1.
2. $888r = \pi$	2.
3. $w = \pi$	3. transitive law of equality (Also known as: quantities that are equal to the same quantity are equal to each other.)

Statement	*Reason*
1. $wz = zw$	1. commutative law of multiplication

Statement	*Reason*
1. $(ab)c = a(bc)$	1. associative law of multiplication

Statement	*Reason*
1. $45(x + y) = 45x + 45y$	1. distributive property

Statement	Reason
1. $2 - 3x = z + 8x$	1.
2. $3x = 3x$	2. reflexive property of equality
3. $2 = z + 11x$	3. If equals are added to equals, the results are equal. (Adding lines 1 and 2 together.)

Statement	Reason
1. $\pi + 6x = 10x$	1.
2. $6x = 6x$	2. reflexive property of equality
3. $\pi = 4x$	3. If equals are subtracted from equals, the results are equal. (Subtracting line 2 from line 1.)

Statement	Reason
1. $8w = 24$	1.
2. $8 = 8$	2. reflexive property of equality
3. $w = 3$	3. If equals are divided by equals, the results are equal. (Dividing line 1 by line 2.)

Statement	Reason
1. $3.87z = 4$	1.
2. $100 = 100$	2. reflexive property of equality
3. $387z = 400$	3. If equals are multiplied by equals, the results are equal.

Statement	Reason
1. $x = z^3$	1.
2. $z^3 = x$	2. symmetric property of equality (If $a = b$, then $b = a$.)

Statement	*Reason*
1. $2x^2 + 18x + 3 = 0$	1.
2. $x = \frac{-18 \pm \sqrt{324 - 4(2)(3)}}{2(2)}$	2. the quadratic formula

Now, if you'd like to use any or all of these super-specific reasons in your geometry proofs, feel free to do so. (I will use "Algebra" to cover all of them.)

One of the chief thrusts of this book is to have fun doing puzzles we call geometry proofs. We are building a geometry theory based on geometrical postulates and not algebra postulates. Later on in your math courses (after calculus) you will encounter an algebra theory in which the postulates will be the commutative, associative, and distributive laws.

The reason that geometry is selected as the first theory that you encounter in your math career is that we can use diagrams and that makes it a lot easier to see what is going on. A couple of chapters from now when we prove one of the top ten theorems of geometry:

implies

you will be happy that we are permitted to make drawings.

It's traveling time now. We will have the opportunity to visit six cities in the next several pages. The first two cities have all the answers listed. The second two have answers for all the odd-numbered questions. Everyone interested in mastering geometry will tackle all the questions that have the answers listed.

In addition, your instructor may assign all of the third or fourth cities as homework. Some instructors occasionally use the last two cities for a quiz or test or as lecture material.

Elmira

1. If A–B–C and B–C–D, must it be true that A, B, C and D are collinear?

2. What is the hypothesis of the theorem: *In △ABC, if AC = BC, then the measure of angle A equals the measure of angle B.*

3. State the converse of the theorem in question two.

4. State the contrapositive of the theorem in question two.

5. We defined △ABC, but we never got around to defining the line segment $\overline{AB}$. Fill in the blank:

Definition: **Line segment** $\overline{AB}$ is the points A and B together with all points C such that __?__ . (Please put this definition on your Definitions page that you are keeping.)

6. Fill in the blank: R is a __?__ of △PQR.

answers

1. Yes. Drawing a picture might help you to see this.

2. AC = BC (or more fully, in △ABC we have AC = BC).

3. In △ABC, if the measure of angle A equals the measure of angle B, then AC = BC.

4. In △ABC, if the measure of angle A does not equal the measure of angle B, then AC ≠ BC. (≠ means *does not equal.*)

5. . . . such that C is between A and B. Or you could have written . . . such that A–C–B.

6. vertex

Parkdale

1. If the coordinate of D is 7 and the coordinate of E is 9 and E is the midpoint of $\overline{DF}$, what can you say about the coordinate of F?

2. Supply the reasons for the proof of the following proposition (Note: a **proposition** is a lightweight or inconsequential theorem.)

Proposition: If A, B, C and D are on ℓ, and if C, D, E and F are on m, then A is on m.

S	R
1. A, B, C and D are on ℓ, and C, D, E and F are on m.	*1.* ?
2. ℓ and m are the same line (and since A is on ℓ, it must be on m.)	*2.* ?

3. Supply the missing statements and/or reasons in the proof of the following proposition.

Proposition: If AC – AB = BC, then A–B–C.

S	R
1. ?	*1.* Given
2. AB + BC = AC	*2.* ?
3. A–B–C	*3.* ?

4. If ℓ and m are concurrent and m and n are concurrent, must ℓ, m, and n be concurrent?

5. If ℓ, m and n are distinct lines that are concurrent and m, n and o are distinct lines that are concurrent, must ℓ, m, n and o be concurrent?

answers

1. The distance from D to E must be the same as the distance from E to F, so the coordinate of F must be 11.

2. Reason 1 is "Given." Reason 2 is "Postulate 1 (One and only one line can be drawn through any two points.)"

3. *1.* AC − AB = BC
   2. Algebra
   3. Def of A–B–C

4. No, there wouldn't have to be. They could be: 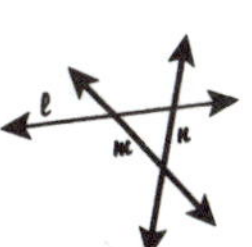

5. In this case they would have to be all concurrent. For fun, here's a proof.

S	*R*
1. ℓ, m, and n are concurrent and m, n and o are concurrent	*1.* Given
2. ℓ, m and n share a point in common. Call it P. m, n and o share a point in common. Call it Q.	*2.* Def of concurrent lines
3. P and Q are the same point (since P is on m and n and Q is also on m and n.	*3.* Postulate one: Only one line can can be drawn through two points.

(The postulate states that *If* you have two points, *then* you can have at most one line through them. The contrapositive of that statement is: *If* you have more than one line containing the point(s), *then* there can't be more than one point.)

4. ℓ, m, n, and o are concurrent.	*4.* Def of concurrent (P is on ℓ, m, n and o.)

Walker

1. AB is the distance between point A and point B. If a is the coordinate of A, and b is the coordinate of B, why isn't b − a equal to AB?

2. If the coordinate of A is a and the coordinate of B is b, and if B is the midpoint of $\overline{AC}$, what is the coordinate of C in terms of a and b?

As you continue in your math career, the questions you are asked to solve will begin to change their character. In elementary school arithmetic problems, you were given 400 questions like 2 + 3 = ?, 4 + 7 = ?, etc. which you could answer as fast as you could write. 400 questions took 400 seconds to answer. Now that you are in geometry the questions may actually take some thought (Heavens!) in order to answer them. When you were a kid, the emphasis was on learning facts: *What's the letter that follows β (beta) in the Greek alphabet?* (answer: γ which is gamma. This is a common question in elementary schools in Greece.) Now the emphasis is on thinking, not memorizing. When I taught college mathematics, all the tests I gave were open-book tests so the students didn't have to memorize anything—but did have to understand a lot.

So in answering this question of finding the coordinate of C in terms of a and b, you shouldn't expect to just answer it off the top of your head. Instead, you might draw a diagram. You might try out various numbers. For example, you might say that a was 7 and that b was 9 and then figure out what the coordinate of C would have to be in this particular case.

3. If P–Q–R and the coordinate of P is 9, could the coordinate of Q be negative?

4. What word having to do with triangles has an "x" in it, but its plural does not have an "x"? (If you are not an English major, you are permitted to read the following hint: *This word with the "x" in it deals with the "corners" of a triangle.*) English majors may continue reading at this point. I have nothing against English majors. One of my daughters majored in English in college. It's just that EMs often enjoy playing word games and I didn't want to spoil their fun by giving them too many hints. EMs that you know might enjoy reading *Joy of Lex.* (That's a real book.)

5. Give an example of an *If ••• then •••* statement that is false. Must its contrapositive also be false?

odd answers

1. Suppose that a were equal to 6 and b were equal to 4. Then b – a would equal –2. In the *Your Turn to Play* in question 1 on p. 20, we wrote that the distance from San Francisco to Yosemite is the same as from Yosemite to San Francisco. Distances are never negative.

3. Yes, this is possible. The coordinate of P is 9. The coordinate of Q might be –50. Since we know that P–Q–R, if the coordinate of Q were –50, then the coordinate of R would have to be –109.

5. Your answer may vary from mine. Here is an *If ••• then •••* statement that is false:

If the sun rises in the east, then the moon is made out of green cheese.

The contrapositive is:

If the moon isn't made out of green cheese, then the sun doesn't rise in the east. This is also a false statement.

An if-then statement and its contrapositive must either both be true or both be false. They are said to be **logically equivalent**.

Danbury

1. If the coordinate of point A is a, and the coordinate of point B is b, then we define AB as the absolute value of the difference between a and b. Translate this English into algebra.

2. Give an example of an if-then sentence which is true, but whose converse is not true.

3. Give an example of an if-then sentence which is true, and whose converse is also true.

4. This is a false statement: *If A, B and C are collinear, then A–B–C.* Is the converse of that statement also false?

5. Let the coordinates of A, B and C be a, b, and c respectively. If A–B–C and if a = 17, b = 23, what are the possible values for c?

odd answers

1. <u>Definition</u> of AB: AB = |b – a|.

(Please put this definition on the Definitions page that you are keeping.)

3. Your answer will probably be different than mine. *If your mother was a duck, then you are a duck* is a true sentence. Its converse is also true: If you are a duck, then your mother was a duck. Definitions are often good places to find if-then sentences whose converses are also true.

5. $c \geq 23$

Jasper

1. If A–B–C and the coordinate of A is 20 and the coordinate of B is 30, what can you say about the coordinate of C?

2. If Q is the midpoint of $\overline{PR}$ and if the coordinate of P is 7 and the coordinate of R is 13, what can you say about the coordinate of Q?

3. State the converse of *If Lambda runs nine miles, she will get sick.*

4. If GH + HI > GH, does this imply that G, H and I are not collinear?

5. Fill in the missing statements and/or reasons in the proof of the following proposition.

<u>Proposition</u>: If points D, E and F have coordinates d, e and f respectively, and if d = 5, e = 7 and f = 9, then D–E–F.

S	*R*
1. Points D, E and F have coordinates d, e and f respectively, and d = 5, e = 7 and f = 9.	*1.* ?
2. DE = |7 – 5| EF = |9 – 7| DF = |9 – 5|	*2.* Def of DE (see answer #1 of the previous City.)
3. DE = 2 EF = 2 DF = 4	*3.* ?
4. DE + EF = DF	*4.* Algebra
5. ?	*5.* Def of D–E–F

Lakota

1. State the contrapositive of *If Fred owns a llama, then Fred will get in trouble.*

2. If B–D–G and the coordinate of B is 8 and the coordinate of G is –5, what numbers could the coordinate of D be equal to?

3. Name the three undefined terms of geometry.

4. In proofs that we have done so far, we have used four different possible reasons. One of them was "Definition." What have been the other three reasons we have used?

5. If I have line ℓ and two different points P and Q on ℓ, what postulate assures us that the coordinates of P and Q will not be equal?

6. Fill in one word in the following definition: Two lines are __?__ if and only if then are not concurrent.

A six-minute walk from the Math building (M) to the campus police station (P) should have taken about six minutes. All Fred had to do was go from point M to point P.

M P

Line segment $\overline{MB}$ with endpoints M and P

How could he miss? In June he would be finishing up his sixth year of teaching college mathematics at this university and so he knew the KITTENS campus like he knew his addition and multiplication tables.

In short, there was no way that he could have any difficulty getting from M to P. And that, in fact, was the case. Six minutes along the straight-line path from the Math building to the police station, Fred was right at the cop's front door.

His body was at the front door, but his mind was off building sand castles in Cyprus.* Actually he was thinking about a very famous book that he had read which was entitled *The Elements*. And Fred, lost in thought, kept walking. Instead of being a line segment, his path became a ray.

M P

Ray $\overrightarrow{MP}$ with endpoint M

The Elements is a geometry book.

That is not quite correct. *The Elements* is *the* geometry book.

That is not quite correct. *The Elements* is *the* math book and it almost wins the award for being the most studied book of any kind in all of Western civilization. It gets the second place award—the Bible winning first place. Except for the Bible, *The Elements* has had more editions published (more than 2000) than any other book ever written. And it's a geometry book!

---

* A favorite expression of my eighth-grade social studies teacher. It's a beautiful image and the alliteration is also pleasing. She used the phrase a lot in class when she noticed that we were daydreaming.

Twenty-three centuries ago (300 B. C.) Euclid (YOU-clid) wrote this best-seller. In the 1800s, a lawyer carried a copy of *The Elements* with him and studied it by candlelight in the evening. Studying Euclid's reasoning in proving geometry theorems influenced this lawyer's presentation of his arguments that he made in the courtroom and in his political speeches. Later this lawyer had his head on a penny and the $5 bill.

Famous student of *The Elements*

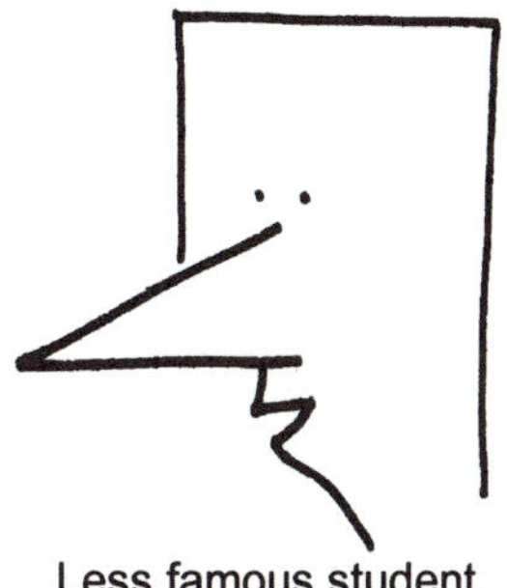

Less famous student of *The Elements*

Euclid's step-by-step proofs starting from his postulates and definitions were something that the world had never seen. *The Elements*, more than just producing the world's first geometrical theory, showed a new way of thinking. A way of thinking—when you think about it—that cut through the haze of fuzzy infantile brain bubbling which is characteristic of two-year-olds, the inebriated, and some politicians.

Some Well-know Examples of Non-Euclidean Thought

The Two-year-old	The Drunk	The Immoral
And why did you take Mommy's perfume and throw it in the toilet?	Your boozing has torn apart your life, alienated your family from you, and ruined your health. Why do you continue?	Is it true that you lied to the American public, cheated on your wife, and abused the power of your office?
The kid's answer:	The response:	The answer (under oath):
"Because."	"I'll quit . . . tomorrow. I promise."	"What do you mean by *is*?"

What Euclid offered was an escape from muddled thought. The reasons that are permitted in a geometry proof come from a small and well-defined list. So far, we have only four reasons that can be used: ① Given, ② Definition, ③ Postulate, and ④ Algebra.

Fred once gave a quiz to his geometry class. He asked them to fill in the reasons in the proof of the proposition: *If points A, B and C have coordinates 5, 8 and 32 respectively, then A–B–C.*

Here's how one student (who will remain unnamed) completed the quiz:

Statement	*Reason*
1. A, B and C have coordinates 5, 8 and 32 respectively.	1. That seems like a good place to start.
2. AB = |8 – 5| BC = |32 – 8| AC = |32 – 5|	2. Because it's true.
3. AB = 3 BC = 24 AC = 27	3. Who am I to argue with that?
4. AB + BC = AC	4. What I am told, I believe.
5. A–B–C	5. This is the conclusion we're looking for.

Your Turn to Play

1. What are the five correct answers to the above quiz?

2. Part of the reason we write out proofs in geometry, giving a legitimate reason for each step, is to make our thought processes very clear. In a short essay (with words and sentences and spelling, etc.), discuss some reasons why people will practice obscurantism (deliberately making things ambiguous or vague or hard to understand).

3. We have never defined the ray $\overrightarrow{MP}$. Since the emphasis in studying geometry is not to memorize "stuff," but to learn to reason and to learn to be creative, give the following a try:

M

P

Ray $\overrightarrow{MP}$ with endpoint M

Before looking at the answer below, create a definition of $\overrightarrow{MP}$. To get you started,
Definition: The **ray** $\overrightarrow{MP}$ is the set of all points C such that. . .

4. In the answer to the previous problem, we mentioned that MM = 0. Back in chapter one (on p. 20) we said that AA = 0 (talking about the distance from San Francisco to San Francisco). In this question we will prove the proposition: AA = 0.

You might be inclined to ask at this point: ***Why prove that? We both know that the distance from some point to itself must be zero.***

That's an important question. A little story might help answer your question. When my daughter Margaret was young (before kindergarten) we sat down to "play chess." I put all the pieces on the board and let her choose which color she wanted to play. Then I made some move with one of my pieces and named it as I moved it: "I'm going to move my pawn to this square." We took turns making moves. She hadn't learned the correct moves for any of the pieces yet; she was just learning the names of the pieces. After a couple of minutes, she looked up at me and asked, "Daddy, how do you win the game?" I explained to her that you win by taking the other guy's king. She reached over and grabbed my king and with a smile that wet both ears she announced, "I win!"

Why do some people play football? Is the whole point just to get the football past the other team's goal line? If so, then why not just arm your team with rifles?

Just saying AA = 0 or just saying, "The base angles of an isosceles triangle are congruent," is like grabbing the other guy's king or shooting all the linebackers.

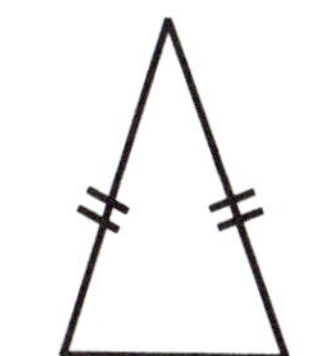

implies

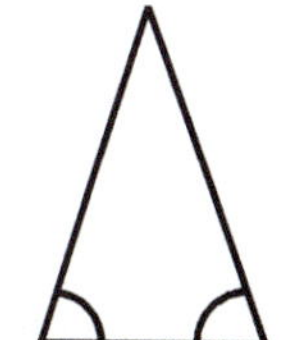

An isosceles triangle is one in which at least two sides are equal in length.

The base angles are conguent if they have the same size.

All of these things are what human beings do. We love to play. And as we grow up, we develop a desire to play the harder games, the ones that are not so easy to win. In kindergarten every kid should succeed 99% of the time. On the other hand, can you imagine what a curse it would be if everything was just given to you? If you had spent years training to be a world-class sumo wrestler, and the only opponents you could find were of Fred's stature?

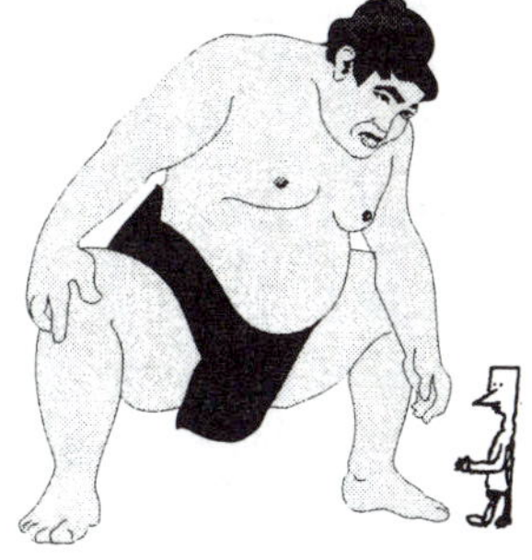

Fill in the reasons in the proof of AA = 0:

S	R
1. There is a unique number associated with the point A. Call it x.	1. ?
2. AA = |x – x|	2. ?
3. AA = 0	3. ?

5. Create a definition of **opposite rays**.

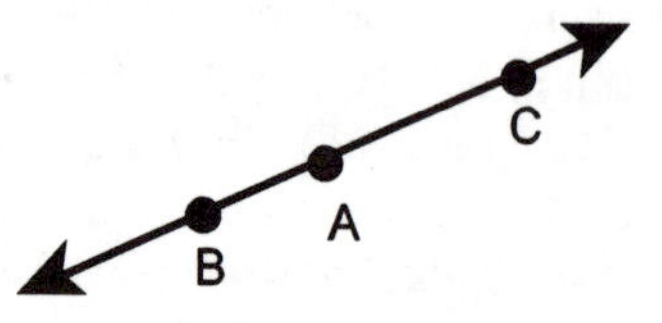

To get you started: Definition: $\overrightarrow{AB}$ and $\overrightarrow{AC}$ are **opposite rays** if and only if. . . .

6. What is wrong with the following definition? $\overrightarrow{AB}$ and $\overrightarrow{AC}$ are **opposite rays** if and only if they point in opposite directions.

1. (1) Given; (2) Def of AB; (3) Algebra; (4) Algebra; (5) Def of A–B–C.

2. My essay will probably differ from yours and you may think of reasons that aren't in my essay why people will practice intentional obscurity.

My essay:

Evaders of Clear Thinking

There are five reasons why people practice obscurantism.

The first is that some people want to impress others by their use of big words like obscurantism. Of course, by showing off, they make it hard for others to understand what they are saying.

Second, there are those obscurantists who want to hide the truth because it might be embarrassing. As Adam illustrated when he responded to God's question, "Where are you?" with his famous answer, "You caught me with my pants down." That was a bit of verbal hiding to obscure the real truth that he had eaten of the forbidden tree.

Third, there are those who are paid to "slant the truth," or to "put our particular spin on the events." Those receiving between $150 and $300 per hour are called lawyers.

Fourth, there are those who hide the truth so that they can puff up their own reputations. They will tell you that the field they specialize in is a field that is only open to the select few because of their superior intelligence, etc. (Often they will get the government to make it difficult for others to enter their speciality by requiring those entering the field to have hard-to-obtain licenses.) I read once that when typewriters were first introduced into business offices, the only ones who "operated" them were males, supposedly because typewriters were so complicated.

Last, are those who don't want to look at reality themselves and in order to do that, they need to assure others of things that really aren't true. "Oh, I can quit smoking anytime I want to." "After Christmas, I'm going on a strict diet." "I don't have a drinking problem since I've never passed out."

3. There are several ways you might have completed that definition. Here are a couple of possibilities.

① The **ray** $\overrightarrow{MP}$ is the set of all points C such that

C is either equal to M or equal to P or

C is between M and P or

P is between M and C.

② The **ray** $\overrightarrow{MP}$ is the set of all points C such that either

C is either equal to M or equal to P or

M–C–P or

M–P–C.

If you wanted to be really tricky, you could shorten the definition even further:

③ The **ray** $\overrightarrow{MP}$ is the set of all points C such that M–C–P or M–P–C.

The set of *all C such that M–C–P* includes the possibility that C is "on top of" M, or to write that more formally, that C equals M.

It's always true that M–M–P. That follows directly from our definition of betweenness. Proposition: M–M–P for any points M and P.

Proof of M–M–P:

On p. 21 we had the definition: G–H–I if and only if GH + HI = GI.

By that definition, M–M–P is true if and only if MM + MP = MP.

Since MM = 0, we know that MM + MP = MP which in turn implies that M–M–P.

4. Reason 1 is the Ruler Postulate (There is a unique number associated with each point). Reason 2 is the Definition of AA. (See p. 49 if you haven't written this on your definitions page.) Reason 3 is Algebra.

5. There are several ways you might have completed the definition. Here are a couple of possibilities:

① $\overrightarrow{AB}$ and $\overrightarrow{AC}$ are **opposite rays** if and only if A is between B and C.

② $\overrightarrow{AB}$ and $\overrightarrow{AC}$ are **opposite rays** if and only if B–A–C.

6. The trouble is that we don't know what "pointing in opposite directions" means. This would have to be a new undefined term (in addition to point, line, and plane).

Any new definitions we create should define the new word in terms of the words we already know. When your little two-year-old asks you, "Daddy/Mommy, what does *lucubration* mean?" you don't answer your kid with, "Lucubration is cogitation or the production of one's oeuvre performed nocturnally." Instead you define lucubration in terms of what the kid already has in his "word bag." You say something like, "Lucubration is studying in the evening."

Fred mentally went through the 465 theorems* and their proofs in *The Elements* and then suddenly realized where he was. He was standing in front of Stanthony's PieOne pizzeria. That's where he would be meeting Betty and Alexander this evening for dinner. Through the window of the pizza shop he could see the cuckoo clock that a soft drink

Drink Sluice! Drink Sluice!

manufacturer had given Stanthony. Since it was only 6:05 A.M., Fred figured he was a tiny bit early. (PieOne opens at 7:30 A.M. for those who like to start their day with Stanthony's famous Breakfast Pizza which has corn flakes on it.)

He turned around and headed back toward the campus police station. *Since I've turned six, I must be getting really absentminded,* he thought to himself. *How could I walk right past the police station?*

He giggled to himself as he made up a little song which he sang to himself as he walked back toward the police station. It was entitled, "♫Another Day, Another Ray♫." Unfortunately, most of the verses of his song have been lost to history. After more than five years of seeing their little math professor walking around campus singing, the students were almost used to the phenomenon. It will be another eight years before his voice will change from soprano to tenor.

Starting at PieOne, which we will call point O, he was heading back to the police station, point P.

Only one verse of his song survives:

♫ Another Day, Another Ray ♫

by Fred Gauss

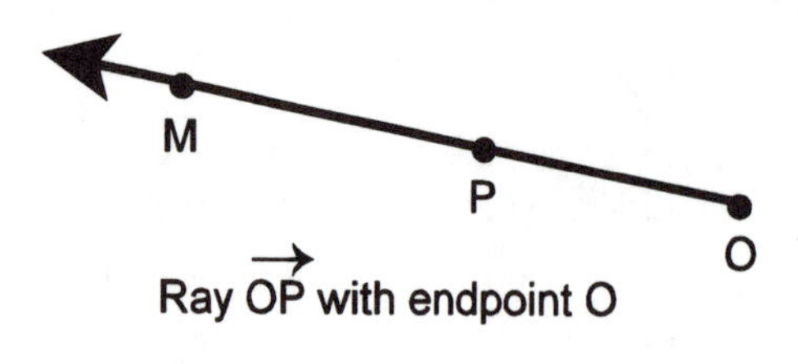

Ray $\overrightarrow{OP}$ with endpoint O

I thought I'd lost my way
When past the cops I stray'd.
Was it that my nucle-
Ic acid's soured or may-
Be I've got feet of clay?

---

* 465 is the actual number of theorems in *The Elements*.

Fred put a lot of heart into singing, hitting a high C at the end of the his verse. These five lines of unforgettable* poetry deserve a little comment.

In English poetry, a (metrical) foot is a bunch of syllables with one of the syllables stressed and the other ones unstressed. The most famous line in Shakespeare's *Hamlet*:

Tŏ bé oř nót tŏ bé

has three stressed syllables and hence has three feet. The stressed syllables are traditionally marked with ′ and the unstressed syllables with ˘. When you write poetry you face two choices: which of four possible feet you will use and how long (in feet) each of your lines will be. Here's a handy pair of charts:

Kinds of Feet

˘′	iambic (eye-AM-bick)
′˘	trochaic (tro-KAY-ick)
˘˘′	anapestic (ana-PES-tick)
′˘˘	dactyllic (dack-TILL-ick)

How Many Feet in a Line

one foot	monometer
two feet	dimeter
three	trimeter
four	tetrameter
five	pentameter
six	hexameter
seven	septameter
eight	octameter

Now you know why English majors have no trouble distinguishing a pentagon from a hexagon.

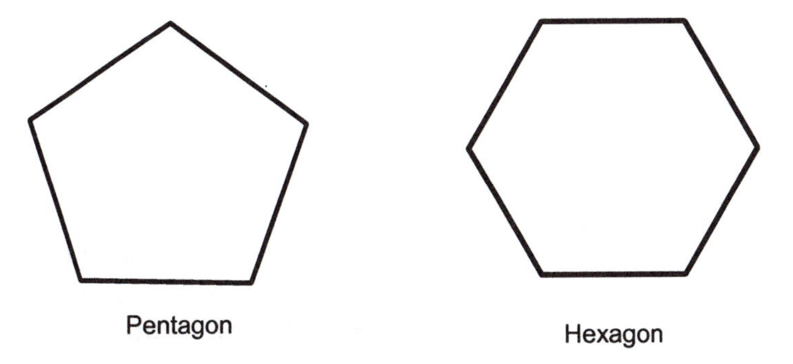

---

* This is irony.

Now with what you know of scansion (the metrical analysis of poems), you see that Fred's "Another Day, Another Ray" is five lines of iambic trimeter.

Fred went nuts making seven words of his song rhyme with *hay*. In order to do that he had to do a little enjambment (having one line run over to the next line). Otherwise, he wouldn't have been able to use *nucleic acid* in his ditty.

Speaking of nucleic acid, did you catch the pun (play on words, a double meaning) when he asked, "Was it that my nucleic acid's soured . . ."?

First meaning: Acids can't sour; they are sour to start with. All acids taste sour—for example, lemon juice or vinegar. Even the word *acid* comes from the Latin word *acere* which means "sour." In German it is even closer. Die Säure means both the acid and the sourness.

Second meaning: Nucleic acids are what make up DNA which carry genetic information. If your DNA is messed up, (gone sour), it means that you are in trouble in a very basic way.

And did you note the literary allusion, "feet of clay," which refers to Daniel's interpretation of a dream of some leader who had a great head (made of gold), wonderful arms and chest (made of silver), neat thighs (made of brass), but when you got down to his feet, they were a mess. They were a combination of iron and clay. (Dan. 2:32) Not the kind of feet you want to have if you like to play basketball. Nowadays, to have feet of clay means that you have a vulnerable point or a defect.

With this analysis of the obvious mental effort that Fred was putting into his "♬Another Day, Another Ray♬," you can see that he might have been a little distracted. In fact, when he turned around to head back to the police station (P), from PieOne (O), he took the wrong path. He headed from point O towards the campus gymnasium (G) and that gave us a look at our first angle. Point O is called the vertex of the angle. In symbols angle POG is written ∠POG.

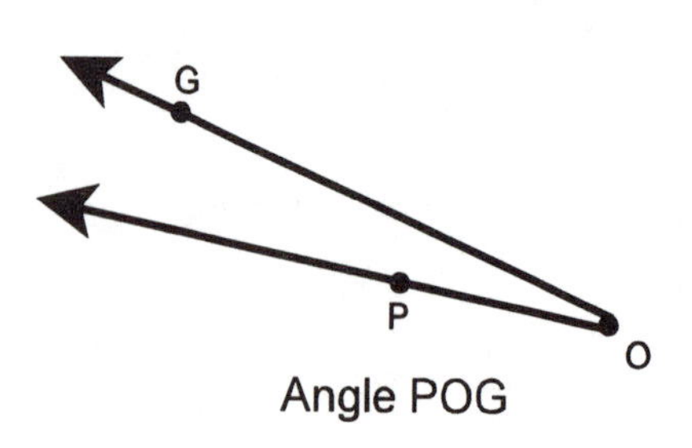

Angle POG

On your definitions page, add: <u>Definition</u> An **angle** is two rays which share a common endpoint.

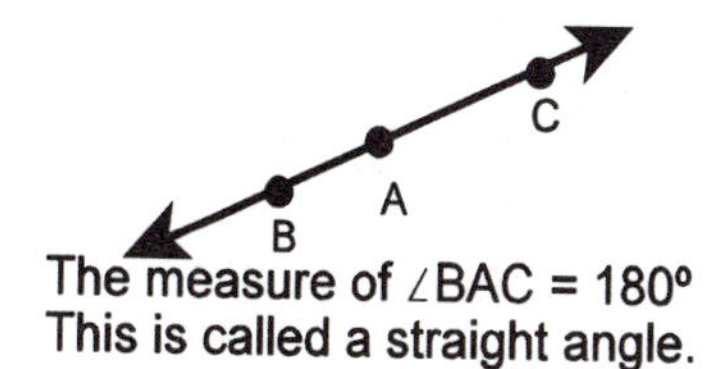

The measure of ∠BAC = 180°
This is called a straight angle.

The biggest possible angle in geometry is formed by opposite rays (which we have already defined). We will say that the measure of that angle is equal to 180°.

The smallest possible angle is formed when the two rays are identical—one lies on top of the other. This would have happened if Fred had been paying attention to where he was walking and had actually headed back from PieOne toward the police station. We will say that the measure of that angle is equal to 0°.

The angle MOP is the same as the angle POM. The important thing is that the point O is mentioned between the M and the P so that it is the vertex of the angle.

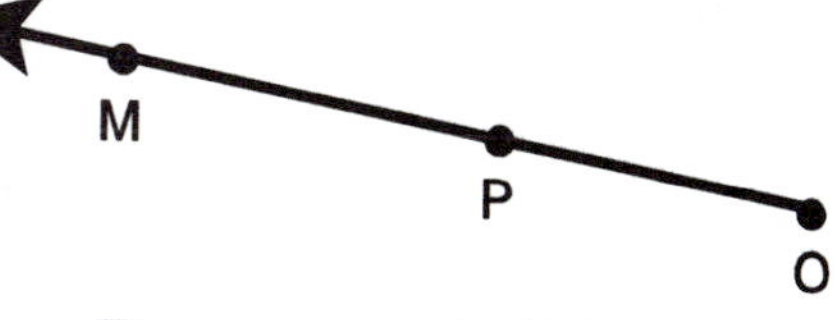

The measure of ∠MOP = 0°

On your symbols page please write "m∠MOP means the measure of ∠MOP." This corresponds to AB indicating the distance between points A and B on the segment $\overline{AB}$.

m∠MOP and AB are numbers.

∠MOP and $\overline{AB}$ are geometrical objects.

In chapter one we had the Ruler Postulate which said that we can match up every point on a line with a real number (which was called the coordinate of that point). We need a similar postulate for angles:

<u>Postulate 3</u>: The **Angle Measurement Postulate**. You can match up every angle with a number between 0 and 180.

This could be rephrased: Every angle has a unique measurement which is some number between 0 and 180 degrees inclusive. (The word "inclusive" means that it includes both 0 and 180 degrees as possibilities. In algebra, this might have been written as $0° \leq x \leq 180°$.)

The angle that Fred made in turning and walking toward the gym was an **acute angle**—its measure is less than 90°.

In symbols, m∠POG < 90°.

Angle POG

A **right angle** is an angle whose measure is equal to 90°. (Please put that on your definitions page. It will be a lot easier to find there than trying to hunt through this book to find the definition.)

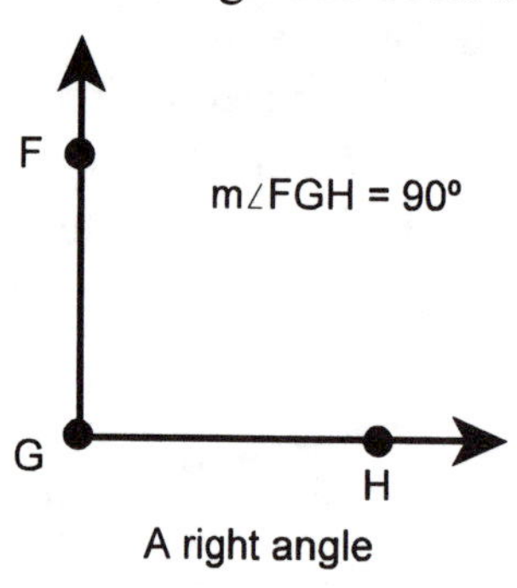

A right angle

We will see lots of right angles in geometry. Often, to indicate that we are looking at a right angle (and not an 89° angle or a 91° angle), we will put a little box at the vertex.

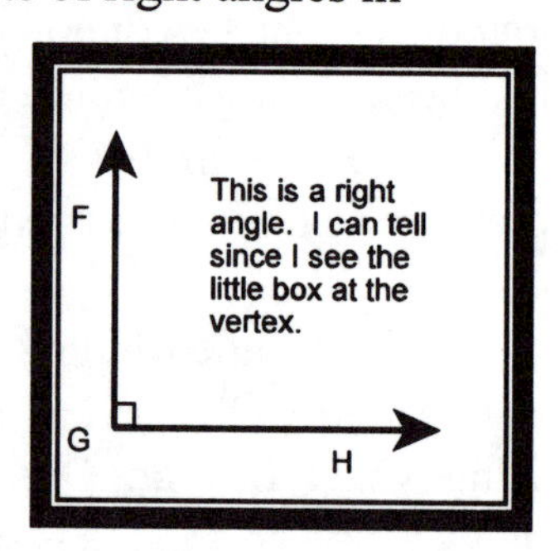

F
This is a right angle. I can tell since I see the little box at the vertex.
G
H

It wasn't three minutes on Fred's new path before he realized he was heading in the wrong direction. He turned toward the police station (point P) and continued his journey. *Thank Goodness no one is watching me make a fool of myself,* he thought to himself. Of course, Fred was unaware of the hundreds of thousands of readers of this book who are watching his every move.

Fred eventually made it to the front of the station.

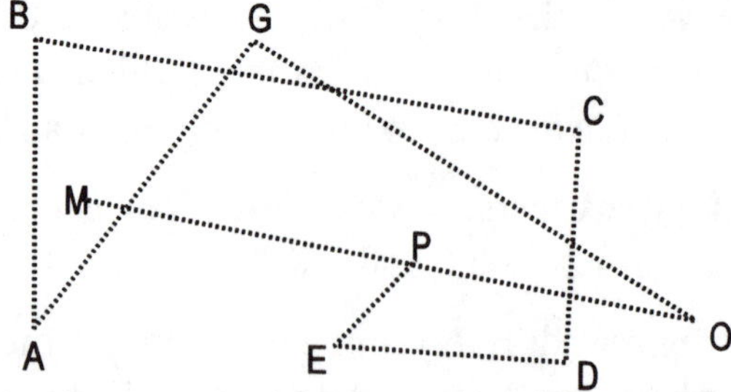

∠CDE looks like it is a right angle. $\overline{CD}$ and $\overline{DE}$ are streets so that isn't very surprising. Many streets meet at right angles. Sometime we name an angle just by its vertex point. ∠CDE could also be called ∠D.

∠G is neither acute nor is it a right angle. Angles like ∠AGO whose measure is greater than 90° are called **obtuse angles**.

Chances are, you own a ruler so that you can measure the length of things. If you want to measure angles, you can go to a drug store or an office supplies store and buy a piece of plastic called a protractor. It is so simple to use that usually no instruction sheet comes with it. You won't need a protractor for geometry, but some people get a dollar's worth of fun out of one of them.

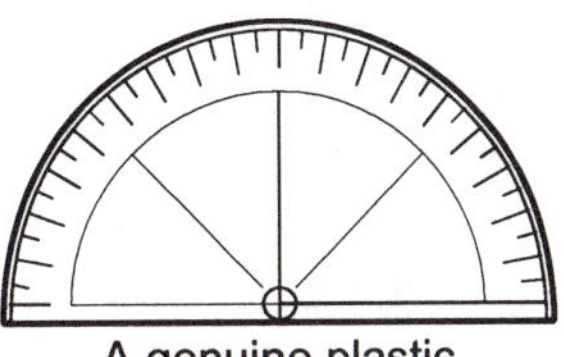

A genuine plastic protractor

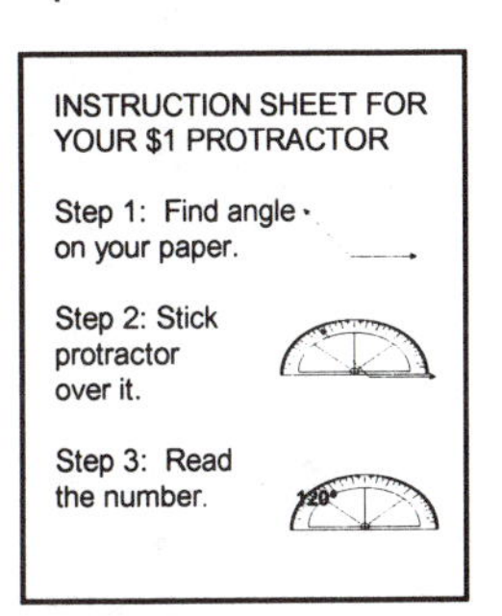

INSTRUCTION SHEET FOR YOUR $1 PROTRACTOR
Step 1: Find angle on your paper.
Step 2: Stick protractor over it.
Step 3: Read the number.
120°

It was quite a police station. Built with federal money, it was the most magnificent structure on campus. Marble floors, solid oak doors, and highly polished brass everywhere announced that THE LAW was going to be enforced here at KITTENS. The only drawback was that crime is almost nonexistent on this campus. Before the current Great Vandalism of the Bocci Ball Lawn, the most recent crime occurred years ago when a visitor to the campus put a quarter in one of the vending machines and it accidentally gave him two candy bars instead of one. He failed to turn in the extra bar to the authorities.

The Millard Fillmore Federally Funded Police Station
Honoring Our 13th President

Fred was always impressed when he stood in front of that billion-dollar building. He walked inside and looked at the directory to find out where he should turn himself in. Of the 130 people listed in the directory (which included federal regulators; federal inspectors; federal compliance officers; federal supervisors who supervised the regulators, inspectors and compliance officers; federal administrators who oversaw the work of the supervisors . . . and one lone part-time campus cop), Fred figured that it was the campus cop he should see.

As Fred made his way down the hall to the broom-closet-sized office of Sergeant Friday, it is now . . .

Your Turn to Play

1. When Euclid wrote *The Elements*, he was working under several handicaps. First, he was writing in a foreign language (Greek) rather than good old English. He didn't even have the regular Roman alphabet with A, B, C, . . . but instead had to write in the Greek alphabet with α (alpha), β (beta), γ (gamma). . . . His second handicap was that he was the first one in any country, in any culture, to write a geometry book with theorems and their proofs. Not knowing about our modern ideas of building a mathematical theory by starting with undefined terms and then building definitions and theorems on top, Euclid tried to define *everything*. His first definition was Δεφινιτιον ονε: Α ποιντ ισ τηατ ωηιχη ηασ νο παρτ which roughly translates as Definition one: A point is that which has no part.

Your job is to criticize Euclid's definition. You may do it in English if you wish.

2. Without a protractor, how would you draw an angle of approximately 45°?

3. Without a protractor, how would you draw an angle of approximately 30°?

4. Euclid defines what we call a "line" in modern geometry as: A breadthless length which lies evenly with the points on itself. Do you forgive him? ☐ yes; ☐ no

5. Do you understand him? ☐ yes; ☐ no

6. If two geometrical figures (two segments, two angles, two triangles, two circles) have the same shape and size, we will want to say that they are congruent. What would be wrong with: Definition: Two triangles are congruent if they have the same shape and size.

7. Complete the following. <u>Definition</u>: $\overline{AB} \cong \overline{CD}$ if and only if. . . .

8. Draw Fred's face and identify some acute, right and obtuse angles.

COMPLETE SOLUTIONS

1. ". . . that which has no part" is really goofy. It is virtually meaningless. Fred's head doesn't have a part.

Fred Gauss
with a part

2. The way I would draw it would be to first draw a right angle and then cut the angle in half (bisect it). Those cute little arcs that I've drawn indicate that the two angles are the same size. One student once suggested you could get a pretty good 45° angle by taking a sheet of paper and folding up a corner.

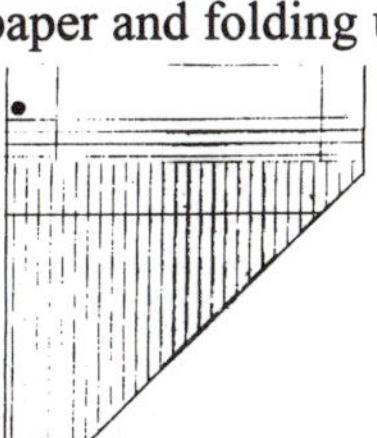

3. I would first draw a right angle and then cut it in thirds (trisect it).

4. First, please note that the word is "breadthless" not "breathless."

A breathless length would be swimming the length of a pool underwater. Second, there are some very famous words about being forgiven to the extent that we forgive others.

5. You might understand him, but I sure can't picture what *lying evenly with the points on yourself* means. Can you imagine being sent to the vice principal's office in high school because you didn't lie evenly with the points on yourself? That is something that Franz Kafka should have put in his book *The Trial*, in which Joseph K., a bank worker, is suddenly arrested and then spends his whole life defending himself and he is never clear what the charges are against him. Definitely nightmare material.

6. You can't define congruent triangles in terms of words that aren't already defined. The "shape" and "size" of triangles have not been defined within our geometry theory. In contrast, we can define: Definition: Two angles, $\angle A$ and $\angle B$, are **congruent** if and only if $m\angle A = m\angle B$. We *do* know what the equality of numbers means, and $m\angle A$ and $m\angle B$ are numbers.

The symbol we use for congruence is $\cong$. We could restate the definition of congruent angles as:

Definition: $\angle A \cong \angle B$ if and only if $m\angle A = m\angle B$.

7. <u>Definition</u>: $\overline{AB} \cong \overline{CD}$ if and only if AB = CD. True story: Yesterday I was at the library checking out a bunch of books—a not unusual act. The librarian and I started to talk about geometry and he commented, "Geometry . . . I took it in school years ago and I never 'got it.' " Two things are true and a third is probably true: ① He never had Fred as his teacher; ② His textbook wasn't the one you are holding in your hands; and ③ He never wrote on his definitions sheet the definitions of congruent angles and congruent segments and he never wrote ≅ on his symbols sheet. Hint! Hint!

8. I'll use an A, R and O to mark some of the acute, right and obtuse angles.

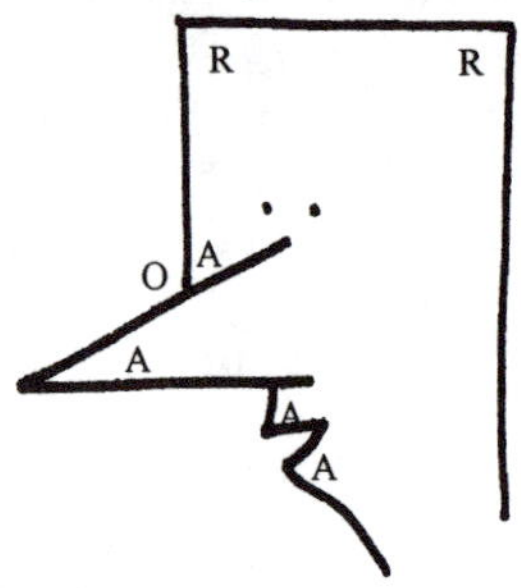

There was a long line of people in front of Sergeant Friday's office. Fred took his place at the end of the line. *A whole lot of crooks must be turning themselves in*, he thought to himself. Everyone in front of him had handfuls of cash. This worried Fred: *I hope they take checks.* Fred began to fill out a check.

FRED GAUSS
ROOM 314, MATH BUILDING
KITTENS UNIVERSITY

321

Date

Pay to the order of *The Police* $

bucks

Kittens Bank

"IT'S IN THE KITTY"

Fred Gauss

Fred recognized Alexander and Betty, but they were too far ahead of him in the line for him to say hello to them. *I wonder what crime they've committed? It must have been something like a parking infraction.*

Fred could see that they were each counting out a lot of money to Sergeant Friday. After they were finished, they headed out a side door, so Fred didn't get a chance to speak to them. It's okay, Fred thought to himself. I'll see them tonight for pizza and we can exchange stories then.

He looked down the hall and saw a Wanted—Dead or Alive poster, but he couldn't read it since it was about 300 feet away.

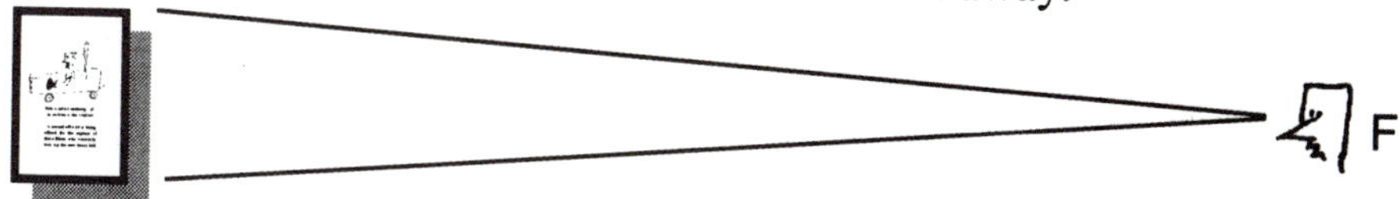

Angle F was an extremely small angle. (Much smaller than is pictured here.) $m\angle F = 0.083°$. That's much smaller than one degree. What we need is a finer unit of measurement for angles. There are 60 **minutes** in a degree. In symbols: $60' = 1°$. To convert from degrees to minutes you multiply by a conversion factor. That's a fraction which is equal to one since its numerator equals its denominator. The conversion factor for changing degrees to minutes is $\frac{60'}{1°}$

Watch it in action: $0.083° \times \frac{60'}{1°} = 4.98'$

We may pretend that the degree marks cancel, leaving the minutes mark:

$$0.083\not{°} \times \frac{60'}{1\not{°}}$$

To convert six pounds into ounces: $6 \text{ lbs.} \times \frac{16 \text{ oz.}}{1 \text{ lb.}}$

To convert 17977 feet into miles: $17977 \text{ feet} \times \frac{1 \text{ mile}}{5280 \text{ feet}}$

To convert 2.0344 miles into feet: $2.0344 \text{ miles} \times \frac{5280 \text{ feet}}{1 \text{ mile}}$

Your Turn to Play

1. To convert days into hours, what conversion factor would you multiply by?
2. To convert minutes into degrees, what conversion factor would you use?
3. On Fred's sixth birthday, how many seconds had he been alive? This is done by first converting years into days, then days into hours, then hours into minutes, then minutes into seconds. You will have four conversion factors.

4. On a bet once, Joe drank 1.27 gallons of Sluice. How many cubic inches was that? (There are 231 cubic inches in a gallon.)

5. If you have really small angles, you may need a unit of measure smaller than minutes. This happens every so often in astronomy. There are 60 **seconds** in a minute of angular measurement. In symbols: 60" = 1'. Convert 23" into degrees.

COMPLETE SOLUTIONS

1. You multiply by $\frac{24 \text{ hours}}{1 \text{ day}}$

2. You use $\frac{1°}{60'}$

3. 6 years × $\frac{365 \text{ days}}{1 \text{ year}} \times \frac{24 \text{ hours}}{1 \text{ day}} \times \frac{60 \text{ minutes}}{1 \text{ hour}} \times \frac{60 \text{ seconds}}{1 \text{ minute}}$

which equals 189,216,000 seconds.

For fun, watch how the units cancel in the above computation:

$$6 \cancel{\text{years}} \times \frac{365 \cancel{\text{days}}}{1 \cancel{\text{year}}} \times \frac{24 \cancel{\text{hours}}}{1 \cancel{\text{day}}} \times \frac{60 \cancel{\text{minutes}}}{1 \cancel{\text{hour}}} \times \frac{60 \text{ seconds}}{1 \cancel{\text{minute}}}$$

and what started out in years is now in seconds.

4. 1.27 gallons × $\frac{231 \text{ cubic inches}}{1 \text{ gallon}}$ which equals 293.37 cubic inches. (If we pay attention to significant digits, we would say he drank 293 cubic inches. The idea of significant digits is explained in *Life of Fred: Advanced Algebra* and in *Life of Fred: Trigonometry*.)

5. 23" × $\frac{1'}{60"} \times \frac{1°}{60'}$ which equals approximately 0.0063888°.

You may have noticed a similarity: 60 seconds in a minute and 60 minutes in a degree seems a lot like 60 seconds in a minute and 60 minutes in an hour, and you may be asking, ***Where did all these 60's come from? I don't like that. I much prefer stuff like the metric system where you multiply by 10's instead. You know, the good ol' decimal system.***

I guess you will have to blame it on the Babylonians (about 2000 B.C.). Instead of a base-10 system (because we have 10 fingers), they used the sexagesimal system. That's base-60. Arithmetic is agony in that system.

Well, you continue in the only math book that you can talk back to, ***why in the world don't we change the system?! Why keep 180° as the measure of a straight angle and 90° as the measure of a right angle? After 4000 years, isn't it time for a little itty-bitty change?***

I agree. And while we're at it, let's get the people in England to drive on the right side of the road. And let's change our crazy typewriters so that the keys are in a much more logical order instead of having QWERTYUIOP across the top row. And let's clean up the funny spellings of so many words in the English language. (Why isn't *yacht* spelled y-o-u-g-h-t like ought and bought? In fact, why isn't *ought* spelled a-w-t? And while we're at it, let's have everyone speak the same language. Would you prefer to learn Chinese or Finnish?) And why does everyone have to say, "How are you?" when you really don't want to hear the real answer? And if you or I were King/Queen of the Universe, we might want to make some other changes: All pizzas would be free. No trees would ever be cut down. The skies would never be cloudy.

But, on second thought. To answer your question why don't we change the silly base-60 system with 60 seconds equaling one minute, we have to note one fact that many visionaries miss:

Proposed Changes Always Involve Costs

To change which side of the road that the English drive on, might involve a lot of fatalities in the first few weeks of the changeover.

To change our QWERTY keyboards would drive a lot of elderly typists nuts.

To change our funny spellings, wud meen that r iballs and awl r old buuks wud need 2 get redun.

To force every pizzeria to serve free pizzas, would mean that they would all go broke and there would be no more pizza shops.

To never cut down a tree would mean that there would be more homeless.

If the skies were never cloudy, it would never rain. (No rain implies no wheat which implies no pizza.)

If we got rid of our peculiar $1° = 60' = 3600''$, then we'd have to throw away our precious one-dollar protractors. (Good news: In calculus we are going to get rid of stupid degrees and adopt radian measurement. Measuring angles in radians will make our calculus formula a lot easier. Bad news: A right angle is equal to approximately 1.5707963 radians.)

After twenty minutes Fred finally made it to the front of the line in front of Sergeant Friday's office. He didn't know what to say. He looked at the policeman and then looked at the long line behind him. There were hundreds of students, faculty, and staff standing behind him, all with lots of cash in their hands.

Finally, Fred blurted out, "Could I speak with you for a moment?"

"Sure, go ahead," Friday responded.

"I mean . . . in private. It's kind of embarrassing."

The policeman frowned a little and then turned and went inside his office. He came out with a large basket and addressed everyone in line behind Fred. "Just put your money in here. Every donation will certainly be appreciated."

Fred didn't understand. It must be new terminology in criminal justice. Why was the policeman calling the fines "donations"?

He took Fred by the arm and took him into his office so that they could talk "in private." As they sat down, Fred noticed a sign on the wall.

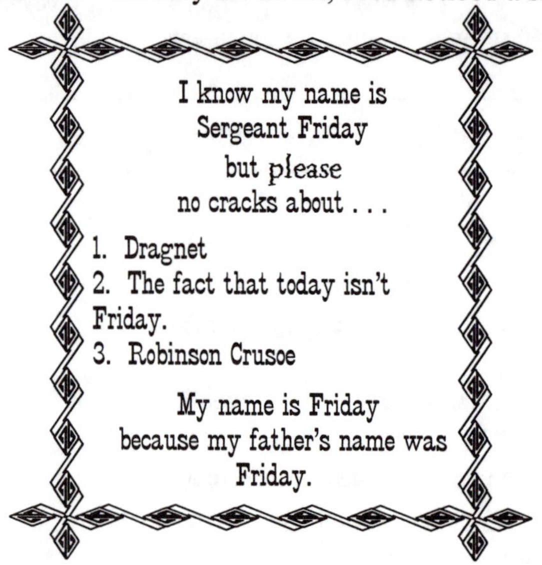

I know my name is
Sergeant Friday
but please
no cracks about . . .
1. Dragnet
2. The fact that today isn't Friday.
3. Robinson Crusoe
My name is Friday
because my father's name was
Friday.

Fred could sympathize with Friday. Fred's last name, Gauss, (rhymes with *house*) has been the butt of many a joke.

"Okay, young man," the officer began, "what have you been up to? Did you break a window with a baseball? Did you steal an apple from the grocery store?"

Fred was shaking his head. He had never played baseball; he was too small for that. He had never stolen anything in his life since he had learned in Sunday School that God didn't want him to do that. That was Law Number 8, as the teacher had explained to him.

"Okay, son, I need to know. We're in the middle of a serious manhunt right now. The first real crime this campus has seen in years. Some bad guys have vandalized our new bocci ball lawn. Did hundreds of dollars worth of damage."

"How much?" Fred asked.

"$638.42 if you must know."

Fred took out his checkbook and filled in $638.42 in the check he had started while he was waiting in line. "Here," he said as he handed the check to the policeman.

"No kid. Contributions to the reward fund go in the basket outside."

"You don't quite understand sir," Fred explained. "I'm the crook. I and my llama ate the grass . . . no I mean my llama, the one I own, ate the grass on the lawn. My llama didn't know any better and I am new at owning and operating a llama and here's the money and I'm sorry and it won't happen again and do I need to bring in the llama to be punished and do I have to go to jail and . . ."

"Hold it, kid. Do you mean that your pet did that damage?"

"Yes, sir. I'm afraid that she did."

"And you are turning yourself in?"

"Of course. I was responsible and the only right thing to do is acknowledge my malefaction."

Friday stood up and instructed Fred, "Wait here."

As Friday left the room, Fred wondered whether he was going to get some handcuffs or maybe a paddy wagon to haul him off to jail.

First Sergeant Friday headed to his policeman's dictionary and looked up MALEFACTION: mal-e-FAK-shun—a no-no; something you're not supposed to do. Then, having figured out what Fred was talking about, he headed outside of his office and held up his hands and announced to the crowd that was donating reward money into the basket, "We caught the crooks. It was a difficult piece of detective work, but we got 'em."

A reporter from *THE KITTEN Caboodle* campus newspaper was there and conducted a mini-interview. Friday explained that he couldn't give out the name of the perpetrator of the crime since the "perp" was a minor. The reporter asked who actually solved the crime. Sergeant Friday, with a blush of confusion, stated that it was Fred Gauss who was the one who identified the malefactor. (Friday had learned that word when he had looked up malefaction.)

An early morning special edition of the campus newspaper headlined that interview.

THE KITTEN Caboodle

The Official Campus Newspaper of KITTENS University — Thursday 6:17 a.m. Special Edition 10¢

Vandals Nabbed.
Gauss Solves the Case

file photo

KANSAS: All the credit for finding the identity of the vandals goes to our own Prof. Fred Gauss according to Sergeant Friday in an exclusive interview with the Caboodle. A special awards ceremony is scheduled for noon. (continued on page 10)

This turned out to be much worse for Fred than if the Caboodle had just reported that Fred and his llama had wrecked the lawn. Now for years afterward his friends would kid Fred about his role in the Great Vandalism caper that he had "solved." Being the crook-who-caught-himself would be very embarrassing.

Meanwhile, while the police officer was out of the room, Fred became increasingly nervous. He pictured being arrested, fingerprinted, and having his mug shot taken.

Friday had a whole bunch of paper clips on his desktop that he had strung together. (KITTENS Cops usually don't have a lot to do given the low crime rate.) Fred undid the clips and arranged them as an angle. Then he made another angle. This was fun and the paper clips would be a lot easier for Sergeant Friday to use since they weren't all tied together.

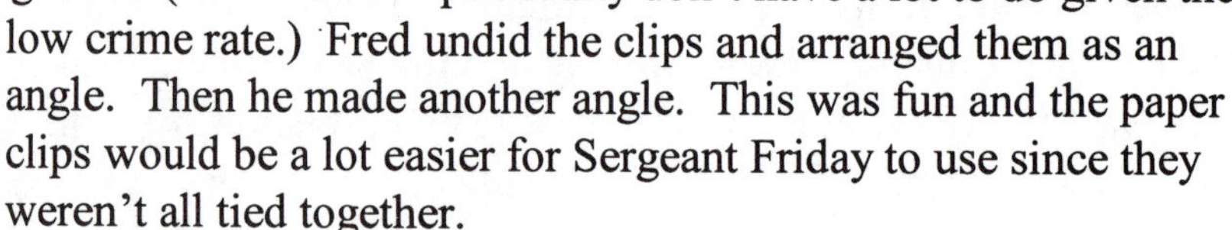

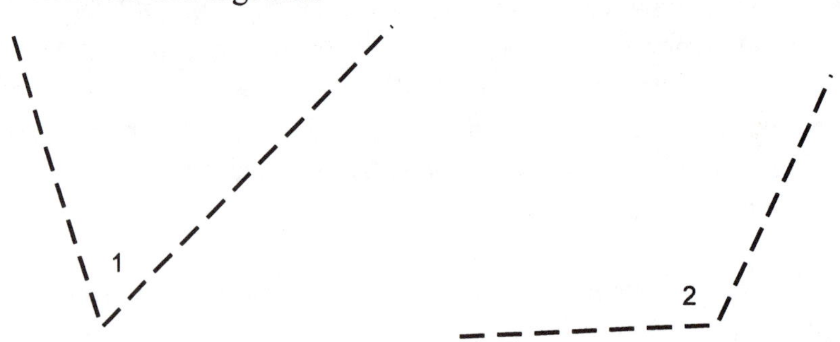

He looked at $\angle 1$ and $\angle 2$ for a moment. Then he thought to himself, *Those angles look like they are supplementary angles.*

Your Turn to Play

1. Here is a pair of supplementary angles:

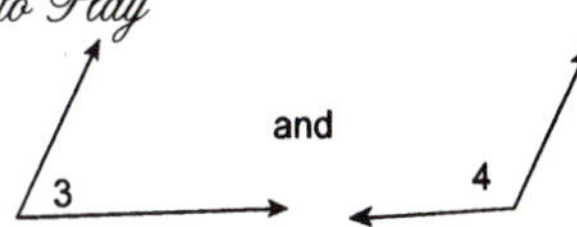

Here are some more pairs of supplementary angles:

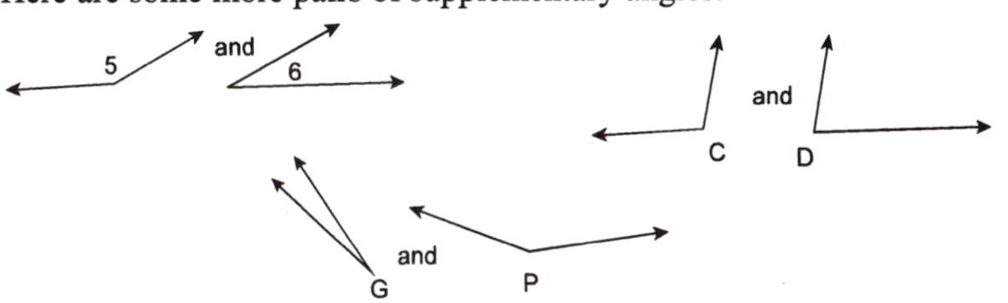

Guess what the definition of supplementary angles is. Here is how the definition would begin: <u>Definition</u>: ∠A and ∠B are supplementary if and only if. . . .

2. If ∠G and ∠P are supplementary and if m∠G = 10°, what is the measure of ∠P?

3. Angle ABC and angle CBD form a linear pair of angles.

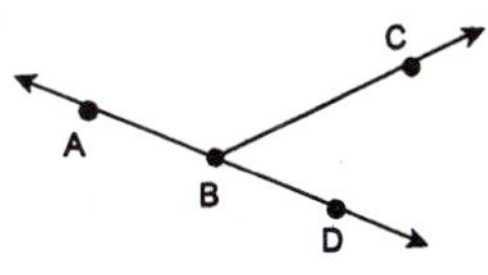

Here are some more examples of linear pairs.

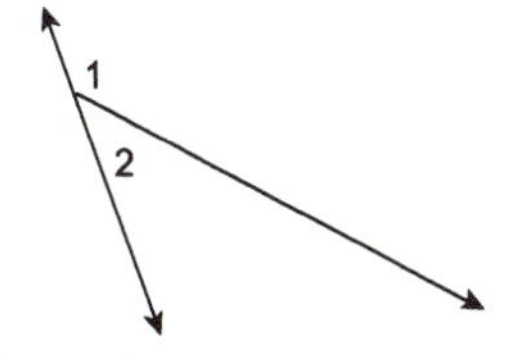

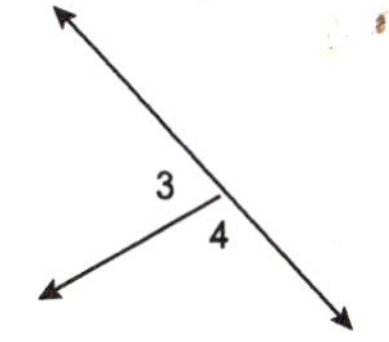

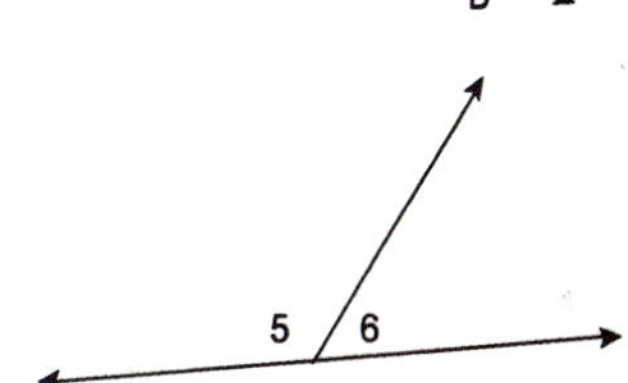

Guess what the definition of a linear pair is. The definition would begin: <u>Definition</u>: ∠ABC and ∠CBD form a linear pair if and only if. . . .

4. Complete the following by filling in the blank with a pentasyllabic* word that begins with "s" and ends with "y": If two angles form a linear pair, then they must always be ________?________.

5. How about the opposite question: If two angles are supplementary, must they always form a linear pair?

6. The two rays that form an angle are sometimes called the **sides** of the angle. So a linear pair could be defined as two angles which share a common side and a common

---

* pentasyllabic = five syllables. Poems with five feet = pentameter and

Pentagon

A student once asked me, "Why is *hexasyllabic* pentasyllabic? Shouldn't it have six syllables?" I could only answer monosyllabically, "I don't know."

vertex and whose other sides are opposite rays. (English always seems harder than math.) For English majors and anyone else who speaks the language, try your hand at creating the definition of vertical angles. ∠1 and ∠3 are vertical angles in the diagram.

Here are the Rules: Pretend you are talking with someone on the telephone. They have read this book up to this point, but they haven't seen the diagram drawn on the right. How would you define vertical angles when all you can do is talk to them and can't draw a picture for them to look at?

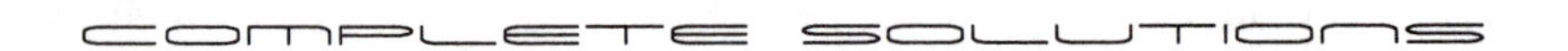

1. Definition: ∠A and ∠B are **supplementary** if and only if m∠A + m∠B = 180°.

2. By algebra, m∠G + m∠P = 180° and m∠G = 10° imply that m∠P = 170°.

3. Definition: ∠ABC and ∠CBD form a linear pair if and only if $\overrightarrow{BA}$ and $\overrightarrow{BD}$ are opposite rays. (Sometimes this definition is continued with ". . . and C does not lie on $\overleftrightarrow{AD}$." Or sometimes the definition is continued with ". . . and $\overrightarrow{BC}$ is not the same as either $\overrightarrow{BA}$ or $\overrightarrow{BD}$.) Any of these three possibilities (either "plain" or with either of the two additions) is fine.

4. I'm going to be totally unfair and not give you the answer to this question.

5. Looking at the four examples of pairs of supplementary angles given in question one, we note that none of them are linear pairs.

6. There are several possible ways to write the definition of vertical angles. Your answer may differ from the ones I suggest. In any event, please place one of the definitions of vertical angles on your definitions page.

Possible definition #1: Two angles are **vertical angles** if they are formed by intersecting lines and they are not a linear pair. (If you need the definition of **intersecting lines,** it is two lines that share a common point.)

Possible definition #2: If the sides of two angles form opposite rays, then the angles are called **vertical angles**.

Possible definition #3: If ∠B is formed by taking the opposite rays of ∠A, then ∠A and ∠B are **vertical angles**.

Fred rearranged the paper clips to make a linear pair of angles.

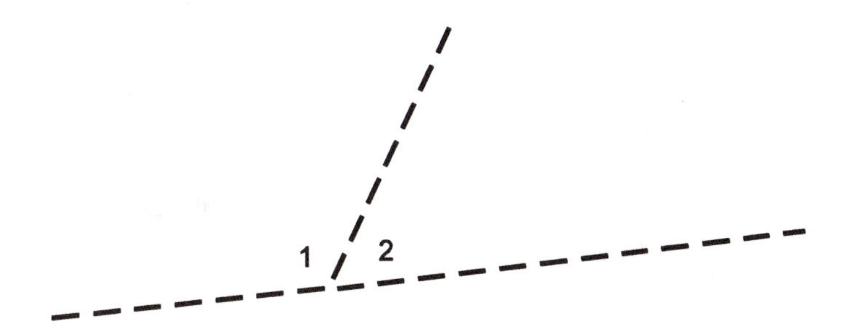

What he noticed was that he couldn't think of a way to prove that $m\angle 1 + m\angle 2 = 180°$. It seemed so obvious. A proof would look something like this:

Statement	*Reason*
1. $\angle 1$ and $\angle 2$ are a linear pair	1. Given
2. ?	2. ?
3. ?	3. ?
(and after a bunch of steps, the last step would be)	
$m\angle 1 + m\angle 2 = 180°$	?

He knew that $m\angle 1$ was some number between 0° and 180°. He knew that by Postulate 3 (the **Angle Measurement Postulate**: you can match up every angle with a number between 0 and 180). He knew that the measure of the straight angle in the diagram was equal to 180° by the definition on p. 59.

He couldn't prove it. He needed to assume it. And since postulates are our beginning assumptions (upon which we base our arguments), we have:

<u>Postulate 4</u>: If two angles form a linear pair, then they are supplementary.

Your Turn to Play

1. What is the hypothesis of Postulate 4?
2. What is the conclusion of Postulate 4?
3. State the contrapositive of Postulate 4. The contrapositive of "If P, then Q" is "If not Q, then not P." In symbols, the contrapositive of P ➜ Q is not-Q ➜ not-P.

4. Prove: If $\angle 1$ and $\angle 2$ form a linear pair, then $m\angle 1 + m\angle 2 = 180°$. (If we were golfing, we'd call this a par 3. Your proof should take about three steps.)

COMPLETE SOLUTIONS

1. and 2. The hypothesis is underlined and the conclusion is double underlined:

If two angles form a linear pair, then they are supplementary.

3. The contrapositive of Postulate 4 is: *If two angles are not supplementary, then they don't form a linear pair.*

4. Proof of: If $\angle 1$ and $\angle 2$ form a linear pair, then $m\angle 1 + m\angle 2 = 180°$.

Statement	Reason
1. $\angle 1$ and $\angle 2$ form a linear pair.	1. Given
2. $\angle 1$ and $\angle 2$ are supplementary.	2. Postulate 4
3. $m\angle 1 + m\angle 2 = 180°$	3. Definition of supplementary angles

There are several things to note about this proof.

♪#1: A good place to begin a proof is by writing down the hypothesis of what you are trying to prove. The reason will be "Given." You also know that the conclusion will be the last step in your proof. So, almost *automatically*, you have two statements and one reason for your proof before you even have to turn on the electricity to the neurons in your brain and begin thinking.

♪#2: The only legal reasons we have (thus far) are Given; a Postulate; Algebra and a Definition.

♪#3: This proof is a good example of a logical train of thought. (Think of hooking all the cars together where the engine is the given.) You started with a linear pair. From there you went to the fact that they were supplementary angles. Then from supplementary angles, you could say that the sum of their measures equals 180°. This is also called a chain of reasoning, since every link attaches to the next link in the chain.

One student once suggested a ZIPPER OF REASONING but the difficulties with zippers is that they sometimes COME APART

Friday returned to his office and sat down next to Fred.

"I don't know how to explain this to you," the sergeant began. He looked down at his desk and noticed that Fred had unhooked all his paper clips and had rearranged them into:

"That's a linear pair of angles," Fred offered, not knowing that silence is the preferred mode when the officer is thinking. But Fred was nervous. The default setting in his brain was: when in doubt, throw in some math. It was a lot better than discussing his possible punishment beyond paying the $638.42 in damages. He was glad that the Constitution prohibited cruel and unusual punishments. He was wondering how much jail time he would have to do.

Friday was still quiet. He had never had to deal with a crook who had also been the good-guy-who-turned-in-the-crook. He played with the paper clips.

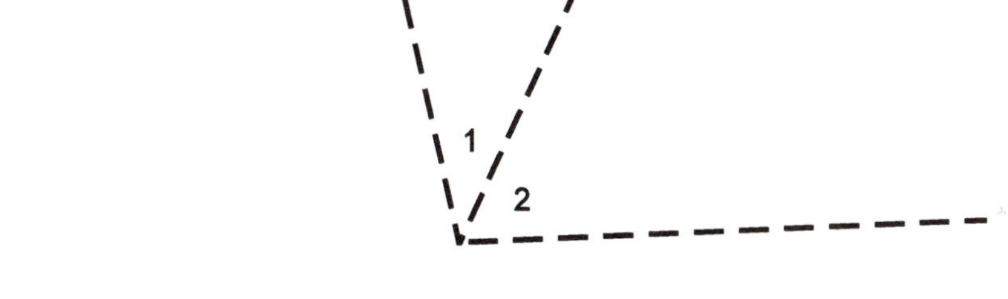

Fred forgot his worries for a moment as he looked at the two adjacent angles.* He realized that he had no way to prove that the measure of angle 1 and angle 2 add up to the measure of the "big angle." The four postulates that he had so far wouldn't prove it.

Postulate 1: Two points determine a line.

Postulate 2: (The Ruler Postulate) The points on a line can be matched up with the real numbers so that every point is matched up with just one number and every number is matched up with just one point.

Postulate 3: (Angle Measurement Postulate) You can match up every angle with a number between 0 and 180.

Postulate 4: Linear pair ➜ supplementary.

So he played with how he would formulate the Angle Addition Postulate which would be Postulate 5.

✶ Definition: Two angles are **adjacent** if they share a common side and a common vertex.

His first attempt was: Postulate 5: For any two angles, ∠AOB and ∠BOC, m∠AOB + m∠BOC = m∠AOC.

Your Turn to Play

1. Writing postulates is not easy. Fred's first attempt was **NOT TRUE**.
In fact, there are three possible diagrams you could draw in which m∠AOB + m∠BOC does not equal m∠AOC. Consider yourself a winner if you can think of any of them.

2. Now for the first Honors Problem of the book. How should the Angle Addition Postulate (Postulate 5) be written so as to avoid the difficulties mentioned in question 1?

1. Fred was thinking of the angles looking like:

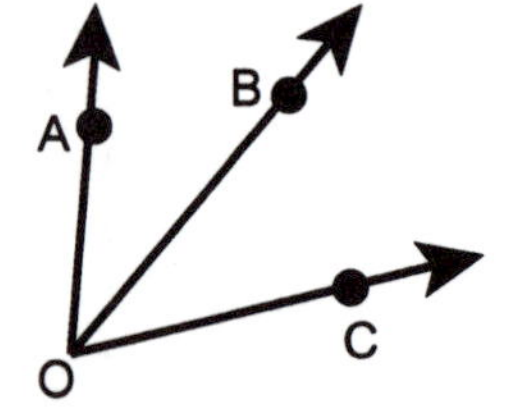

But what if the diagram were

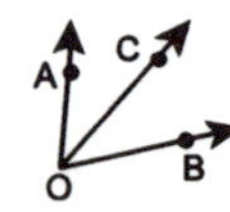

Then m∠AOB + m∠BOC ≠ m∠AOC.

The second possible difficulty would be if we were trying to add together two obtuse angles. The result would be a measure greater than 180°. In geometry* all angles will have a measure between 0° and 180° inclusive.

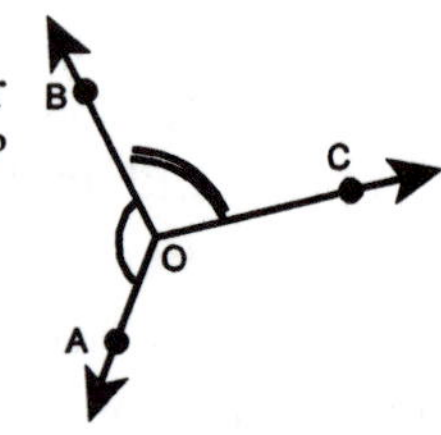

The third possible difficulty would be if we were working in space (in three dimensions) instead of just on a plane.

Imagine a three-legged stool. In this case, m∠AOB = m∠BOC = m∠COA. But everything we are doing in geometry will be on a single plane (in two dimensions) so this third possibility really isn't a possibility. This is plane geometry. (Almost at the end of this book, in chapter 12, we will do a little solid geometry (three dimensions).

---

* In trigonometry we will expand things a bit and have angles with a measure of 3349994872°.

2. Hey. This is an Honors Problem. The fun is in trying to figure it out, not in just reading the answer.

"Son," Friday began. "You know that what you did was wrong."

Fred nodded.

"And you know that you'll have to pay for what your llama did."

Fred nodded again and held out his $638.42 check again. This time the officer took it. It was a lot of money for Fred—more than a month's wages. (When he had turned six, KITTENS increased his salary from $500/month to $600/month.)

"And you know that you're never supposed to let that happen again."

Fred wasn't sure whether he was supposed to nod his head or shake it, so he said, "I won't, sir."

"And since you confessed your crime, the law says that your payment is sufficient punishment."*

Sergeant Friday was trying to look as stern as he could, but he couldn't hold it in any longer. First a little drop of sweat formed on his forehead. Then his tense lips began to tremble a little. Then a crack of a smile. Then a giggle. Then he burst out laughing. Uncontrollably.

Fred didn't know what to think.

(When this book is made into a movie, the role of Sergeant Friday will be one that will be sought after by many famous movie stars. The actor who plays this part well is almost assured of winning an Oscar for Best Supporting Actor as he goes through all of the emotions from sternness and anger through hilarity.)

After about five minutes, Friday settled down a little bit and uttered, "I'm sorry" and then laughed a little longer.

---

* The apocryphal law reads: KANSAS STATE PENAL CODE §349.90, PARAGRAPH C, SUBSECTION M2, SUBPARAGRAPH 89, AS AMENDED JULY 14, 1908, FOR THE WANTON DESTRUCTION OF UNIVERSITY BOCCI BALL LAWNS IN WHICH THE MALEFACTOR IN CONJUNCTION WITH A PET LLAMA DO WILLFULLY CHEW, EAT, OR MUNCH SAID LAWN, THE PENALTY WILL BE THIRTEEN YEARS IN THE STATE PRISON AND A FINE EQUAL TO THE AMOUNT NECESSARY TO RESTORE SAID LAWN. IN THE EVENT THE PERPETRATOR TURNS HIMSELF IN, THE JAIL TIME WILL BE WAIVED.

Wiping the tears from his eyes, he got up and went outside again and came back with the basket of money that they had been collecting all morning. It was the reward money. There was $70,000 in the basket.

He handed it to Fred.

In an Oscar-winning performance, Fred went from bafflement through incredulity to guffaw (missing giggle, snicker and titter completely).

As Fred left the police station $70,000 – $638.42 = $69,361.58 richer than when he had arrived there, he thought to himself, Life sure is funny. Sometimes when you're expecting the worst, good things happen, and sometimes when the skies seem clear, lightning strikes. He thought about a joke a student had told him: "What makes God laugh?" Answer: "Hearing our plans."

In geometry proofs there's some sense of order. Each step follows from the previous steps. In life it's not quite that way. On this side of the grave, you never know. I could lose this whole basket of money to a robber or a have a heart attack and the medical bill could be $69,361.58.

Instead of worrying about robbers or heart attacks, Fred amused himself by proving some easy geometry theorems in his head.

Theorem 3: Supplements of congruent angles are congruent.

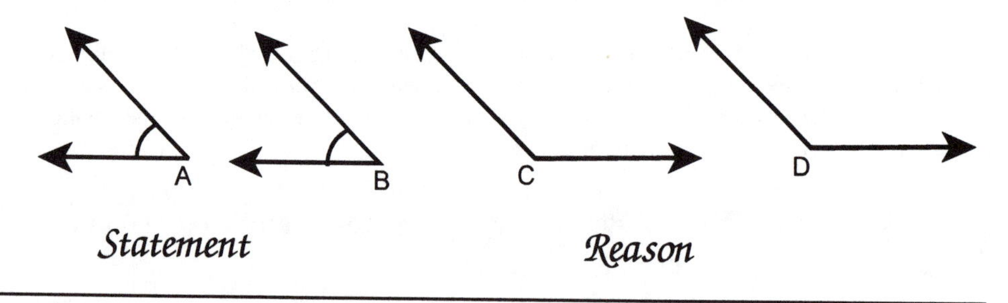

Statement	*Reason*
1. $\angle A \cong \angle B$	1. Given
2. $\angle A$ and $\angle C$ are supplementary. $\angle B$ and $\angle D$ are supplementary.	2. Given
3. $m\angle A = m\angle B$	3. Def of congruent angles
4. $m\angle A + m\angle C = 180°$ $m\angle B + m\angle D = 180°$	4. Def of supplementary angles
5. $m\angle C = m\angle D$	5. Algebra
6. $\angle C \cong \angle D$	6. Def of congruent angles

Six steps! That is the longest proof you have ever seen. Fred could have made it five steps long if he had combined steps 1 and 2 together:

Statement	Reason
1. ∠A ≅ ∠B ∠A and ∠C are supplementary. ∠B and ∠D are supplementary.	1. Given

Either way is fine. The only two rules for creating proofs are that every statement have a (proper) reason and the last statement is the thing that you want to prove. Have you seen the rules for golf or football or any of the board games you may have received as a birthday present? They are pages and pages long. The rules for doing math proofs are only two little tiny sentences: every step has a reason and stop when you get to the end.

In math, things tend to be short and juicy. We've only proved three official theorems thus far and look what page we're on. Theorem 1 (If P is the midpoint of $\overline{TF}$, then PF = ½ TF) took only three steps to prove. Theorem 2 (On any line ℓ, there are five different points on ℓ) also took only three steps.

The difficulty comes when you read the six-line proof of Theorem 3 as if you were reading *The Red Badge of Courage* or *Moby Dick*. Speed reading is not quite appropriate. Even what you might consider slow-speed reading isn't what may give you success in geometry. If you read the proof on the previous page, saying, "I get it. I get it. I get it. . . ," that still might have been a tad fast. The real test is whether or not you can close this book right now and prove that supplements of congruent angles are congruent. "Auswendig lernen"* is the way my German teacher used to express it. "C" students in geometry should be able to prove each of the theorems of geometry—to reproduce the proofs that have been given in this book, auswendig. D students should be able to follow the proofs that the C student gives. A and B students will be able to *create* proofs. The same would be true in a music composition class, *mutatis mutandis*.**

---

* Auswendig lernen = to learn something by heart

** That's a common Latin expression meaning, "after making the appropriate changes."

When you are reading history, say the eleven-volume *The Story of Civilization* by Will and Ariel Durant, if you skip a couple of pages, who is going to know? If you get the mumps and miss two weeks in October in your psychology class, you probably won't be "lost" when you come back to class. The weeks that you missed might have been when the teacher was talking about the physiology of the brain and now the teacher is talking about B. F. Skinner.

It's in math courses that students talk about getting l-o-s-t. That is because math builds on itself. To understand what is going on at any point, you need to understand what happened yesterday.

For example, in Fred's proof of two pages back, he went from

3. $m\angle A = m\angle B$
4. $m\angle A + m\angle C = 180°$
   $m\angle B + m\angle D = 180°$

and by Algebra declared in step 5 that $m\angle C = m\angle D$. If your beginning algebra is foggy, then you know what the weather report is for geometry.

Fred arrived at the bank just as it was opening. KITTENS Bank opens at 6:35 a.m. to accommodate all the early birds. Fred was naturally an early morning person and had assumed that everyone else was too. In his eight o'clock class he once cheerily repeated the saying, "The early bird gets the worm. That's what every mother bird tells her children." One sleepy-eyed student retorted, "And the mother worm tells her kids to stay in bed or they'll get eaten!"

The teller smiled and greeted Fred, "And what can we do for you today?"

Fred carefully put the basket on the counter and announced, "I'd like to make a deposit." And then he wondered to himself whether or not he was being slightly redundant. After all, he wasn't putting the money on the counter just to be admired.

The bank president was going to award Fred the "Depositor of the Year" medal, but since this was the largest deposit that KITTENS Bank had received in the last two years, they gave him the "Depositor of the Past & Present Year" medal.

After the deposit, the bank officers had trouble figuring out what to do with that much money. It wouldn't all fit in their safe.*

It was now 6:45 a.m. and time to get back to his office and make sure that Lambda was doing okay. In the three minutes it would take for him to get back to his office in the math building, he amused himself by proving the **Vertical Angle Theorem**.

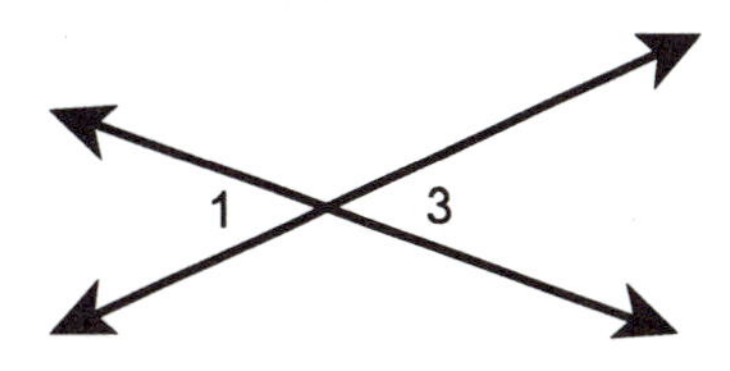

You could probably state the theorem even before you have seen it for the first time. It begins:

<u>Theorem 4</u>: Vertical angles are ____?_____ .

What would you guess goes in the blank? Let's make it a multiple-choice question. Vertical angles are . . .

A) raspberry flavored
B) often found in grocery stores
C) nudists
D) stars above the River Don as it empties into the Sea of Azov
E) congruent

If you guessed "often found in grocery stores" you were close. That is true and you should receive partial credit for your answer.** However, this would make a very difficult theorem to prove in geometry.

---

* Here's what the bank looked like after Fred's deposit.

** Some teachers who teach a combination Geometry & Photography course, often make the assignment: Go to the grocery store and take pictures of four vertical angles you find there. Students are frequently amazed at the number of vertical angles they find. They had never noticed them before. Much of education is like that. You learn to see things with "new eyes."

To start the proof, Fred drew the diagram on the blackboard in his mind and looked at it. "What's the given?" is almost always a good place to start. Next to the diagram he wrote:

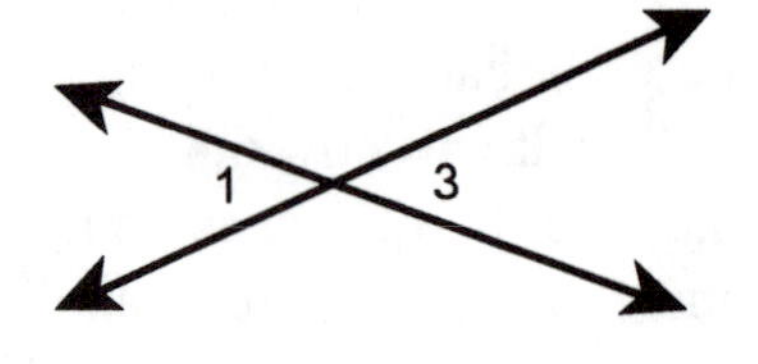

Statement	*Reason*
1. ∠1 and ∠3 are vertical angles.	1. Given

What's next? If you are stuck, one thing to do is to look over your Definition sheet, your Theorem sheet, your Postulate sheet. The only place that *vertical angles* is mentioned is on your Definition sheet: "Two angles are vertical angles if they are formed by intersecting lines and they are not a linear pair."

See any linear pairs?

It can't hurt.* We'll make that step 2.

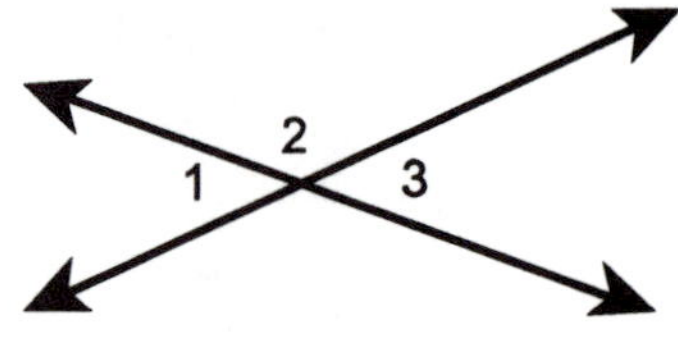

Statement	*Reason*
1. ∠1 and ∠3 are vertical angles.	1. Given
2. ∠1 and ∠2 form a linear pair. ∠2 and ∠3 form a linear pair.	2. Def of linear pair

---

* Even if we don't use that step, it doesn't violate any rules to throw it in. The only two rules in doing proofs is that every statement have a proper reason and the last statement is what you want to prove. When you have done that, you have a proof.

If you want a "pretty proof," you can always go back and ~~cross out~~ any steps that you didn't use.

If you want a long proof, you can always toss in extra steps like:

Statement	Reason
$18^2 = 324$	Algebra

Linear pair, Linear pair, Linear pair, Linear pair, Linear pair, Linear pair, Linear pair, Linear pair, does that ring a bell?
Turn to your list of postulates.

Postulate 1: Two points determine a line.
Postulate 2: (The Ruler Postulate) The points on a line can be matched up with the real numbers so that every point is matched up with just one number and every number is matched up with just one point.
Postulate 3: (Angle Measurement Postulate) You can match up every angle with a number between 0 and 180.
Postulate 4: **Linear pair** → supplementary.
Postulate 5: (Angle Addition Postulate) m∠AOB + m∠BOC = m∠AOB. (This postulate is still under construction. It's wrong as it currently stands.)

Postulate 4 looks like the only one that applies. We'll use it.

Statement	*Reason*
1. ∠1 and ∠3 are vertical angles.	1. Given
2. ∠1 and ∠2 form a linear pair. ∠2 and ∠3 form a linear pair.	2. Def of linear pair
3. ∠1 and ∠2 are supplementary. ∠2 and ∠3 are supplementary.	3. Postulate 4: If two angles form a linear pair, then they are supplementary.

Supplementary, Supplementary, Supplementary, Supplementary, Supplementary, Supplementary, Supplementary, Supplementary, Supplementary, Supplementary, Supplementary, does that turn on a light bulb?

On your Definitions Sheet:

Definition: ∠A and ∠B are **supplementary** if and only if m∠A + m∠B = 180°.

It can't hurt. It will make a nice step 4.

Statement	*Reason*
1. ∠1 and ∠3 are vertical angles.	1. Given
2. ∠1 and ∠2 form a linear pair. ∠2 and ∠3 form a linear pair.	2. Def of linear pair
3. ∠1 and ∠2 are supplementary. ∠2 and ∠3 are supplementary.	3. Postulate 4: If two angles form a linear pair, then they are supplementary.
4. m∠1 + m∠2 = 180° m∠2 + m∠3 = 180°	4. Def of supplementary angles.

Those are equations in line 4. If you subtract the second one from the first you get: $m\angle 1 - m\angle 3 = 0°$. Transpose the $m\angle 3$ to the right side of the equation: $m\angle 1 = m\angle 3$. That would make a nice step 5:

5. $m\angle 1 = m\angle 3$ 5. Algebra

Does that finish the proof?

Answer: No it doesn't. You are supposed to stop when your last line is the thing you want to prove. You were trying to prove $\angle 1 \cong \angle 3$. You have (on line 5) $m\angle 1 = m\angle 3$. Getting from $m\angle 1 = m\angle 3$ to $\angle 1 \cong \angle 3$ isn't too hard:

Statement	*Reason*
1. $\angle 1$ and $\angle 3$ are vertical angles.	1. Given
2. $\angle 1$ and $\angle 2$ form a linear pair. $\angle 2$ and $\angle 3$ form a linear pair.	2. Def of linear pair
3. $\angle 1$ and $\angle 2$ are supplementary. $\angle 2$ and $\angle 3$ are supplementary.	3. Postulate 4: If two angles form a linear pair, then they are supplementary.
4. $m\angle 1 + m\angle 2 = 180°$ $m\angle 2 + m\angle 3 = 180°$	4. Def of supplementary angles.
5. $m\angle 1 = m\angle 3$	5. Algebra
6. $\angle 1 \cong \angle 3$	6. Def of $\cong$ ∠s

"∠s" is the common abbreviation for "angles." The abbreviation of *abbreviation* is abbrev.

And the plural of Δ is ∆s.

Any kind of knowledge
gives a certain
amount
of power.
—Aimee Buchanan

Camas

1. We have two different words in geometry that describe statements that are proved. What are they?

2. Name the four possible kinds of things that can be used as reasons in a proof. (This is the current list. We will soon expand that list to five items.)

3. We never did define $\overline{AB}$. In chapter one, we just said that $\overline{AB}$ was the line segment with endpoints A and B. (It was just too early to start throwing all kinds of definitions at you.) Now, however, is a good time to define $\overline{AB}$. Please do so.

4. Did we ever define $\overleftrightarrow{AB}$, the line through points A and B?

5. Prove: *If $\overrightarrow{PQ}$ and $\overrightarrow{PR}$ are opposite rays, then Q–P–R.* (This is a pretty short proof. Please attempt it before you look at the answer below.)

6. At KITTENS University, they changed over from the quarter system to the semester system. Every three units under the old system now became two units under the new system. 3 old units = 2 new units.

Darlene had 130 units under the quarter system. Showing your work and circling your conversion factor, find how many new semester units she would have.

answers

1. Theorem and proposition. A proposition is a lightweight theorem.

2. Given, postulate, definition, and algebra.

3. Your definition may differ slightly from mine. That's okay. <u>Def</u>: $\overline{AB}$ consists of points A and B and all points C on $\overleftrightarrow{AB}$ such that A–C–B.

4. We didn't need to. It was one of the three undefined terms.

5.

S	*R*
1. $\overrightarrow{PQ}$ and $\overrightarrow{PR}$ are opposite rays.	*1.* Given
2. Q–P–R	*2.* Def of opposite rays

6. 130 old units × $\frac{\text{2 new units}}{\text{3 old units}}$ = 86⅔ new units.

Hamlet

1. Which of these four possible kinds of reasons in a proof—given, postulate, definition or algebra—is superfluous (more than is required) and is just there for our convenience?

2. Theorem 3 reads: Supplements of congruent angles are congruent. Sometimes it's helpful to know that supplements of *the same* angle are congruent. Supply the missing statements and reasons in the proof of this new statement.

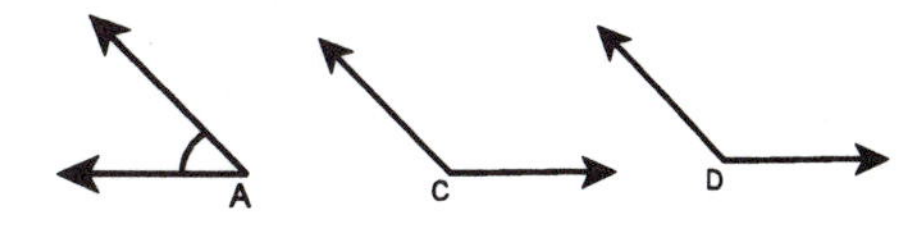

S	*R*
1. ∠A and ∠C are supplementary. ∠A and ∠D are supplementary.	*1.* ?

2. m∠A = m∠A	*2.* ?
3. ∠A ≅ ∠A	*3.* ?
4. ?	*4.* Theorem 3

3. One (or more) of the following statements is not possible. Which one(s)?

① PQ = $\sqrt{7}$

② the coordinate of A is –8

③ BC ÷ DE = 4

④ $\overline{FG} + \overline{GI} = \overline{FI}$

⑤ HH = 0

⑥ JK = –3

4. You have had some time to think about the first Honors problem which was presented on p. 76. Now it is time to write Postulate 5, the Angle Addition Postulate. Recall, the incorrect statement was: *For any two angles, ∠AOB and ∠BOC,* m∠*AOB* + m∠*BOC* = m∠*AOC.*

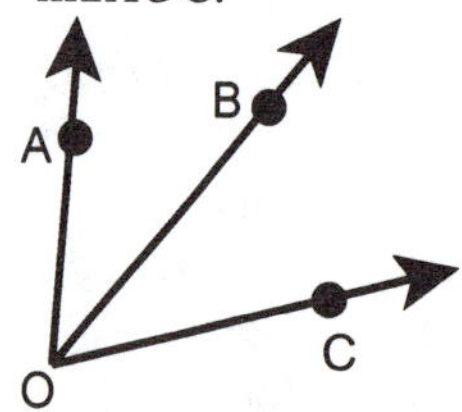

answers

1. We could do all our proofs and never use any definition. Every time, for example, that we encountered ΔABC, we could replace that with its definition, "the geometrical object consisting of three non-collinear points A, B and C and the line segments $\overline{AB}$, $\overline{BC}$ and $\overline{CA}$." It wouldn't take long for our proofs to become incredibly wordy.

2.

S	*R*
1. ∠A and ∠C are supplementary. ∠A and ∠D are supplementary.	*1.* Given
2. m∠A = m∠A	*2.* Algebra
3. ∠A ≅ ∠A	*3.* Def of ≅ ∠s
4. ∠C ≅ ∠D	*4.* Theorem 3

The biggest surprise is the reason for step 4. It is a previously proven theorem. This is our newest permissible kind of thing we may use as a reason. It is a lot like using a definition as a reason—both of them are superfluous. Instead of writing, "Theorem 3," I could have just inserted the entire proof of Theorem 3 in its place.

Theoretically, whenever you see a proof, you could remove every reason that was a definition or a previously proven theorem and you would obtain a proof resting just on the original undefined terms and postulates. It might turn out to be horribly long and virtually unreadable.

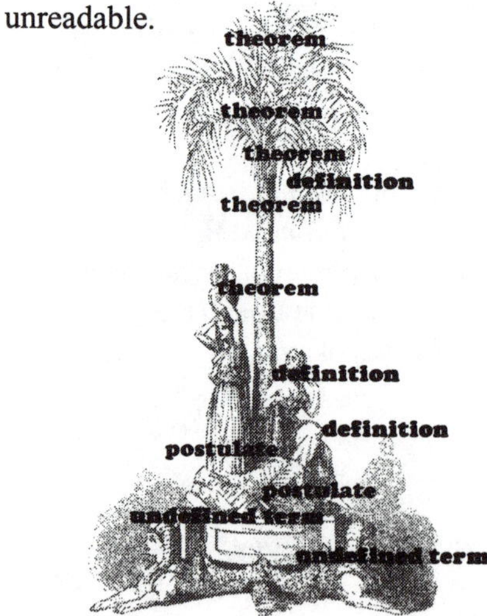

The proofs of theorems can rest on previously proven theorems. It makes things a lot nicer.

The proof of, "Supplements of the same angle are congruent," followed directly from Theorem 3, which is "Supplements of congruent angles are congruent." Such a statement is called a **corollary**. (CORE-o-larry) On your Theorems page, write: Corollary to Theorem 3: Supplements of the same angle are congruent.

3. ④ and ⑥. You can't add line segments together and the distance between points is never negative.

4. Somehow we need to make $\overrightarrow{OB}$ lie between $\overrightarrow{OA}$ and $\overrightarrow{OC}$, but we have never defined betweenness for rays. Here's one possible (correct) formulation of Postulate 5: (Angle Addition Postulate) For any two angles, $\angle AOB$ and $\angle BOC$, if A–B–C, then $m\angle AOB + m\angle BOC = m\angle AOC$.

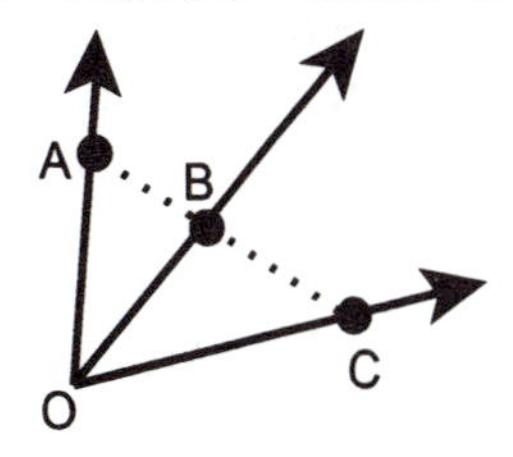

There are other ways to avoid the three possible difficulties that were discussed in problem 1 on p. 76. You may have discovered one of the other ways. That is okay.

Having said all that, we are going to keep things simple. When we do proofs in this book, we will skip the official requirement that A–B–C. Instead, if we have an $\angle AOC$ and if point B is "inside" of $\angle AOC$ as you look at the diagram, then by the Angle Addition Postulate, we will write that $m\angle AOB + m\angle BOC = m\angle AOC$.

It's nice to have a little freedom.

In contrast to algebra where there is only one correct answer to $3x = 12$, the postulates, definitions, and proofs in geometry can often be done in several different ways. In the next chapter, we will discuss three completely different proofs that the base angles of an isosceles triangle are congruent.

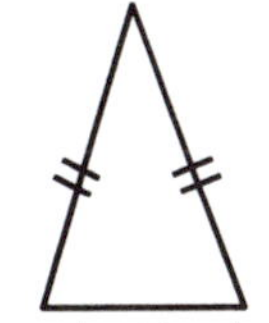

implies

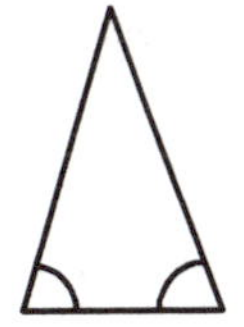

An isosceles triangle is one in which at least two sides are equal in length.

The base angles are conguent if they have the same size.

Oversimplifying a little bit, algebra is `rigid and mechanical` with one way to do things. Geometry offers *creative opportunities*. It's the difference between making a blueprint of a bridge and making a painting of a bridge.

This is not to say that geometry is better than algebra. It is a matter of taste. Some people like to make blueprints. Others love to sketch.

Santa Barbara

1. We can now prove "Theorem 4: Vertical angles are congruent" in just four steps (instead of the original six steps) since we are now allowed to use previously proven theorems. (Previously proven theorems include previously proven propositions and corollaries.) Here's the diagram. Please supply the four-step proof. (Hint: the first three steps are identical to our original proof.)

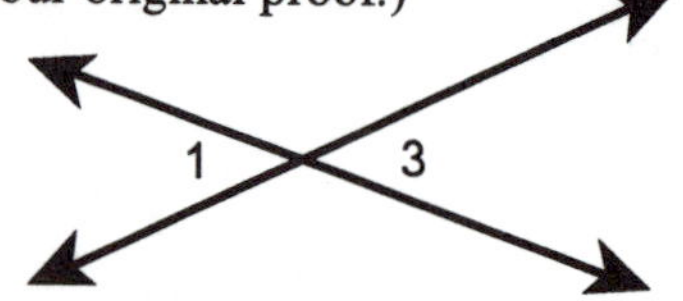

2. Name the five permissible kinds of reasons one may use in proving geometry theorems.

3. If $\angle R$ is obtuse, what can we say about the possible value of $333(m\angle R)$?

4. The advertisement at PieOne read: "Six pizzas for 75 frogskins!" Compute how many frogskins five pizzas would cost. Circle your conversion factor.

5. Look at the picture and then complete the following definition.

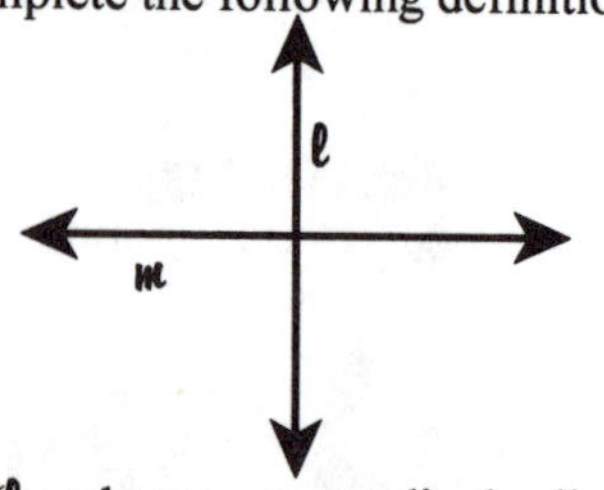

(ℓ and m are perpendicular lines)

Definition: Two lines are **perpendicular** if and only if. . . .

6. We define perpendicular rays:

Definition: $\overrightarrow{AB} \perp \overrightarrow{CD}$ if and only if $\overleftrightarrow{AB} \perp \overleftrightarrow{CD}$. With that hint, it isn't hard to define perpendicular segments. Do it.

odd answers

1.

S	*R*
1. $\angle 1$ and $\angle 3$ are vertical ∠s.	*1.* Given
2. $\angle 1$ and $\angle 2$ form a linear pair. $\angle 2$ and $\angle 3$ form a linear pair. *[Insert a "2" in the diagram at this point.]*	*2.* Def of linear pair
3. $\angle 1$ and $\angle 2$ are supplementary. $\angle 2$ and $\angle 3$ are supplementary.	*3.* Def of supplementary ∠s
4. $\angle 1 \cong \angle 3$	*4.* Corollary to Theorem 3

3. Since $90° < m\angle R \le 180°$, then multiplying by 333, we obtain: $29970° < 333(m\angle R) \le 59940°$.

5. There are several different ways to complete this definition. Perhaps the easiest is: Definition: Two lines are **perpendicular** if and only if they form right angles.

If $\overleftrightarrow{AB}$ and $\overleftrightarrow{CD}$ are perpendicular, we symbolize this as $\overleftrightarrow{AB} \perp \overleftrightarrow{CD}$.

Upsala

1. If you know that $m\angle ABC = 0°$, then must it be true that either B–A–C or B–C–A?

2. By what reason would I be able to assert that there exists an angle whose measure is equal to 8.9963°?

3. If $493(m\angle P) = 43877°$, must $\angle P$ be acute?

4. Complete the following definition.
Definition: Angle A is a right angle if and only if. . . .

5. Prove the proposition: *If two $\angle$s form a linear pair and are $\cong$, then they are right angles.*

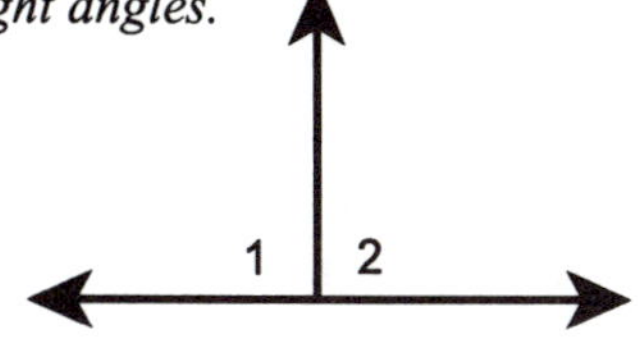

6. Convert 2° into seconds. Show your conversion factors.

7. Angle bisectors will come in handy when we work with triangles in the next chapter. The angle bisector of $\angle AOC$ looks like:

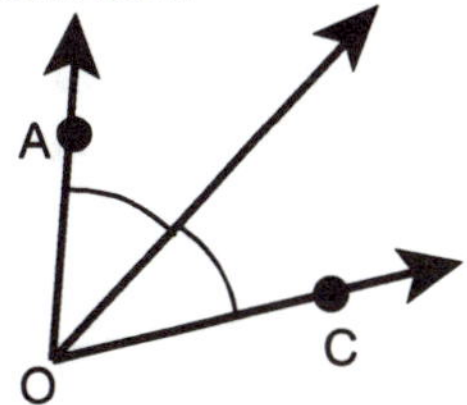

Complete the definition:
Def: $\overrightarrow{OB}$ is the **angle bisector** of $\angle AOC$ if and only if. . . .

8. Does every angle have a supplement?

odd answers

1. yup.

3. Just barely. Dividing both sides of the equation by 493, we obtain $m\angle P = 89°$.

5.

S	*R*
1. $\angle 1$ and $\angle 2$ are a linear pair. $\angle 1 \cong \angle 2$	*1.* given
2. $m\angle 1 = m\angle 2$	*2.* Def of $\cong$ $\angle$s
3. $\angle 1$ and $\angle 2$ are supplementary	*3.* Post. 4
4. $m\angle 1 + m\angle 2 = 180°$	*4.* Def of supplementary $\angle$s
5. $m\angle 1 = 90°$ $m\angle 2 = 90°$	*5.* Algebra
6. $\angle 1$ and $\angle 2$ are right angles	*6.* Def of right angle

7. Def: $\overrightarrow{OB}$ is the **angle bisector** of $\angle AOC$ if and only if
$m\angle AOB = m\angle BOC$.

(If we want to prevent the kind of troubles we had with the Angle Addition Postulate, we should also toss into the definition the requirement that A–B–C. That will prevent the backward angle bisector pictured above.)

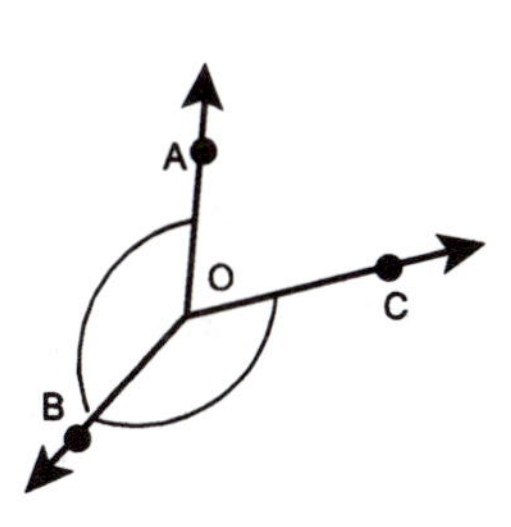

Industry

1. If the coordinate of point P is 5 and the coordinate of point Q is –4, find PQ.

2. Fill in the blank with one word: If $m\angle DEF = 180°$, then $\angle DEF$ is called a _____?_____ angle.

3. If A–B–C, then must it be true that C–B–A?

4. Prove that if $\angle A$ and $\angle B$ are right angles, then they must be congruent.

5. If ∠C and ∠D are both acute angles, must ∠C ≅ ∠D?

6. If ∠C and ∠D are both acute angles, must m∠C + m∠D > 90º?

7. If ∠C and ∠D are both acute angles, must m∠C – m∠D > 0?

8. If ∠C and ∠P are supplementary angles, then we say that ∠P is the **supplement** of ∠C.

If ∠C and ∠D are both acute angles, and if ∠P is the supplement of ∠C, and if ∠Q is the supplement of ∠D, and if m∠C > m∠D, then must it be true that m∠Q > m∠P?

Kalona

1. If A–B–C and the coordinates of A and C are 6 and 345 respectively, what can you say about the coordinate of B?

2. Prove: *If ∠2 and ∠3 form a linear pair and if ∠1 and ∠3 are supplementary, then ∠1 ≅ ∠2.*

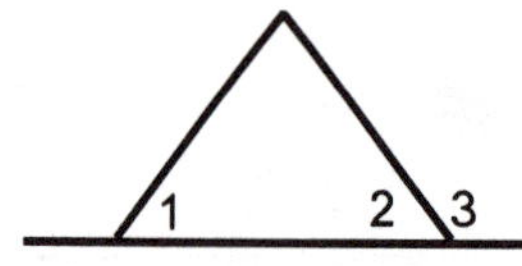

3. One (or more) of the following statements is *not* possible. Which one(s)?

① $\overrightarrow{AB} > \overrightarrow{CD}$

② –3.98 < EF

③ line ℓ and line *m* are different lines and they intersect at points P and Q and PQ = 4

④ line ℓ and line *m* are different lines and they intersect at points P and Q and PQ = 0

⑤ AB = 8, BC = 2, AC = 6

4. Find a pair of supplementary angles on Fred's face.

5. Convert 17' to degrees.

6. Convert 17' to seconds.

7. Some of the following are false. For each false statement, draw a picture to show that it is false.

① If two angles are adjacent, then they are a linear pair.

② If two adjacent ∠s are ≅, then their common side is the angle bisector of the angle formed by the sides that the angles do not share in common.

③ Supplements of angles which are not congruent must also not be congruent.

Chapter Three
Triangles

Fred finished proving the Vertical Angle theorem just as he got to the door of his office/home. He hesitated. *Should I knock?* he wondered. *After all, llamas may want a little privacy.* Upon further reflection he decided that if he knocked he might wake her up and that would not be very good.

He quietly opened the door and tiptoed in. It was 6:48 A.M. and it was still pretty dark outside, but he didn't turn on the room lights. Instead he sat at his desk and turned on a small table lamp. Then he could work on preparing his math lectures for the day without disturbing Lambda.

He looked at his teaching schedule for the day:

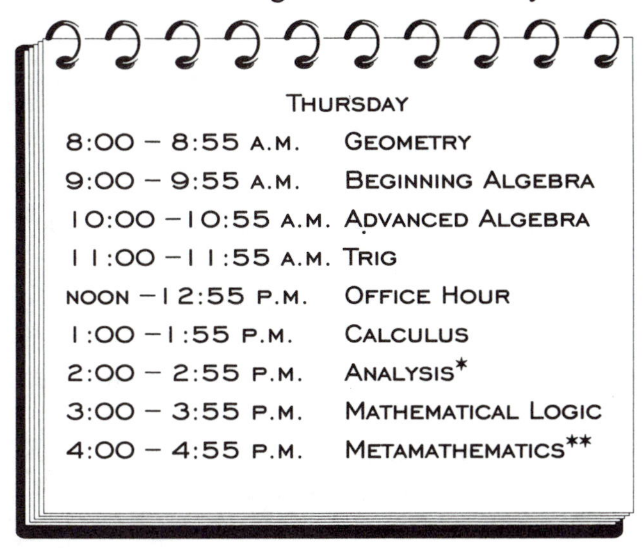

THURSDAY
8:00 – 8:55 A.M. GEOMETRY
9:00 – 9:55 A.M. BEGINNING ALGEBRA
10:00 – 10:55 A.M. ADVANCED ALGEBRA
11:00 – 11:55 A.M. TRIG
NOON – 12:55 P.M. OFFICE HOUR
1:00 – 1:55 P.M. CALCULUS
2:00 – 2:55 P.M. ANALYSIS*
3:00 – 3:55 P.M. MATHEMATICAL LOGIC
4:00 – 4:55 P.M. METAMATHEMATICS**

---

* Analysis, roughly speaking, is the theory behind calculus.

** Metamathematics, again roughly speaking, studies mathematics itself. In geometry, we create a theory dealing with points and lines. In metamathematics, we create a theory dealing with the things of mathematics itself such as definitions, theorems, and proofs. In metamathematics, there are theorems about . . . theorems! It is a graduate course which you take after you have a bachelor's degree in math.

Fred just loved teaching. The administration only required three hours per day in the classroom, but didn't object when Fred asked for permission to teach the other classes without pay.

And today in geometry (his favorite course), he was going to start lecturing on triangles (his favorite part of geometry). The proofs about triangles were clean and neat.

He looked across the room in the semi-darkness toward what he called "Lambda's office" and hoped that she was resting well. *Maybe 18 miles was a little long for our first jog*, he reflected.

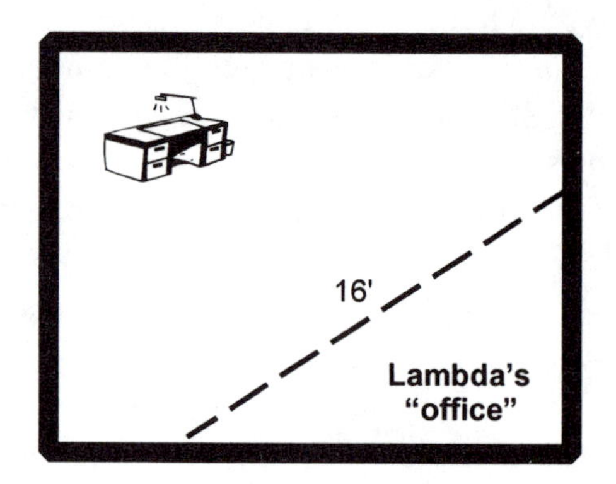

Layout of Fred's Office

Fred had constructed her nest using some fencing that he had found in the general storage closet in the math building. The fencing formed the longest side (called the **hypotenuse***) of the **right triangle** (that's a triangle with a right angle) which was her part of Fred's office. The shorter two sides of a right triangle are called the **legs**. Since many of his students often visited Fred during his office hours, the use of Lambda's office as an example of a right triangle would be a perfect illustration to use in his geometry lecture today.

He pulled his digital camera out of a desk drawer. Fred mumbled to himself, *A picture of a happy sleeping llama inside of a right triangle will be the perfect thing to shine on the wall of the lecture hall. Everyone will surely remember what a right triangle is. I'm sure everyone loves my little "llamakins."*

---

* pronounced high-POT-en-noose. The hypotenuse of a right triangle is sometimes described as the side opposite the right angle. It has absolutely nothing to do with being high because pot is in use.

He turned on the camera and walked slowly over to the edge of the fence. Just then, the sun began to shine through his office window. I'll call this picture, "Dawn Breaks Over Lambda." He pointed the camera.

He couldn't see her through the camera's viewfinder. He took the camera away from his eye and looked.

No Lambda.

Maybe she's under all the paper towels I put in her nest. He put his hand through the fence and carefully felt around in the two feet of bedding.

No Lambda.

Fred raced over to the light switch and turned on the overhead fluorescents. He pulled the fence away and kicked the towels around. He stepped in the bowl of Sluice and then slipped on some of the llama drool.

No Lambda.

"LAMBDA!" Fred cried out.

Fred ran to his desk and slipped again. He was covered with pieces of paper towels. He flipped open the campus phone directory and then dialed the campus police: 007.

No answer. Sergeant Friday has probably gone home to take a nap. He isn't used to having to get up so early in the morning to solve crimes. He dialed Friday's home number: 007½.*

Mrs. Friday answered the phone with a cheerful, "Good morning!"

Fred announced who he was (which is always a polite thing to do when you make a phone call) and then asked to speak with her husband.

---

* Five years ago when Fred first moved into his office, he noticed that the phone only had whole numbers (= {0, 1, 2, . . .}) on the dialing pad. He jokingly mentioned this to Joe who was one of his students. Joe took Fred seriously and began a campaign to liberate the phone system. He marched in front of the phone company with a sign, "The Whole Nos. Ain't Good Enough!" After weeks of enduring Joe's protest, the phone company installed minus signs on the key pads of all the phones so now you could dial integers (= {. . . −3, −2, −1, 0, 1, 2, 3, . . .}).

Unsatisfied, Joe wrote an editorial for *THE KITTEN Caboodle*, "No Half-Way Measures! Give Us More Numbers!" which ended with the stirring line, ". . . and as for me, give me the rational numbers or give me death!" (Joe liked to use a lot of exclamation points.) After several more weeks of pressure, the phone company again relented and installed virgules on every phone on campus. (A virgule is a diagonal stroke that some people call a slash. "/") Now people on campus could dial any rational number (= any number that can be written as x/y where x and y are integers and y ≠ 0.)

"I'm afraid Frank has gone to bed. He had a tough case to solve this morning and came home exhausted."

Fred explained that this was an emergency.

In a few moments, a sleepy voice announced, "This is Friday. Can I help you."

The gravity of the situation overwhelmed Fred. He started crying. "My . . . my Lambda. I can't find her. She's not here." As he said that, he realized what must have happened. It was no longer a matter of reporting a lost-and-found case. "Sir, I think my pet has been llama-napped."

"I'll be right over," Friday told him, and then, since he had heard it so often in the movies, he uttered the famous lines, ". . .and don't touch anything. We may have to dust for fingerprints."

After Friday hung up the phone, he turned to his wife and said, "Poor kid. His pet llama—you know, the one that wrecked the bocci ball lawn—has escaped. I gotta go help find it."

Fred dutifully sat in his chair waiting for the police officer to arrive. Occasionally, he would blow his nose on one of the paper towels that was stuck to him.

Friday's trip from his home (F) to the math building (M) was almost as circuitous as the one Fred had made earlier from the math building to the police station. First he headed from his home (F) to the Edison phonograph store (E) to pick up a record he had ordered. Then to the Five & Dime (D) to buy some penny candy. At that point he realized that he had forgotten to take the lunch that his wife had packed for him, so he headed back home (F). From there to the shoe-repair store (S) to pick up his spectators which he was having resoled. Then to the camera store (C) to pick up some Polaroid® film. A quick trip back home (F) to pick up his camera for his Polaroid® film and then he hurried to the crime scene at the math building (M).*

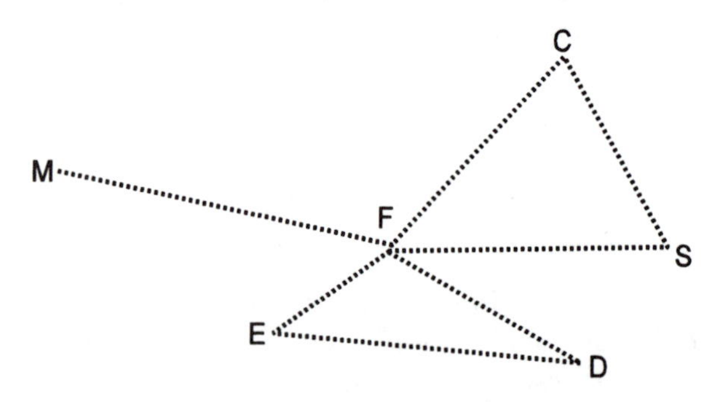

* To find out how old you are, count the number of things (n) you didn't recognize in this paragraph. Your age = 60 – 10n.

Your Turn to Play

1. There are two triangles in the route that Friday took. Neither is a right triangle. Name the obtuse triangle and give a definition of an obtuse triangle. It will begin, Definition: A triangle is an **obtuse triangle** if and only if. . . .

2. Name the acute triangle in Friday's route.

3. Name the largest number that will make the following statement true: Every triangle has at least ___?___ acute angles.

4. Define **acute triangle**.

5. Fill in the correct number: Every triangle has at most ___?___ obtuse angle(s).

6. When we look at the *angles* of a triangle, we can classify triangles as right triangles, acute triangles, or obtuse triangles. When we look at the *sides* of a triangle there are two mutually exclusive possibilities. A triangle is either isosceles or scalene. Twice in chapter two we defined an isosceles triangle. Complete the following: Definition: An **isosceles triangle** is one in which. . . .

7. Any triangle that is not isosceles is called a scalene triangle. Without using the word "isosceles," define a scalene triangle.

COMPLETE SOLUTIONS

1. ΔFED is an obtuse triangle. Definition: A triangle is an **obtuse triangle** if and only if it contains an obtuse angle.

2. ΔFSC is an acute triangle.

3. Every triangle must have at least two acute angles. (It might have three acute angles, but it must have at least two.)

4. Def: An **acute triangle** is a triangle in which all of the angles are acute. (It would be silly to say that it is a triangle which has an acute angle, since every triangle has at least two acute angles.) You should be able to fill in the blanks mentally in the following definitions.

An ___?___ triangle has an ___?___ angle.

An ___?___ triangle has three ___?___ angles.

5. If you try and draw a triangle with two obtuse angles, things get mighty weird. Even attempting to draw a triangle with two right angles can often get one in a lot of difficulty. (In chapter 11½ we will create such a triangle! Mathematics, like life, can contain an occasional surprise or two—an understatement.)

6.

An **isosceles triangle** is one in which at least two sides are equal in length, or the definition could read: A triangle is **isosceles** if and only if it has at least two congruent sides.

7. Def: A **scalene triangle** is a triangle in which no two sides are congruent.

It was a quarter after seven and Fred was getting a little antsy waiting for Sergeant Friday. At eight he had to be at Archimedes Hall to present his beginning algebra lecture. He was sitting very still, being careful not to get fingerprints on anything. This seemed a bit silly since he had lived in his office for more than five years.

When Friday entered Fred's office, he was shocked. Fred sat at his desk chair (on three phone books to give him enough height) with paper towels plastered all over him. The room smelled strongly of llama. The floor was damp from an overturned bowl of Sluice. It was sticky because of the $C_{12}H_{22}O_{11}$ (sugar) contained in the Sluice. (Sluice's motto: "A pound in every pint."*) Fred's eyes were very red.

"Explain what happened, son."

"No Lambda."

"Could you give me a few more details?"

"She's not here. She was here. She's not here anymore."

Friday scratched his head. There wasn't much to go on. It was a true mystery. He tried a different tack: "Did anyone see her leave the room?"

"Yes."

"Who?"

"Lambda. No. I'm not sure about that. If the llama-nappers blindfolded her before they took her, then she wouldn't have seen herself leave the room."

"Now, son, how do you know that someone kidnaped her?"

---

* When Sluice changed over to metric, their motto was ruined. "A kilogram in every liter" just didn't have the same alliteration.

Fred was silent. Couldn't think of an answer. His brain had done what many computers often do—it had crashed.

Sergeant Friday urged Fred to get to the bathroom and clean himself up.

Then Friday spotted a tiny clue. There were drag marks in the llama drool from the Lambda's nest out the office door. Following the marks, he found they led down the hall, then down two flights of stairs, out across the parking lot and to the dumpster. Friday knew he was "hot on the trail" when he saw four stiff cold llama legs sticking out of the top of the dumpster. He pushed the legs over and shut the lid.

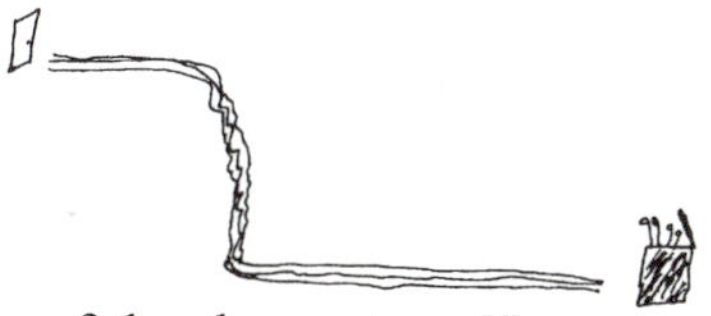

Back in the building, he knocked on the door marked: *Samuel P. Wistrom, Chief Educational Facility Math Department Building KITTENS University, Inspector/Planer/Remediator for offices 225–324.* "Hi Sam. What's with the dead llama down in the dumpster?"

"Llama?" Samuel responded. "Don't know nothing about no llama. Don't even know what one of them things is. Only thing I do know about is that mule that Professor Fred had. You know it up and died. Didn't have the heart to leave it there in Fred's office and have Fred find it all dead and everything. So I took it down and stuck it in the dumpster. Besides, that durn thing stunk so much. I'm sure Fred will be glad to be rid of it."

Friday thanked Sam and headed down the hallway to give Fred the bad news. This was toughest part of being a cop.

Fred returned from the bathroom after having washed his face and putting on a new clean bow tie. The bow tie was important since his first class was 45 minutes away and he often liked to wear a bow tie when he taught.

"Son, I've got bad news for you," Friday announced.

"Do you want the reward money back?"

"No. That's yours to keep. You earned it."

"Have you drawn up a list of suspects for the crime?"

"No. I haven't thought of anybody who may have committed a crime."

"Did you put out an all-points bulletin to search for my pet?"

"No. It wouldn't make any sense since I'm the only cop on campus."*

"Well, do you think there's any hope?"

At this point, the Sgt. Frank Friday just didn't know what to say. He didn't have the heart to tell the six-year-old that his favorite pet was dead. He chickened out and told Fred to go see the chaplain at the university chapel. He said that the chaplain knew all the details. Both Friday and Fred worshiped at the chapel each Sunday, so the sergeant knew that he wasn't sending Fred to a stranger.

As Fred hurried across campus, Friday phoned Reverend Fosdick and told him everything. He concluded his call, "I just don't have the heart to tell this six-year-old . . . you know . . . when I look into his little eyes, I just can't."

Fosdick assured the officer, "Don't worry. I'll take care of telling him. Each of us has our points of bravery. You have a job that calls for physical bravery—pointing your gun at violent criminals. That's something that would scare me a lot."

Friday blushed a little as he hung up the phone. As a KITTENS campus cop, he didn't even own a gun. He didn't need one. He knew that Fosdick was actually a much braver man than he was. He had heard stories of how Fosdick had served as an army chaplain down in Texas and how he had been called many times to bars to help soldiers who had drunk too much get back to base. These soldiers were often depressed and sometimes violent. The bartenders who called Fosdick knew that if they called the police instead, the soldier would probably be arrested and receive a black mark on his military record.

As Fred entered the chapel, he was met by Reverend Fosdick. They went and sat on a pew together. In words appropriate for a six-year-old, he explained to Fred that Lambda was "no longer with us," and that his pet was "with Jesus as you and I shall be someday." Fred had the premonition

---

* Sharp-eyed readers may be wondering about the police cars that were patrolling all over campus in chapter one. They were members of the Kansas National Guard who were called out by the Kansas governor after he declared a state of emergency. (Destruction of university bocci ball lawns = state of emergency.) After the crime was solved, they were demobilized leaving Sergeant Friday as the only law enforcement officer on campus.

that he was going to be told that his pet had died. The assurance that she was in a happy place now was a comfort to him. He looked forward to the day that he would see Lambda again.

(Although llamas are not specifically mentioned in the Bible, Fosdick's words to Fred were based on the description of the life to come in the Bible's next-to-last chapter in which "He will wipe away all tears. There will be no more death, no more grief or pain." If Fred needed a llama to make that come true, then Fred would have a llama.)

As Fred left the chapel to head toward Archimedes Memorial Lecture Hall where he taught his classes, he looked back at the chapel.

The stained-glass windows on the side of the building were all the same size and shape. In a word, those triangles were congruent.

Your Turn to Play

1. What would be wrong with: *Definition: Two triangles are congruent if and only if they are the same size and shape* ?

2. Here are two triangles that are congruent. What angle in the second triangle matches up with angle G in the first triangle?

3. What side in the second triangle corresponds to $\overline{GH}$?

4. If two triangles are the same shape and size, then we know that corresponding angles must be congruent, and corresponding sides must be congruent. This will be the basis for our definition. Fill in the blank: <u>Def</u>: ΔABC is **congruent** to ΔDEF if and only if

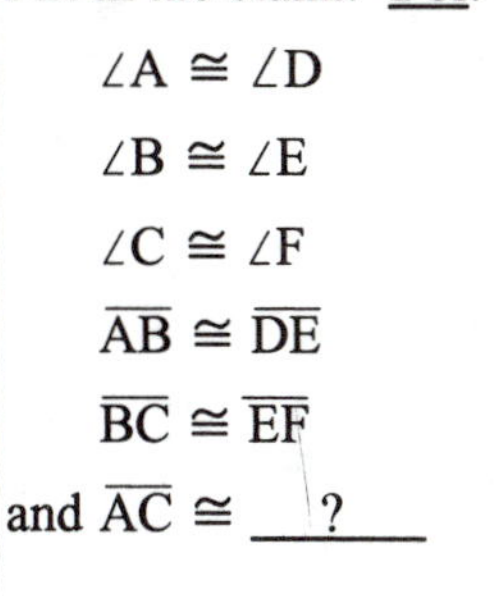

$\angle A \cong \angle D$

$\angle B \cong \angle E$

$\angle C \cong \angle F$

$\overline{AB} \cong \overline{DE}$

$\overline{BC} \cong \overline{EF}$

and $\overline{AC} \cong$ ___?___

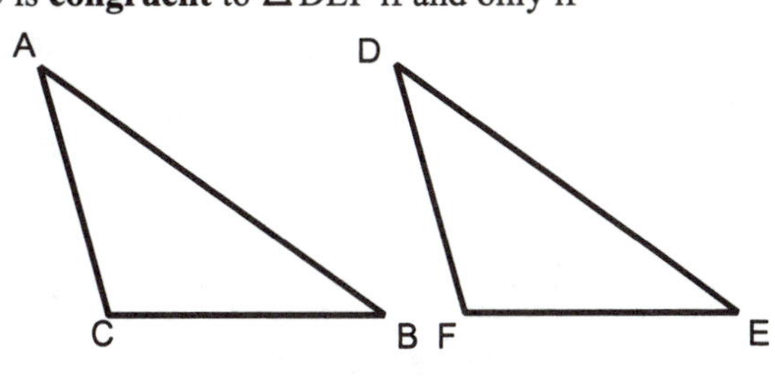

---

5. When we write that $\Delta ABC \cong \Delta DEF$, the order in which the vertices of the triangle are listed is important. Even without the diagrams of the triangles, if you are given that $\Delta PQR \cong \Delta JMT$, you know that $\angle P$ matches up with $\angle J$, etc. Fill in the four missing congruences: Given $\Delta PQR \cong \Delta JMT$, we know that

$\angle P \cong \angle J$, $\angle Q \cong \angle M$, . . .

6. Fill in the missing reason in the following proof:

Statement	*Reason*
1. $\Delta LMN \cong \Delta HIJ$	1. Given
2. $\angle M \cong \angle I$	2. ?

COMPLETE SOLUTIONS

1. Officially speaking, we don't know what "size" and "shape" mean. They aren't in the list of undefined terms (point, line, and plane) and they haven't been previously defined.

2. $\angle H$ and $\angle P$ match up with each other since they both look like they are nearly right angles. $\angle I$ and $\angle S$ match up with each other since they are the smallest angles. That leaves $\angle Z$ to match up with $\angle G$.

3. $\overline{GH}$ is opposite the smallest angle (which is $\angle I$). $\overline{PZ}$ is opposite the smallest angle in the second triangle ($\angle S$).

4. $\overline{DF}$

5. $\angle R \cong \angle T$, $\overline{PQ} \cong \overline{JM}$, $\overline{QR} \cong \overline{MT}$, $\overline{PR} \cong \overline{JT}$.

6. Def of $\cong$ $\triangle$s

Fred noticed a new poster on the campus bulletin board.

The new toppings (which for some reason PieOne called flavors) didn't sound very appealing to Fred, but the thought of not having a crust around the edge was fascinating. There is, of course, crust underneath.

He had just eaten last night, so he wasn't quite hungry yet. (It takes something more than an 18-mile early-morning jog to awaken hunger in Fred.)

But what did attract his attention was the task of phoning in an order for a triangular pizza. In the old days of circular pizzas, you'd just say, "I'd like an eight-inch pizza." Now, in order to specify the exact size and shape, you might say, "I'd like a Tri-Peetz with

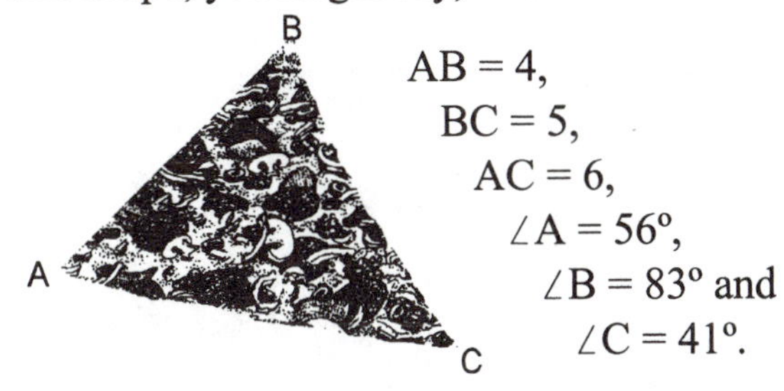

Then you knew, by definition of congruent triangles, that the size and shape of pizza that they would make would be the same as the one you were thinking of. Two triangles are congruent if all three pairs of corresponding sides are congruent and all three pairs of corresponding angles are congruent.

That's overkill, thought Fred. By the time I've told them what the lengths of the three sides of the triangle are, I have said enough. If they make a pizza with sides equal to 4, 5 and 6 inches, it has gotta be the same size and shape as the one in my head.

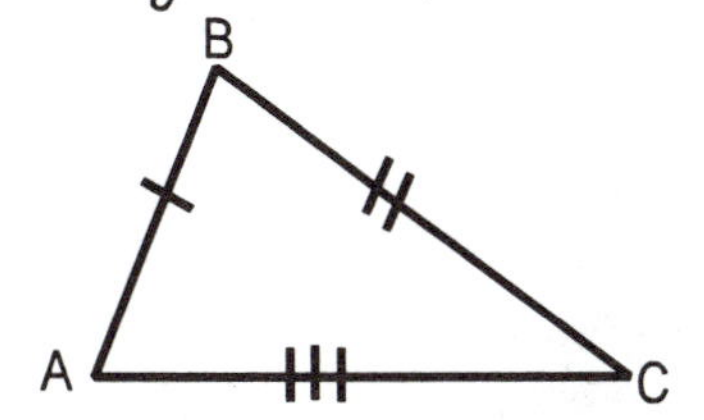

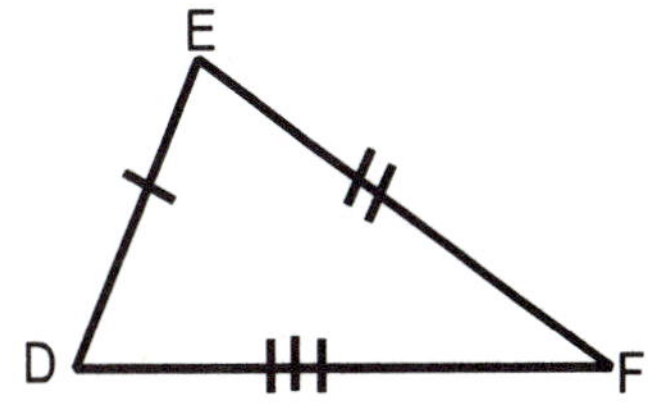

<u>Postulate 6</u>: If AB = DE, BC = EF, and AC = DF, then $\triangle ABC \cong \triangle DEF$.

This is called the Side–Side–Side Postulate. Or the S.S.S. Postulate. Or just SSS. It says that if all three corresponding pairs of sides are congruent, then the triangles are congruent. And that, in turn, by the definition of congruent triangles, means that all three corresponding pairs of angles are congruent.

Your Turn to Play

1. Could we have an SS postulate? If we ask for a pizza with one side equal to 4" and the second side equal to 5", would that specify the size and shape of the Tri-Peetz that we would receive?

2. Could we have an AAA postulate. If we order a pizza with angles of 41°, 56° and 83°, have we said enough to fix the size and shape of the triangular pizza?

3. What about SAS? If we ask for a Tri-Peetz with one side equal to 4" and a second side equal to 6" and the *included* angle (between the two sides) equal to 56°, have we said enough?

4. What about SSA? If we order ΔABC with AB = 4", BC = 5" and $\angle C = 41°$, do we know, for sure, what we'll receive?

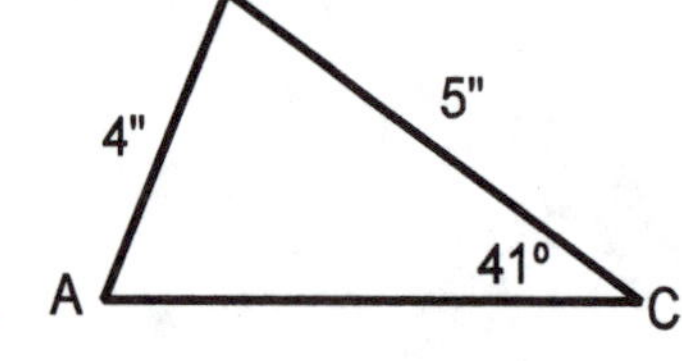

5. Finally, how about ASA? If we specify two angles and the *included* side (between the angles), have we said enough to completely determine the size and shape of the triangle?

COMPLETE SOLUTIONS

1. and have two pairs of corresponding sides congruent, but the triangles aren't the same size and shape. Fred used to draw an alligator on the blackboard. Its "lips" didn't change length when it opened its mouth.

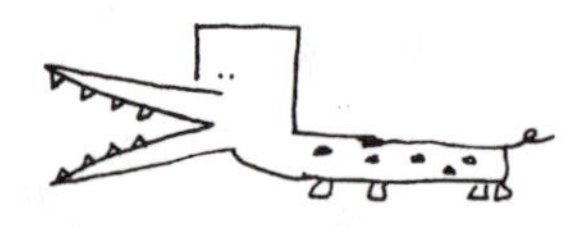

2. Ask photographers about AAA. When they do enlargements of photographs in the darkroom, the angles of the triangle don't change, but the size of the triangle does.

wallet-size photo

regular size

3. SAS is a winner. Stating two sides and the included angle is enough to completely determine what the triangle will look like. After all, how many ways are there to complete this triangle?

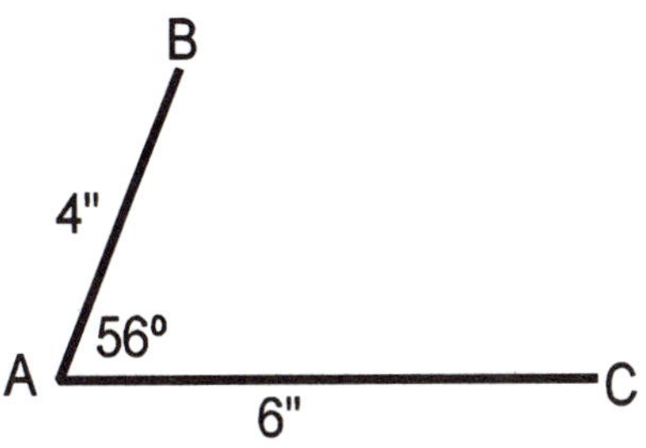

This gives us our second triangle-congruence postulate.

Postulate 7: (SAS) If AB = DE, AC = DF, and $\angle A \cong \angle D$, then $\Delta ABC \cong \Delta DEF$.

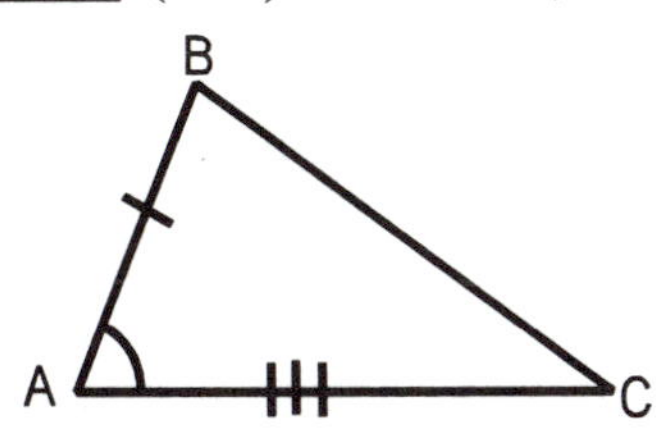

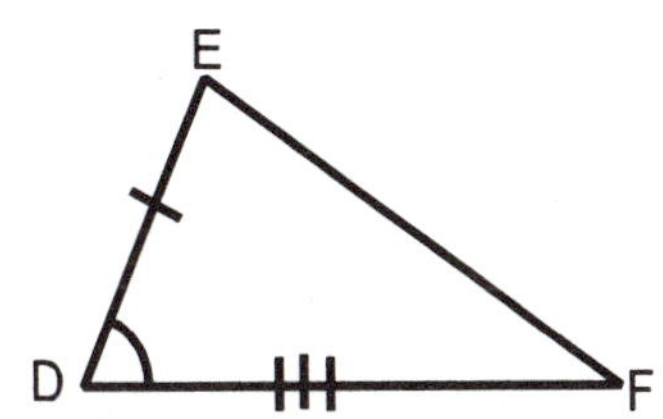

4. SSA doesn't work. There are often two possibilities. In specifying AB = 4", BC = 5" and $\angle C = 41°$,

we might have gotten a pizza that looked like

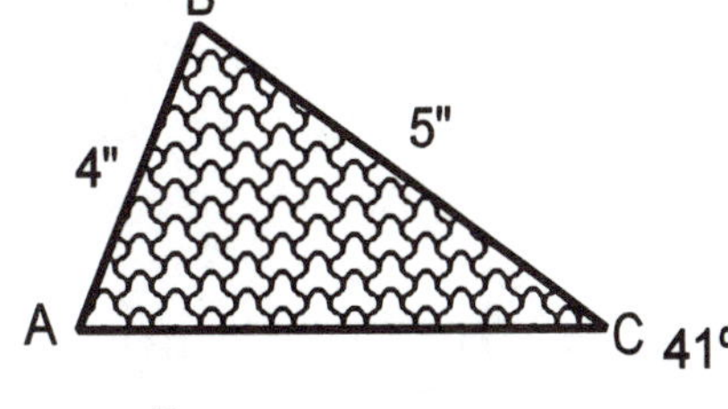

or we might have received

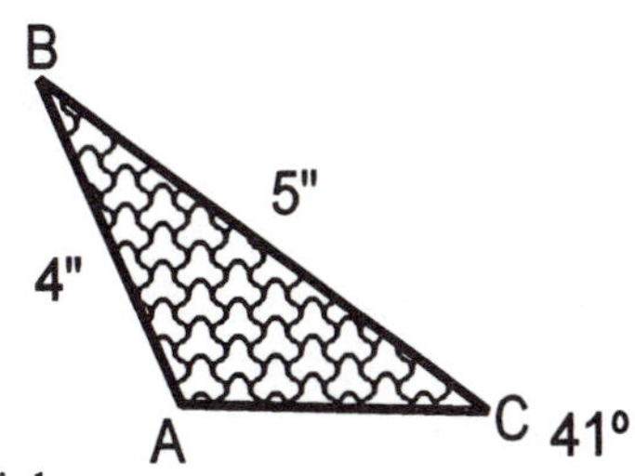

The way to remember that SSA doesn't work is to think of the word in the dictionary that means both donkey and a stupid or stubborn person. One does not wish to make an SSA of oneself.

5. With the ASA (which one of my students pointed out were the initials of the American Sunbathing Association), we have a triangle that looks like

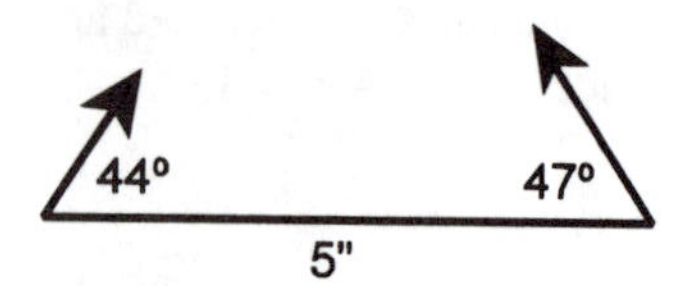

There is only one way to finish drawing that triangle. Its size and shape are completely determined. Any two triangles that have two angles with measures of 44° and 47° and an included side whose length is 5" must be congruent.

This gives us our third triangle-congruence postulate:

Postulate 8: (ASA) If $\angle A \cong \angle D$, $\angle C \cong \angle F$, and $AC = DF$, then $\triangle ABC \cong \triangle DEF$.

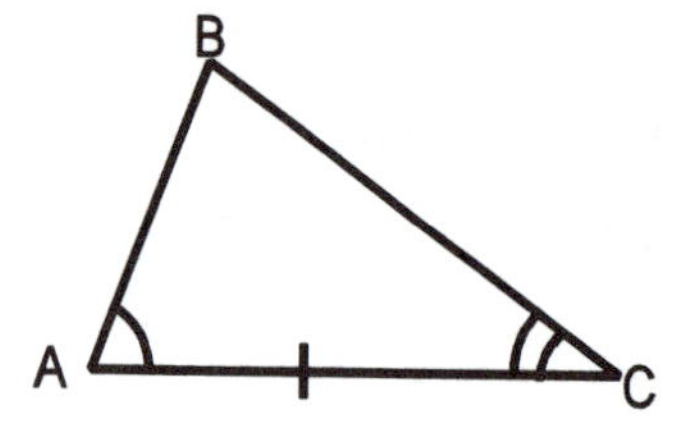

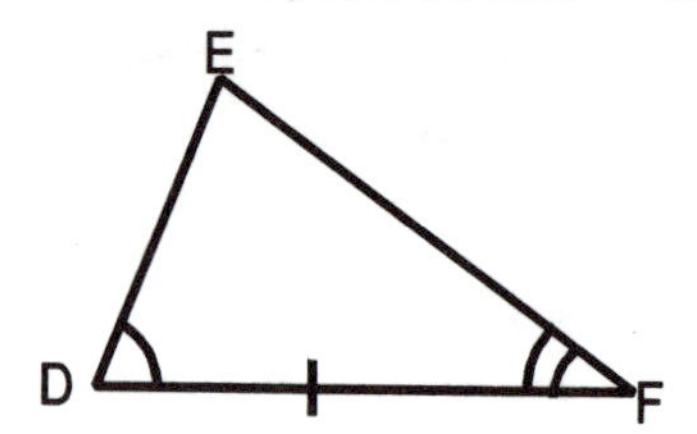

These three new postulates can be referred to as Postulate 6, Postulate 7, and Postulate 8, but much more commonly they are called SSS, SAS, and ASA.

Fred laughed to himself. There was finally a use for a protractor: If you worked at PieOne and someone called in an order for a 44°–5"–47° pizza, you would need that piece of plastic in order to cut the correct angles. (However, if you don't work for PieOne, you still probably don't need one.)

Fred met Joe and Darlene who were also heading to Archimedes Memorial Lecture Hall. This was not really surprising since they were taking Fred's eight o'clock geometry course. They liked this eight o'clock class for three reasons: ① they were learning geometry; ② they enjoyed hearing Fred's stories; and ③ they also liked sitting next to each other in class and holding hands.

Joe was on his cell phone talking with Stanthony at PieOne. "Can we order a Kissing Tri-Peetz?" This was an invention of Darlene's and so Joe had to explain it to Stanthony.

"To make a Kissing Tri-Peetz, first you have to make a big X on the pizza. The X stands for a kiss just like when you get a letter in the mail with an X just before the signature, it means a kiss.

"And, then you take the X," Joe continued,

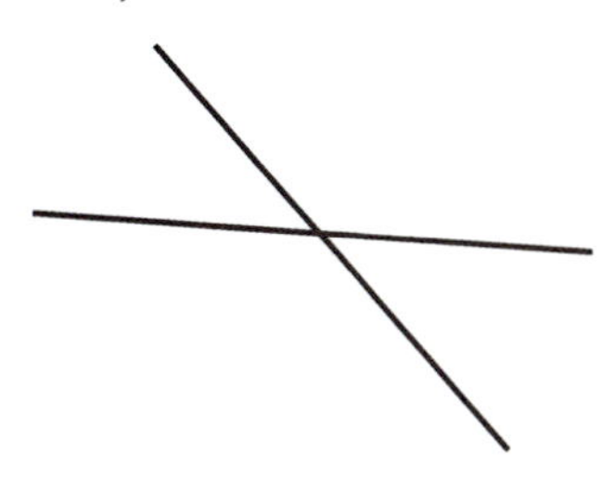

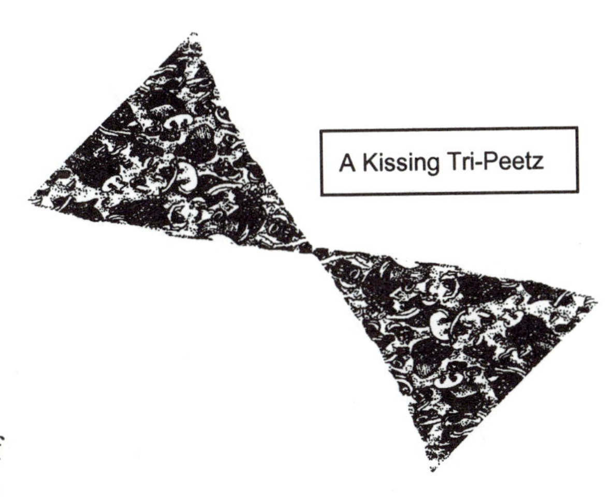

"and chop the third sides off to make the sides that touch the kiss equal." What Joe meant was:

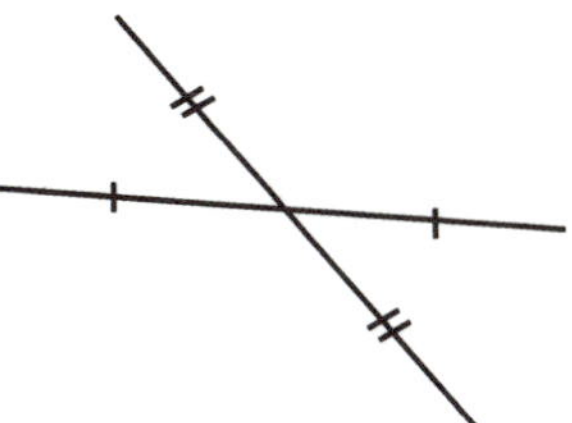

although he didn't express it very clearly.

After the third sides were chopped off:

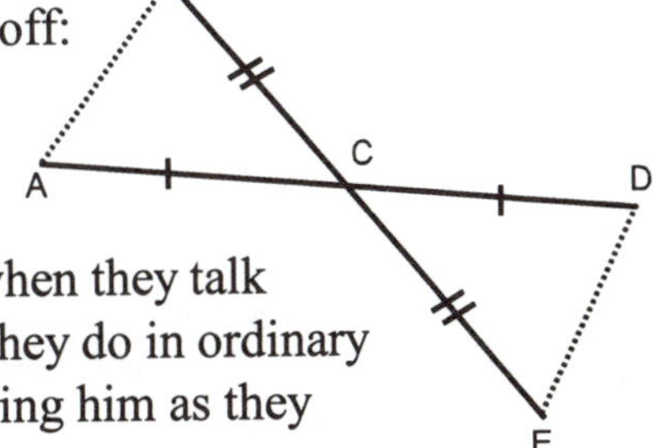

Joe (and a few other individuals) when they talk on a cell phone, speak a little louder than they do in ordinary conversation. Fred couldn't help overhearing him as they walked to class together.

After Joe hung up (or switched off or whatever you do when you terminate a cell-phone call), Darlene leaned over and kissed him. "In honor of our Kissing Tri-Peetz." Joe was kind of embarrassed, being kissed in front of others.

Fred, in order to cover Joe's obvious discomfort, switched the subject a bit. "You know, there's a geometry proposition right there in the Kissing Tri-Peetz that you invented Darlene."

[To the reader: With what is marked on the diagram above, please attempt to prove that the triangles are congruent, **before** you see Fred's proof on the next page. Note that if you write $\triangle ABC \cong \triangle CDE$ you have the letters in the wrong order.]

Here's Fred's proof:

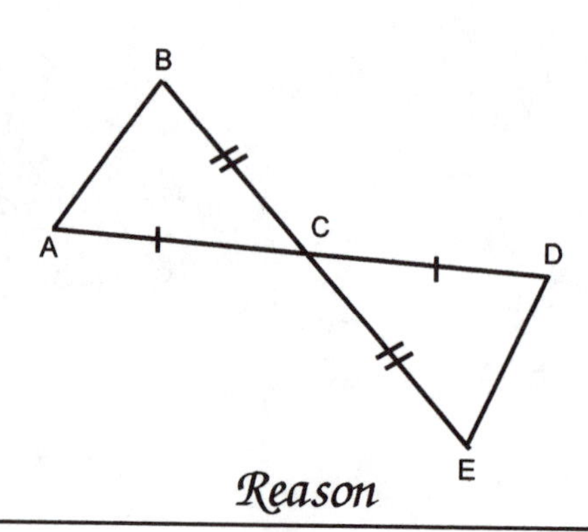

Statement	*Reason*
1. $\overline{AC} \cong \overline{CD}$, $\overline{BC} \cong \overline{CE}$	1. Given
2. $\angle ACB \cong \angle DCE$	2. Vertical Angle theorem (Vertical angles are congruent.)
3. $\triangle ABC \cong \triangle DCE$	3. ASA

Since those two triangular pieces are congruent, they have equal areas and so both Joe and Darlene will be happy with their respective pieces. (*Congruent triangles have equal areas* is a postulate we will have when we get to areas in chapter seven. But we won't mention that yet since it's too early.)

Wait a minute! you, the reader, exclaim. **You just did! You can't get away with that.**

I just did.

But why did you do that? Couldn't you wait till chapter seven to start talking about area. Everyone knows that area belongs in chapter seven and not in chapter three.

There's a reason.

Yes. And do tell me, what is that reason?

Do you remember in chapter one when I mentioned that if we have an isosceles triangle, then the base angles must be equal?

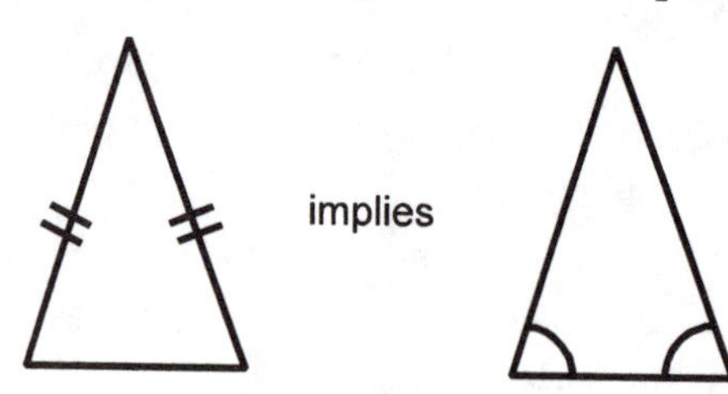

Do I remember!! You mentioned it four times so far. You did it on pp. 37, 45, 53 and 106. Who could forget?

My point precisely. The Isosceles Triangle Theorem is probably one of the top five theorems of geometry. Now you already know it before we will get around to officially introducing it and proving it.

And when will that ever happen? How many more sneak previews before we actually see it?

In round numbers, the answer to your question is 0.

1. State the Isosceles Triangle Theorem.
2. What is the hypothesis of the ITT?
3. What is the conclusion of the ITT?
4. Here is a proof of the ITT (as it's sometimes called). Fill in the missing parts.

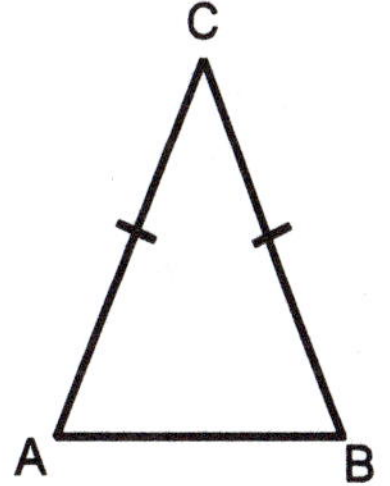

Statement	*Reason*
1. $\overline{AC} \cong \overline{BC}$	1. ?
2. AM = MB	2. Definition of midpoint
3. $\overline{AM} \cong \overline{MB}$	3. ?
4. CM = CM	4. ?
5. $\overline{CM} \cong \overline{CM}$	5. ?
6. $\Delta ? \cong \Delta BCM$	6. ?
7. $\angle A \cong \angle B$	7. ?

The line segment $\overline{CM}$ was not in the original diagram. The fancy name for such an extra segment that we draw in during the course of the proof is an **auxiliary segment**. Of course, there are **auxiliary lines** and **auxiliary rays** and **auxiliary** anythings that you draw in during the course of a proof.

5. The converse of the statement, "If a hexagon has six sides, then moderate exercise makes you feel better," is the statement, "If moderate exercise makes you feel better than a hexagon has six sides." What is the converse of the ITT?

6. The converse of a true statement need not necessarily be true. For example, the converse of the true statement, "If I smoke a pack a day, then I will be dead two centuries from now," is the false statement, "If I'm dead two centuries from now, then I smoked a pack a day." What about the converse of the ITT? Draw some pictures and then make a guess as to whether or not it is true.

COMPLETE SOLUTIONS

1. The Isosceles Triangle Theorem: The base angles of an isosceles triangle are congruent.

2. and 3. The hypothesis is underlined and the conclusion is double underlined:

If a triangle is isosceles, then the base angles are congruent.

4.

Statement	Reason
1. $\overline{AC} \cong \overline{BC}$	1. Given
2. AM = MB	2. Definition of midpoint
3. $\overline{AM} \cong \overline{MB}$	3. Definition of congruent segments
4. CM = CM	4. Algebra
5. $\overline{CM} \cong \overline{CM}$	5. Definition of congruent segments
6. $\triangle ACM \cong \triangle BCM$	6. SSS
7. $\angle A \cong \angle B$	7. Definition of $\cong$ ⚠

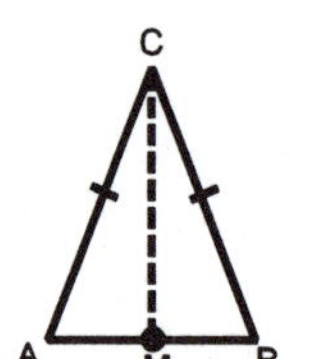

5. The converse of the Isosceles Triangle Theorem is *If the base angles of a triangle are congruent, then the triangle must be isosceles.*

6. If you drew the pictures, then you probably came to the correct conclusion that the converse of the ITT is also true.

You did it! You crossed the *Pons Asinorum.*

I did what? you exclaim.

You crossed the *Pons Asinorum*—that's what they call the ITT.

Yeah, I got that. What's the Latin mean in plain English?

The Isosceles Triangle theorem and its proof is the traditional dividing line between those who will make it in geometry and those who won't. *Pons* means "bridge." The *asinorum* are those that don't make it over the bridge. The modern translation of *asinorum* is something like, "unlit lightbulb."

asinorum crossing bridge

There is a creek that runs through the KITTENS campus. To get to Archimedes Memorial Lecture Hall, Fred, Darlene, and Joe had to cross a little bridge.

As they crossed, Fred was reminded of a famous geometry theorem. Since Darlene and Joe were busy holding hands and talking about what Fred called "gushy stuff," he decided not to interrupt them.

With a piece of chalk he wrote out a second proof of the Isosceles Triangle theorem on the floor of the bridge. You, as reader, may take your choice as to which one you'd like to read about.

Choice #1 Takes one minute

Darlene: Did you hear about Alexander and Betty? They've set the wedding date for this summer. Isn't that romantic?

Joe: Yeah. I guess. I hope they're happy.

Darlene: Oh, I'm sure they are. Going to a wedding is so exciting. Being married to the one you love is what everyone always dreams about. Here in this bride's magazine that I subscribe to is a description of what a modern bride wears. Would you like to look at it?

Joe: Maybe later. I'm chewing gum right now and it's difficult to concentrate on reading.

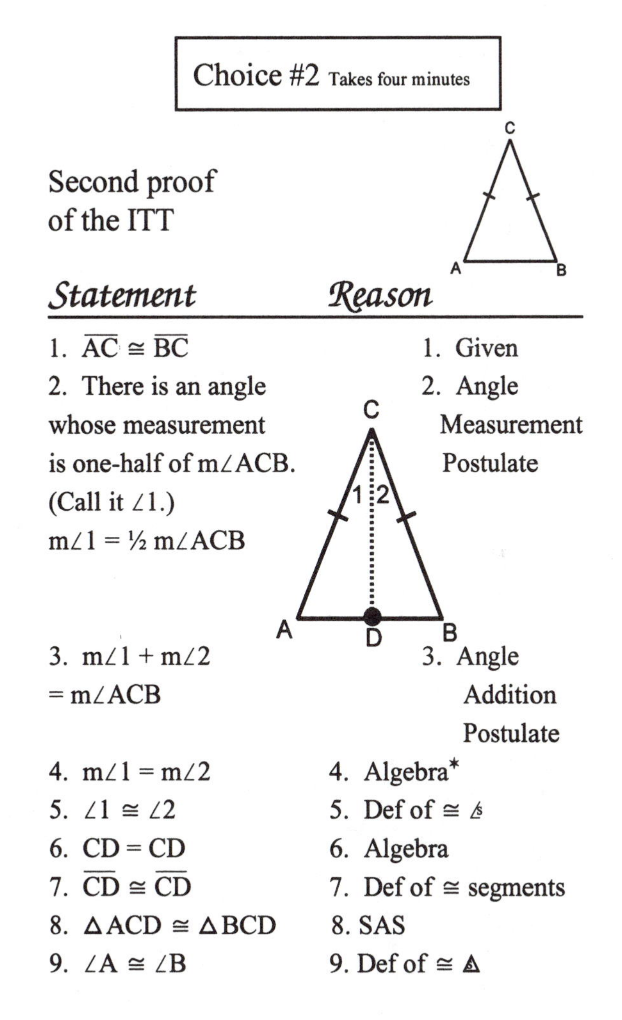

Choice #2 Takes four minutes

Second proof of the ITT

Statement	*Reason*
1. $\overline{AC} \cong \overline{BC}$	1. Given
2. There is an angle whose measurement is one-half of $m\angle ACB$. (Call it $\angle 1$.) $m\angle 1 = \frac{1}{2}\, m\angle ACB$	2. Angle Measurement Postulate
3. $m\angle 1 + m\angle 2 = m\angle ACB$	3. Angle Addition Postulate
4. $m\angle 1 = m\angle 2$	4. Algebra*
5. $\angle 1 \cong \angle 2$	5. Def of ≅ ∠s
6. CD = CD	6. Algebra
7. $\overline{CD} \cong \overline{CD}$	7. Def of ≅ segments
8. $\triangle ACD \cong \triangle BCD$	8. SAS
9. $\angle A \cong \angle B$	9. Def of ≅ △s

✶ The equations in steps 2 and 3 ($m\angle 1 = \frac{1}{2}\, m\angle ACB$ and $m\angle 1 + m\angle 2 = m\angle ACB$) by algebra yield $m\angle 1 = m\angle 2$. It's the same as saying that $x = \frac{1}{2}z$ and $x + y = z$ imply that $x = y$.

Cambria

1. A triangle is **equilateral** if and only if all three sides are congruent. If $\triangle ABC$ and $\triangle DEF$ are both equilateral, must it be true that $\triangle ABC \cong \triangle DEF$?

2. A triangle is **equiangular** if and only if all three of its angles are congruent. On scratch paper draw an equiangular triangle and estimate the size of each of the angles.

3. The proof of the proposition that if a triangle is equilateral, then it is equiangular doesn't take many steps. Fill in the missing parts.

S	*R*
1. $\triangle ABC$ is equilateral.	*1.* ?
2. $\overline{AB} \cong \overline{BC} \cong \overline{CA}$	*2.* ?
3. $\angle A \cong \angle B \cong \angle C$	*3.* ?
4. $\triangle ABC$ is ?	*4.* ?

4. If in some proof we have the statement that $\triangle DEF \cong \triangle GHI$, there are five possible reasons that might be used to justify that statement. Name them.

5. Given the diagram as marked, prove that $\angle B \cong \angle D$.

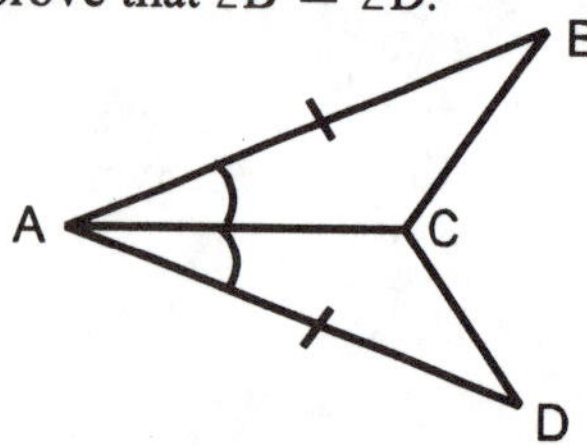

6. May an obtuse triangle be equilateral?

answers

1. They would be the same shape but not necessarily the same size.

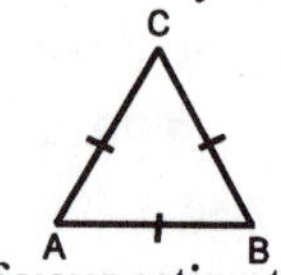

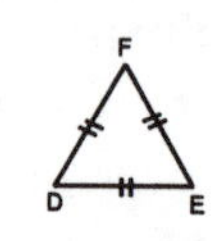

2. If your estimate came near 60°, you are doing good.

3. *1.* Given
2. Def of equilateral $\triangle$
3. Isosceles Triangle Thm
4. . . . equiangular
4. Def of equiangular $\triangle$

4. It might be because $\triangle DEF \cong \triangle GHI$ was given. It might be, "Definition of congruent triangles," which could be used if we knew (in a previous line) that $\overline{DE} \cong \overline{GH}$, $\overline{EF} \cong \overline{HI}$, $\overline{DF} \cong \overline{GI}$, $\angle D \cong \angle G$, $\angle E \cong \angle H$, and $\angle F \cong \angle I$. The other three possible reasons are SSS, SAS, ASA.

5.

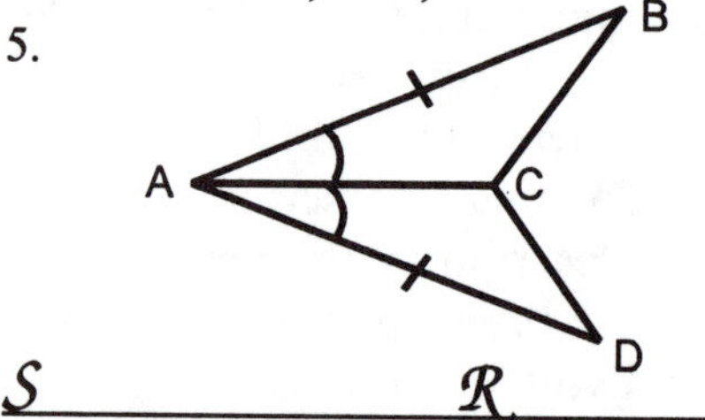

S	*R*
1. $\angle BAC \cong \angle DAC$	*1.* Given
2. $\overline{AB} \cong \overline{AD}$	*2.* Given
3. $AC = AC$	*3.* Algebra
4. $\overline{AC} \cong \overline{AC}$	*4.* Def of $\cong$ segments
5. $\triangle BAC \cong \triangle DAC$	*5.* SAS
6. $\angle B \cong \angle D$	*6.* Def of $\cong$ ⚠

6. Drawing some pictures, we note that the side opposite the obtuse angle is always larger than either of the other two sides.

Parker

1. A computer was once asked to prove the Isosceles Triangle Theorem. It came up with an intriguing proof. It wasn't the first proof which we gave on p. 108 in which we drew an auxiliary line that was a **median** (the segment in a triangle whose endpoints are a vertex of the triangle and the midpoint of the opposite side).

Nor was the proof that the computer came up with one in which we drew an auxiliary line that was an angle bisector.

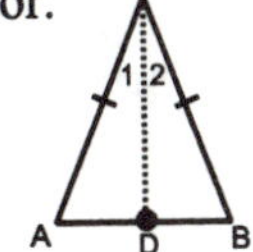

The computer's proof didn't use any auxiliary lines at all.

Here is the computer's proof. Fill in the missing reasons.

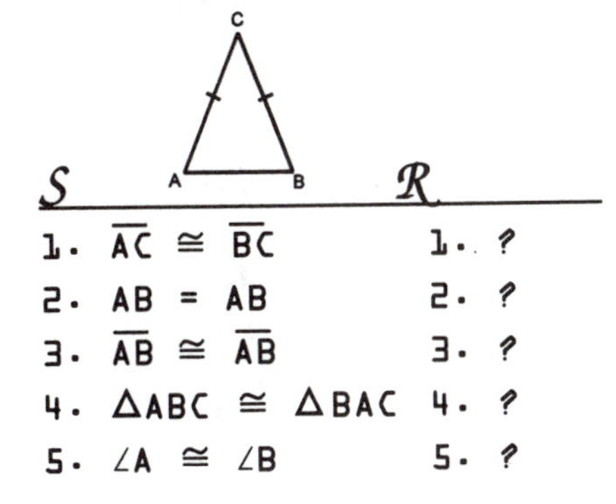

S	*R*
1. $\overline{AC} \cong \overline{BC}$	1. ?
2. AB = AB	2. ?
3. $\overline{AB} \cong \overline{AB}$	3. ?
4. $\triangle ABC \cong \triangle BAC$	4. ?
5. $\angle A \cong \angle B$	5. ?

2. Using the computer's pancake flipping approach (from the previous problem), prove the converse of the ITT.

3. In $\triangle$GHI, if we know that GH ≠ GI, must the triangle be scalene?

4. Given $\angle A \cong \angle 1$ and $\angle 2 \cong \angle C$, prove that $\overline{AD} \cong \overline{CD}$.

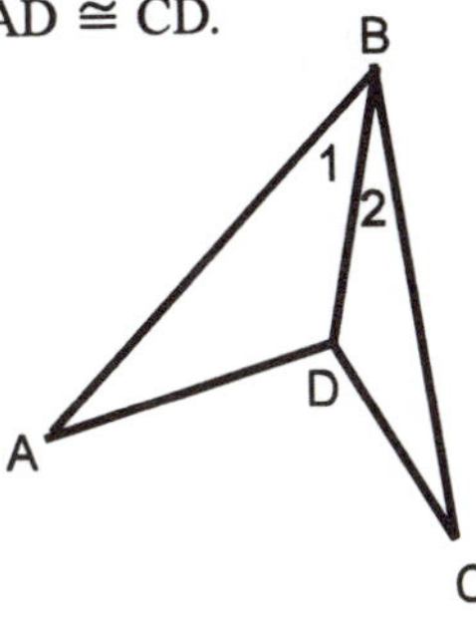

answers

1.
1. Given
2. Algebra
3. Def of ≅ segments
4. SSS
5. Def of ≅ △s

Step 4 is the creative step that the computer came up with. It had taken the given triangle and had mentally made a second copy of it. The second copy was flipped over like a pancake.

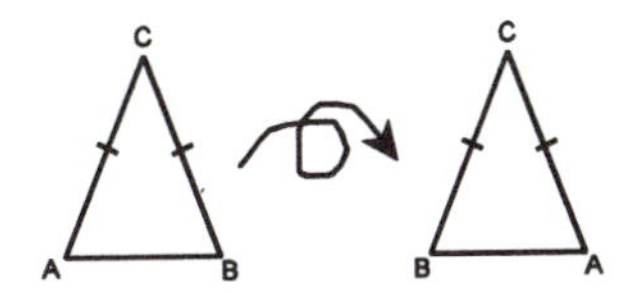

Looking at these "two" triangles, $\angle A$ in the first triangle corresponds with $\angle B$ in the second triangle. It's all very tricky (you know how computers are).

The neat thing was that this proof of the ITT was very short and didn't involve auxiliary lines at all.

2. The converse of the ITT: If the base angles of a triangle are congruent, then the triangle must be isosceles.

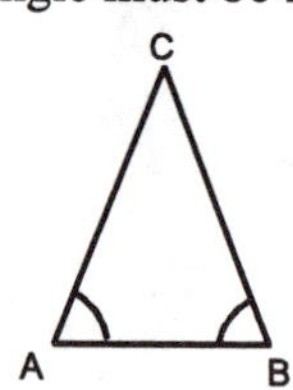

S	R
1. ∠A ≅ ∠B	1. Given
2. AB = AB	2. Algebra
3. $\overline{AB} \cong \overline{AB}$	3. Def of ≅ segments
4. △ABC ≅ △BAC	4. ASA
5. $\overline{AC} \cong \overline{BC}$	5. Def of ≅ △s
6. △ABC is isosceles	6. Def of isosceles

3. No.

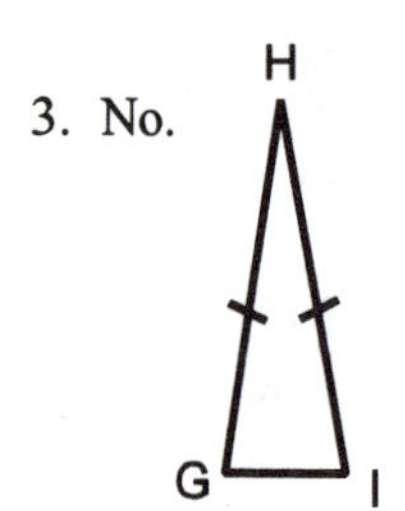

4.

S	R
1. ∠A ≅ ∠1 and ∠2 ≅ ∠C	1. Given
2. △DAB and △DBC are isosceles.	2. Converse of the ITT (which was proved in problem 2 above).
3. $\overline{AD} \cong \overline{BD}$ $\overline{CD} \cong \overline{BD}$	3. Def of isosceles
4. AD = BD CD = BD	4. Def of ≅ segments
5. AD = CD	5. Algebra
6. $\overline{AD} \cong \overline{CD}$	6. Def of ≅ segments

Quincy

1. Can an equilateral triangle be scalene?

2. If AC = DF and ∠A ≅ ∠D, what other piece of information would we need in order to prove that △ABC ≅ △DEF? (There are two possible answers.)

3. Given the figure as marked, prove that ∠E ≅ ∠W.

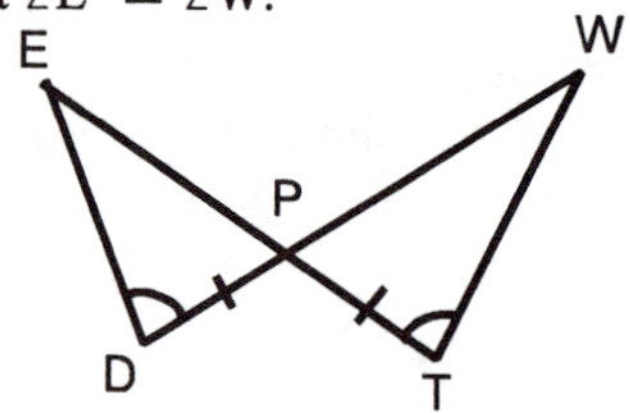

4. Darlene is standing at point D and is looking across the river to Joe (point J). She wants to throw him a sandwich she has made for him. Her problem is that she isn't sure if she can throw it that far.

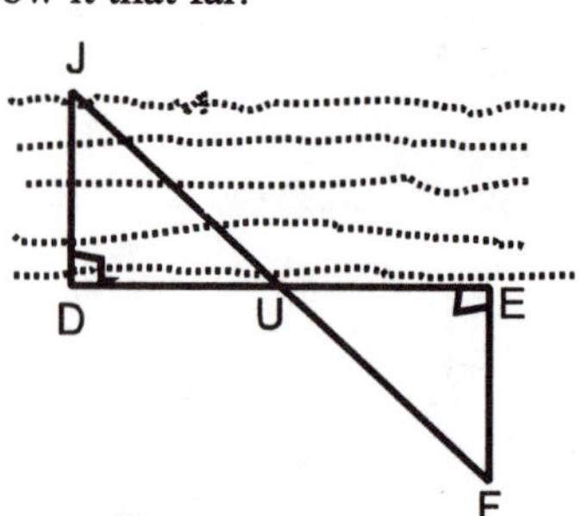

She turns right and walks a dozen paces to point U and sticks her umbrella in the ground. Then she walks another dozen paces to point E.

(That makes DU = UE.) Then she turns right again. With her back to the river she flings the sandwich as hard as she can. It lands at point F. She walks to the sandwich and picks it up (and dusts the dirt off of it). She looks at Joe and notices that he is in a direct line with the umbrella. She now does a short geometry proof that $\overline{JD} \cong \overline{FE}$ which establishes that she can indeed heave the sandwich across the river to Joe. Fill in the missing parts of her proof:

S	R
1. DU = UE ∠D and ∠E are right angles.	*1.* Given
2. $\overline{DU} \cong \overline{UE}$	*2.* ?
3. m∠D = 90° m∠E = 90°	*3.* ?
4. ?	*4.* Algebra
5. ∠D ≅ ∠E	*5.* ?
6. ∠JDU ≅ ∠FEU	*6.* ?
7. △UDJ ≅ △?	*7.* ?
8. ?	*8.* ?

odd answers

1. A scalene triangle is one in which no two sides are congruent. An equilateral triangle is one in which all the sides are congruent. The answer is a definite "no."

3.

S	R
1. $\overline{DP} \cong \overline{TP}$, ∠D ≅ ∠T	*1.* Given
2. ∠DPE ≅ ∠TPW	*2.* Vertical Angle Thm.
3. △DPE ≅ △TPW	*3.* ASA
4. ∠E ≅ ∠W	*4.* Def of ≅ ⚠s

Van Buren

1. Does the ITT apply to obtuse triangles?

2. Prove that if △ABC ≅ △DEF and △DEF ≅ △GHI, then △ABC ≅ △GHI. (This property is true for lots of different relations. For example, it's true for equality:

If x = y and y = z, then x = z.

It's true for >:

If a > b and b > c, then a > c.

It's true for "redder than."

If tomatoes are redder than radishes, and radishes are redder than bowling balls, then tomatoes are redder than bowling balls.

This property is called the **transitive law**. The transitive law also applies to "smoother than," "taller than," and "richer than." But it doesn't apply to "is in love with." In fact, if A is in love with B and B is in love with C, A may not like C very much at all.

3. What is the hypothesis of the converse of the Isosceles Triangle theorem?

4. The converse of an *If ••• then •••* statement exchanges some or all of the hypothesis with some or all of the conclusion. With the ITT the hypothesis has only one part (an isosceles triangle) and the conclusion has only one part (the base ∠s are ≅.) In that case there is only one possible converse.

However if the *If ••• then •••* statement has two parts to the hypothesis, there are several ways (three, in fact) of making a converse. ① You could switch the first part of

the hypothesis with the conclusion.
② You could switch the second part of the hypothesis with the conclusion.
③ You could switch both parts with the conclusion.

Take the (true) *If ••• then •••* statement: *If my roses are red and it's raining, then my roses are wet*, and form the three possible converses.

Here is one possible converse to help you get started: *If my roses are wet and it's raining, then my roses are red.*

5. Given the diagram as marked, fill in the missing parts in the proof that $\angle Q \cong \angle S$.

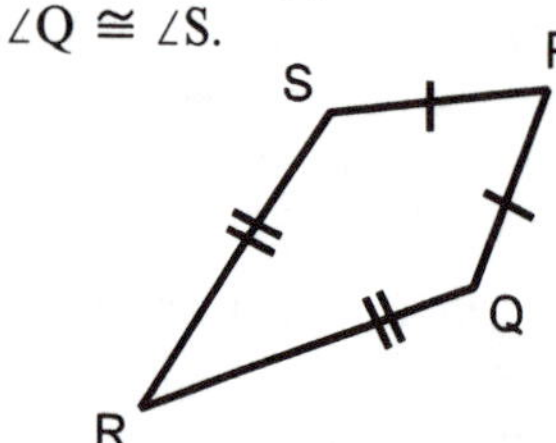

S	*R*
1. ?	*1.* Given
2. There exists a unique line through S and Q.	*2.* ?

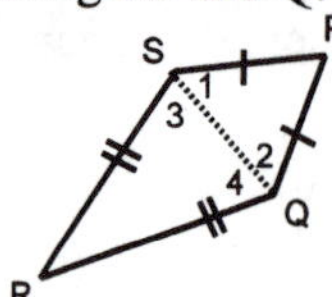

3. $\angle 1 \cong \angle 2$ $\angle 3 \cong \angle 4$	*3.* ?
4. $m\angle 1 = m\angle 2$ $m\angle 3 = m\angle 4$	*4.* ?
5. $m\angle 1 + m\angle 3 = m\angle PSR$ $m\angle 2 + m\angle 4 = m\angle PQR$	*5.* ?
6. ?	*6.* Algebra
7. $\angle PSR \cong \angle PQR$	*7.* ?

6. There are often several different ways to prove a particular theorem. We have proved the Isosceles Triangle Theorem three different ways (so far).

Prove the previous problem, this time drawing the auxiliary segment $\overline{PR}$. You will find that the proof is easier than the one shown in problem 5.

odd answers

1. This is the same as asking if an obtuse triangle can have two congruent sides. The answer is yes.

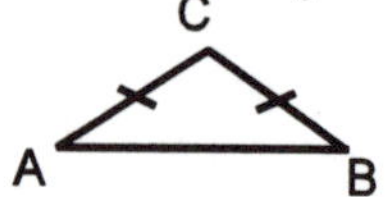

3. The converse of the ITT is: If the base angles of a triangle are congruent, then the triangle is isosceles. The hypothesis is: If the base angles of a triangle are congruent.

5. *1.* $\overline{PS} \cong \overline{PQ}$, $\overline{RS} \cong \overline{RQ}$

2. Postulate 1 (Two points determine a line.)

3. ITT

4. Def of ≅ ∠s

5. Postulate 6 (Angle Addition Postulate)

6. $m\angle PSR = m\angle PQR$

7. Def of ≅ ∠s

Wallace

1. Can an acute triangle be equilateral?

2. Must an equilateral triangle be an acute triangle?

3. If AC = DF, $\angle A \cong \angle D$ and BC = EF, can we prove that $\Delta ABC \cong \Delta DEF$?

4. Prove that all right angles are congruent. Namely, show that if $\angle A$ and $\angle B$ are right angles, they must be congruent.

5. In ΔABC, in each case classify the triangle as either acute, right, or obtuse.

   case ①: $m\angle A = 12°$, $m\angle B = 39°$, $AC = 4$

   case ②: AC = BC = AC

   case ③: $m\angle A = m\angle B = 89°$

Jarbridge

1. Is every equilateral triangle an isosceles triangle?

2. If $\Delta LMN \cong \Delta PQR$, what angle in ΔLMN is congruent to $\angle R$?

3. Given $\overline{AD} \cong \overline{JD}$ and $\overline{DC} \cong \overline{DH}$, prove that $\angle A \cong \angle J$.

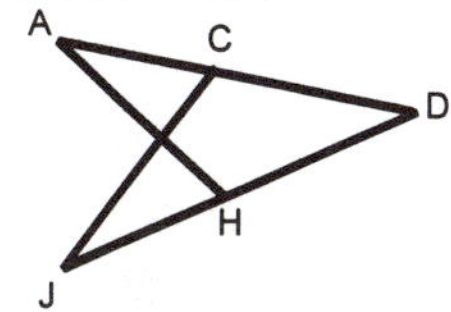

4. You are given two points. You wish to draw an auxiliary line through them. What reason would be used to justify the existence of such a line?

5. Given $\angle 1 \cong \angle 2$ and $\angle 3 \cong \angle 4$, prove that $\angle B \cong \angle D$.

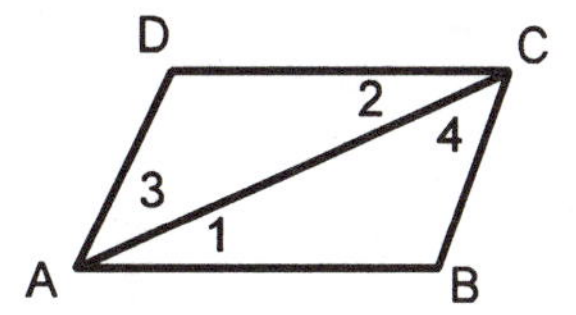

[A small note about the Latin on p. 108: There was no room on that page to put a footnote of apology.

As everyone knows who's read *Latin Made Simple* by Rhoda A. Hendricks: "Because Latin is a highly inflected language, the ending of a Latin word is of primary importance and must be considered as carefully as the base of the word." (p. 9)

That animal is an asinus (in the nominative singular). In the plural genitive (possessive) it is asinorum.

It's the same as amicus (the friend) and amicorum (of the friends).

So it really was an asinus that was crossing the pons.

Chapter Four
Parallel Lines

From the campus clock tower came the announcement that it was 8 A.M. As the bell sounded its last note, the geometry class quieted. Fred stood in front of the lectern since he was not tall enough to be seen if he stood behind it.

The students never knew what to expect from their teacher. All they were certain of was that it wouldn't be just a list of postulates and definitions and theorems copied out of old lecture notes onto the blackboard. At the very least, he would make each piece of mathematics come alive by relating it to the larger context of the world.

Not a good place for Fred to stand

"Today we shall compose an opera. In fact, each of you will make your own composition."

Most of the class gasped. Only two of them were smiling at this point: Chad, who was a music major, and Darlene, who was sitting in the back of the classroom with Joe. Darlene was already thinking about the opera-length pearls she would wear to the opening of her new opera and how she might get Joe to buy them for her.

Joe raised his hand. "Professor Gauss, I've got a question. How do you do an opera? I mean, like, is it notes and stuff or is it also with words?"

Chad, meanwhile, had already selected his dramatic theme (a remake of Aristophanes' *Lysistrata** set in 1916 in Kansas) and was busy sketching out the key signatures, orchestral balance, and various melodic themes.

---

* This is one of the ten must-read classical works that Victor Davis Hanson and John Heath recommend in their book *Who Killed Homer?: The Demise of Classical Education and the Recovery of Greek Wisdom*. Readers of *Life of Fred: Geometry* will find *Lysistrata* understandable (unless they try to read it in the original Greek). In *Lysistrata* the women of Greece decide they want to end the Peloponnesian War that their husbands are fighting. They barricade themselves and announce that they are going on a sex strike. Twenty-four hundred years ago, these women were telling the men that they had a choice: Make love or make war.

From Joe's question, Fred knew that he had some explaining to do. "First of all, to write an opera, you should draw five parallel lines and put a treble clef on the left side.

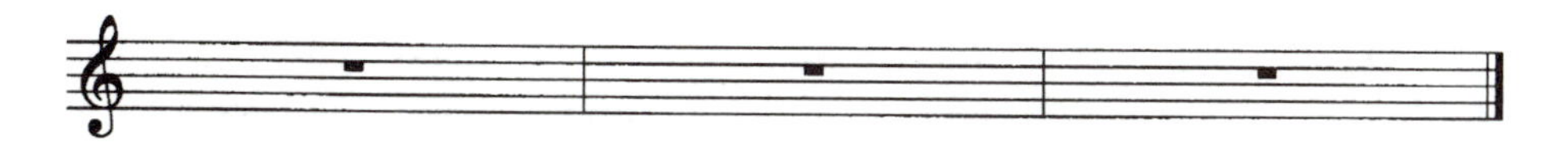

"Then you have to think of a name for your opera. I'm going to call mine *Boat Ride on Lake Tiberius.*✶ You have to start with some notes and end with some notes. Every opera has notes at each end."

Joe wrote down, "Notes at each end."

"My first chord will be A-minor ☹ and my last chord will be A-major ☺. The ride will start out all stormy, but will end happily.

Boat Ride on Lake Tiberius

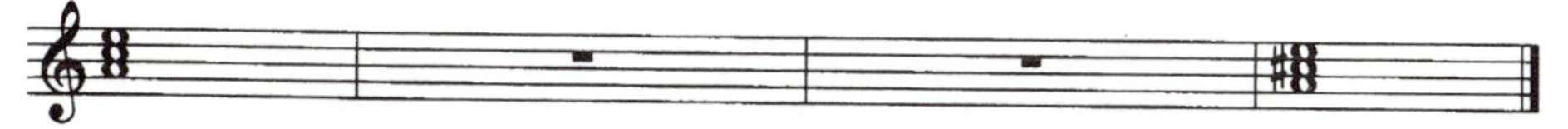

"Now all you have to do to finish your opera is add some notes in the middle and throw in some words."

No one in the history of teaching opera writing had made the whole process so easy to understand.

But Joe had one difficulty in starting to write his opera. It is a difficulty that many composers have when they buy paper without preprinted music staffs.

When Darlene looked over at Joe's work, she said, "It's good that you have five lines, but those lines have to be parallel."

<u>Definition</u>: Two lines are **parallel** if and only if they don't intersect (don't contain a point in common).

✶ This lake was named after the Roman emperor Tiberius. In modern Hebrew they call it Lake Kinnerert. It's also known as the Sea Of Galilee.

Almost every definition seems to contain the phrase "if and only if". What this means is that if we have parallel lines then we know that they don't intersect AND if we know that two lines don't intersect, then we know that they are parallel.

It gets tiring for mathematicians to write "if and only if" all day long.✶ The standard mathematical abbreviation for that phrase is iff.

Definition: Two lines are **parallel** iff they don't intersect.

In chapter one (on p. 31), we said that everything in this book (except chapter 12) lies on a single plane. Everything is **coplanar**. If we were working in three dimensions, then our definition would have to state that two lines are parallel iff they are coplanar and don't intersect.

In space, two lines might not intersect and yet not be parallel. Think of two rocket ships passing each other in outer space.

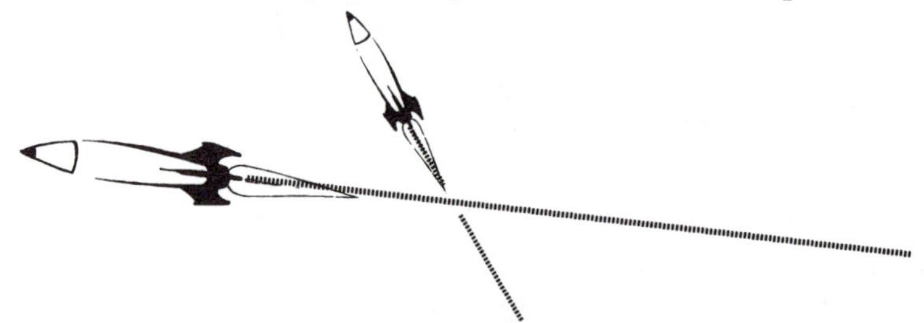

These lines are called **skew lines**.

Darlene took Fred's idea of boats and wrote her opera:

✶ Mathematicians are a lazy lot. Although doing mathematics is considered hard, it is a lot easier to make $50 doing math than it is flipping burgers, or teaching English in junior high school or being a soldier in a combat zone. (I just repeated myself.)

And many people find mathematics fun, except when it is taught in some schools. (In those schools even sex ed becomes a chore.) People everywhere enjoy doing puzzles, whether they are chess puzzles, or logic puzzles ("Name all the people in the world who are not descendants of people who are not your ancestors") or situations in Quake II® where you are trying to figure out how to overcome some bad guys who are shooting at you. Logical thinking can be pleasurable.

Joe looked over at Darlene's work and compared it with the class notes that he had taken. "You gotta have notes at each end. Otherwise, it's not an opera."

"I wasn't done yet." She finished her opera:

She looked into Joe's eyes and said, "You know, parallel lines must be awfully lonely."

"Why is that?"

She squeezed his hand. "They never meet."

Joe thought for a moment. "That's cool. Maybe I could tell Professor Fred that. When he writes a book about parallel lines, he could call it *The Lonely Lines*." Joe never seemed to understand what Darlene was really trying to tell him.

Your Turn to Play

1. If we were working in space, (in three dimensions), would this be a good definition of skew lines? Definition: Two lines are **skew** iff they are not coplanar.

2. We are going to look at the two different parts of iff.

The *If ••• then •••* statement "Jeanette MacDonald is a soprano if roses are red." might be written in logic as R ➜ S. (We are letting "roses are red" = R and "Jeanette MacDonald is a soprano" = S.)

We translate *S if R* as R ➜ S. (This could also be written as S ⬅ R.)

Now look at the *••• only if •••* statement "Jeanette MacDonald is a soprano *only if* roses are red." This has a different meaning than the sentence in the previous paragraph. This means that Jeanette MacDonald is a soprano *implies that* roses are red. It means that S ➜ R.

This question will ask you to guess something. You are to guess what symbol from logic looks like that can be used to fill in the blank in the following statement.

Jeanette MacDonald is a soprano *if and only if* roses are red is written in logic as S __?__ R.

COMPLETE SOLUTIONS

1. Take two pencils and hold them up in the air. Pretend the pencils are lines. The minute that you touch them together, the two lines will have to lie in the same plane. The minute that you make your pencils parallel to each other, then they will have to also lie in the same plane. If the pencils are neither intersecting nor parallel, then they will have to be skew. Skew lines are lines that are not coplanar is a good definition.

2. If "S *if* R" is written as S ← R,
and if "S *only if* R" is written as S → R,
then it might seem natural to write "S *iff* R" as S ↔ R, which is exactly what they do in logic classes.

"Is it nine o'clock yet?" Joe whispered to Darlene.

"No silly. If it were nine o'clock, Fred would have stopped lecturing. You know he never runs over the appointed hour since it makes students nervous. Listen. You can still hear him talking. Hence, it isn't nine o'clock yet."

Darlene's argument is an example of an **indirect proof**. All the proofs we have done so far in geometry have been direct proofs. We start with the given and then use postulates, definitions, previously proven theorems, and algebra to work our way to the desired conclusion.

But sometimes there is no direct way to prove something and you need to resort to an indirect proof.

Some people consider an indirect proof an act of desperation. That's because you begin an indirect proof by saying, "Suppose I'm wrong." It is sometimes called betting the farm—betting everything.

If you are wrong, you may find yourself slightly naked and homeless. If you are right, then the beginning assumption of "Suppose I'm wrong" will lead you to an impossibility. It will lead you to some contradiction. Then you can rejoice. That means that your "Suppose I'm wrong" caused the contradiction. That means that the "Suppose I'm wrong" was itself incorrect. That means, in effect, that you are right and you don't have to be slightly naked and homeless.

Darlene wanted to prove to Joe that it wasn't yet nine o'clock. If you put Darlene's argument in Statement-Reason form, it might look like:

Statement	*Reason*
1. Assume for the sake of argument that it is nine o'clock.	1. Beginning of an indirect proof.
2. Then Fred would not now be lecturing.	2. Fred never lectures after the lecture hour has ended.
3. Fred is now lecturing.	3. Just listen to him. He's writing on the blackboard and talking.
4. It's not nine o'clock.	4. Contradiction in steps 2 & 3. Therefore the assumption in step 1 is not true.

We have five types of reasons we can give in geometry:

① Given
② Postulate
③ Definition
④ Previously proven Theorem and
⑤ Algebra.

We are going to add two more items that you can use as reasons in a proof:

⑥ Beginning of an indirect proof
⑦ Contradiction in steps ___ and ___ and therefore the assumption in step ___ is not true.

The reasons "⑥ Beginning of an indirect proof" and "⑦ Contradiction" always come as a matched pair in a proof. If you start with the assumption that "I'm wrong," you had better end with a mess (a contradiction, an impossibility, a snafu*).

Using an indirect proof, here is how Joe could prove that he didn't forget his car key this morning. (He often does.)

---

* snafu (pronounced sna-FOO) is a word that originated during WWII. According to polite dictionaries snafu stands for Situation Normal: All Fouled Up.

Joe would begin by assuming the opposite of what was true.

Statement	*Reason*
1. Assume I did forget my car key this morning.	1. **Beginning of an indirect proof.**
2. Then I couldn't start my car.	2. Need my car key to start my car.
3. Then I couldn't drive my car.	3. If a car isn't started, you can't drive it.
4. I couldn't have gotten to the bakery on the edge of town which was having its Once-A-Year-Leftover-Valentine's-Cake Sale that is only held on February 23rd of each year.	4. There are no buses that go out near that bakery and it is too far to walk from Joe's apartment and Joe is too cheap to ever use a taxi.
5. If I didn't get there today, I couldn't have bought the heart cake.	5. The bakery doesn't deliver. You have to be there to buy that cake.
6. If I didn't buy that cake, I couldn't have brought it here to give to you.	6. You can't give what you don't have.
7. I am giving you that heart cake right now.	7. Look.
8. I didn't forget my car keys this morning.	8. **Contradiction in steps 6 and 7 and therefore the assumption in step 1 is false.**

The important reasons are in **bold face**.

When Darlene looked at the cake, she wasn't very happy. "There's a piece missing!"

An awkward silence ensued as Joe wondered what was bothering Darlene. He knew that it was a big cake, and Darlene has always been dieting. He explained, "I gave a small piece to Nancy."

"Nancy! That . . . that . . . that . . . person!" For some reason, Darlene couldn't stand the thought of Nancy. It might be because Nancy sometimes attracted Joe's attention.

"You don't like me anymore?" Joe moaned.

"Listen lunkhead, I love you. And I can prove it."

Darlene's proof would have been an indirect proof:

Statement	*Reason*
1. Assume that I didn't love you.	1. **Beginning of an indirect proof.**
2. You would be dead right now since I would have killed you for giving me that half-eaten week-old Valentine's cake that you gave Nancy the first bite out of.	2. Jealousy is capable of murder.
3. You aren't dead.	3. You are breathing.
4. I love you.	4. **Contradiction in steps 2 and 3 and therefore the assumption in step 1 is false.**

"It must be nine o'clock," Joe announced.

"How do you figure that?" Darlene asked. She knew that it was only around 8:15 since Fred had just started his eight o'clock geometry class.

"I have an indirect proof that it must be nine o'clock. Would you like to see it?"

"Sure," said Darlene. She wanted to encourage every feeble effort he made at engaging in rational thought.

Statement	*Reason*
1. Assume that it isn't nine o'clock.	1. Beginning of an indirect proof.

That was a good beginning for an indirect proof. Joe was assuming the opposite of what he was trying to establish. His proof continued:

2. If it weren't nine, then my watch wouldn't say it was nine. 	2. Joe has a high-quality watch that he bought yesterday from some stranger in the hall. The stranger had told Joe that it was a top-of-the-line Rolex™ which he let Joe have for only $4 "since I have more of them than I can use."
3. My watch says nine.	3. Look.
4. It is nine o'clock.	4. Contradiction in steps 2 and 3 and therefore the assumption in step 1 is false.

There are two requirements for a valid proof: ① Every statement has a reason which justifies it, and ② the last line is what you wanted to prove. In Joe's proof one of his statements has a very shaky reason. Joe's watch has read nine o'clock ever since he bought it.

"And so today," Fred began, "we need to introduce a new postulate. It's called the Parallel Postulate. Without it, we wouldn't be able to do many of our proofs concerning parallel lines."

Joe wrote in his notes, "Parallel Postulate," looked at it for a moment, and then raised his hand. "I bet I can guess what it is."

Fred nodded to Joe indicating that Joe should make his guess.

"I bet it is: Parallel lines are lines that don't intersect."

Your Turn to Play

1. Is that the Parallel Postulate? (Even without having seen the Parallel Postulate, you can answer that question.)
2. Is this the Parallel Postulate: "Two lines are parallel iff they are not concurrent"?

COMPLETE SOLUTIONS

1. The *definition* of parallel lines is that the lines don't intersect. When we write that line ℓ is parallel to line m, we know by definition that ℓ and m do not intersect.

2. The definition of concurrent lines is lines that all share a common point. To say that two lines are parallel iff they don't share a common point is just another restatement of the definition of two parallel lines.

In order to save a little effort, we symbolize "line ℓ is parallel to line m" as $\ell \parallel m$.

"That's a true statement, Joe. Parallel lines never intersect, but the Parallel Postulate is a little different." Fred wrote on the blackboard:

Postulate 9: (The Parallel Postulate) If you have a line ℓ and a point P not on ℓ, then there is at most one line through P that is parallel to ℓ.

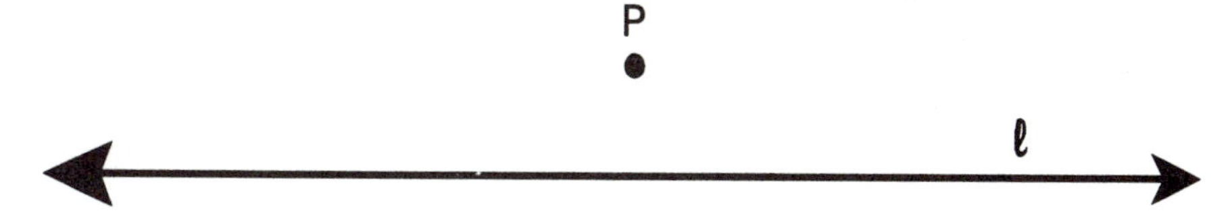

Then Fred completed the diagram by drawing the line through P that was parallel to ℓ.

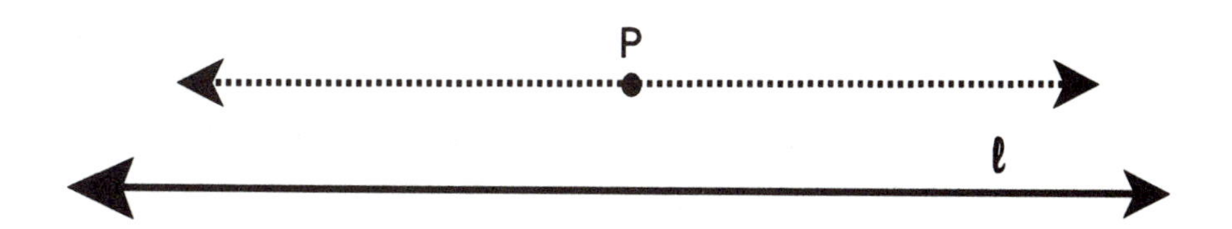

What the Parallel Postulate says is that you can't have two lines through P that are both parallel to ℓ."

Joe wrote in his notes:

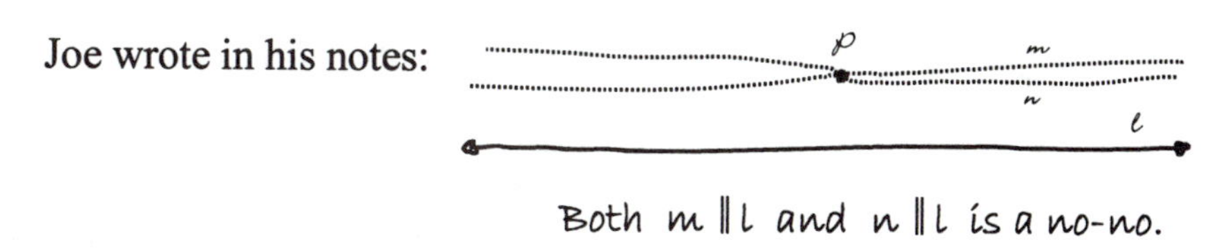

p
m
n
l
Both m ∥ l and n ∥ l is a no-no.

. . . and he was right.

Joe raised his hand again. "That's so obvious. Everybody knows that. Why do we have to say it? In fact, couldn't we just prove it and then it would be a theorem instead of a postulate?"

Fred tackled Joe's two questions one-at-a-time.

✱ *Why do we have to say it? Everyone knows it's true.*

Fred's answer: "Do you remember our first postulate which says that two points determine a line—that there is exactly one line that you can draw through any two distinct points? That was obvious also. We want our beginning assumptions, our postulates, to be obvious. We use them as the foundation stones for the theory we're building." *

✱ *Why don't we just prove the Parallel Postulate?*

Fred's answer: "That would be desirable. Mathematical theories with very few beginning assumptions and lots of theorems are considered more elegant. The dream of some mathematicians is to create a theory in which there is only one postulate, and from that derive all of geometry, algebra, set theory, calculus, and metamathematics. If they could start with a simple postulate such as "A = A" and from that derive the calculus formula $\int \sec^2 \theta \, d\theta = \tan \theta + C$, that would be quite an accomplishment.

"Ever since Euclid wrote his version of the Parallel Postulate in *The Elements*, mathematicians have been trying to prove it using the other postulates. It looks so *provable*. It sounds like a theorem."

At this point Joe got all excited. He would be the one to prove the Parallel Postulate.

He whipped out a fresh sheet of binder paper and wrote at the top:

✱ Recall that famous illustration of what a mathematical theory looks like. At the bottom are the undefined terms (point, line, and plane) and the assumed statements (the postulates). Upon that base we build the definitions and theorems.

It should be noted that some geometry books don't use a palm tree to illustrate what a mathematical theory looks like. It is silly not to use a palm tree.

A building is not as good a metaphor since buildings don't grow.

A maple tree won't work because it's deciduous (the leaves fall off in winter). You wouldn't want your theorems falling off your theory.

Joe's Proof of the Parallel Postulate

I will prove that if you have a line l and a point P not on l, then there is at most one line through P that is parallel to l.

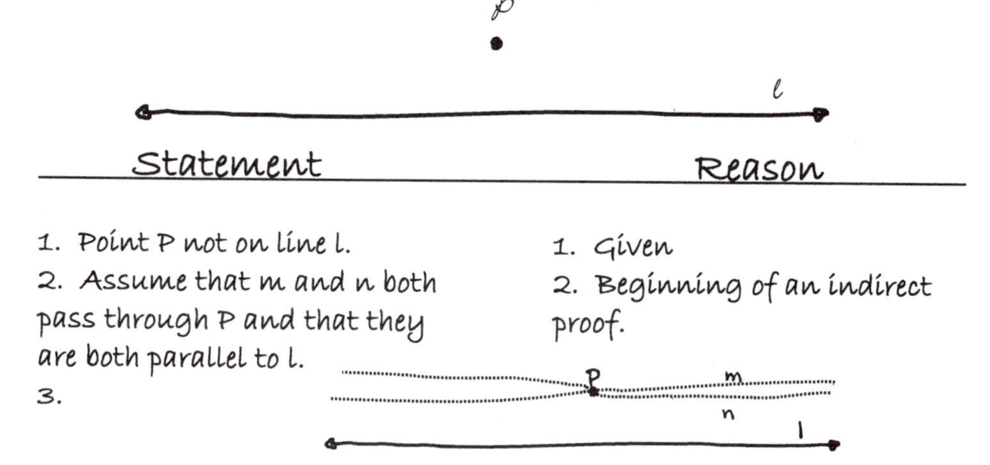

Statement	Reason
1. Point P not on line l.	1. Given
2. Assume that m and n both pass through P and that they are both parallel to l.	2. Beginning of an indirect proof.
3.	

At this point Joe wanted to write that it just wasn't possible for both $m \parallel l$ and $n \parallel l$. Anybody could see that. But he knew that if he played by the rules for proving things in mathematics, the only possible reasons he could give are: ① *Given;* ② *Postulate;* ③ *Definition;* ④ *Previously proven theorem;* ⑤ *Algebra;* ⑥ *Beginning of an indirect proof;* and ⑦ *Contradiction in steps ___ and ___ and therefore the assumption in step ___ is false.* None of those reasons would give him any kind of contradiction which he needed.

Fame and fortune had almost seemed within his grasp. He went back to listening to Fred.

". . . and hundreds of mathematicians from 300 B.C. to A.D. 1800 searched for the proof of the Parallel Postulate. Twenty-one centuries of searching ended in the early 1800s. Three mathematicians each independently arrived at a most surprising conclusion. They established that it couldn't be proved. Some of them even set up geometries where the Parallel Postulate was false."*

✶ We'll look at those non-Euclidian geometries in chapter 11½. They are weird. At least I think so when I see triangles with two right angles in them.

Joe balled up his piece of binder paper and made "two points" into the wastebasket.

Your Turn to Play

1. Prove that if two different lines (m and n) are both parallel to ℓ, then $m \parallel n$.

2. In algebra, the symbol $\neq$ meant "not equal to."

In working with sets, $\in$ means "is a member of," so that, for example, a $\in$ {a, d, g}. And $\notin$ means "is not a member of," so that, for example, b $\notin$ {a, d, g}.

In algebra, to indicate "is not less than" you can write $\not<$ or you can write the more familiar $\geq$. They mean the same thing.

What is the symbol for "is not parallel to"?

3. An **exterior angle** of a triangle is easy to define using a diagram.

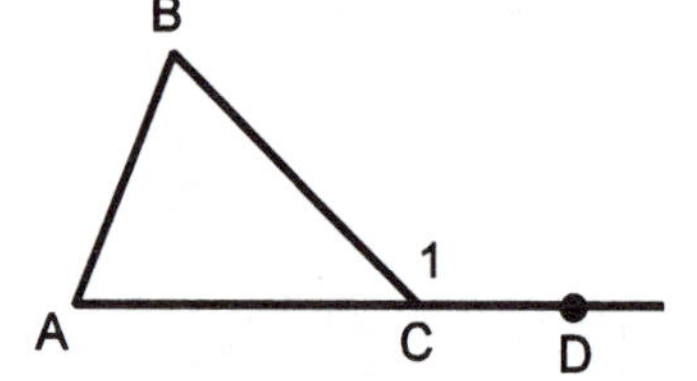

$\angle 1$ is an exterior angle of $\triangle$ABC.

It is not so easy to put that definition into words.

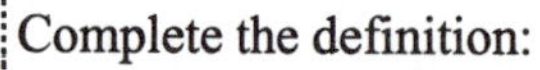

Complete the definition:

An **exterior angle** of $\triangle$ABC is . . . .

4. How many exterior angles does a triangle have? (Hint: the answer isn't 1 or 3.)

5. In the above diagram, $\angle$A and $\angle$B are called the **remote interior angles** to $\angle 1$. Every exterior angle will have two remote interior angles. Draw a picture, if possible, in which a remote interior angle has a measure larger than the exterior angle it is related to.

6. On the basis of your answer to the previous question, make a guess as to what the Exterior Angle Theorem states.

7. The Exterior Angle Theorem is probably one of the top ten theorems in plane geometry. It doesn't seem like much when you first read it, but it will unlock the door to almost all of our parallel lines theorems.

It's proof is—how shall we say this?—wonderful. Filled with wonder. Beginning geometry students often think, "How did someone ever think of this proof?"

Luckily for us, someone did. And we have it as a gift. The same is true of many things in our world today. How did someone ever figure out how to make pizza? Or zippers? Or butter? These are all gifts of our civilization. Since it's hard to thank most of the contributors directly since they are currently dead, perhaps the best that we can do is simply to be grateful. (In the extreme case, we might even contribute something of our own to our civilization. Perhaps a buttered pizza that comes in a zippered package.)

Here's the proof of the Exterior Angle Theorem with most of the steps supplied. Please fill in the missing steps.

To prove that $m\angle B < m\angle BCD$

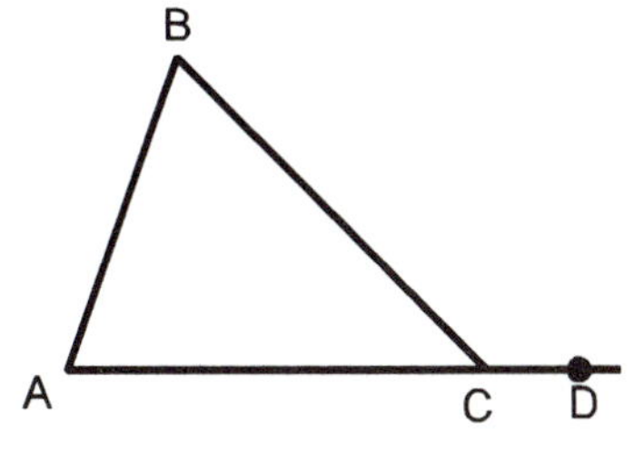

Statement	*Reason*
1. ΔABC and A–C–D	1. Given
2. A, C and D are collinear (on the same line)	2. Def of A–C–D
3. M is the midpoint of $\overline{BC}$.	3. Timeout! In chapter one, we introduced the definition of midpoint, but we never proved that every segment has a midpoint. Here's a quick proof in a box. If you were to cut out this box and paste it on the previous page, then the reason for step 3 would just read, "Midpoint Theorem."

The Midpoint Theorem: Every segment $\overline{BC}$ has a midpoint M.

Proof:

Statement	Reason
1. $\overline{BC}$	1. Given
2. B has coordinate b and C has coordinate c	2. Ruler Postulate: Every point has a number (a coordinate) associated with it.
3. There is a point M on $\overleftrightarrow{BC}$ with coordinate $(b + c)/2$	3. Ruler Postulate which also states that every number has a point associated with it.
4. $BM = \lvert(b + c)/2 - b\rvert$ $MC = \lvert(b + c)/2 - c\rvert$	4. Definition of distance (p. 49)
5. BM = MC	5. Algebra
6. M is a midpoint of $\overline{BC}$.	6. Definition of midpoint

4. BM = MC	4. ?
5. $\overline{BM} \cong \overline{MC}$	5. ?

6. There exists a line through A and M — 6. Postulate: Two points determine a line.

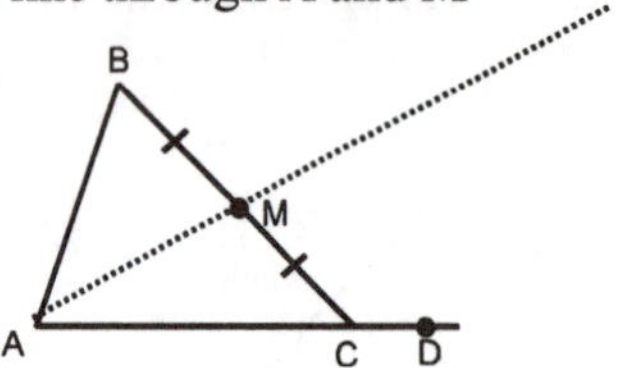

7. Let the coordinate of A be a.
Let the coordinate of M be m.
Let E be the point with coordinate 2m − a — 7. Ruler Postulate: Every point has a number associated with it and to every number there is a point.

8. ME = |(2m − a) − m|
AM = |m − a| — 8. Definition of distance

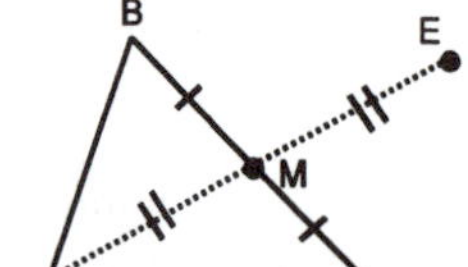

9. ME = AM — 9. Algebra (You can see why algebra was a prerequisite for studying geometry.)

10. There exists a line through C and E. — 10. ?

11. ∠AMB ≅ ∠EMC — 11. Vertical Angle Theorem

12. △AMB ≅ △EMC — 12. ?

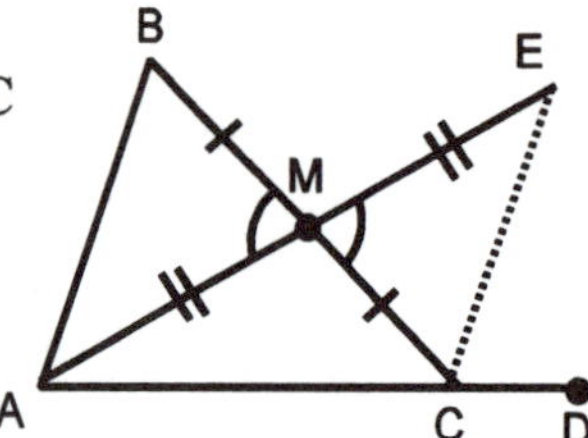

13. ∠B ≅ ∠ECM — 13. Definition of ≅ △s

14. m∠B = m∠ECM — 14. Definition of ≅ ∠s

15. Now at this point every other geometry book that I know of says something to the effect, "Lookie! Since m∠B is equal to m∠ECM and since m∠ECM is less than m∠MCD, then by algebra we have what we want to prove. Namely, m∠B is less than m∠MCD." For 99.932% of all students, this is just fine. After all, a 15-step geometry proof is plenty long. We have established that the exterior angle of a triangle is greater than a remote interior angle.

But for the 0.068% of you, we need to talk about one little sticky point. Let me put it in tiny type so that everyone else can easily skip over it. The question is: How do we know that m∠ECD > 0? You can't just say, "Look at the diagram." That's not one of the seven permissible reasons.

We can argue that $m\angle ECD > 0$ with an indirect proof. Suppose that $m\angle ECD$ were equal to zero.

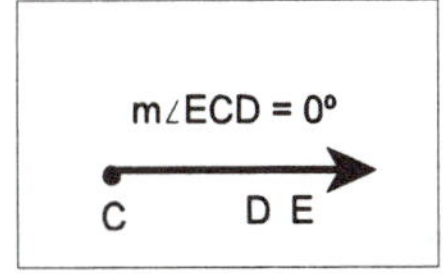

So under the assumption that $m\angle ECD = 0$, we have that E is collinear with A, C and D. Since two points determine a line, if A and E are on the line through A, C and D, then M must also be on the line through A, C, E and D. Now we have that A, M, C, E and D are collinear. If M and C are both on the line through A, M, C, E and D, then B must be on the line through A, M, C, E and D. Thus A, B, M, C, E and D are collinear. Where's the contradiction? Easy. We have that A, B and C are collinear, but we are given $\triangle ABC$ and the definition of a triangle is that the three vertices are not collinear. With that contradiction we now can say that $m\angle ECD$ is not equal to zero. Now the 0.068% of you can join the rest of us as

. . . we announce Q.E.D.

"Q.E.D." is how to indicate the end of a proof in mathematics. It doesn't stand for Quack E. Duck, but for the Latin phrase *quod erat demonstrandum* which roughly translates as, ". . . which is what we were trying to demonstrate". It's a nice finishing touch on long proofs.

In some math books, instead of Q.E.D., they write a little black box ■ to mark the end of proofs. On the blackboard a lecturer may write ⊠.

COMPLETE SOLUTIONS

1.

Statement	Reason
1. $m \parallel \ell$ and $n \parallel \ell$ m and n are different lines.	1. Given
2. Assume that m is not parallel to n.	2. Beginning of an indirect proof
3. m and n intersect at some point P.	3. Definition of parallel lines
4. P is not on ℓ.	4. Definition of parallel lines (We know from step 1 that $m \parallel \ell$ so they can't have a point in common.)
5. There is at most one line through P that is parallel to ℓ.	5. Parallel Postulate

6. $m \parallel n$

6. Contradiction between steps 1, 3 and 5, and therefore the assumption in step 2 is false.

2. ∦ I bet you guessed this one correctly.

3. Definition: An **exterior angle** of Δ ABC is ∠BCD where A–C–D.

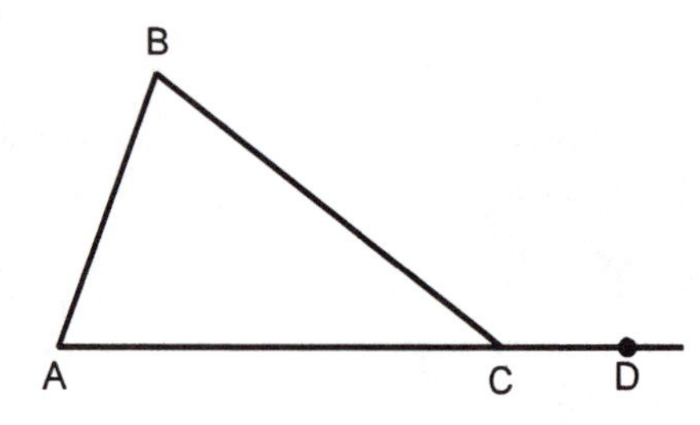

4.

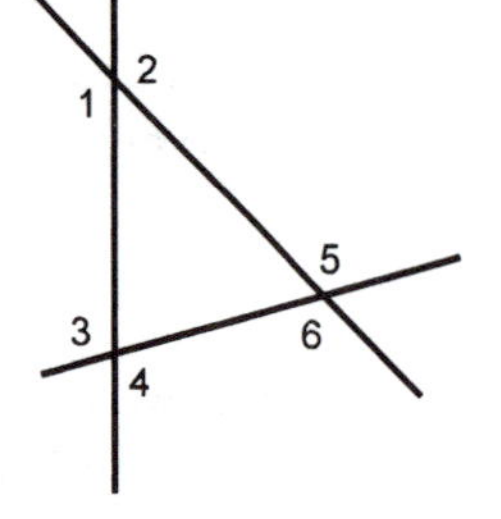

. . . which makes three pairs of vertical angles.

5. This is difficult to do. Really tough. In fact, saying that it is quite hard is an understatement. Even Fred can't do it.

6. Theorem: (The Exterior Angle Theorem) The measure of an exterior angle of a triangle is greater than the measure of either remote interior angle.

7.
4. Def of midpoint
5. Def of congruent segments
8. Postulate: Two points determine a line
10. SAS

Joe realized that he had just thrown away his last piece of binder paper. He had no more paper on which to take notes.

He carefully formulated his Plan of Action: He would quietly get out of his chair, head to the wastebasket, retrieve the balled-up paper, go back to his seat, unfold and smooth out the paper, cross off the "Joe's Proof of the Parallel Postulate," and use the bottom half of the sheet to take notes.

Joe had seen lots of commando movies where the soldiers would create their Plans of Action, and he thought it would be cool to have everything worked out in advance before he acted.

The Plan of Action worked. Almost perfectly. When Joe unfolded the binder paper, he didn't see "Joe's Proof of the Parallel Postulate" anywhere on the paper.

Joe had taken the wrong piece of paper out of the wastebasket.

Hey Snowy,
We knock over the KITTENS Bank at a quarter after nine today. Here is how to get there.
Your pal,
C.C.

On the back of the sheet was a crudely drawn map.

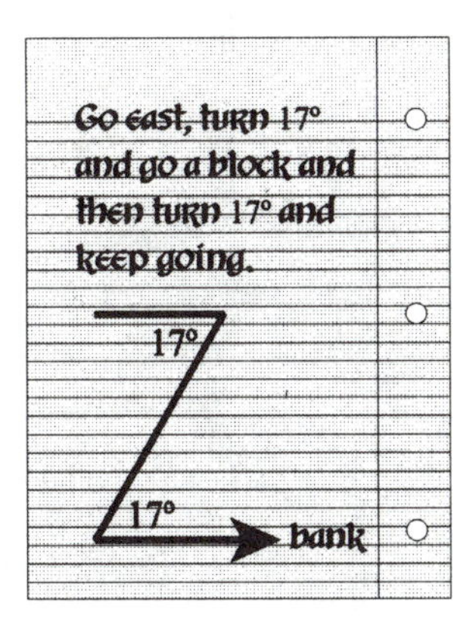

Joe looked at his watch. In 15 minutes there would be a bank robbery right there on the KITTENS campus. Without even creating a new Plan of Action, Joe stood up and told everyone that there was going to be a bank stickup and asked if anyone wanted to come with him and "see the show."

The entire class arose en masse and dashed to the bank, many of them filling out withdrawal slips as they ran.

Fred stood before an empty classroom. He walked over to Joe's desk and looked at the map that Joe had left behind.

When we look at things, we often see what we are ready to see. When Joe had seen the paper, he saw the excitement of a bank robbery. When Fred saw the paper, he saw two lines cut by a transversal and a pair of alternate interior angles.

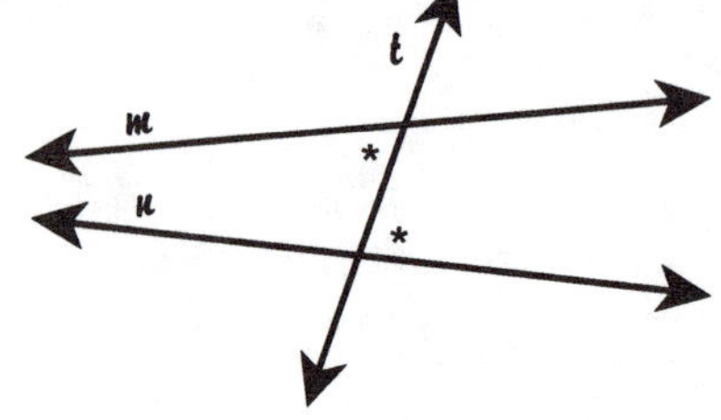

Lines *m* and *n* are cut by the **transversal** *t*. A transversal is a line that intersects two (or more) lines at different points. The two angles marked by * are alternate interior angles.

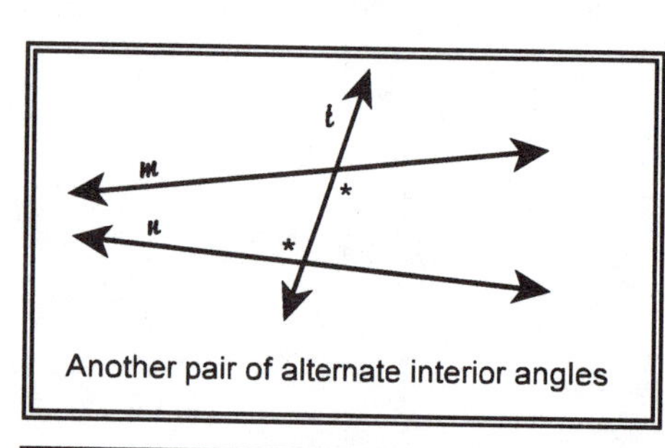

Another pair of alternate interior angles

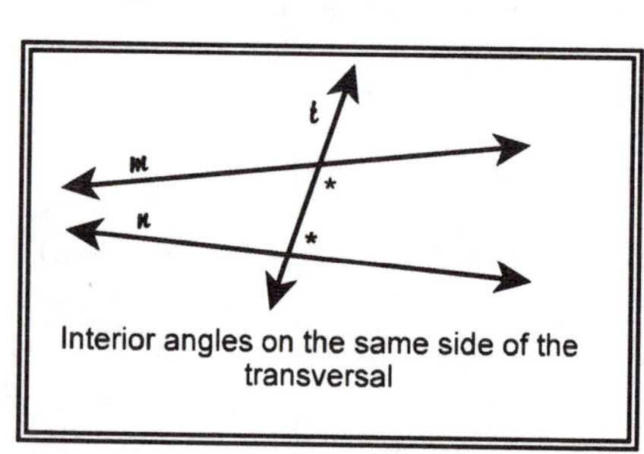

Interior angles on the same side of the transversal

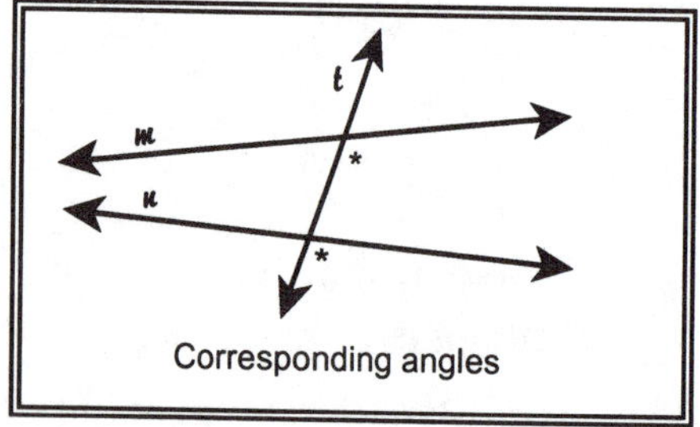

Corresponding angles

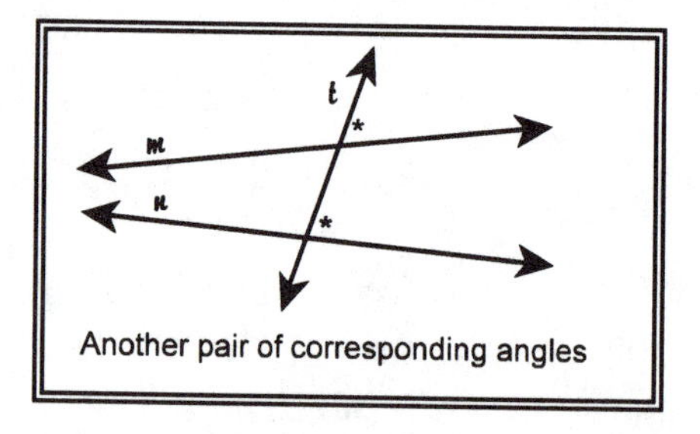

Another pair of corresponding angles

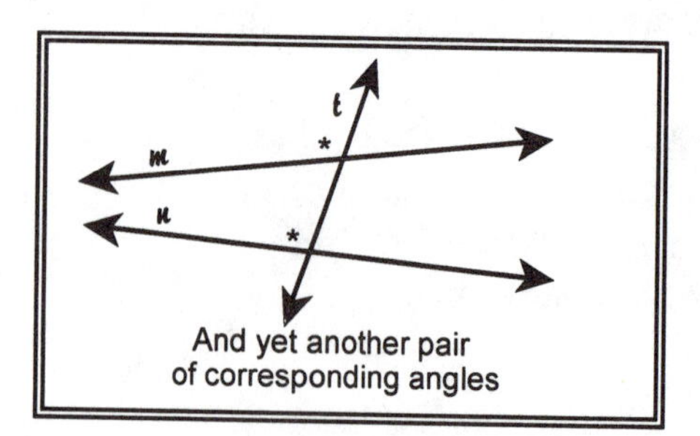

And yet another pair of corresponding angles

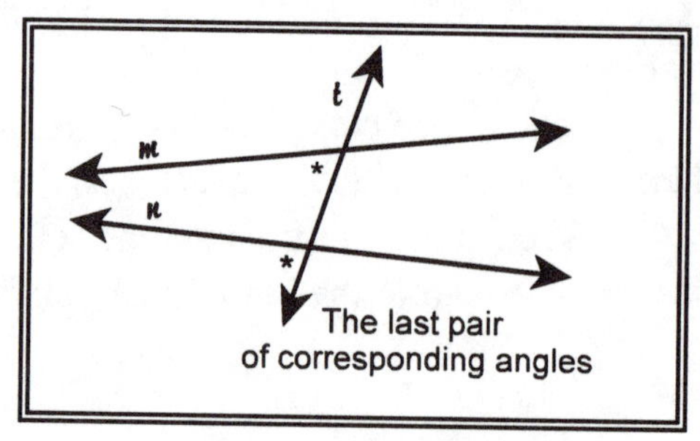

The last pair of corresponding angles

When Fred looked at the map he saw two alternate interior angles that were special. They happened to be congruent alternate interior angles.

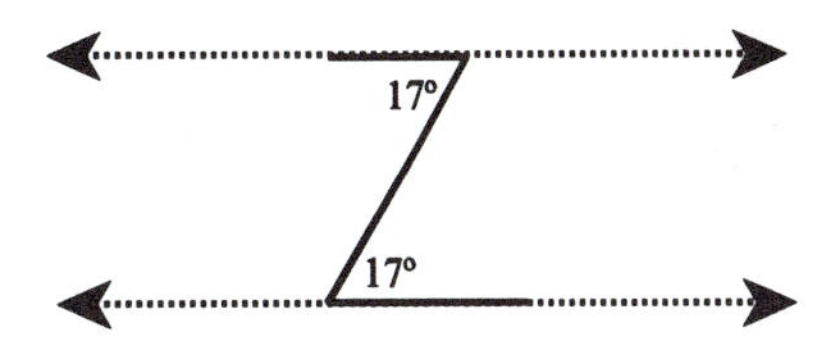

Those lines on the map that looked parallel *are* parallel.

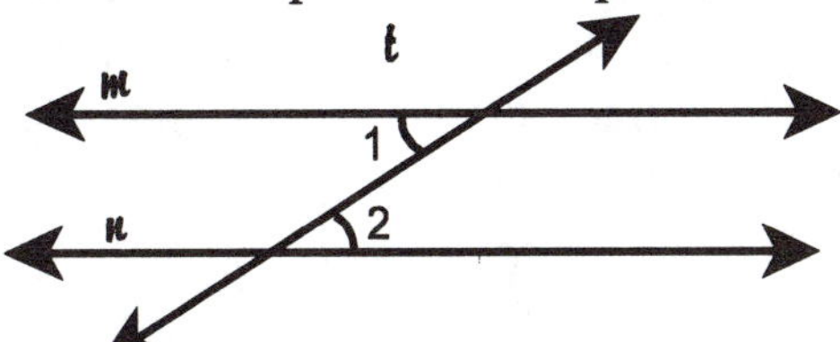

<u>Theorem</u>: If two lines are cut by a transversal and the alternate interior angles are congruent, then the lines are parallel.

or you could say

If t is a transversal to m and n and if $\angle 1 \cong \angle 2$, then $m \parallel n$.

or you could say

Congruent alternate interior angles imply parallel lines.

which is sometimes abbreviated as

AI ➔ P

or with even more abbreviation

AIP

The proof of AI ➔ P is short and sweet.

Sweet

Short & Sweet

Statement	*Reason*
1. t is a transversal to m and n and forms congruent alternate interior angles. $\angle 1 \cong \angle 2$	1. Given
2. Assume $m \nparallel n$. 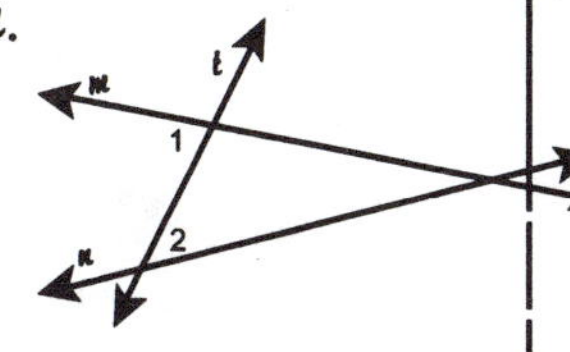	2. Beginning of an indirect proof.
3. $m\angle 1 > m\angle 2$	3. Exterior Angle Theorem
4. $m \parallel n$	4. Contradiction in steps 1 and 3 Q.E.D.

Of course, the picture that we drew in the middle of the proof is goofy. It portrays an impossible situation if we know that $\angle 1 \cong \angle 2$.

The word had spread quickly across the KITTENS campus and everyone was at the bank to see the big event (after first drawing out their money). When Sergeant Snow and C.C. Coalback arrived at 9:15 A.M. for the heist, a long line of students stretched out the front door and down the street.

"We could cut in line," Snow suggested.

"Nah. We got all the time in the world," Coalback assured him.

"We could wear masks like bank robbers do in the movies," Snow suggested.

"You're full of suggestions today, aren't you? Look. It's obvious we don't want to tip them off why we're here. If we wore masks then everyone would know we're the bad guys. We don't want that. Hence we shouldn't wear masks." Coalback had offered his partner a very nice indirect proof.

"But C.C.! We gotta wear some kind of disguise. We can't have our faces out there in plain sight."

After a few minutes of discussion Coalback finally gave in. "Okay. You be in charge of our disguises. Just make sure that we blend in. We don't want to be sticking out like sore thumbs."

Sergeant Snow, who had been doing a little reading about the fauna of the Great Plains, knew exactly what was common in Kansas. The two men took their places at the end of the line of students.

They were cleverly disguised.

As we wait for the line to move forward, let's take a little break and knock off some corollaries to the AI ➜ P Theorem.

Your Turn to Play

1. Have we established that *If you have a line* ℓ *and a point P not on* ℓ*, then there is at least one line through P that is parallel to* ℓ?

2. Why is the following not true? Given a point P and a line ℓ, there is always at least one line through P that is parallel to ℓ.

3. What is a corollary?

4. Fill in the missing parts in the proof of *If you have a line* ℓ *and a point P not on* ℓ*, then there is at least one line through P that is parallel to* ℓ.

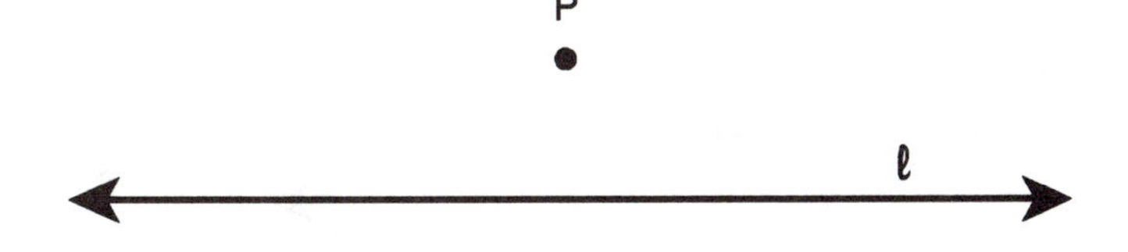

Statement	Reason
1. ?	1. Given
2. There are points Q and R on ℓ. (Say that Q corresponds to the number 5 and R to the number 18. The numbers 5 and 18 are arbitrary. Choosing any two different numbers gives us two different points which is what we wanted.)	2. Ruler Postulate
3. m∠PQR is equal to some number, x°, between 0° and 180°.	3. ?
4. There is an angle such that m∠QPS = x°.	4. Angle Measurement Postulate
5. $\overleftrightarrow{SP} \parallel \ell$	5. ?

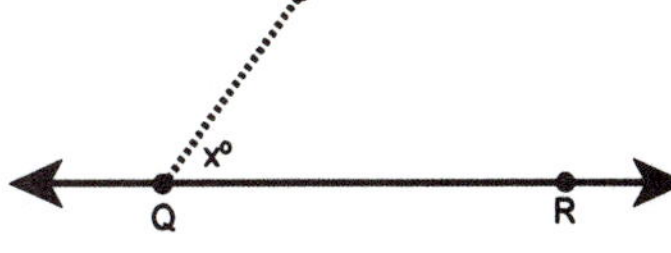

Q.E.D.

5. The PAI Theorem is also a corollary of the AIP Theorem. Make a guess as to what the PAI Theorem states.

6. Fill in the missing parts of the proof of the PAI Theorem.

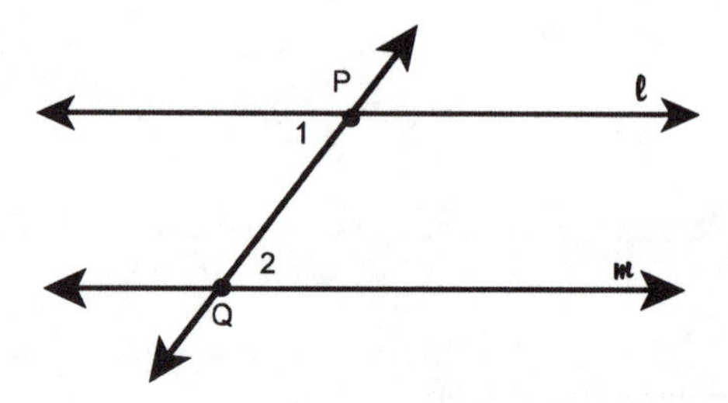

Statement	Reason
1. $\ell \parallel m$ and t is a transversal to ℓ and m.	1. Given
2. Assume that $m\angle 1 \neq m\angle 2$.	2. ?
3. There is some $\angle QPR$ such that $m\angle 2 = m\angle QPR$.	3. Angle Measurement Postulate

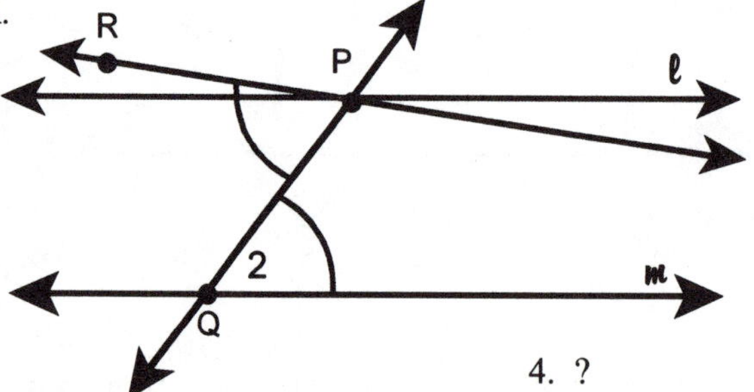

Statement	Reason
4. $\overleftrightarrow{PR} \parallel m$	4. ?
5. There is at most one line parallel to m which passes through point P.	5. ?
6. ?	6. Contradiction in steps 1, 4 and 5 and therefore the assumption in step 2 is false.

■

7. The CAP Theorem ("If two lines cut by a transversal form congruent corresponding angles, then the lines are parallel") is a corollary of the AIP Theorem. Prove the CAP Theorem.

Once you note that $\angle 1$ and its vertical angle are congruent, the proof will almost write itself.

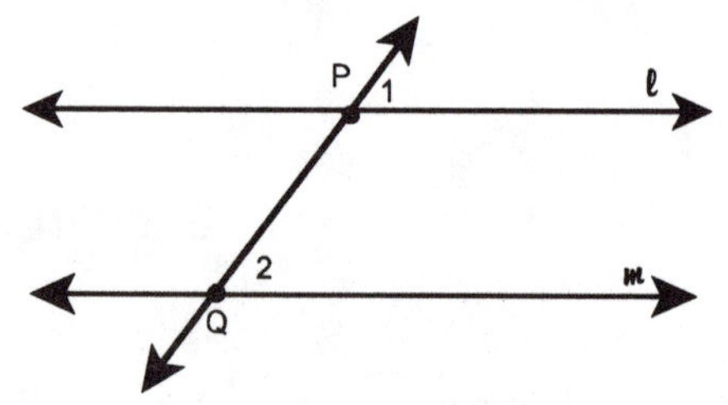

COMPLETE SOLUTIONS

1. That would be easy to confuse with the Parallel Postulate which states that there is *at most* one line through P that is parallel to ℓ. No. We haven't talked about the existence of at least one line through P that is parallel to ℓ.

2. If P were on ℓ, then it would be really tough to find a line through P that was parallel to ℓ.

3. We partially defined the word *corollary* in chapter two. A **corollary** is a theorem. It is a theorem that follows directly/easily from a previously proven theorem. The corollaries that we prove in this *Your Turn to Play* will all be derived from the AIP Theorem.

4. 1. A line ℓ and a point P not on ℓ.

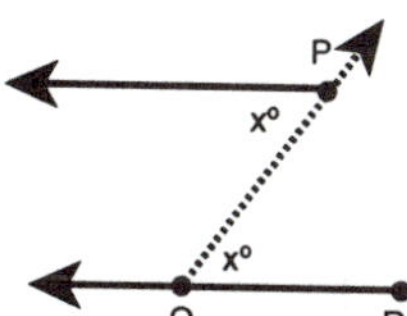

3. The Angle Measurement Postulate which states that to every angle there corresponds a number between 0° and 180° (and also to every number between 0° and 180° there corresponds an angle).

5. If two lines are cut by a transversal and the alternate interior angles are congruent, then the lines are parallel. (a.k.a. AI ➜ P)

5. The PAI Theorem states that if two parallel lines are cut by a transversal, then the alternate interior angles must be congruent. The PAI Theorem is both a converse and a corollary to the AIP Theorem. (PAI is a converse because it is obtained by switching parts of the hypothesis and the conclusion of the AIP. It is a corollary because its proof follows directly from the AIP.)

6.

2. Beginning of an indirect proof

4. AI ➜ P

5. Parallel Postulate

6. $m\angle 1 = m\angle 2$

7. Proof of the CAP Theorem

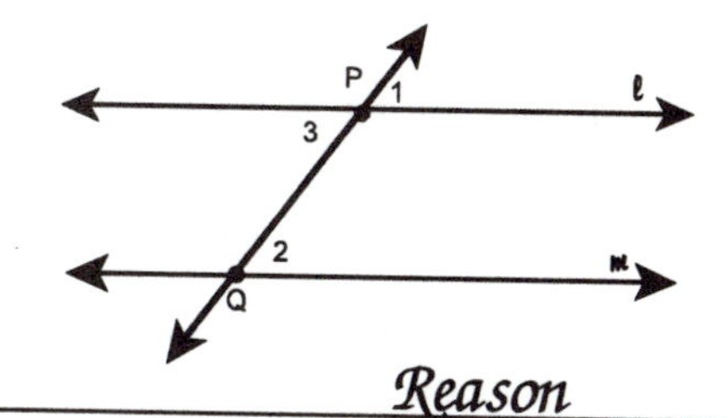

Statement	*Reason*
1. ℓ and m are cut by a transversal and form congruent angles. ($\angle 1 \cong \angle 2$)	1. Given
2. $\angle 1$ and $\angle 3$ are vertical $\angle$s.	2. Def of vertical $\angle$s
3. $\angle 1 \cong \angle 3$	3. Vertical Angle Theorem

4. $\angle 2 \cong \angle 3$	4. Timeout! If $\angle 2 \cong \angle 1$ and $\angle 1 \cong \angle 3$, it seems pretty obvious that $\angle 2$ should be congruent to $\angle 3$. But you can't use "Obvious" as a reason in a proof. We didn't prove this little proposition back in chapter two when we were looking at angles because you would have wondered why we were dealing with such a trivial and obvious fact. But now we need it. Here's another quick proof-in-a-box. If you were to cut out this box and paste it anywhere in chapter two, then the reason for step 4 in the proof we are now doing would read, "The Transitive Property of Congruent Angles Theorem."*

The Transitive Property of Congruent Angles Theorem

Statement	Reason
1. $\angle A \cong \angle B$ and $\angle B \cong \angle C$	1. Given
2. $m\angle A = m\angle B$ and $m\angle B = m\angle C$	2. Def of $\cong$ $\angle$s
3. $m\angle A = m\angle C$	3. Algebra
4. $\angle A \cong \angle C$	4. Def of $\cong$ $\angle$s

5. $\ell \parallel m$	5. AI → P Theorem

The line moved forward and the buffalo shuffled forward.**

Those customers who had already withdrawn all their money from the bank borrowed chairs from a nearby classroom and brought them to the bank. They sat down and waited for the robbery to begin.

Snow was beginning to sweat profusely. The buffalo disguises that he had rented from Buffalo Rental ("MADE FROM REAL BUFFALOS." SM) were hot and heavy. As everyone knows, even when it's snowing outside, you don't have to invite your pet buffalo into your house to keep warm.

---

* In algebra, we had the transitive property for equality: if x = y and y = z, then x = z. Some relations (such as = and ≅) are transitive. Others are not. "Is a descendent of" is transitive. If Pat is a descendent of Chris and Chris is a descendent of Robin, then Pat is a descendent of Robin. "Is the mother of" is not transitive. "Is taller than" is transitive.

** I wonder if some famous poet like Ogden Nash ever wrote lines which included the phrase, ". . . the buffalo shuffaloed"?

Snow removed most of his suit and just kept the head. He thought that no one would be able to identify him if his face remained covered.

As he approached the teller he said, "Muffx, yuommf lufm umf," and then realized he couldn't be understood. He opened the mouth of the mask and the words came out much more clearly. "I'd like to make a withdrawal."

The teller smiled. "Everyone seems to be doing that today. I hear that there's a big bank robbery planned, Sergeant Snow."

"How'd you know my name?"

The teller pointed to his name tag that was sewn onto his shirt. It was an old habit that he had picked up in the army.

"Okay Mr. Teller. This really isn't a withdrawal. This is it. This is a stickup!"

The audience cheered. It was 9:15 and the robbery was really happening.

Coalback was tired of his partner's ineptitude. He pushed Snow aside and took charge. "Okay.✶ Stick up your hands and give me all the money in the till."

"I was wondering," the teller responded, "may I ask a couple of quick questions? First, I'm having some difficulty reaching into the till when both of my hands are in the air."

Snow and Coalback walked away from the teller and reformulated their Plan of Action. When Coalback came back, he told the teller that it was more important that he get the money out of the till than to keep his hands up. Snow went and sat down next to Joe and Darlene in the audience. (Coalback had told him to do that so that he wouldn't make any more silly mistakes and foul up the robbery.)

"You said that you had two questions," Coalback said.

The teller nodded. "I don't mean to sound impertinent, but when I've watched movies like "Bonnie & Clyde," the bank robbers always

---

✶ All robbers tend to start their sentences with "Okay." It seems to relieve some tension and clear their throats.

seemed to have pistols. You don't seem to possess any weapons. Why should I obey your orders?"

That was a good question.

Blushing under his buffalo costume, he turned to Snow. "Didn't you bring any guns?"

"No I didn't, Colonel Coalback. I was in charge of our disguises."

Joe wrote down in his notebook, *Col. Coalback*, right underneath *Sgt. Snow*. He was making notes on the robbery in case he was called upon to testify. He turned to Darlene and said, "Maybe, someday, I could write a book entitled *The Great KITTENS Bank Robbery.*"

Darlene told Joe that robberies like this are never in books—it would be too silly.

Snow turned to Joe. "Excuse me. Do you happen to have a pistol that I could borrow for a moment?"

Joe shrugged his shoulders and then he remembered that Darlene carried a snub-nosed .38 special in her purse.* He reached into her purse and pulled out the gun so that he could hand it to Snow.

When Darlene yelled at Joe, "WHAT ARE YOU DOING!!!???" he involuntarily squeezed the trigger.

The bullet ricocheted off the face of the bank vault, then off a marble column and landed gently back in Joe's hand. Since Darlene had loaded her pistol with tracer bullets, you could see the neat triangle that the bullet made.

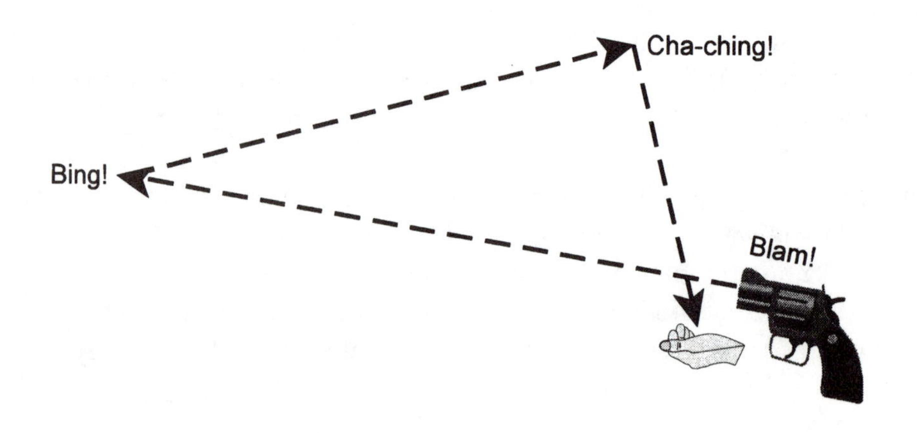

* Darlene had a concealed weapons permit, so that was legal.

Joe was holding a smoking pistol in one hand and a hot bullet in the other. "Ouch!" he exclaimed and dropped the bullet to the floor.

"Ouch!" he exclaimed again as Darlene smacked the back of his head with her hand.

"Ouch!" again as Snow stood up and tossed off his buffalo head. A horn of the head landed on Joe's foot.

Snow held his hands up in the air. "Don't shoot, mister! I've got a wife and kids, and he (pointing to Coalback) made me do it, and this was all a mistake, and I needed the money to pay for the vet for the cancer operation on my kids' kitty cat whom they love dearly, and I'm not responsible since I had a rotten childhood, and my horoscope told me I had to do it or the moon in Sagittarius would hex me."*

Before Joe could utter another "Ouch!" the robbers lay down on the floor in felony-arrest position with their arms outstretched.

Darlene took her pistol back from Joe, put it into her purse, took out her cell phone and called Sergeant Friday.

When Friday arrived, the crowd greeted him with 'rest them! 'rest them!" Friday turned to Darlene. "Could I please borrow your gun?"

Not knowing what else he should do after he completed the arrest, he took lots of pictures,

This saved the day.

What he looked like

View out bank window

and then he got out his plastic protractor and measured the angles in the Blam!—Bing!—Cha-ching! triangle.
The Bing! angle measured 37°.
The Cha-ching! angles measured 92°.
The Blam! angle measured 51°.

Cha-ching!
Bing!
Blam!

---

* Lies. All lies.

To make his report complete, he added the angles: 37° + 92° + 51°, and they came out to an even 180°. He knew that in an equiangular triangle, the sum of the measures of the angles would add up to 180°, but

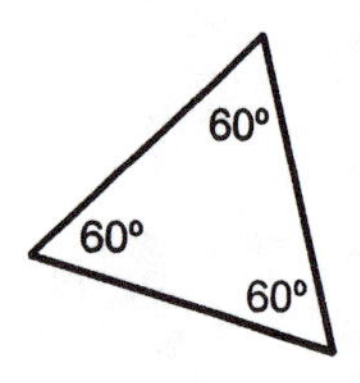

he was surprised that the angles of this tracer-bullet triangle also added up to 180°. He wrote "Important Clue" next to his computation. He was sure that whoever read his report would be greatly impressed by his discovery of a second triangle with the angles-add-to-180° property.

However, most people who read Friday's crime scene report remained unimpressed. In fact, some of them laughed at his "Important Clue" notation. Everyone who has ever studied geometry knows that the sum of the measures of the angles of *any* triangle always add to 180°. It's a fact of nature.

The proof is one of the most creative in all of geometry. First, you start out with any triangle:

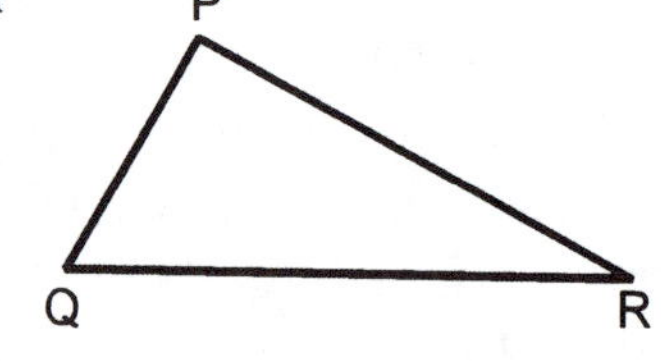

The question is, "How in the world do you show that $m\angle P + m\angle Q + m\angle R$ equals 180°? Before you look at the proof below, please play with this a few moments and see if you can discover the proof on your own. *Please.* Your delight will be increased if you make that effort.

<u>Theorem</u>: The sum of the angles of any triangle add to 180°.

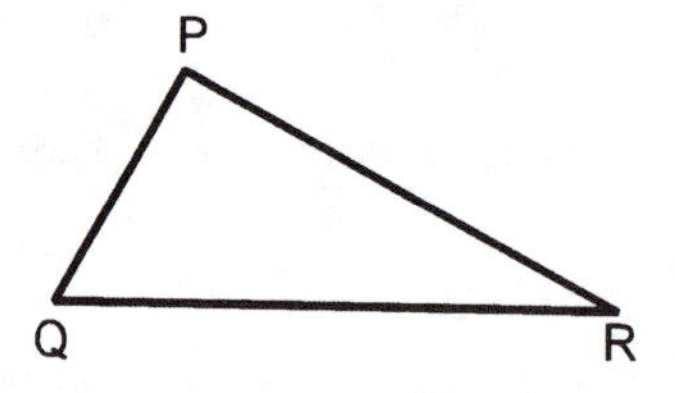

We prove it by first drawing a line through P which is parallel to $\overleftrightarrow{QR}$. We know that such a line exists by the theorem we proved in *Your Turn to Play* on p. 137 in problem 4. (The Parallel Postulate won't help us here since it says that there is *at most* one line through P that is parallel to $\overleftrightarrow{QR}$.)

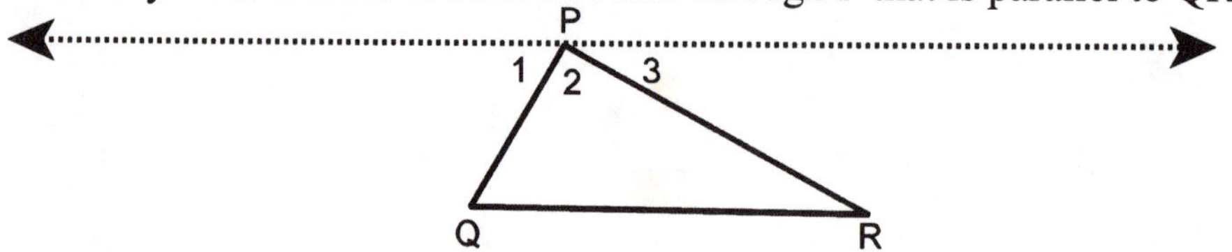

By P ➜ AI, we have that $\angle Q \cong \angle 1$ and also that $\angle R \cong \angle 3$. Since $m\angle 1 + m\angle 2 + m\angle 3$ equals 180°, it must be true that $m\angle Q + m\angle 2 + m\angle R$ also equals 180°.

Neat, isn't it?

This was a big news day for the Caboodle. Even though all the students had witnessed the attempted robbery, they each bought a copy of the 9:30 edition of the paper. It would be something that they could show their grandchildren some day and say, "I was there."

THE KITTEN Caboodle

The Official Campus Newspaper of KITTENS University

Thursday 9:30 a.m. Edition 10¢

Sgt. Friday Nabs Bank Robbers Singlehandedly

KANSAS: Our campus policeman was on the job this morning. Exhibiting great bravery, he foiled the first robbery ever attempted at the KITTENS Bank. Responding to an urgent call from a concerned student, Darlane, who was at the scene of the crime, Friday arrived at the bank shortly after gunfire had erupted and took masterful control of the situation. (continued on p. 14)

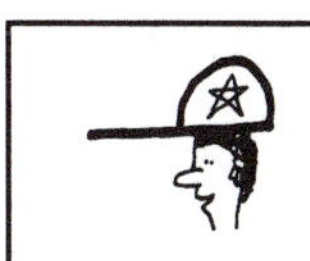

Our Campus Cop

"When they weren't looking I got the drop on them."

Some of the students were amazed at the inaccuracies in the newspaper report.

Darlene was more than just amazed. "Those mush brains! They spelled my name wrong." She was also angry that the newspaper didn't give the proper credit to her boyfriend. She told him, "You know, Joe, you were mighty brave in that crisis. I'm so proud of you. You knew exactly what to do and you went ahead and did it."

His head still hurt from where she had hit him.

Your Turn to Play

We need to tie up a couple of loose ends regarding the CAP Theorem and the Exterior Angle Theorem and then we can find out what Fred was doing during all the excitement of the bank robbery.

1. Back on p. 139, problem 7, we proved the CA → P Theorem ("If two lines cut by a transversal form congruent corresponding angles, then the lines must be parallel.") Its converse, the PCA Theorem ("If two parallel lines are cut by a transversal, then the corresponding angles are congruent") is also true.

Fill in the missing parts of the proof of the P → CA Theorem.

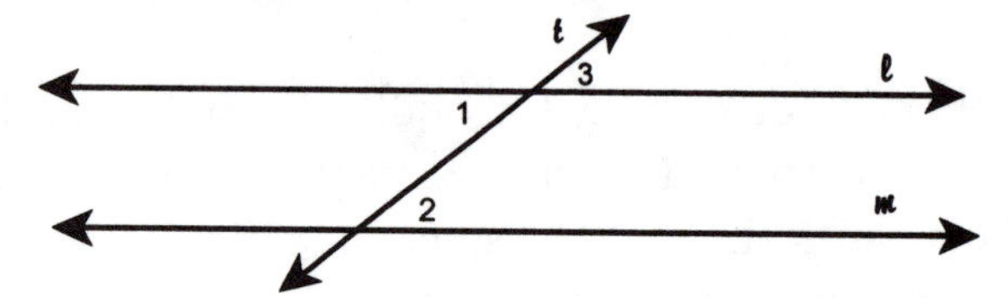

Statement	Reason
1. $l \parallel m$ and t is a transversal.	1. Given
2. $\angle 1 \cong \angle 2$	2. ?
3. $\angle 1 \cong \angle 3$	3. ?
4. $\angle 2 \cong \angle 3$	4. The Transitive Property of Congruent Angles Theorem which we proved about six pages ago. (If $\angle A \cong \angle B$ and if $\angle B \cong \angle C$, then $\angle A \cong \angle C$.)

2. The Exterior Angle Theorem stated that an exterior angle is greater than either remote interior angle. We have only shown that $m\angle B < m\angle BCD$. How could we show that $m\angle A$ is also less than $m\angle BCD$.

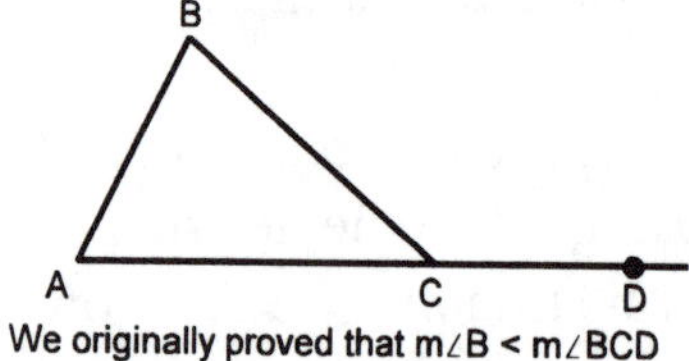

We originally proved that $m\angle B < m\angle BCD$

The diagram we used in that proof looked like:

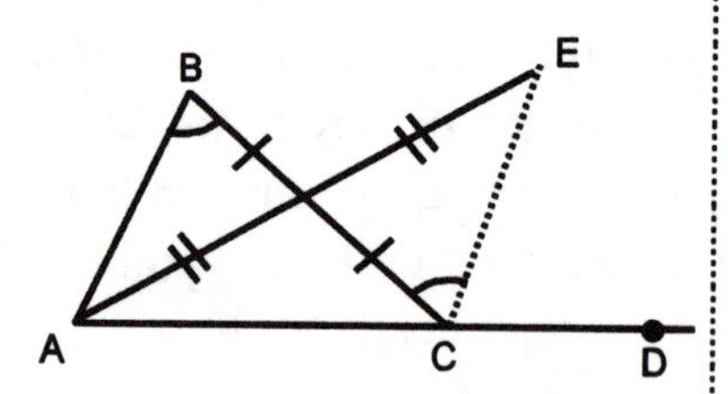

It's not clear from that diagram how $m\angle A$ could be shown to be less than $m\angle BCD$. This is a question that art majors may enjoy. Redraw the diagram to illustrate the proof that $m\angle A < m\angle BCD$. (You need not write out the proof.) The new diagram will use the midpoint of $\overline{AC}$ instead of the midpoint of $\overline{BC}$ and the vertical angle to $\angle BCD$.

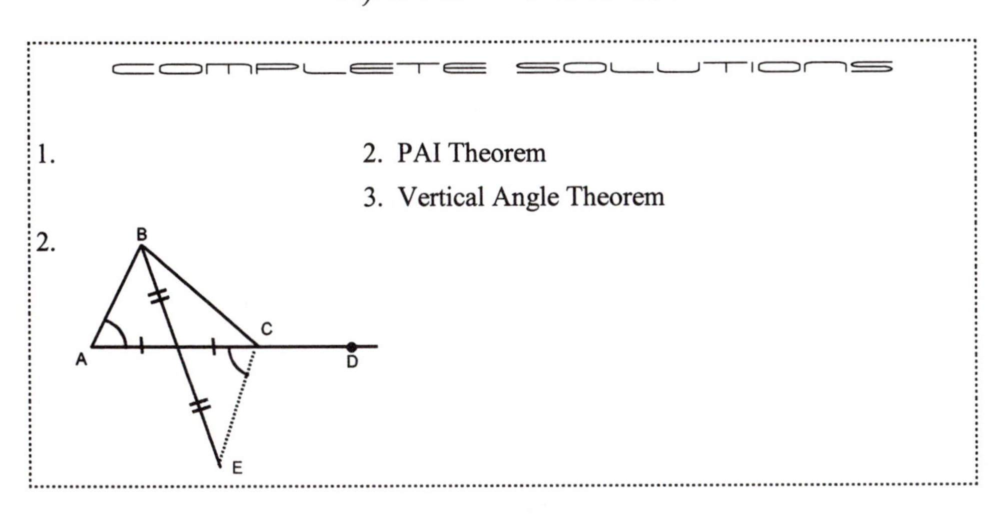

It was pretty quiet in Fred's classroom with all his students gone. Since all the chairs in the room were too big for him, he sat down on the floor and took some mail out of his briefcase. He kicked off his shoes and crossed his left ankle over his right knee. After all that jogging this morning, being barefoot felt good. Besides, after he read each letter he could file them between his toes according to a filing systems that he had invented.

Letter has math question for me to answer.

Fan mail. Write thank you letter.

Bill from bookstore.

Junk mail.

Fred always immediately answered the mail that contained math questions. He always wrote thank you notes on the day that he received fan mail. And, in fact, he always paid his bills on the day that he received them. He was that kind of guy.

He read his first piece of fan mail. *Dear Sir, May I call you Fred? Your teaching method of relating everyday life to mathematics makes such a difference!*

The other ladies at the Wednesday Bridge Club and I think you are soooooooo cute! Could you please send us an autographed picture of yourself? We're thinking of starting a chapter of the Fred Gauss Math Club right here in our town.

Affectionately, Ms. D Commonsizerinski

p.s. My brother, T Commonsizerinski and his wife, L Commonsizerinski, have a lovely little daughter, P Commonsizerinski, who has just finished her high school studies and will be heading to KITTENS University for the spring semester. If you see her, please tell her "Hi!" from Auntie D.

In novelists' parlance, this is adumbration if there ever was adumbration.

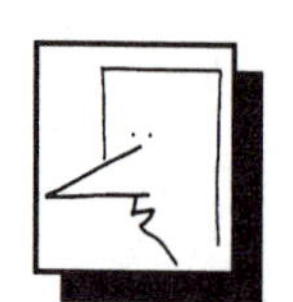

Fred's drawing of himself

Since Fred didn't have a photograph, he drew a picture of himself (which looked surprisingly like any photograph of him) and wrote a note of appreciation to accompany his drawing.

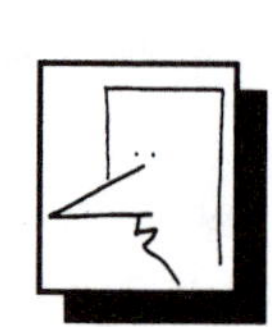

Actual photo of Fred

As he filed Ms. D Commonsizerinski's letter between his toes, he thought to himself, This looks a lot like the diagram for the Exterior Angle Theorem.

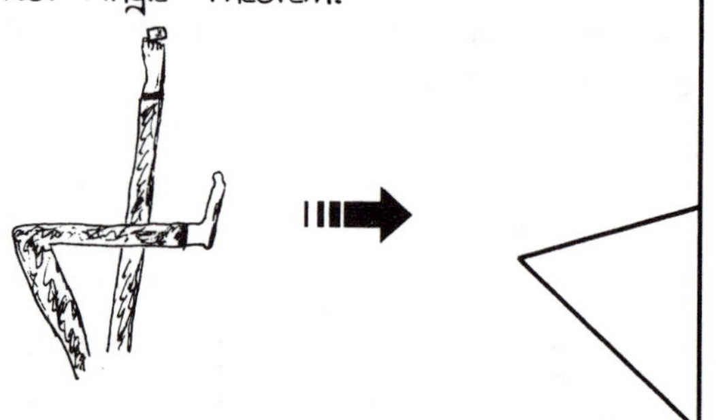

It's too bad that I only had a chance to prove the weak form of the Exterior Angle Theorem: that ∠1 is greater than either ∠2 or ∠3.

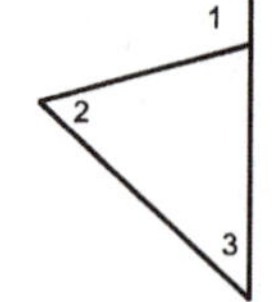

I wanted to show my class the strong form. It's much nicer. The strong form says that the exterior angle (∠1) is not only greater than either of the interior angles, but is actually equal to the sum of the two remote interior angles (∠2 + ∠3).

Proving the weak form took 15 steps. There are two different proofs of the strong form and both of them are shorter than 15 steps.

Fred split his mind down the middle and proved the strong form twice, using one half of his brain for each proof. (Both halves of Fred's brain are mathematical.)

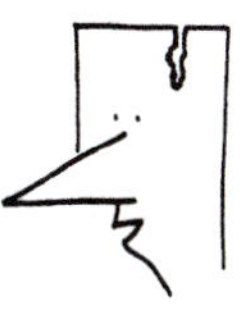

First Proof

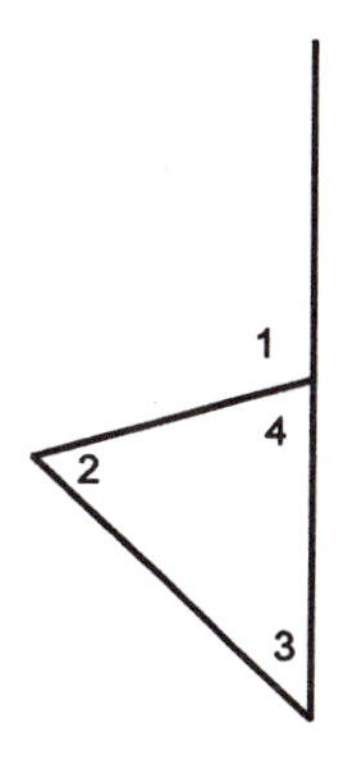

Angles 2, 3 and 4 add up to 180°.
Angles 1 and 4 add up to 180°.
Then by Algebra we're done:

$m\angle 1 = m\angle 2 + m\angle 3.$

Second Proof

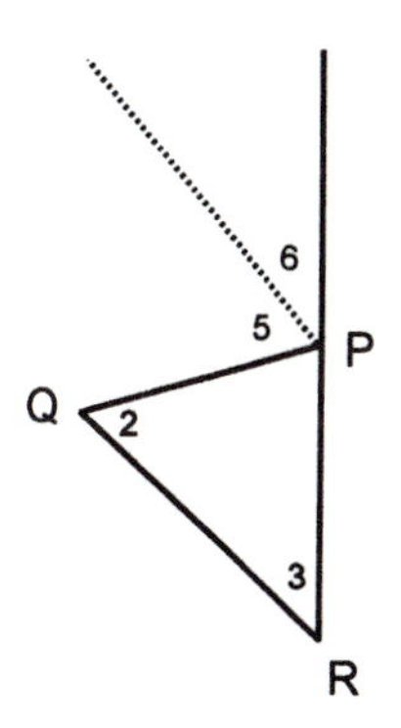

We draw a line through P that is parallel to $\overleftrightarrow{QR}$.
$\angle 2 \cong \angle 5$ by PAI*
$\angle 3 \cong \angle 6$ by PCA**

* the transversal is $\overleftrightarrow{PQ}$ ** the transversal is $\overleftrightarrow{PR}$

These proofs are in paragraph form rather than in statement-reason form. This is the way that mathematicians carry proofs around in their heads. Then, if they need to present the proof on the blackboard, they translate the paragraph form into the statement-reason form and fill in all the mechanical details as they write. There is a lot less to memorize if you are a math major than a . . .

❀ biology major where you have to remember all the names of all those bugs and the difference between *mitosis* (which is the normal way that cells divide) and *meiosis* (which is a good word for getting rid of vowels in Scrabble® and which is cell division in which the resulting cells each have only half of their goodies—in preparation for birds-and-bees stuff.)

❀ foreign language major where you have to memorize at least 1500 words for a basic vocabulary. How do you say *French Toast* in Russian?

❀ art major where you have to discuss the chiaroscuro (pronounced key-R-eh-skoor-oh) in hundreds of paintings and tell (a) who painted it; (b) in what year; and (c) its content.

Quick quiz: Do this for

Answer: This is a detail from Titian's *The Presentation in the Temple*, 1534–1538, in which he painted Mary ascending the steps in the Temple. One reason it took him so long to create this painting is that the original is 306 inches wide. We have reduced it somewhat in order to fit in this book.

Here is a translation of the second proof of the strong form of the Exterior Angle Theorem into statement-reason form:

Statement	*Reason*
1. $\triangle$PQR with $\angle$QPS as an exterior angle.	1. Given
2. There exists a line $\overleftrightarrow{PT}$ through P which is parallel to $\overline{QR}$.	2. Theorem: If you have a line ℓ and a point P not on ℓ, then there is at least one line through P that is parallel to ℓ.

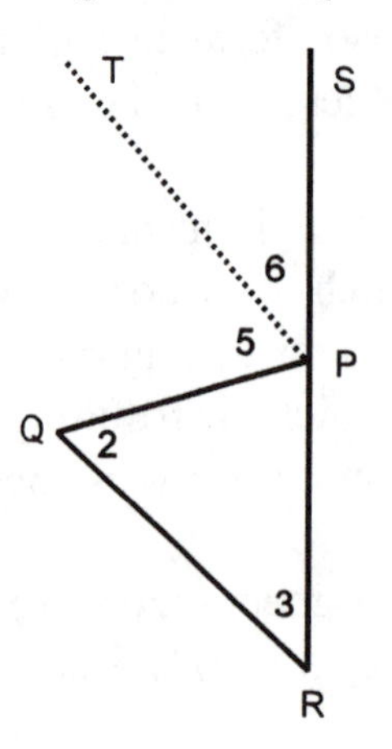

3. $\overleftrightarrow{PQ}$ is a transversal to lines $\overleftrightarrow{PT}$ and $\overleftrightarrow{QR}$. $\overleftrightarrow{PR}$ is a transversal to lines $\overleftrightarrow{PT}$ and $\overleftrightarrow{QR}$.	3. Definition of transversal
4. $\angle 2 \cong \angle 5$	4. P ➜ AI (If two parallel lines are cut by a transversal, then the alternate interior angles are congruent.)
5. $\angle 3 \cong \angle 6$	5. P ➜ CA (If two parallel lines are cut by a transversal, then the corresponding angles are congruent.)
6. $m\angle 2 = m\angle 5$ $m\angle 3 = m\angle 6$	6. Definition of $\cong$ $\angle$s
7. $m\angle 5 + m\angle 6 = m\angle QPS$	7. Angle Addition Postulate
8. $m\angle 2 + m\angle 3 = m\angle QPS$	8. Algebra

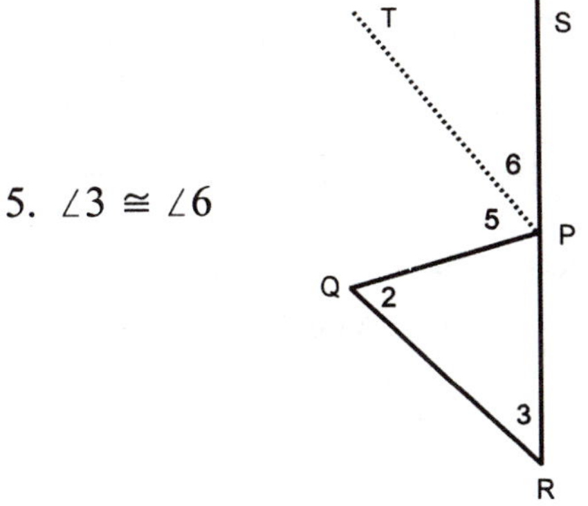

⊠

Elmo

1. Fill in the missing parts of the proof that no triangle can contain two right angles.

S	*R*
1. ΔABC	*1.* ?
2. Assume $\angle$A and $\angle$CBA are right $\angle$s.	*2.* ?
3. $\angle 2$ and $\angle 3$ form a linear pair.	*3.* ?
4. ?	*4.* Postulate 4 (If two angles form a linear pair, then they are supplementary.)
5. m$\angle 2$ + m$\angle 3$ = 180º	*5.* ?
6. m$\angle 2$ = 90º and m$\angle 1$ = 90º	*6.* Def of right angle (from step 2)
7. m$\angle 3$ = 90º	*7.* Algebra
8. m$\angle 3$ > m$\angle 1$	*8.* ?
9. $\angle$A and $\angle$CAB are not both right angles.	*9.* Contradiction in steps 6, 7 and 8

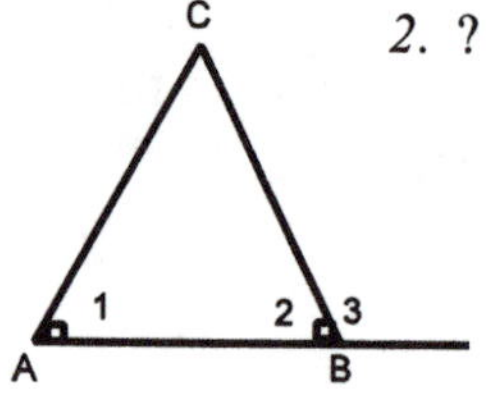

2. Fill in the one missing reason in the proof of the theorem: If P is not on ℓ, then there is at most one perpendicular from P to ℓ.

S	*R*
1. P is not on ℓ.	*1.* Given
2. Assume there are two perpendiculars from P to ℓ.	*2.* Beginning of an indirect proof
3. ΔPQR is a triangle with two right angles.	*3.* Def of $\perp$ lines
4. ΔPQR can't have two right angles.	*4.* ?
5. There are not two perpendiculars from P to ℓ.	*5.* Contradiction in steps 3 and 4

3. This will test your ability to think in three dimensions: If two lines intersect, must they be coplanar?

4. Let B = "This poem is beautiful," and T = "This poem contains truth," and if T ➜ B, then do we put "if" or "only if" in the blank space:

This poem contains truth __?__ this poem is beautiful.

5. If you were trying to prove that you are alive using an indirect proof, what would be the assumption step?

answers

1.
1. Given
2. Beg. of indirect proof
3. Def of linear pair
4. $\angle 2$ and $\angle 3$ are supplementary
5. Def of supplementary
8. (Weak form of) the Exterior Angle Theorem

2. The theorem we just proved in the previous problem.

3. With any two intersecting lines, you can always find a plane that contains them. Since we don't have any postulates concerning planes yet, we can't prove this now.

4. *Implies* (➜) can be translated as *only if.*

5. You would say, "Assume that I am dead." Then you would do something that dead people can't do (such as kissing).

Harlan

1. Is *If my car is out of gas, it won't run* the converse of *My car won't run if it is out of gas*?

2. The Vertical Angle Theorem reads: *Vertical angles are congruent.* Is its converse also a true statement?

3. In any proof, what is the first word in the statement which corresponds to the reason *Beginning of an indirect proof*?

4. Two angles are **complementary** iff their measures add to 90°. Prove that the acute angles of a right triangle are complementary.

5. If $\overline{AC} \cong \overline{BC}$ and if $\overline{DE} \parallel \overline{AB}$, prove that $\angle 1 \cong \angle 2$.

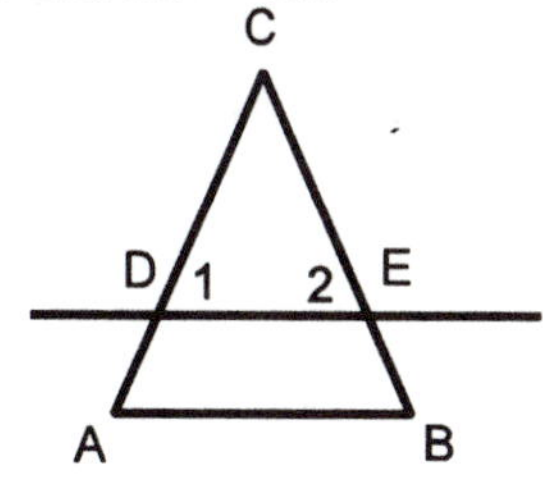

6. This is the picture of a house. It is an equilateral triangle on a square. Find the angle marked "?"

answers

1. The converse of an *If ••• then •••* sentence is obtained by interchanging one or more parts of the hypothesis with one or more parts of the conclusion. The hypothesis of both sentences is *my car is out of gas* and the conclusion of both sentences is *it won't run*. Since nothing has been interchanged, the answer is no.

2. The Vertical Angle Theorem in *If ••• then •••* form reads *If two angles are vertical angles, then they are congruent.* Its converse would be *If two angles are congruent, then they are vertical angles.* This isn't true.

3. The first word has invariably been "assume."

4.

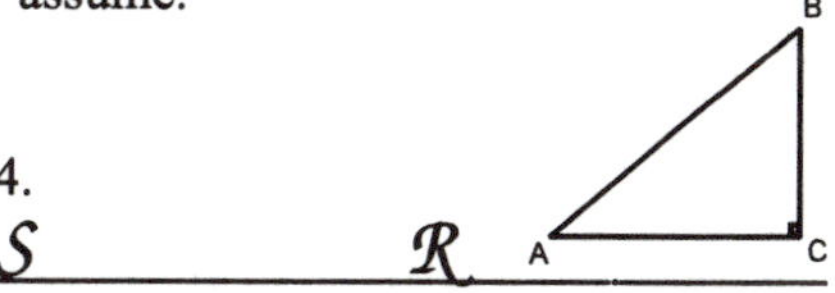

S	*R*
1. △ABC in which ∠C is a right angle.	*1.* Given
2. m∠C = 90°.	*2.* Def of right angle
3. m∠A + m∠B + m∠C = 180°.	*3.* Thm: The sum of the ∠s of a △ = 180°
4. m∠A + m∠B = 90°	*4.* Algebra
5. ∠A and ∠B are complementary.	*5.* Def of complementary

5.

S	*R*
1. $\overline{AC} \cong \overline{BC}$, $\overline{DE} \parallel \overline{AB}$	*1.* Given
2. ∠A ≅ ∠B	*2.* ITT
3. ∠1 ≅ ∠A and ∠2 ≅ ∠B	*3.* P ➜ CA
4. ∠1 ≅ ∠2	*4.* Transitive Property of Congruent Angles Theorem

6. The angle we're looking for is part of an isosceles triangle whose largest angle is 90° + 60°. m∠? = 15°

Tarrytown

1. Sometimes in giving a proof, we have shortened the reason "Contradiction in steps 6 and 7 and therefore the assumption in step 2 is false" to just "Contradiction in steps 6 and 7."

How could you mechanically find out what step in the proof was false? In other words, how might you instruct a computer to find what line of the proof was false?

2. Prove that the measures of an equilateral triangle are equal to 60°.

3. Given $\overline{AB} \parallel \overline{DE}$ and C bisects (cuts into congruent segments) $\overline{BD}$. Prove that C bisects $\overline{AE}$.

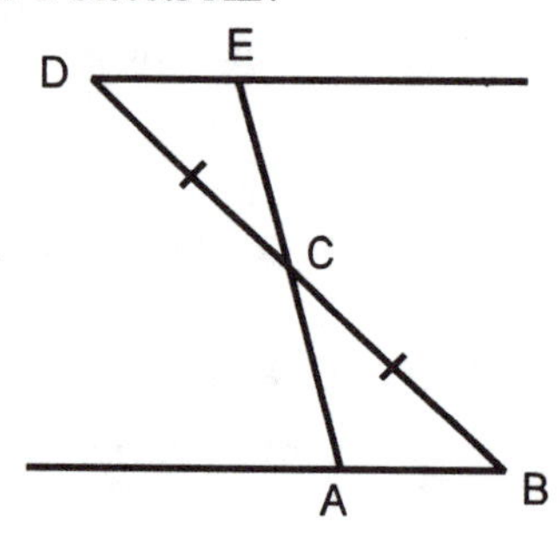

4. Describe a triangle in which all six of its exterior angles are congruent.

5. Prove that if one pair of alternate interior angles is congruent, then the other pair must also be congruent. Do the proof without using the PAI Theorem.

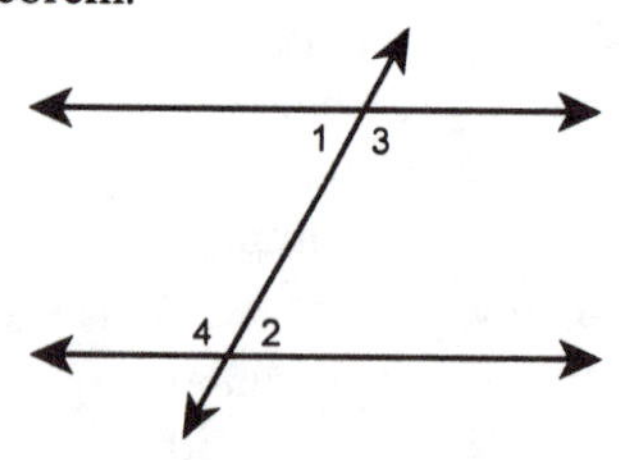

odd answers

1. Just tell the computer to look upward in the proof for the word *Assume*. That will be the line that is false. Another way of doing it would be to tell the computer to look for the line with the reason *Beginning of an indirect proof.*

Sometimes, when the location of the contradiction is obvious, some writers of proofs just write *Contradiction.*

And some writers who really don't want to write the big long word *Contradiction* just write the symbol #. You can't get much shorter than that.

3.

S	*R*
1. C bisects $\overline{BD}$ and $\overline{AB} \parallel \overline{DE}$.	*1.* Given
2. $\overline{BC} \cong \overline{CD}$	*2.* Def of bisects
3. $\angle ACB \cong \angle ECD$	*3.* Vert ∠ Thm
4. $\angle CDE \cong \angle CBA$	*4.* PAI
5. $\Delta CAB \cong \Delta CED$	*5.* ASA
6. $\overline{AC} \cong \overline{EC}$	*6.* Def of ≅ ▲s
7. C bisects $\overline{AE}$	*7.* Def of bisects

5.

S	*R*
1. $\angle 1 \cong \angle 2$	*1.* Given
2. $\angle 1$ and $\angle 3$ form a linear pair. $\angle 2$ and $\angle 4$ form a linear pair.	*2.* Def of linear pair
3. $\angle 1$ and $\angle 3$ are supplementary ∠s. $\angle 2$ and $\angle 4$ are supplementary ∠s.	*3.* Postulate 4 (Linear pair ➜ supp.)
4. $\angle 3 \cong \angle 4$	*4.* Theorem: (Supplements of ≅ ∠s are ≅ .)

Zeigler

1. If, in some step of a proof, you have $\angle A \cong \angle B$, state seven possible reasons that might have been used to justify that step. Of course, which one of these reasons is appropriate will depend on what has happened earlier in the proof.

2. Complete the proof of *If two angles of one triangle are congruent to two angles of a second triangle, then the third angles of each triangle must be congruent.*

(The AA ➜ AAA Theorem)

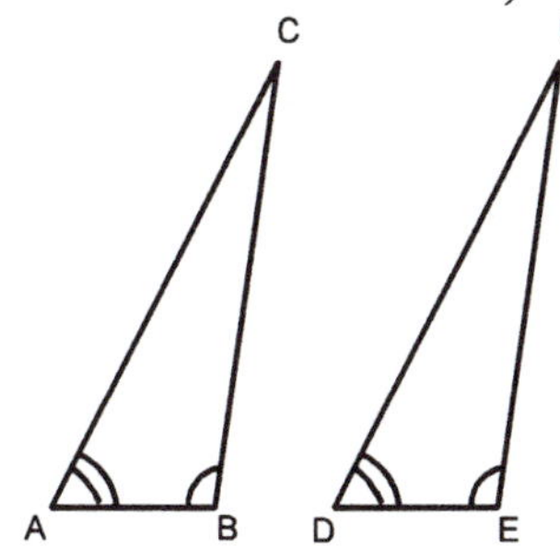

S	*R*
1. $\triangle ABC$ and $\triangle DEF$, $\angle A \cong \angle D$, $\angle B \cong \angle E$.	*1.* Given
2. $m\angle A = m\angle D$ and $m\angle B = m\angle E$	*2.* Def of $\cong$ $\angle$s
3. $m\angle A + m\angle B + m\angle C = 180°$ $m\angle D + m\angle E + m\angle F = 180°$	*3.* Theorem: The sum of the measures of a $\triangle$ equals 180°.
4. [Two more steps to go.]	

3. Prove the AAS Theorem: In triangles ABC and DEF, if $\angle A \cong \angle D$, $\angle B \cong \angle E$, and $\overline{BC} \cong \overline{EF}$, then the triangles are congruent.

4. If in some proof we have the statement that $\triangle DEF \cong \triangle GHI$, there are now six possible reasons that might be used to justify that statement. Name them.

5. [This is fairly similar to problem 3 in Tarrytown.] Without using the Vertical Angle Theorem, prove that if $\overline{AB} \parallel \overline{DE}$ and if C bisects (cuts into congruent segments) $\overline{BD}$, then C bisects $\overline{AE}$.

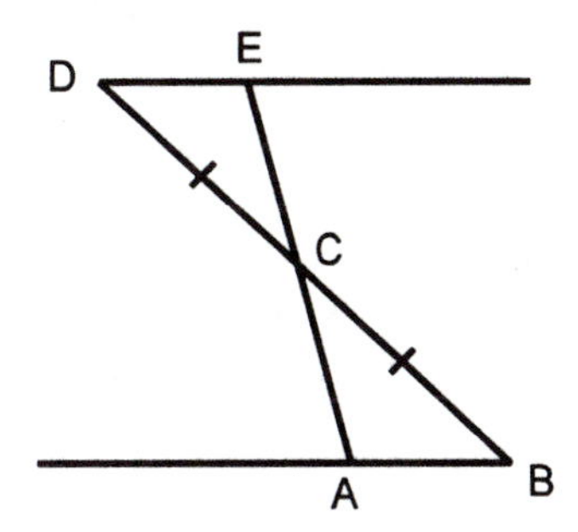

odd answers

1. a) Def of congruent angles
   b) Given
   c) Def of congruent triangles
   d) PCA (If two parallel lines are cut by a transversal, then corresponding angles are congruent.)
   e) PAI
   f) Isosceles Triangle Theorem (The base angles of an isosceles triangle are congruent.)
   g) Transitive Property of Congruent Angles Theorem

3.

S	*R*
1. $\angle A \cong \angle D$, $\angle B \cong \angle E$ and $\overline{BC} \cong \overline{EF}$	*1.* Given
2. $\angle C \cong \angle F$	*2.* AA ➜ AAA Thm (which we just proved in the previous problem)
3. $\triangle ABC \cong \triangle DEF$	*3.* ASA

5.

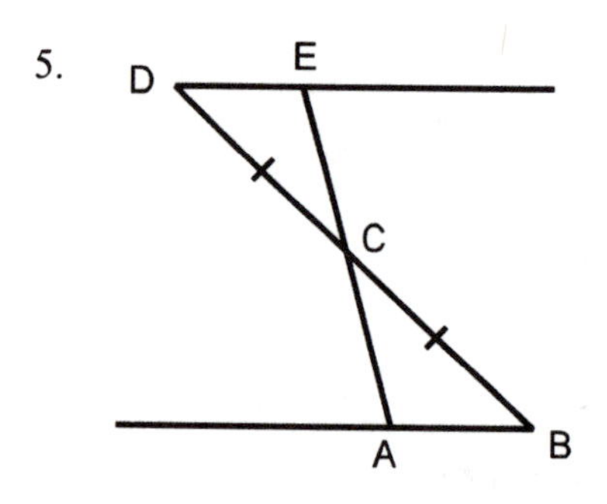

S	*R*
1. C bisects $\overline{BD}$ and $\overline{AB} \parallel \overline{DE}$	*1.* Given
2. $\overline{BC} \cong \overline{CD}$	*2.* Def of bisects
3. $\angle CED \cong \angle CAB$, $\angle CDE \cong \angle CBA$	*3.* PAI
4. $\triangle CAB \cong \triangle CED$	*4.* AAS (which we just proved in problem 3)
5. $\overline{AC} \cong \overline{EC}$	*5.* Def of $\cong$ $\triangle$s
6. C bisects $\overline{AE}$	*6.* Def of bisects

Dane

1. The watch that Joe bought was really a dud. It didn't run. It always read 9 o'clock. It is a true statement that *A watch is a dud if it ever tells the incorrect time.*

Is the converse of that statement true?

2. Given $\overline{DE} \cong \overline{DF}$ and $\overline{DH} \parallel \overline{EF}$. Prove that $\overrightarrow{DH}$ bisect $\angle GDF$.

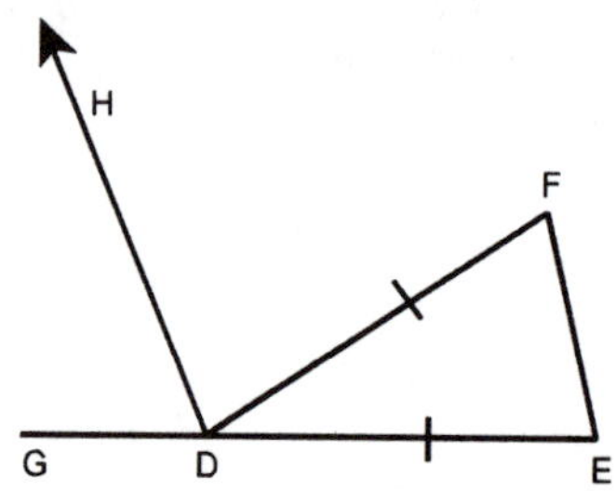

3. Is it possible to have a triangle with exactly three of its six exterior angles congruent?

4. In $\triangle ABC$, $m\angle A = 89°$, can $\triangle ABC$ be obtuse?

5. If $\overline{AC}$ bisects $\angle DAB$ and if AB = BC, prove that $\overline{AD} \parallel \overline{BC}$.

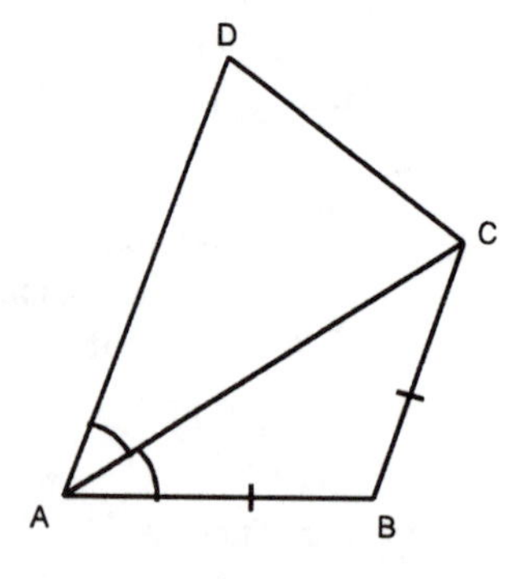

Lamar

1. Which one (or more) of the following alternatives is/are a converse to *If the sky is blue and I've recently been kissed, then I have a song in my heart.*

A) I have a song in my heart if the sky is blue and I've been recently kissed.

B) If I've recently been kissed and have a song in my heart, then the sky is blue.

C) If I have a song in my heart, then the sky is blue and I've recently been kissed.

D) The sky is blue if I've recently been kissed and there is a song in my heart.

2. Fill in the missing parts of the proof of *If $\overline{AC}$ and $\overline{BD}$ bisect each other, then $\overline{AB} \parallel \overline{CD}$ and $\overline{AD} \parallel \overline{BC}$.*

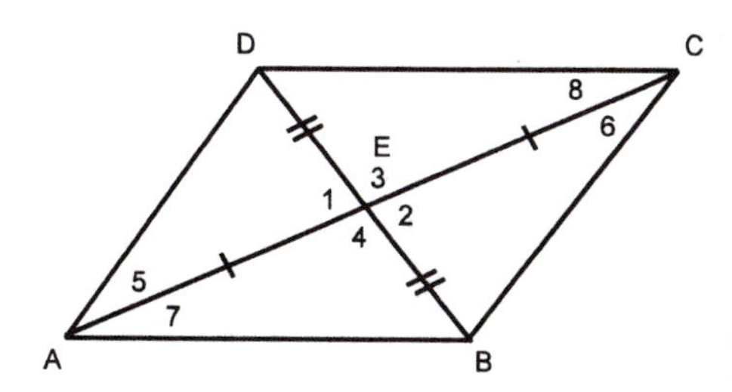

S *R*

1. $\overline{AC}$ and $\overline{BD}$ bisect each other.
1. Given

2. $\overline{AE} \cong \overline{EC}$ and ?
2. Def of bisect

3. $\angle 1 \cong \angle 2$ and ?
3. ?

4. $\Delta AED \cong \Delta CEB$ and $\Delta DEC \cong$? (Remember to write the letters in the correct order so that they correspond to DEC.)
4. ?

5. $\angle 5 \cong \angle 6$ and ?
5. Def of $\cong$ $\triangle$s

6. $\overline{AB} \parallel \overline{CD}$ and $\overline{AD} \parallel \overline{BC}$
6. ?

3. Is it possible to have a triangle with exactly four of its exterior angles congruent to each other? If it is possible, describe the triangle. If it is not, explain why it is not possible.

4. Given $\overline{DH} \parallel \overline{EF}$ and $\overrightarrow{DH}$ bisects $\angle GDF$, prove that ΔDEF is isosceles.

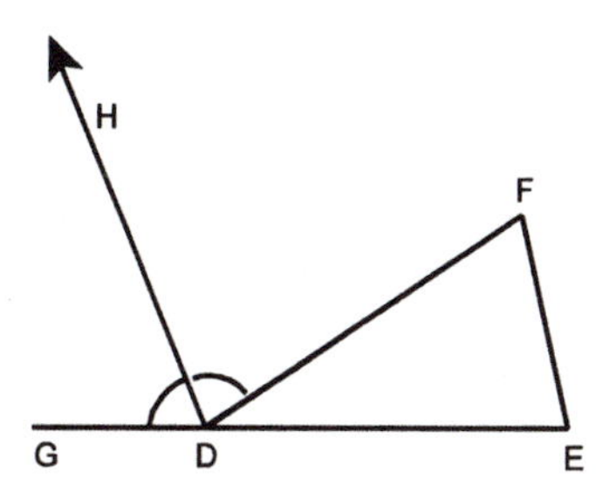

5. In ΔABC, if $m\angle C = 90°$, can ΔABC be an acute triangle?

6. Which of the following is *not* sufficient to prove two triangles are congruent?

SSS
SAS
SSA
AAS
AAA

Chapter Five

Perpendicular Lines

Fred was in his own happy little world: barefoot, sitting on the floor, answering his mail, and thinking about two proofs of the strong form of the Exterior Angle Theorem. He had his health, and he knew that he was loved by God. He had a wonderful job teaching at the University, thousands of dollars in the bank, and he would be seeing his friends Alexander and Betty for pizza tonight.

He was unaware of another pleasant fact: that two men who had given him such problems in his life, Sergeant Snow and Col. C.C. Coalback, had just been arrested.

He was also unaware of a girl who had just walked into the classroom and was looking down at him.

"Hi. Is that my Auntie's letter you have stuck between your toes?"

Fred was used to being asked questions. It was part of his life as a teacher. Often he had been asked why he put letters between his toes after he read them, and he was happy to explain that this was a logical way to sort the mail.

The KITTENS administration was used to seeing pictures of their most famous professor in national magazines, but they had some misgivings when they saw a cover photograph of Fred in *Modern Teachin'* sitting on the floor with letters sticking out of his toes. The university president sent him a new "handy-dandy" (his words) metal letter sorter in hopes that Fred would use it and not embarrass the school. Fred sent the president a thank-you note explaining that although it was dandy (first-rate), it wasn't handy (convenient). He always had his toes, but lugging around something that was 6% of his body weight was a bother. (Several teachers around the country tried out Fred's technique, but they found it was difficult to hold more than a couple of letters at a time in their toes. Fred had the advantage of five years of practice.)

Fred has always had a problem with conversational English. He listens too hard to the question. He had just been asked, "Is that my Auntie's letter you have stuck between your toes?" He wasn't stuck in answering why he used his toes, but he wondered if he had to prove that it was her Aunt's letter.

He was baffled. He needed more information. He was going to ask, "Are you her niece?" That would have given him enough information to answer her question. As he looked up at her, he began, "Are you. . . ."

That's all he could say.

Something new had entered into the life of Fred.

When Ms. D Commonsizerinski had written that she had a "lovely little niece, P Commonsizerinski," she wasn't exaggerating. This was the loveliest example of a female Homo sapiens that Fred had ever seen. He would have gasped except that his heart was in his throat.

Fred's thoughts raced. Her eyes, her hair, her smile.

Before Fred stood P Commonsizerinski, also age six. Fred, who had been proclaimed the Lecturer of the Century®, couldn't put together a spoken sentence.

She pointed to the letter wondering if he hadn't understood her. Fred's toes involuntarily spread apart, and her Aunt's letter fell to the ground. He sputtered, "They're clean. My toes. I showered after I jogged this morning." But he really hadn't showered this morning as he normally did after jogging. Things had been so unusually busy that his usual routine was interrupted. But Fred really wasn't lying since he thought he was telling the truth. The problem was that his brain had gone into DEFAULT MODE and was capable of only the most basic functions:

✓ inflate lungs
✓ beat heart
✓ do math
✓ produce enzymes
✓ recall basic facts (name, age, and usual routines).

To break the awkwardness she changed the subject. "My aunt tells me that when you first came here to teach that there were only 98 students on campus, and now there are more than 6,000. And she said that your teaching has attracted so many students."

Fred blushed some more. He stood up and gazed at her. It was puppy love at first sight. She was just his height and had the most beautiful red curly hair.

When she noticed that Fred had uttered only two words, she tried a different approach to get to know him. "I'm new to this campus. Would you care to walk with me and show me around?"

Fred had fallen . . . hard.

He was half way to the door when she pointed out to him that he had forgotten his shoes and socks. He rushed back and debated with himself whether he should take the time to put them on or just stuff them in his briefcase. P turned and looked at the math that Fred had written on the blackboard. He sat down and put on his shoes (and put his socks in the briefcase).

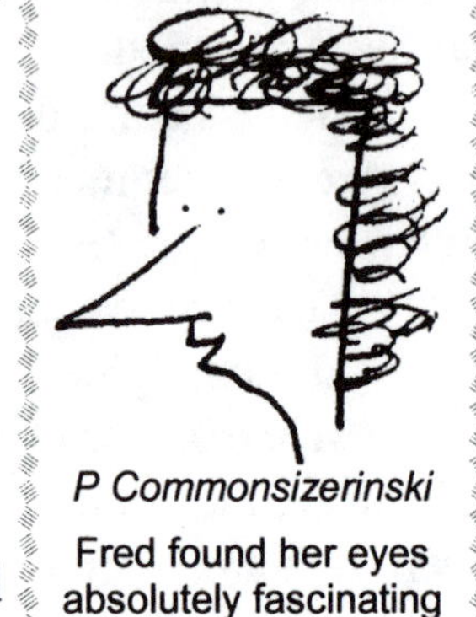

P Commonsizerinski

Fred found her eyes absolutely fascinating

He thought about asking her if she were going to take any math courses, but he realized that if he tried asking, all that would probably come out of his mouth would be the words, "Are you."

"I was wondering. Is there any room in your one o'clock calculus class? I finished all my high school studies a couple of years ago, but my folks said I had to stay at home until I was six. My sixth birthday was just this last Friday."

Fred nodded vigorously. He was afraid to speak.

Finally they were in the hallway of the Archimedes Memorial Lecture Hall. They headed west down the hallway, then south. When they took another left turn, Fred figured out how he could get his tongue unstuck. He would talk about mathematics.

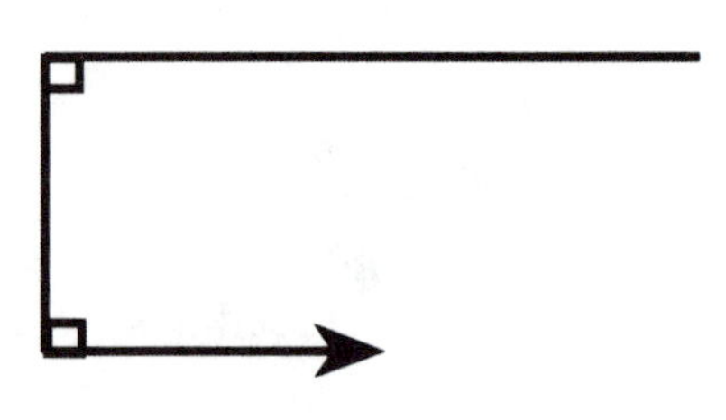

"You know, if two lines are perpendicular to the same line, they are parallel. It's a theorem in geometry."*

✶ Two whole sentences and no sputtering. This illustrates the old adage: On your first date it's important that you know some geometry theorems.

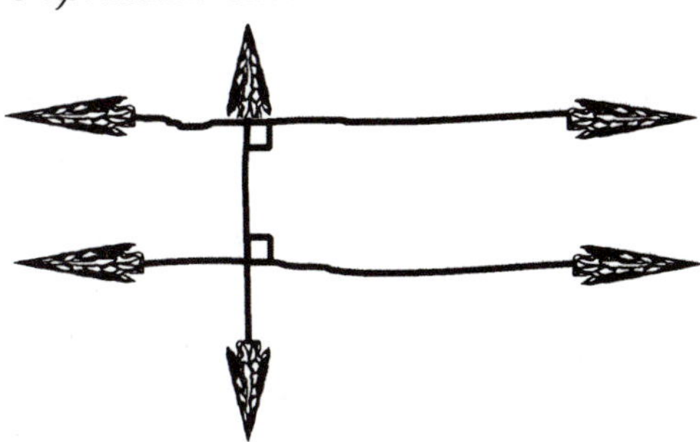

They ducked into an empty classroom and Fred drew on the blackboard. He made the arrowheads very fancy.

"Oh yes," she said. "The perpendicular lines form right angles and all right angles are congruent. Then it's one easy step from that. Just use the AIP Theorem and you're done. But I personally like its sister theorem: If a line is perpendicular to one of two parallel lines, then it's perpendicular to the other." She drew on the blackboard.

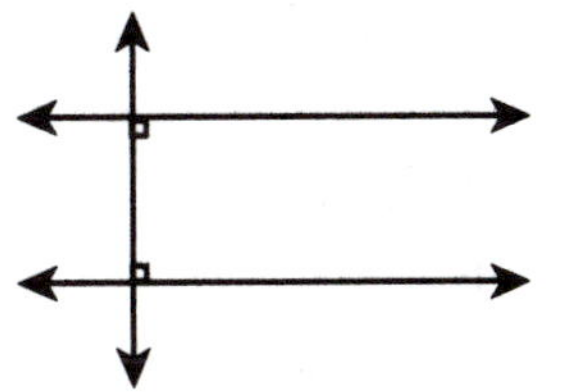

Her lines were a lot straighter than Fred's.

"The proof is just as easy," she announced. "This time it's a corollary of the PAI Theorem."

Your Turn to Play

1. Fill in the missing reasons in the proof of *If a line is perpendicular to one of two parallel lines, then it is perpendicular to the other.*

Statement	Reason
1. $\ell \parallel m$ and $\ell \perp n$	1. Given
2. $\angle 1$ is a right angle.	2. ?
3. $\angle 1 \cong \angle 2$	3. ?
4. $m\angle 1 = 90°$	4. Def of right angle
5. $m\angle 1 = m\angle 2$	5. Def of $\cong$ $\angle$s
6. $m\angle 2 = 90°$	6. ?
7. $\angle 2$ is a right angle	7. ?
8. $n \perp m$	8. ?

⊠

2. Are these two theorems converses of each other?

<u>Theorem</u>: *If two lines are perpendicular to the same line, they are parallel.*
and
<u>Theorem</u>: *If a line is perpendicular to one of two parallel lines, then it is perpendicular to the other.*

3. As Fred and P were walking down the hallway, when they made their first turn to head south, Fred was reminded of the proposition: *From every point on a line, there is a perpendicular to that line through that point.* (In contrast, at that moment P was wondering where Fred's socks were.)

Fill in the missing parts of the proof.

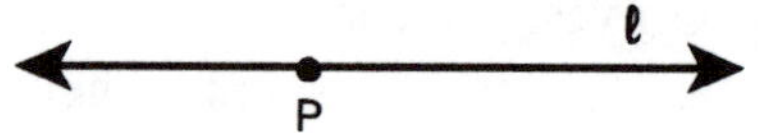

Statement	Reason
1. Point P is on line ℓ.	1. ?
2. There is another point Q on ℓ.	2. Theorem 2 on p. 41
3. There is an ∠QPR such that m∠QPR = 90°.	3. Angle Measurement Post.
4. ∠QPR is a right angle.	4. ?
5. ?	5. Def of ⊥ lines

This is called "erecting a perpendicular to ℓ through P."

4. It is a lot harder to establish that you can drop a perpendicular from P to ℓ than it is to show that you can erect a perpendicular to ℓ through P when P is on ℓ.

<u>Theorem</u>: Given a point P not on ℓ, there is at least one perpendicular from P to ℓ.

Fill in the missing parts of the proof.

• P

Statement	Reason
1. Point P, not on line ℓ.	1. Given
2. Let Q and R be points on ℓ.	2. Theorem 2 on p. 41

• P

Statement	Reason
3. Assume that there does not exist a line that is perpendicular to ℓ through point P. This means, in particular, that $\overleftrightarrow{PQ}$ is not perpendicular to ℓ.	3. ?
4. There is an $\angle RQS$ which has a measure equal to $m\angle PQR$.	4. Angle Measurement Post.
5. There is a point T on $\overrightarrow{QS}$ such that QT = QP.	5. Rule Postulate

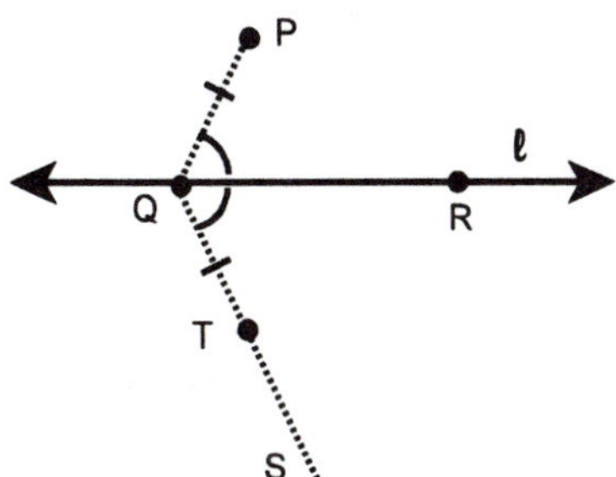

Statement	Reason
6. There is a unique line through P and T.	6. ?

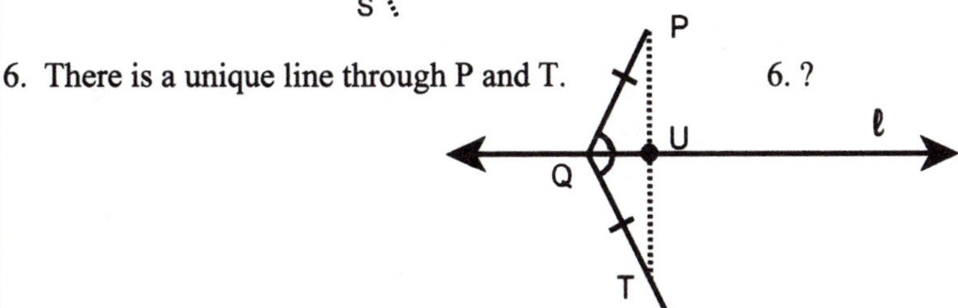

Statement	Reason
7. QU = QU	7. ?
8. $\overline{QU} \cong \overline{QU}$	8. ?
9. $\Delta PQU \cong \Delta TQU$	9. ?
10. $\angle 1 \cong \angle 2$	10. ?
11. ?	11. Definition of $\cong$ $\angle$s

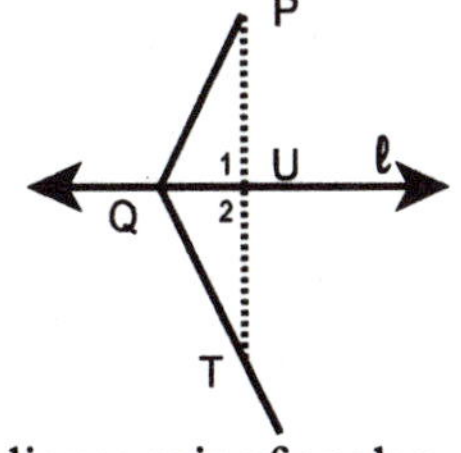

Statement	Reason
12. $\angle 1$ and $\angle 2$ form a linear pair of angles.	12. Definition of linear pair
13. $\angle 1$ and $\angle 2$ are supplementary.	13. ?
14. $m\angle 1 + m\angle 2 = 180^\circ$	14. ?
15. $m\angle 1 = 90^\circ$	15. Algebra (steps 11 and 14)
16. $\angle 1$ is a right angle.	16. Definition of right angle
17. $\overleftrightarrow{PT} \perp \ell$	17. ?
18. There does exist a line through P which is perpendicular to ℓ.	18. Contradiction in steps 3 and 17 and hence the assumption in line 3 is false.

Q.E.D.

Eighteen steps! That deserves its Q.E.D. It would have been a lot shorter if we had made a little preparation by first proving the theorem *If two lines intersect and form a linear pair of congruent angles, then the lines are perpendicular.* That would have knocked lines 11, 12, 13, 14, 15 and 16 out of the proof.

This theorem-in-preparation-for-proving-an-upcoming theorem is called a **lemma**. Lemmas are like John the Baptist in that they prepare the way.

We now have four words for statements that are proven:

- theorem: *any statement which is proven*
- proposition: *a lightweight (trivial) theorem*
- lemma: *a theorem proven in preparation for the proof of a big theorem to come*
- corollary: *a theorem which follows easily from a big theorem*

Darlene once explained to Joe that if marriage were considered the big theorem, then dating is the lemma and having children are the corollaries.

Joe was not at all comfortable with that. He told her that it was much better to think of the big theorem as the Super Bowl and all the games that led up to it were the lemmas and the party afterwards was the corollary.

5. In the previous problem we showed that: Given a point P not on ℓ, there is *at least* one perpendicular from P to ℓ. To show that there is *at most* one line through P that is perpendicular to ℓ is an easy corollary of a theorem we proved in the previous chapter.

If we begin the (indirect) proof by assuming that there are two lines through P which are perpendicular to ℓ, we will have a triangle with two right angles in it. Find the theorem which prohibits this.

COMPLETE SOLUTIONS

1.

2. Def of $\perp$ lines
3. PAI Theorem
6. Algebra
7. Def of a right angle
8. Def of $\perp$ lines

2. The converse of the first theorem is: If two lines are parallel, then they must both be perpendicular to some line. This is not the second theorem.

3.

1. Given
4. Def of a right angle
5. $\overrightarrow{PR} \perp \ell$

4.

	3. Beginning of an indirect proof
	6. Two points determine a line. (Postulate 1)
	7. Algebra (The fact that $a = a$ for all a is called the **reflexive law of equality** in algebra.)
	8. Def of ≅ segments
	9. SAS
	10. Def of ≅ ⚠
11. $m\angle 1 = m\angle 2$	
	13. If two angles form a linear pair, then they are supplementary. (Postulate 4)
	14. Def of supplementary angles
	17. Def of ⊥ lines

5. It is the first problem in Elmo on p. 152

Fred just stood there admiring the straight lines that P had drawn. He had read that motor skills develop more quickly in girls than in boys, and this was living proof. He made a mental note to obtain a yardstick from the KITTENS supply room so that the lines he drew on the blackboard would be neater in the future.

Everything about her enthralled him. When he looked into her eyes, he thought of Postulate 1. Her nose was a perfect acute angle. She knew geometry. What more could Fred ask for in a sweetheart?

P looked at her watch. "Oh my! It's almost ten. I've got to get to my Plays of Sophocles class. Today we're starting *Antigone*. It's not as well known as Oedipus Rex, but it's cool. I've already read it. The play says that when everyone around you has got their values all screwed up, then you've got to decide: Abandon what you know is right, or stand fast and get creamed in the process."

Fred was having trouble listening. He was in a panic, worrying that he'd never see her again. "Are you free tonight? A couple of my friends, Betty and Alexander, and I are meeting at PieOne tonight at 6. They've got great Tri-Peetz there. Can I pick you up at 5:45?" The words came pouring out of Fred. He was surprised at how much he could say when he really needed to speak.

"Sure. I'd be delighted." She hurried off.

Fred realized that he also had a class at ten o'clock. As he walked out of the empty classroom and down the hall toward his giant lecture room where he did his teaching, he noticed that there were very few students in the hallway.

He picked a Caboodle out of the trash can.

THE KITTEN Caboodle

The Official Campus Newspaper of KITTENS University — Thursday 9:45 a.m. Edition 10¢

Classes Canceled!!! Pres Says: No Classes Today or Tomorrow

"We must ensure," our university president declared in an exclusive Caboodle interview (continued on p. 24)

KANSAS: Starting at 10 a.m. today all classes have been canceled by order of the president of the University. Classes will resume on Monday.

When asked why this unprecedented double holiday was proclaimed, our president explained that because of the need to promote adequate attendance at the various awards ceremonies that have been scheduled, he was ensuring that there would be no conflicts between the ceremonies and the students' education.

Here is the Schedule of Ceremonies for today and tomorrow. Be sure to attend!

• Noon today. Special Awards Ceremony honoring Prof. Fred Gauss for his bravery in solving the notorious Bocci Ball Lawn Vandalism case.

• 1 p.m. today. Special Awards Ceremony honoring Sergeant Friday for his bravery in apprehending the two criminals in the notorious KITTEN Bank Robbery.

• 2 p.m. Special Awards Ceremony honoring Samuel P. Wistrom, Chief Educational Facility Math Department Building Janitor, for bravery in cleaning up the (continued on p. 6)

By 10:01 A.M. there was a traffic jam at the exits of the campus.

* Note the skis on top of every car. Many of the students decided they would rather be off to Mt. Sunflower (the highest point in Kansas, elevation 4039 feet) and enjoy a four-day weekend holiday than attend the awards ceremonies.

The campus store ran an emergency special sale on ski equipment. The ads were plastered all around the campus:

A right triangle and all you know is the hypotenuse and one leg, thought Fred. Is that enough? It would be, if telling me the length of the hypotenuse (H) and the one leg (L) would completely determine the size and shape of the triangle. It would be enough if two right triangles had to be congruent if their hypotenuses and one pair of corresponding legs were congruent.

Fred thought for a moment (0.000000000326 seconds) and came up with:

<u>Theorem</u>: (The Ski Boot Satchel Theorem*) Two right triangles are congruent if their hypotenuses and one pair of corresponding sides are congruent.

---

* A lot of people don't call it by that name. They call it the Hypotenuse-Leg Theorem or the H-L Theorem.

Wait a minute! you exclaim. ***We already went through that. Remember back on p. 103 in problem 4 you say that SSA doesn't work. You even made the cute comment,*** "One does not wish to make an SSA of oneself." ***Remember that? This* H-L *stuff that you're talking about now has got SSA written all over it.***

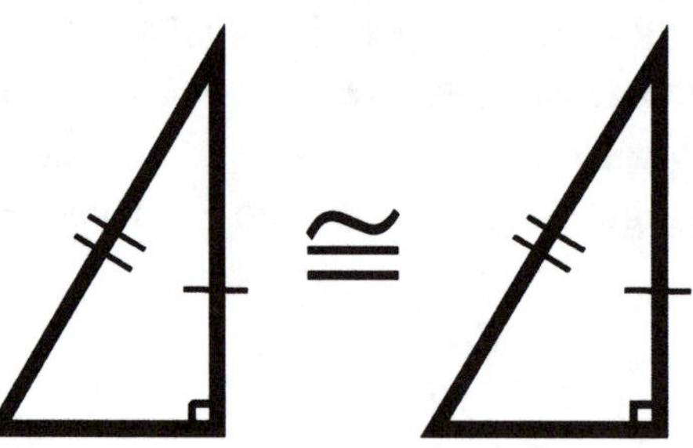

Well, yes and no. It . . .

Stop! None of this political talk. I want to know. Just spit it out. Does SSA work?

No.

That's better. Enough of this namby-pamby* chatter. Now admit that the* H-L *Theorem is just a hoax.

No.

Whadda you mean "No"?

I thought *that* was pretty definite. I mean that the H-L Theorem is not a hoax. It is really true. May I explain?

Sure. It's your book.

In general, SSA doesn't work—knowing two sides and a (non-included) angle isn't enough to make the two triangles congruent.

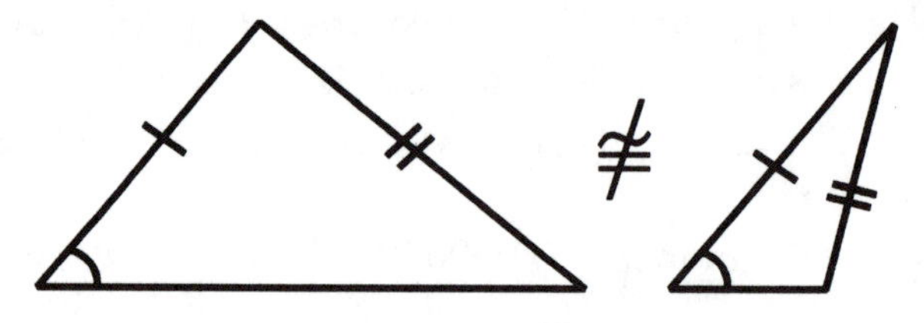

But in the special case in which the "A" (the angle) is a right angle, then the triangles have to be congruent, just like the big heavy picture you drew at the top of this page.

Okay buster. Prove it.

✶ namby-pamby = can't make up your mind

It is kind of tricky proving the Hypotenuse-Leg Theorem. I start with two triangles:

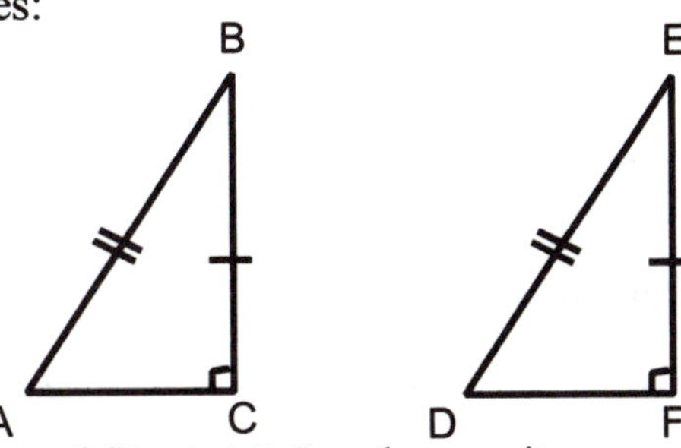

. . . and then I sort of flip ΔABC and move it over next to ΔDEF:

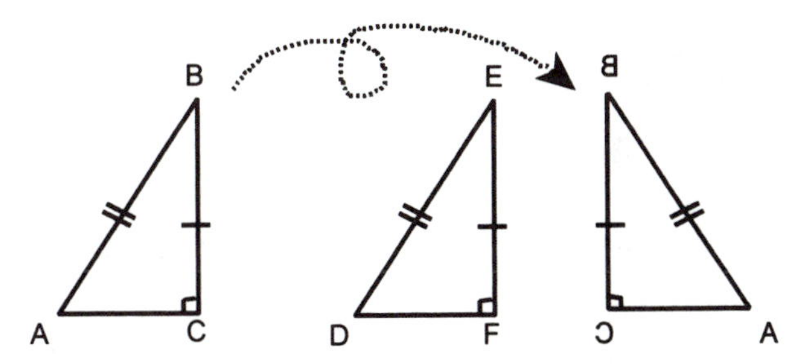

. . . and then I jam them together:

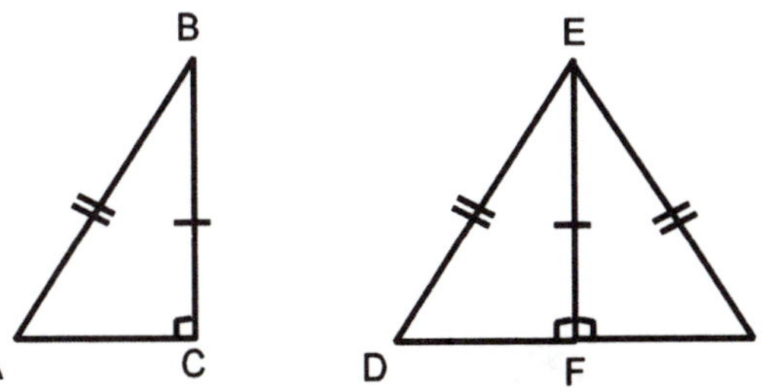

"Kind of . . . sort of flip . . . jam them together"—Get a grip on yourself man! This is a geometry book. You've told me a hundred times that you can't use "flipping" and "jamming" as reasons in a proof. You are limited to the official seven reasons:

① Given
② Postulate
③ Definition
④ Previously proven theorem
⑤ Algebra
⑥ Beginning of an indirect proof
⑦ Contradiction in steps ___ and ___ and therefore the assumption in step ___ is false.

Now play by the rules and give me a proof of the H-L Theorem.

Okay. I just thought that you might like to know my Plan of Action.

Theorem: (The H-L Theorem) Two right triangles are congruent if their hypotenuses and one pair of corresponding legs are congruent.

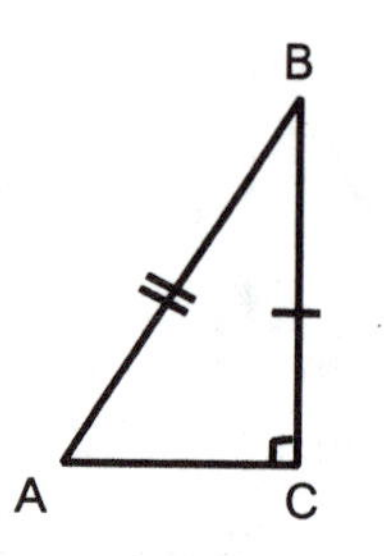

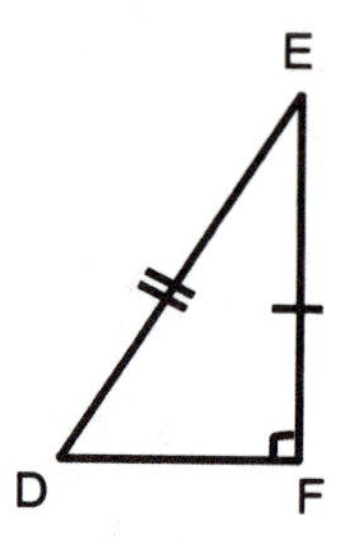

Statement	Reason
1. $\triangle ABC$ and $\triangle DEF$ are right triangles. $\overline{AB} \cong \overline{DE}$ and $\overline{BC} \cong \overline{EF}$.	1. Given
2. There is exactly one line through D and F.	2. Postulate: Two points determine a line.
3. There is a point G on $\overleftrightarrow{DF}$ such that FG = AC.	3. Ruler Postulate
4. There is exactly one line through E and G.	4. Postulate: Two points determine a line.

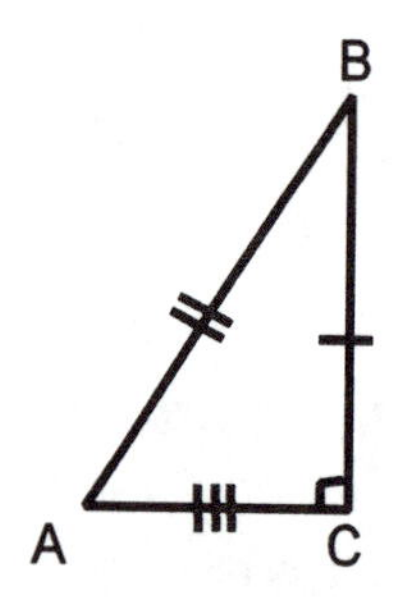

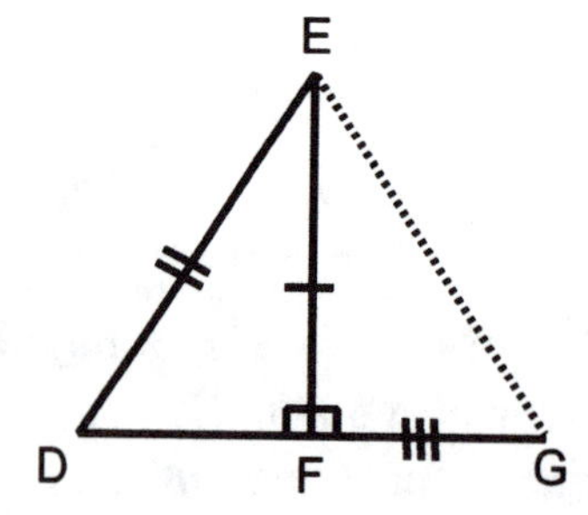

Hold it! You didn't move $\triangle ABC$. You just copied it.

Mea culpa.

Hey! I don't speak Italian!

It's Latin, but in common usage in this country. *Mea culpa* means that I admit my guilt.

I accept your confession. Now get on with the proof.

I need to prove a lemma first. It will make the proof of the H-L Theorem a lot shorter.

Okay. Prove your lemma . . . and then get back to proving the* H-L *Theorem.

Lemma: If two lines intersect and form one right angle, then all four angles are right angles and they are all congruent to each other.

Proof: ***Wait! Stop! This is too easy. Save it for Cities or something. If $\angle 1$ is a right angle then $\angle 3 \cong \angle 1$ by the Vertical Angle Theorem. And in order to show that $\angle 2$ is a right angle just note that $m\angle 1 = 90^o$ and that $\angle 1$ and $\angle 2$ form a linear pair, etc. Get back to the main proof please.***

Okay.

Statement	*Reason*
5. $\angle 1$ and $\angle 2$ are both right angles. $\angle 1 \cong \angle 2$ 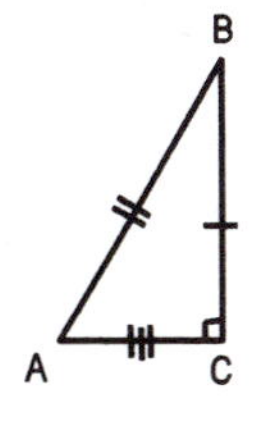 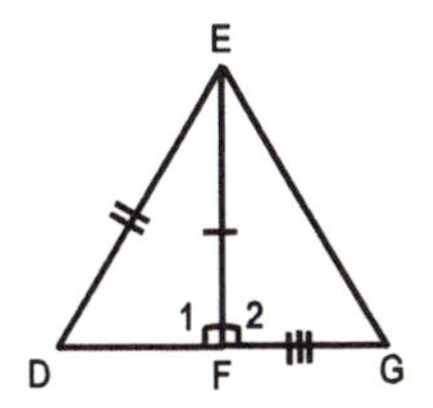	5. Lemma: If two lines intersect and form one right angle, then all four angles are right angles and they are congruent to each other.
6. $\Delta ABC \cong \Delta GEF$	6. SAS
7. $\overline{AB} \cong \overline{GE}$	7. Def of $\cong$ ⚠
8. $\angle D \cong \angle G$	8. Isosceles Triangle Theorem
9. $\Delta DEF \cong \Delta GEF$	9. AAS
10. $\Delta ABC \cong \Delta DEF$	10. Transitive Property of Congruent Triangles Theorem

⊠

But Fred wasn't interested in skiing right now. If it weren't for the mathematics in the Ski Boot Satchel, he wouldn't have given it a second thought. He was thinking about his sweetheart. A thousand thoughts crossed his mind. How could someone be that lovely? Why is her name "P"? Does she like me? What's her favorite song? How did she finish high school before the age of six? Do I need to get a tuxedo for tonight?

He suddenly realized that he knew virtually nothing about this whole romance thing. The only fact that he had learned in his short life was that if you are separated from your true love, you get in a boat and search till you find him/her. Last week, he had read that in Longfellow's poem *Evangeline*. When Fred had read it, he couldn't figure out why Evangeline had spent all those years looking for her guy. Now, a week later, after he had met P Commonsizerinski, he understood. Sometimes you read great literature and then later in life it will suddenly become vibrantly alive as its vision matches the events of your life.*

Next to the poster for the Ski Boot Satchels was another advertisement. Here was something that talked directly about romance. As lawyers like to say, this was *on point*.

Get the girl of your dreams!
Use Ralph's Mouth Dust Powder
AVAILABLE IN CONVENIENT ONE- AND TWO-QUART SIZES

Fred had his checkbook with him so he turned and headed to the campus store. He knew that he was a "two-quart" kind of guy.

Then he turned back to look at the picture of Cupid again. Perfect! he thought to himself. That arrow is a super illustration of a perpendicular bisector of a segment.

---

* English teachers often explain to their students that they should read the great works of literature only if they have a future, and only if they want to live a full life. Otherwise, reading the backs of cereal boxes is just fine.

Definition: Line ℓ is the perpendicular bisector of $\overline{AB}$ iff it is perpendicular to $\overline{AB}$ and it bisects $\overline{AB}$ (passes through the midpoint of $\overline{AB}$.)

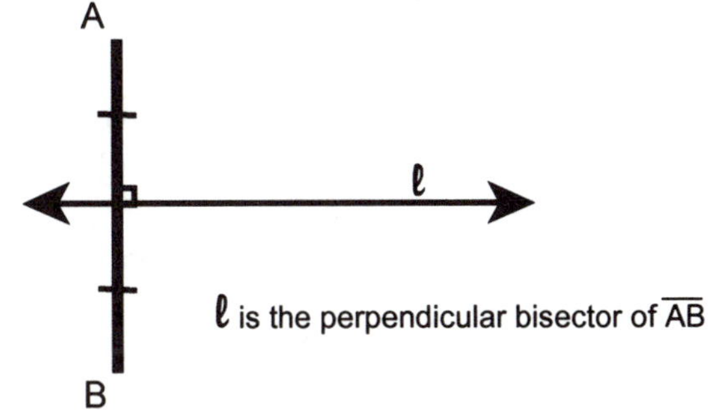

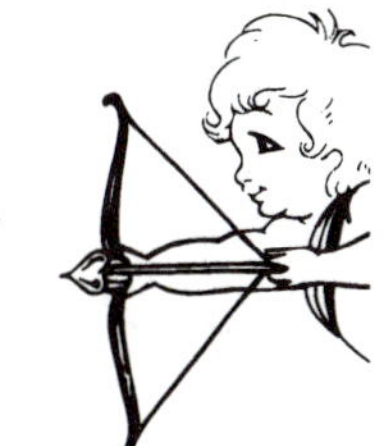

Fred looked at the string in Cupid's bow. From Cupid's left fingers the distance to each endpoint of the bow was the same. In fact, *any point* on the perpendicular bisector was equidistant (the same distance) from the endpoints of the bow.

Theorem: (Cupid's Bow Theorem*) Any point on the perpendicular bisector of a segment is equidistant from the endpoints of the segment.

Your Turn to Play

1. Prove it.
2. What is the converse of the Cupid's Bow Theorem?
3. Is the converse true? If so, prove it. If not, give an example of when it is false.
4. When Joe was studying perpendicular bisectors, he came up with a most astounding theorem. He "proved" that every triangle is isosceles! Where is his error?

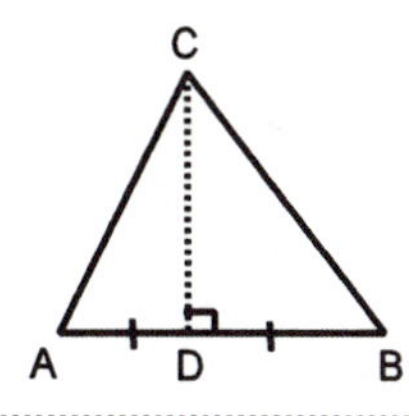

Joe's proof ➡ Start with any triangle ABC. That will be the given. Then drop the perpendicular bisector from vertex C to side $\overline{AB}$. Since $\overline{CD}$ is the $\perp$ bisector, D is the midpoint and therefore $\overline{AD} \cong \overline{BD}$. $\overline{CD} \cong \overline{CD}$. The two triangles are congruent by SAS. Finally, $\overline{AC} \cong \overline{BC}$ by definition of congruent triangles. ☺ Q.E.D.

✶ Hardly anyone except Fred calls this theorem by that name. The cognoscenti might call it something like Half of the Perpendicular Bisector Theorem. (*cognoscenti* = people who are "in the know," people who know a lot in a particular field. Normally pronounced con-yeah-SHIN-tea. An alternate pronunciation is cog-neah-SHIN-tea.)

COMPLETE SOLUTIONS

1.

Statement	*Reason*
1. ℓ is the perpendicular bisector of $\overline{AB}$ and P is any point on ℓ.	1. Given
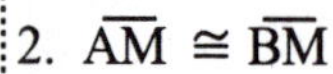 2. $\overline{AM} \cong \overline{BM}$	2. Def of $\perp$ bisector
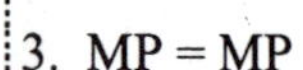 3. MP = MP	3. Algebra
4. $\overline{MP} \cong \overline{MP}$	4. Def of $\cong$ segments
5. $\angle$AMP is a right angle	5. Def of perpendicular
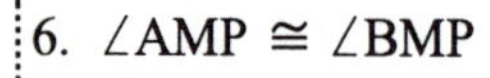 6. $\angle$AMP $\cong$ $\angle$BMP	6. Lemma: If two lines intersect and form one right angle, then all the angles are congruent
7. $\triangle$AMP $\cong$ $\triangle$BMP	7. SAS
8. $\overline{AP} \cong \overline{BP}$	8. Def of $\cong$ $\triangle$s

2. The converse is: *If any point is equidistant from the endpoints of a segment, then it must lie on the perpendicular bisector.*

3. Converses of true statements need not necessarily be true. (For example, the converse of this true statement is false: *If I have just eaten a large combination pizza, then I am full.*) But in the case of Cupid's Bow Theorem, its converse is also true. Here is a proof.

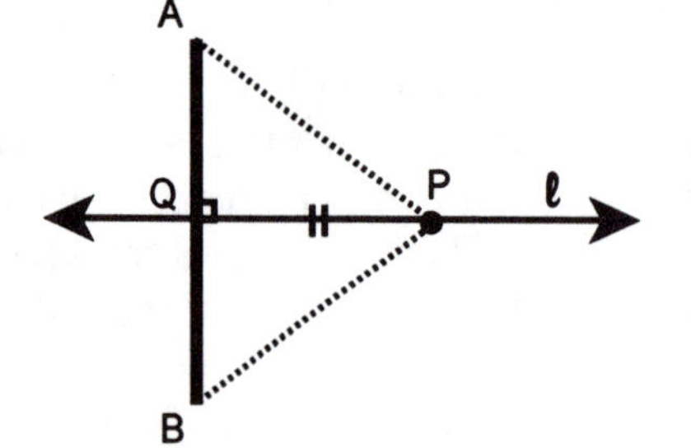

Statement	*Reason*
1. $\overline{AP} \cong \overline{BP}$	1. Given
2. $\overleftrightarrow{PQ} \perp \overleftrightarrow{AB}$	2. Thm: Given a point not on a line, there is at least one perpendicular from that point to the line.
3. PQ = PQ	3. Algebra
4. $\overline{PQ} \cong \overline{PQ}$	4. Def of $\cong$ segments
5. $\angle$AQP and $\angle$BQP are right angles.	5. Def of $\perp$ lines
6. $\triangle$AQP $\cong$ $\triangle$BQP	6. Hypotenuse-Leg Theorem
7. $\overline{AQ} \cong \overline{BQ}$	7. Def of $\cong$ $\triangle$s
8. $\overline{PQ}$ is the $\perp$ bisector of $\overline{AB}$	8. Def of $\perp$ bisector

Sharp-eyed readers may have noticed that the proof didn't cover all the possibilities. We are proving that *any* point that is equidistant from the endpoints of $\overline{AB}$ must lie on the perpendicular bisector. There is one point that we overlooked: What if P lies *on* $\overline{AB}$? Then we couldn't have those neat triangles which we could then prove to be congruent. All we would have is

But this case is even easier to deal with. On p. 162 in problem 3, we proved that you can erect a perpendicular to a line from any point on the line. And that perpendicular will be the perpendicular bisector of $\overline{AB}$ since it passes through the midpoint of $\overline{AB}$.

Sometimes there is more than one case that you need to examine in order to complete a proof.

Since the Cupid's Bow Theorem and its converse are both true, the two theorems are often combined into one:

Theorem: (The Perpendicular Bisector Theorem) A point is on the perpendicular bisector of a segment iff it is equidistant from the endpoints of the segment.

The important word

4. If Joe had put his proof in statement-reason form, it would have been fun to see what reason he would have supplied for his dropping a perpendicular bisector from C to $\overline{AB}$.

We do have a theorem that allows us to drop a perpendicular from a point P to a line ℓ. That can always be done.

We do know that you can always draw a **median** (which is the segment joining the vertex of a triangle to the midpoint of the opposite side). This takes two steps. First, by the Midpoint Theorem, we know that $\overline{AB}$ has a midpoint, and second, the line through C and M exists by Postulate 1 (Two points determine a line).

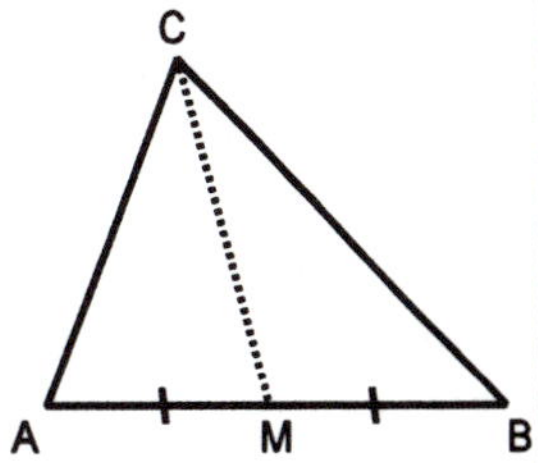

But you can't draw a line and say that it is *both* a median and is perpendicular. In general, these are two different lines.

Fred headed off to the campus store with his checkbook in his hand.

Cambridge

1. Given a point P not on line ℓ. Make a guess as to how we will define the distance from P to ℓ.

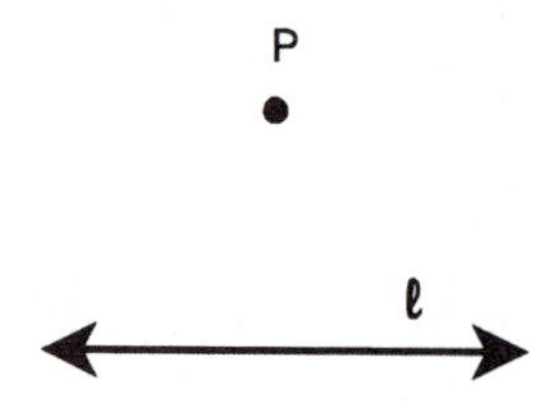

2. Prove that any point on an angle bisector is equidistant from the sides of the angle.

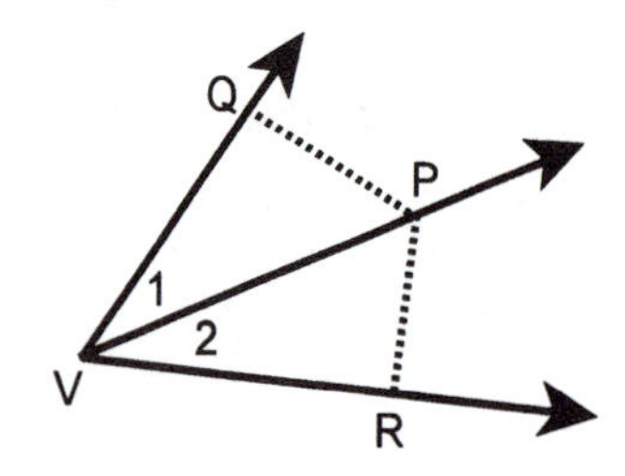

3. In $\triangle ABC$ drop a perpendicular from C to $\overleftrightarrow{AB}$ and let Q be the foot of the perpendicular. Must A–Q–B?

4. In $\triangle DEF$ drop a perpendicular from F to $\overleftrightarrow{DE}$ and let Q be the foot of the perpendicular. Also suppose that we know that D–Q–E. What can we say about the size of $\angle D$?

5. Prove that if $m \parallel n$ and if ℓ intersects m, and if $\ell \neq m$ (which means that ℓ and m are not the same line), then ℓ must intersect n.

answers

1. There are many points on ℓ and thus many candidates for our definition of the **distance from P to ℓ**, but there is one point that seems to work best. It's called the foot of the perpendicular from P to ℓ. (Definition: The **foot** of the perpendicular from P to ℓ is the point where the perpendicular from P to ℓ intersects ℓ.) We define the distance from P to ℓ as the distance between P and the foot of the perpendicular from P to ℓ.

2.

S *R*

1. $\overrightarrow{VP}$ is the angle bisector of $\angle QVR$.
1. Given

2. $\angle 1 \cong \angle 2$ — *2.* Def of $\angle$bisector

3. $\overline{PQ} \perp \overline{VQ}$ and $\overline{PR} \perp \overline{VR}$
3. Def of distance from a point to a line

4. $\angle VQP$ and $\angle VRP$ are right angles
4. Def of $\perp$

5. $VP = VP$ — *5.* Algebra

6. $\overline{VP} \cong \overline{VP}$ — *6.* Def of $\cong$ segments

7. $\triangle VQP \cong \triangle VRP$
7. AAS Theorem

8. $\overline{PQ} \cong \overline{PR}$ — *8.* Def of $\cong$ $\triangle$s

9. P is equidistant from $\overleftrightarrow{VQ}$ and $\overleftrightarrow{VR}$.
9. Def of equidistant

3. No. Q need not necessarily be between A and B. It might be the case that, for example, A–B–Q.

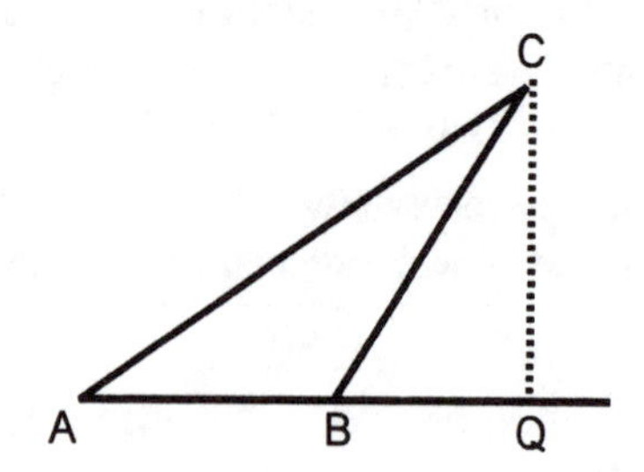

4. $\angle D$ could not be an obtuse angle.

5.

S *R*

1. $m \parallel n$ and ℓ intersects m at point P, and $\ell \neq m$. *1.* Given

2. Assume ℓ does not intersect n. *2.* Beginning of an indirect proof

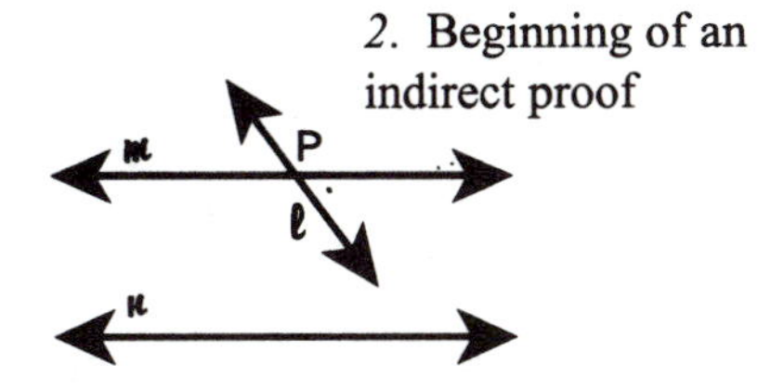

3. $\ell \parallel n$ *3.* Def of ∥ lines

4. $\ell = m$ (equality between two geometrical objects means that they are the same object.) *4.* Parallel Postulate

5. ℓ intersects n. *5.* Contradiction in steps 1 and 4, and therefore the assumption in line 2 is false.

Emerson

For all the problems in the City, use the following diagram.

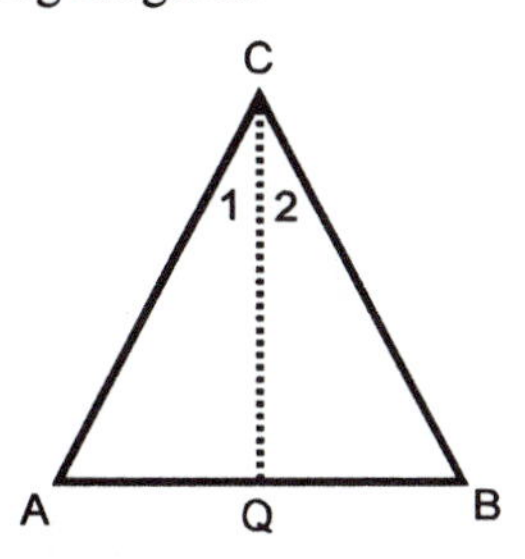

Here are four statements:

i) ΔABC is isosceles (AC = BC)

ii) $\overline{CQ}$ is a median

iii) $\overline{CQ} \perp \overline{AB}$

iv) $\overline{CQ}$ bisects $\angle$ACB ($\angle 1 \cong \angle 2$)

In each of the following describe how you would . . .

1. Prove that i and ii imply iii.
2. Prove that i and iii imply iv.
3. Prove that ii and iii imply iv.
4. Prove that ii and iii imply i.
5. Prove that iii and iv imply i.

Describing how you would prove something is called giving an **outline of a proof** or **sketching a proof.** When you sketch a proof, you give an overview of the proof with enough details that "anybody" could fill in the steps to make a formal statement-reason proof. (By "anybody" we mean any C or better geometry student.)

answers

1. To show that isosceles and median will imply that $\overline{CQ}$ is ⊥, we note that $\overline{CQ} \cong \overline{CQ}$ and the have the triangles congruent (ΔACQ and ΔBCQ) by SSS.

Then the two angles down at Q are congruent by definition of ≅ ⚠.

Since the angles down at Q are also a linear pair, they are right angles (by the theorem we proved in problem 5 on p. 89).

2. To show that isosceles and perpendicular imply that $\overline{CQ}$ is an angle bisector.

With $\overline{CQ}$ congruent to itself, we have ≅ ⚠ by H-L.

$\angle 1 \cong \angle 2$ by definition of ≅ ⚠.

3. To show that a median which is also perpendicular will imply that $\overline{CQ}$ is an angle bisector.

The angles down at Q are congruent by the lemma on p. 171.

With $\overline{CQ}$ ≅ to itself, we have ≅ ⚠ by SAS. Then $\angle 1 \cong \angle 2$ by definition of ≅ ⚠.

4. To show that a median which is also perpendicular will imply that the triangle is isosceles.

With $\overline{CQ}$ congruent to itself, we have $\cong$ △s by SAS. Then $\overline{AC} \cong \overline{BC}$ by definition of $\cong$ △s.

5. To show that if the angle bisector is also perpendicular to the opposite side of the triangle, then the triangle is isosceles.

The angles down at Q are congruent by the lemma on p. 171.

The △s are congruent by ASA.

$\overline{AC} \cong \overline{BC}$ by definition of $\cong$ △s.

Some things to notice:

#1: We used a lot of different reasons to establish congruent triangles in the above proofs (SSS, H-L, SAS, and ASA), depending on what was given.

#2: We didn't exhaust all the possible theorems. There are seven others:

i and iv → iii
ii and iv → iii
i and ii → iv
i and iii → ii
i and iv → ii
iii and iv → ii
ii and iv → i.

(If any teachers ever need additional test questions, here they are.)

#3: Can you imagine combining all of the above into a single theorem? It might read: *Any two of i, ii, iii, and iv are sufficient to establish the other two.*

#4: Worse yet, can you imagine proving that theorem?

Santa Clara

1. Prove that any point that is equidistant from the sides of an angle is on the angle bisector.

2. We proved the Hypotenuse-Leg Theorem. If there were such a thing as the Hypotenuse-Angle Theorem, what would it say?

3. Why don't we have an Hypotenuse-Angle Theorem?

4. If the dotted lines are perpendiculars and are also congruent, prove that ΔABC is isosceles.

(Hint: Look at ΔABD and ΔBAE.)

5. The four-sided figure ABCD is a squashed square (better known as a rhombus).

We are given that AB = BC = CD = DA. Prove that the diagonals (the dotted lines) are perpendicular to each other.

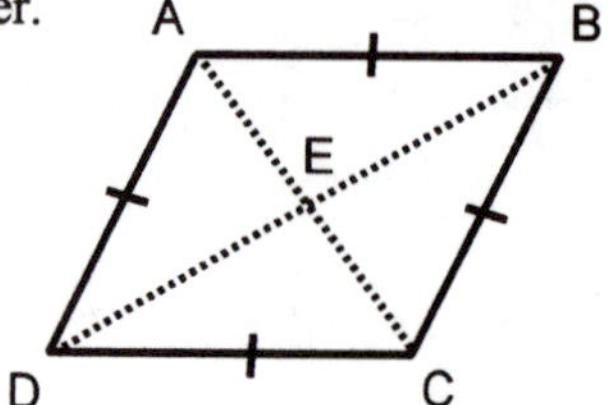

1.

S	*R*
1. P is equidistant from the sides of the angle.	*1.* Given
2. $\overline{PQ} \perp \overline{VQ}$ and $\overline{PR} \perp \overline{VR}$	*2.* Def of distance from a point to a line
3. $\overline{PQ} \cong \overline{PR}$	*3.* Def of equidistant
4. VP = VP	*4.* Algebra
5. $\overline{VP} \cong \overline{VP}$	*5.* Def of congruent segments
6. $\triangle VPQ \cong \triangle VPR$	*6.* H-L Theorem
7. $\angle 1 \cong \angle 2$	*7.* Def of $\cong$ △s
8. $\overrightarrow{VP}$ is the angle bisector of $\angle QVR$.	*8.* Def of $\angle$ bisector

3. We already have a more general theorem that includes Hypotenuse-Angle as a special case. It is the AAS Theorem.

5.

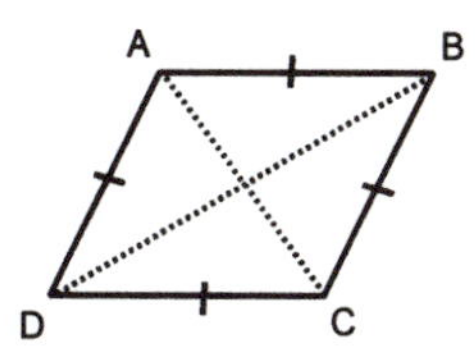

S	*R*
1. AB = BC = CD = DA	*1.* Given
2. A and C are on the $\perp$ bisector of $\overline{BD}$	*2.* Perpendicular Bisector Theorem
3. $\overline{AC}$ is the $\perp$ bisector of $\overline{BD}$	*3.* Postulate (Two points determine a line.)

Vancouver

1.

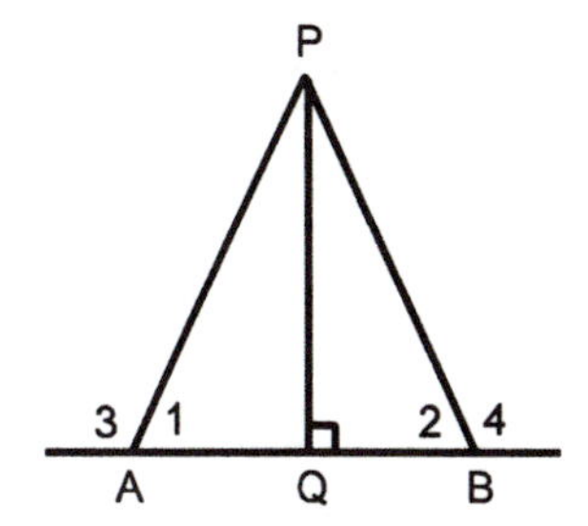

If $\overline{PQ}$ is the perpendicular bisector of $\overline{AB}$, prove that $\angle 3 \cong \angle 4$.

2. We proved the Hypotenuse-Leg Theorem. If there were such a thing as the Leg-Leg Theorem, what would it say?

3. Why don't we have a Leg-Leg Theorem?

4. Prove that in an isosceles triangle, the segment from the vertex which is perpendicular to the opposite side is also a median.

5. Here's a question for potential art majors. Suppose that during a bank robbery the crook keeps his two pistols at right angles to each other with one pistol pointing at teller A and the other at teller B.

He starts out near teller A and walks to teller B. By doing a little sketching, determine what the shape of the curve of his path must be.

odd answers

odd answers

1.

$\mathcal{S}$ $\mathcal{R}$

1. $\overline{PQ}$ is the perpendicular bisector of $\overline{AB}$.	*1.* Given
2. $\overline{PA} \cong \overline{PB}$	*2.* Perpendicular Bisector Theorem
3. $\angle 1 \cong \angle 2$	*3.* Isosceles Triangle Theorem
4. $\angle 1$ and $\angle 3$ are a linear pair. $\angle 2$ and $\angle 4$ are a linear pair.	*4.* Def of linear pair
5. $\angle 1$ and $\angle 3$ are supplementary. $\angle 2$ and $\angle 4$ are supplementary.	*5.* Linear Pair Postulate
6. $\angle 2 \cong \angle 4$	*6.* Theorem: Supplements of congruent angles are congruent.

3. In some sense we already do, since if we ever needed Leg-Leg, we could use SAS.

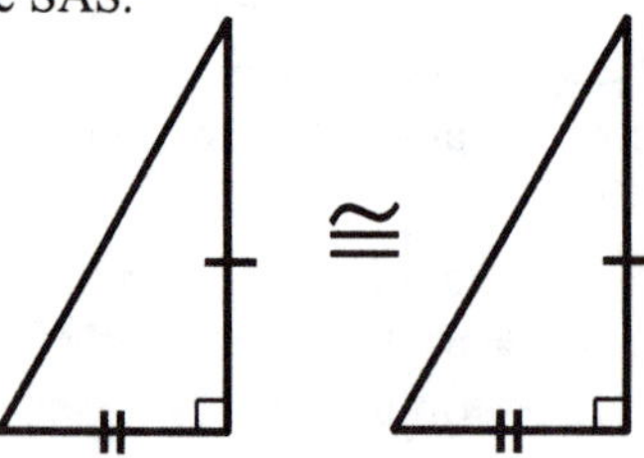

5.

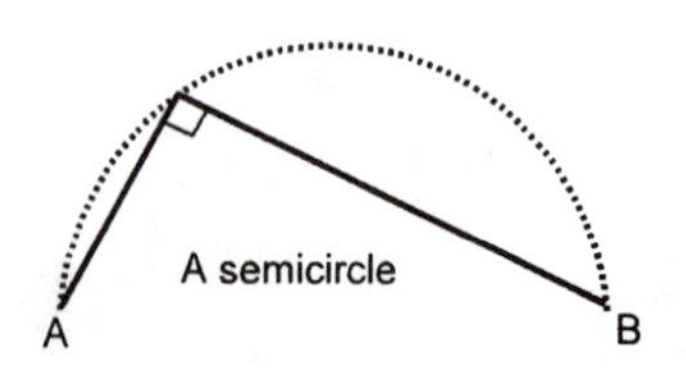

Zell

1. Is this true? *If two points, P and Q, are each equidistant from the endpoints of* $\overline{AB}$*, then* $\overleftrightarrow{PQ}$ *is the perpendicular bisector of* $\overline{AB}$. If it is true, prove it. Otherwise, find a counterexample to show that it is false.

2. If $\angle A$ and $\angle B$ are right angles, prove that $\angle A \cong \angle B$.

3. Is the converse to problem 2 true? If it is, prove it. If it is not, give a counterexample to show that it is false.

4. If $\angle 1$ is a right angle, prove that $\angle 3$ is right angle.

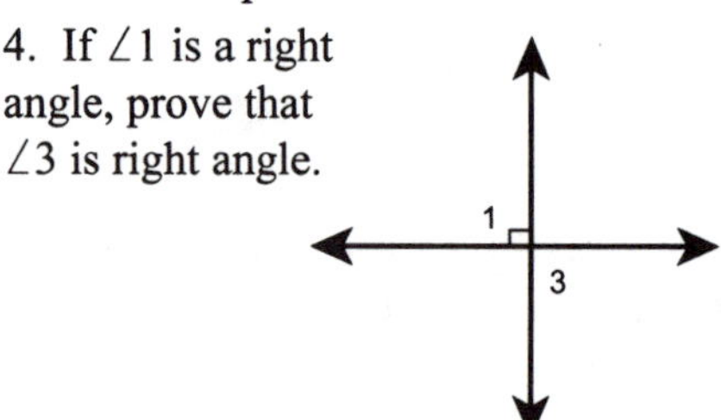

5. A tough question: If $\overline{CQ}$ is a median and $\overline{CQ}$ is also the angle bisector of $\angle ACB$, prove that ΔABC is isosceles.

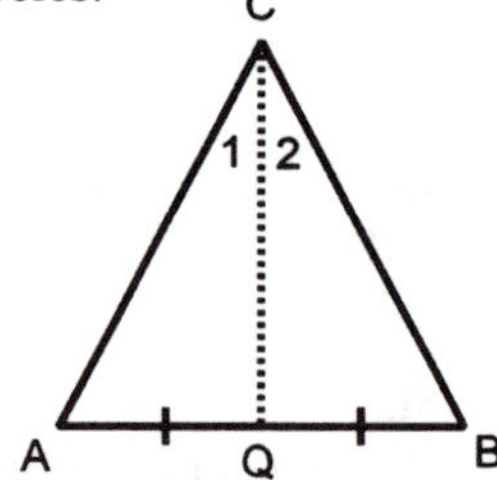

Please try it first on your own. If you can't figure it out after five minutes, you may look at the hint given at the end of the next page.

Kane

1. Prove that if $\angle 1$ is a right angle, then $\angle 2$ is also a right angle.

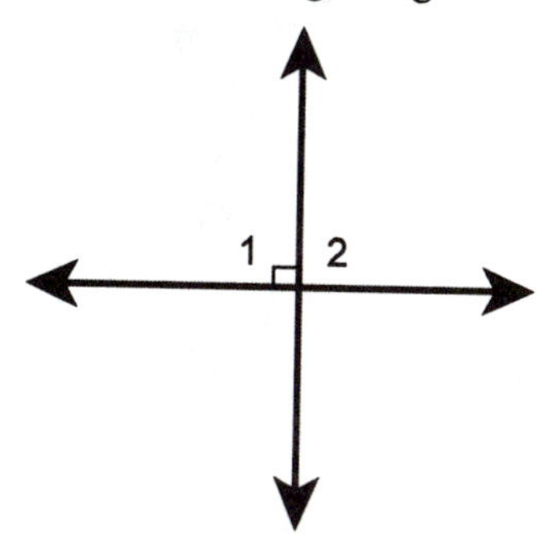

2. Suppose you were reading a very short math book that contained only one theorem, one corollary, and one lemma. In what order would these three appear in the book?

3. Prove that in an isosceles triangle, the angle bisector is perpendicular to the opposite side.

4. What's wrong with: *If two parallel lines are cut by a transversal, then the alternate interior angles are equal.*

5. If P and Q are two different points on ℓ, and if the distance from P to m is equal to the distance from Q to m, what can you say about ℓ and m?

6. A tough problem: Prove what you said in problem 5 is true.

(Hint to problem 5 of Zell: Drop perpendiculars from Q to $\overline{AC}$ and $\overline{BC}$ and use the theorem that says that any point on an angle bisector is equidistant from the sides of the angle.)

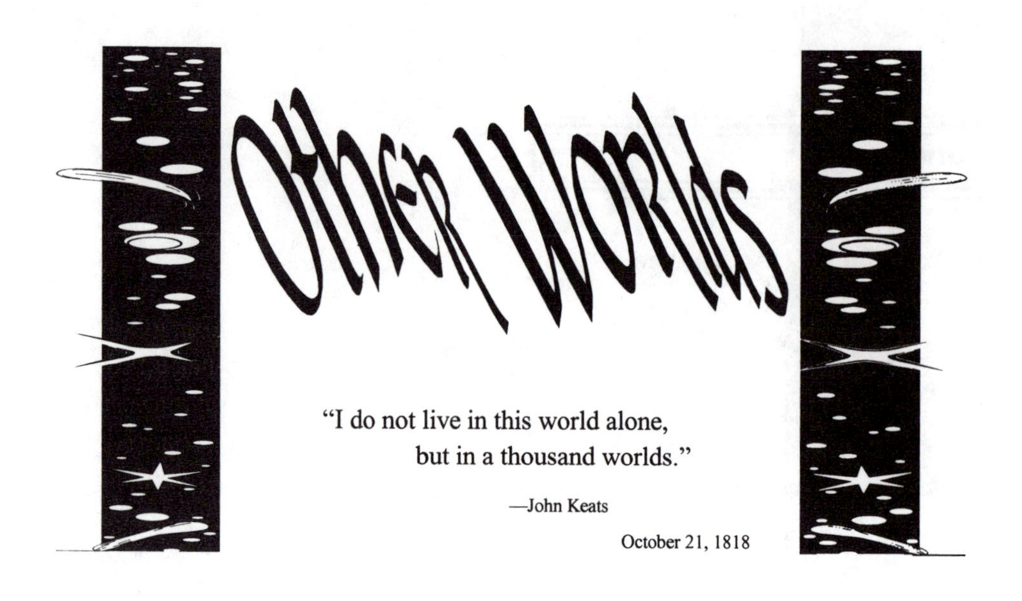

Other Worlds
"I do not live in this world alone,
but in a thousand worlds."
—John Keats
October 21, 1818

Chapter 5½
Chain the Gate

As I was walking to work one morning several years ago. . .

Wait! Stop! What's going on? you exclaim. ***Where's chapter six? We just finished chapter five. The next number is six. Nobody in the world has a chapter 5½.***

If I may correct your English just a little bit. Perhaps you meant to say, "Nobody else in the world has a chapter 5½."

Mea culpa—to use your words. Wait a minute! Why am I apologizing? You're the one who stuck this 5½ thing in the book. Please tell me what's happening.

This is a bonus chapter. A gift to you. Geometry is much bigger than just plain ol' plane geometry. In this Other Worlds chapter, I give you a glimpse of some of the rest of geometry.

And if I want just "plain ol' plane geometry?"

Then just turn to chapter six right now. I've made it easy to spot these interlarded chapters (there are six of them) by putting them in a different type face. They are easy to skip over.

But what if I don't want to skip over them?
Then you just let me continue telling my story.
Okay. Fire away.

As I was walking to work one morning several years ago, I noticed a large gate in a chain-link fence. Behind the fence was a bunch of yellow school buses. The gate was locked with a long chain and a big padlock. (This was to keep the buses from escaping.)

If you had the key, you could free the buses.

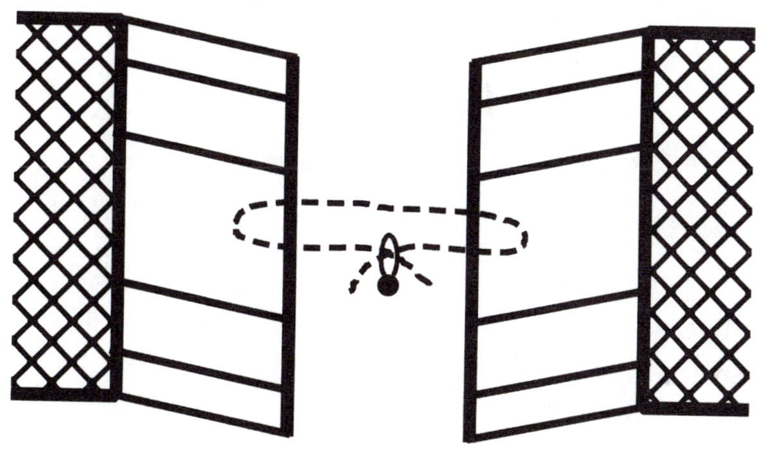

As I continued walking, I thought about the missile silos where it takes two different keys (in two different) locks) in order to launch the ICBM. They give the keys to two different people so that no one person can fire the rocket and start WWIII.

I wondered if that could be done with the one chain and two different locks. Could you lock the chain with the two locks so that both locks would need to be unlocked in order to open the gate?

That was a cinch.

If the locks were named P and Q, then to open the gate you would need to open P and Q. In logic they write this as P & Q.

Then I thought about the case where you would want the unlocking of either lock to open the gate. Suppose the holder of the key to P works Monday through Friday and the holder of the key to Q works on the weekend.

My first solution was to interlock the locks:

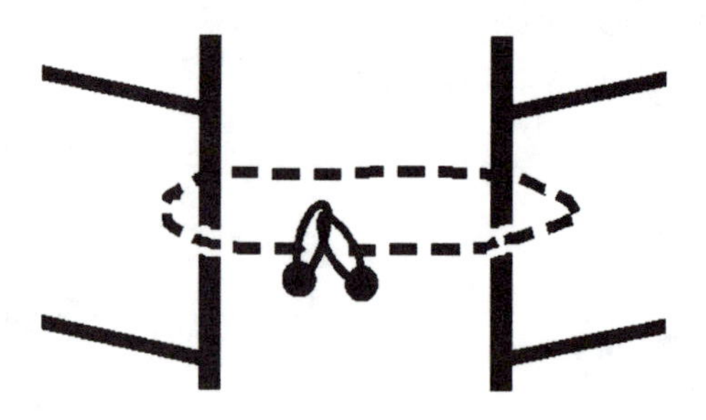

That was easy till I looked more closely at those heavy-duty locks that they use. They had such thick hasps that you couldn't lock two of them together.

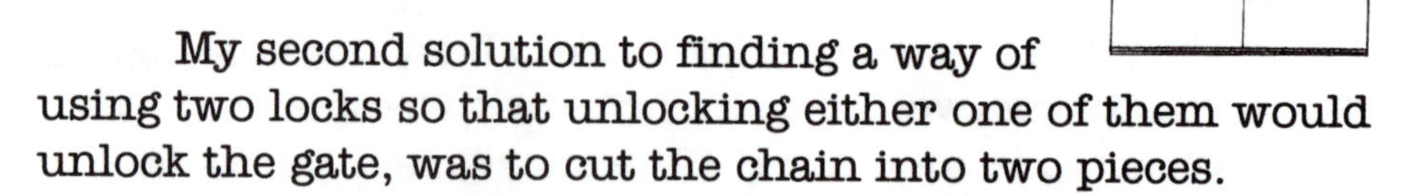

My second solution to finding a way of using two locks so that unlocking either one of them would unlock the gate, was to cut the chain into two pieces.

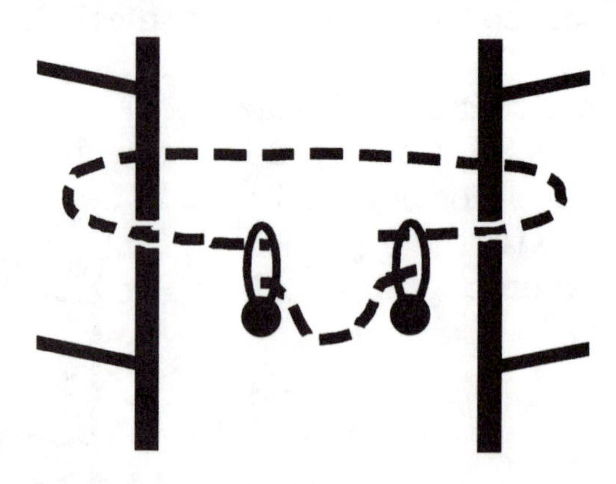

But the chain was too heavy to cut.

I found a third solution in which I didn't interlock the locks and I didn't cut the chain.

Puzzle Question

(The answer is at the end of this chapter so you won't accidentally see it until you want to.)

#1. Figure out how to chain the gate with two padlocks so that opening either lock will open the gate.

Opening either P or Q (or both of them) will open the gate. In logic this is written $P \vee Q$. This is called the "non-exclusive or." It is the kind of "or" that mathematicians use.

Unfortunately, English majors and policemen often use the "exclusive or." When a cop tells you, "Stop or I'll shoot!" you understand: one or the other, *but not both.* It is often very distressing if you stop and also get shot.

When I asked my daughter (when she was about three years old), "Would you like the chocolate cake or the cherry pie?" she answered, "Yes" (indicating that she wanted both). Knowing that I am a mathematician, she assumed I was using the non-exclusive or.

In legal language where things are supposed to be very precise, when they want to indicate the non-exclusive or, they will write "and/or." PENAL CODE: §5.037: KILL A POLICEMAN AND/OR SMOKE A JOINT IS SUFFICIENT REASON TO HAVE YOU EXECUTED.

They could have written: PENAL CODE: §5.037: (KILL A POLICEMAN $\vee$ SMOKE A JOINT) IS SUFFICIENT REASON TO HAVE YOU EXECUTED.

And since "➜" is the symbol for implies, they could have written: PENAL CODE: §5.037: (KILL A POLICEMAN $\vee$ SMOKE A JOINT) ➜ YOU WILL BE EXECUTED.

By the time I had walked to work, I realized that I was in the middle of creating a new mathematical theory. It had its undefined terms (chain, gate, . . .), and its postulates

(① You can't interlock two locks. ② You can't cut the chain. etc.) and its theorems:

Theorem: (P) You can lock the gate with one lock.

Theorem: (P & Q) You can lock the gate with two locks so that you need to open both locks in order to open the gate.

Theorem: (P ∨ Q) You can lock the gate with two locks so that opening either one (or both) of them will open the gate.

Some things to note before we continue:

♪#1: Did you ever wonder where math comes from? You don't wander outside and find math growing on trees.

$x^0 = 1$

Math is a product of human thought. Just walking to work, I found myself creating a new piece of mathematics. Humans (and perhaps some day soon, their mechanical children*) are the only source of mathematics in the entire physical universe. Of course, when we die and meet the Ultimate Mathematician, one of the (minor) joys of the life hereafter will be to look into His answer book.

♪#2: Have you ever been to a science fair? Many of them allow mathematical entries. Judges are often very impressed by mathematical submissions since so few students choose that part of the scientific world for a project. Extending this chain-the-gate math theory beyond what is written here might not be that hard to do. For example, finding the way to chain the gate with three locks so that either P alone or both Q and R together could open the gate is a corollary of what we've done so far. P ∨ (Q & R) Your booth could have chains and locks for people to play with.

♪#3: I actually really truly honestly did see a chained gate as I was walking to work and actually really truly honestly did begin to create this new mathematics as I walked. No one was giving me a salary.

✶ Our mechanical children = computers

Chapter 5½ Chain the Gate

In the next couple of weeks I continued to play with this chain-the-gate theory. I figured out a way to lock the gate with three locks, P, Q and R, so that unlocking any two of them would unlock the gate. That theorem might be symbolized: (P & Q) ∨ (Q & R) ∨ (P & R). I'll put the answer to this near the end of chapter six so that you won't accidently see it until you want to. We'll call it *Puzzle Question #2.*

The hardest one that I worked on and solved was P ∨ Q ∨ R: given three locks so that unlocking any one of them would unlock the gate. I'll put the answer to this at the end of chapter seven so that you won't accidently see it until you want to.

Answer to Puzzle Question # 1

#1. How to chain the gate with two padlocks so that unlocking either lock will open the gate.

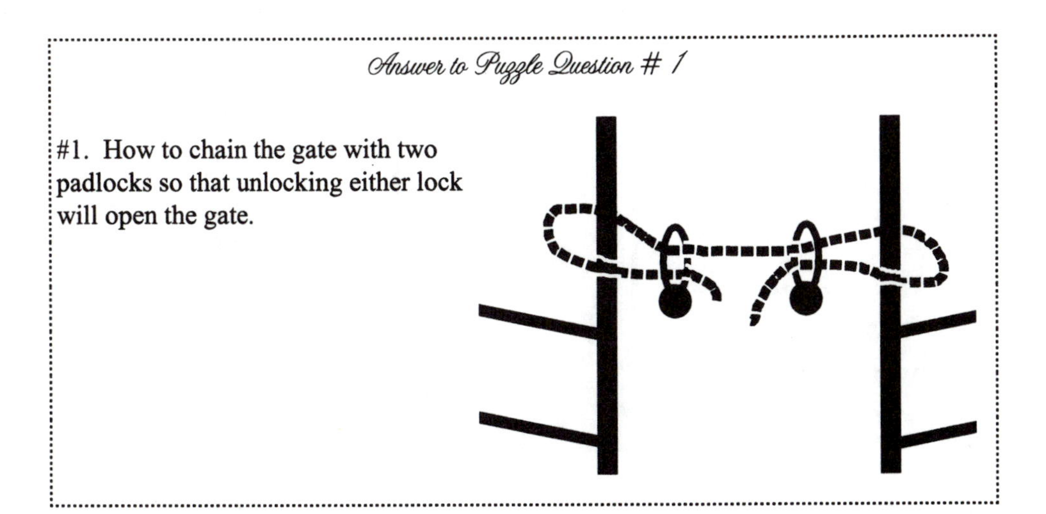

Chapter Six
Quadrilaterals

Fred was in a hurry since there wasn't that much time before he would be picking up P for their first date. So many things for him to do before 5:45. At the top of his list was to get to the campus store and pick up Ralph's Mouth Dust Powder. He walked across the quad.

Many college campuses have a quad. That's short for quadrangle, an open space surrounded by buildings. A quadrangle is a figure with four angles and four sides.

In contrast, a quadrilateral is a figure with four sides and four angles. Many people find them difficult to tell apart.

Just as triangles come in many varieties (acute, right, obtuse, scalene, isosceles, equilateral . . .), so quadrilaterals also can be classified in many different ways.

A Scrapbook of Quadrilaterals

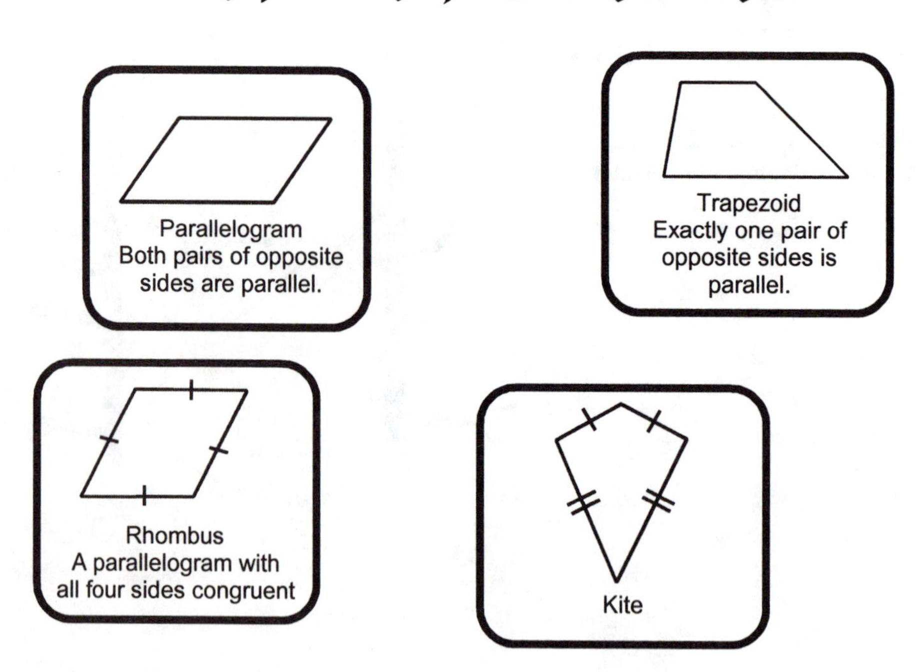

Parallelogram
Both pairs of opposite sides are parallel.
Trapezoid
Exactly one pair of opposite sides is parallel.
Rhombus
A parallelogram with all four sides congruent
Kite

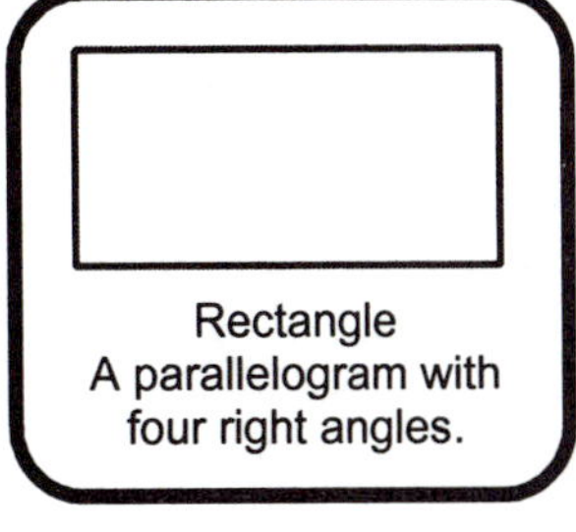

Rectangle
A parallelogram with four right angles.

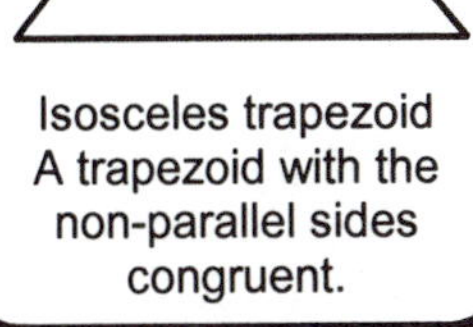

Isosceles trapezoid
A trapezoid with the non-parallel sides congruent.

A plain old quadrilateral with nothing special to brag about.

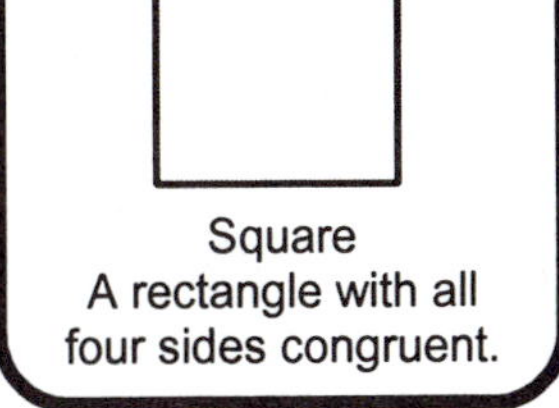

Square
A rectangle with all four sides congruent.

There aren't *that* many different kinds of quadrilaterals. I bet you could name more kinds of dogs. There's the Labrador retriever, the German shepherd, and the Dachshund. A more complete list might include: Affenpinscher, Afghan Hound, African Wild Dog, Ainu Dog, Airedale Terrier, Akbash Dog, Akita Inu, Alapaha Blue Blood Bulldog, Alaskan Husky, Alaskan Klee Kai, Alaskan Malamute, Alopekis, Alpine Dachsbracke, American Bandogge Mastiff, American Black and Tan Coonhound, American Blue Gascon Hound, American Bulldog, American Cocker Spaniel, American Crested Sand Terrier, American Eskimo Dog, American Foxhound, American Hairless Terrier, American Mastiff, American Pit Bull Terrier, American Staffordshire Terrier, American Staghound, American Toy Terrier, American Water Spaniel, American White Shepherd, Anatolian Shepherd Dog, Anglos-Françaises, Anglos-Françai Grand, Anglos-Français de Moyenne Venerie, Anglos-Françaises de Petite Venerie, Appenzell Mountain Dog, Ariegeois, Armant, Aryan Molossus, Argentine Dogo, Arubian Cunucu Dog, Australian Bandog, Australian Bulldog, Australian Cattle Dog, Australian Kelpie, Australian Shepherd, Australian Stumpy Tail Cattle Dog, Australian Terrier, Austrian Brandlbracke, Austrian Shorthaired Pinscher, Azawakh, Banjara Greyhound, Barbet, Basenji, Basset Artesien Normand, Basset Hound, Bavarian Mountain Hound, Beagle, Beagle Harrier, Bearded Collie, Beauceron, Bedlington Terrier, Bedouin Shepherd Dog, Belgian Griffons, Belgian Mastiff, Belgian Shepherd Groenendael, Belgian Shepherd Laekenois, Belgian Shepherd Malinois, Belgian Shepherd Tervuren, Belgian Shorthaired Pointer, Belgrade Terrier, Bergamasco, Berger des Picard, Berger des Pyrénées, Berger Du Languedoc, Bernese Mountain Dog, Bichon Frise, Bichon Havanais, Bichon/Yorkie, Billy, Black and Tan Coonhound, Black Forest Hound, Black Mouth Cur, Black Russian Terrier, Bleus De Gascogne, Bloodhound, Blue Heeler, Blue Lacy, Bluetick Coonhound, Boerboel, Bohemian Terrier, Bolognese, Border Collie, Border Terrier, Borzoi, Boston Terrier, Bouvier des Flanders, Bouvier de Ardennes, Boxer, Boykin Spaniel, Bracco Italiano, Braque D' Ariege, Braque D' Auvergne, Braque Du Bourbonnais, Braque Dupuy, Braque Saint-Germain, Braques Françaises, Brazilian Terrier, Briard, Brittany Spaniel, Briquet, Broholmer, Brussels Griffon, Bull Boxer, Bull Terrier, Bulldog, Bullmastiff, Cairn Terrier, Cajun Squirrel Dog, Canary Dog, Canaan Dog, Cane Corso Italiano, Canis Panther, Canoe Dog, Cão da Serra da Estrela, Carlin Pinscher, Caravan Hound, Carolina Dog, Carpathian Sheepdog, Catahoula Bulldog, Catahoula Leopard Dog, Catalan Sheepdog, Caucasian Ovtcharka, Cardigan Welsh Corgi, Cavalier King Charles Spaniel, Central Asian Ovtcharka, Cesky Fousek, Cesky Terrier, Chart Polski, Chesapeake Bay Retriever, Chien D' Artois, Chien De L' Atlas, Chiens Francaises, Chihuahua, Chin, Chinese Chongqing Dog, Chinese Crested, Chinese Foo Dog, Chinese Shar-Pei, Chinook, Chippiparai, Chortaj, Chow Chow, Cirneco Dell 'Etna, Clumber Spaniel, Cockapoo, Cocker Spaniel, Collie, Combai, Continental Toy Spaniel, Corgi, Coton De Tulear, Cretan Hound, Croatian Sheepdog, Curly-Coated Retriever, Cypro Kukur, Czechoslovakian Wolfdog, Czesky Terrier, Dachshund, Dalmatian, Dandie Dinmont Terrier, Danish Broholmer, Danish/Swedish Farm Dog, Danish Chicken Dog, Deutsche Bracke, Deutscher Wachtelhund, Dingo, Doberman Pinscher, Dogo Argentino, Dogue Brasileiro, Dogue de Bordeaux, Dorset Olde Tyme Bulldogge, Drentse Patrijshond, Drever, Dunker, Dutch Shepherd Dog, Dutch Smoushond, East-European Shepherd, East Russian Coursing Hounds, East Siberian Laika, Elkhound, English Bulldog, English Cocker Spaniel, English Coonhound, English Foxhound, English Pointer, English Setter, English Shepherd, English Springer Spaniel, English Toy Spaniel, Entelbucher Sennenhund, Epagneul Francais, Epagneul Pont-Audemer, Epagneuls Picardies, Eskimo Dog, Estonian Hound, Estrela Mountain Dog, Eurasier, Farm Collie, Fauves De Bretagne, Feist, Field Spaniel, Fila Brasileiro, Finnish Hound, Finnish Lapphund, Finnish Spitz, Flat-Coated Retriever, Foxhound, Fox Terrier, French Brittany Spaniel, French Bulldog, French Mastiff, Galgo Espanol, Gascons-Saintongeois, German Hunt Terrier, German Longhaired Pointer, German Pinscher, German Sheeppoodle, German Shepherd Dog, German Shorthaired Pointer, German Spitz,

German Wirehaired Pointer, German Wolfspitz, Giant Schnauzer, Glen of Imaal Terrier, Golden Retriever, Goldendoodle, Gordon Setter, Gran Mastin de Borinquen, Grand Anglo-Français, Great Dane, Great Pyrenees, Greater Swiss Mountain Dog, Greek Hound, Greek Sheepdog, Greenland Dog, Greyhound, Griffon Nivernais, Griffons Vendéens, Groenendael, Grosser Müünsterlläänder Vorstehhund, Guatemalan Bull Terrier, Hairless Khala, Halden Hound, Hamiltonstovare, Hanoverian Hound, Harlequin Pinscher, Harrier, Havanese, Hawaiian Poi Dog, Hellenikos Ichnilatis, Hellenikos Poimenikos, Hertha Pointer, Himalayan Sheepdog, Hokkaido Dog, Hovawart, Hygenhund, Hungarian Greyhound, Hungarian Kuvasz, Hungarian Puli, Husky, Ibizan Hound, Icelandic Sheepdog, Inca Hairless Dog, Irish Glen Imaal Terrier, Irish Red and White Setter, Irish Setter, Irish Staffordshire Bull Terrier, Irish Terrier, Irish Water Spaniel, Irish Wolfhound, Italian Greyhound, Italian Spinoni, Jack Russell Terrier, Japanese Spaniel, Japanese Spitz, Japanese Terrier, Jindo, Kai Dog, Kangal Dog, Kangaroo Dog, Kanni, Karabash, Karakachan, Karelian Bear Dog, Karelian Bear Laika, Karelo-Finnish Laika, Keeshond, Kelb Tal-Fenek, Kemmer Feist, Kerry Beagle, Kerry Blue Terrier, King Charles Spaniel, King Shepherd, Komondor, Kooikerhondje, Koolie, Krasky Ovcar, Kromfohrlläänder, Kugsha Dog, Kunming Dog, Kuvasz, Labradoodle, Labrador Husky, Labrador Retriever, Lagotto Romagnolo, Lakeland Terrier, Lancashire Heeler, Landseer, Lapinporokoira, Lapphunds, Larson Lakeview Bulldogge, Latvian Hound, Leonberger, Leopard Cur, Levesque, Lhasa Apso, Lithuanian Hound, Llewellin Setter, Louisiana Catahoula Leopard Dog, Lowchen, Lucas Terrier, Lundehund, Lurcher, Magyar Agar, Mahratta Greyhound, Majestic Tree Hound, Maltese, Malti-Poo, Manchester Terrier, Maremma Sheepdog, Markiesje, Mastiff, McNab, Mexican Hairless, Mi-Ki, Middle Asian Ovtcharka, Miniature Australian Shepherd, Miniature Bull Terrier, Miniature American Eskimo, Miniature Pinscher, Miniature Poodle, Miniature Schnauzer, Mioritic Sheepdog, Moscow Toy Terrier, Moscow Vodolaz, Moscow Watchdog, Mountain Cur, Mountain View Cur, Mucuchies, Mudhol Hound, Mudi, Neapolitan Mastiff, Nebolish Mastiff, Mutt, Nenets Herding Laika, New Guinea Singing Dog, New Zealand Huntaway, Newfoundland, Norbottenspets, Norfolk Terrier, North American Miniature Australian Shepherd, Northeasterly Hauling Laika, Northern Inuit Dog, Norwegian Elkhound, Norwegian Buhund, Norwich Terrier, Nova Scotia Duck-Tolling Retriever, Olde Boston Bulldogge, Old Danish Bird Dog, Old English Mastiff, Old English Sheepdog, Old-Time Farm Shepherd, Olde English Bulldogge, Original English Bulldogge, Otterhound, Owczarek Podhalanski, Papillon, Pashmi Hound, Patterdale Terrier, Pekepoo, Pekingese, Pembroke Welsh Corgi, Perdigueiro Portugueso, Perdiguero de Burgos, Perdiguero Navarro, Perro Cimarrón, Perro de Pastor Mallorquin, Perro de Presa Canario, Perro de Presa Mallorquin, Perro Ratonero Andaluz, Peruvian Inca Orchid, Petit Basset Griffon Vendéen, Petit Brabancon, Pharaoh Hound, Pit Bull Terrier, Plott Hound, Podengo Portuguesos Grande, Podengo Portuguesos Méédio, Podengo Portuguesos Pequeno, Podenco Ibicenco, Pointer, Poitevin, Polish Hound, Polski Owczarek Nizinny, Polski Owczarek Podhalanski, Pomeranian, Poodle, Poos, Porcelaine, Portuguese Podengo Pequeno, Portuguese Water Dog, Portuguese Rabbit Dog, Potsdam Greyhound, Prazsky Krysavik, Presa Canarios, Pudelpointer, Pug, Puli, Pumi, Pyrenean Mastiff, Pyrenean Mountain Dog, Queensland Heeler, Rafeiro do Alentejo, Rajapalyam, Rampur Greyhound, Rastreador Brasileiro, Rat Terrier, Redbone Coonhound, Rhodesian Ridgeback, Rottweiler, Rough Collie, Rumanian Sheepdog, Russian Bear Schnauzer, Russian Harlequin Hound, Russian Hound, Russian Spaniel, Russian Tsvetnaya Bolonka, Russian Wolfhound, Russo-European Laika, Saarlooswolfhond, Sabuesos Espanoles, Sage Ashayeri, Sage Koochee, Sage Mazandarani, Saint Bernard, Saluki, Samoyed, Sanshu Dog, Sarplaninac, Schapendoes, Schillerstovare, Schipperke, Schnauzers, Schnoodle, Scotch Collie, Scottish Deerhound, Scottish Terrier (Scottie), Sealydale Terrier, Sealyham Terrier, Segugios Italianos, Shar-Pei, Shetland Sheepdog (Sheltie), Shiba Inu, Shichon, Shih-Tzu, Shika Inus, Shikoku, Shiloh Shepherd, Siberian Husky, Siberian Laikas, Silken Windhound, Silky Terrier, Simaku, Skye Terrier, Sloughi, Slovensky Cuvac, Smalandsstovare, Small Greek Domestic Dog, Smooth Collie, Smooth Fox Terrier, Soft Coated Wheaten Terrier, South Russian Ovtcharka, Spanish Mastiff, Spanish Water Dog, Spinone Italiano, Springer Spaniel, Stabyhoun, Staffordshire Bull Terrier, Staghound, Standard American Eskimo, Standard Poodle, Standard Schnauzer, Stephens Stock, Stichelhaar, Strellufstover, Styrian Roughhaired Mountain Hound, Sussex Spaniel, Swedish Lapphund, Swedish Vallhund, Swiss Shorthaired Pinscher, Swiss Laufhunds, Tahltan Bear Dog, Taigan, Tasy, Teddy Roosevelt Terrier, Telomian, Tenterfield Terrier, Tepeizeuintli, Thai Ridgeback, The Carolina Dog, Tibetan Mastiff, Tibetan Spaniel, Tibetan Terrier, Titan Terrier, Tosa Inu, Toy American Eskimo, Toy Fox Terrier, Toy German Spitz, Toy Manchester Terrier, Toy Poodle, Transylvanian Hounds, Treeing Tennessee Brindle, Treeing Walker Coonhound, Tuareg Sloughi, Tyroler Bracke, Valley Bulldog, Vasgotaspets, Victorian Bulldog, Villano de Las Encartaciones, Vizsla, Volpino Italiano, Vucciriscu, Weimaraner, Welsh Corgi, Welsh Sheepdog, Welsh Springer Spaniel, Welsh Terrier, West Highland White Terrier, West Russian Coursing Hound, West Siberian Laika, Westphalian Dachsbracke, Wetterhoun, Wheaten Terrier, Whippet, White German Shepherd, Wirehaired Fox Terrier, Wirehaired Pointing Griffon, Wirehaired Vizsla, Wolf Hybred, Xoloitzcuintle, Yorkshire Terrier, and Yugoslavian Hounds.

My favorite

Now doesn't that make the classification of quadrilaterals seem a bit easier?

Your Turn to Play

1. One thing you can do with quadrilaterals is to play the game called "Every XXX is a YYY."

For example, every square is a rectangle.

How many can you name? (Omit kites from this game since it is not clear whether kites can be rhombuses.)

2. The harder game to play is called "No XXX is a YYY." One example is no parallelogram is a trapezoid. There is only one other easy example. Name it.

3. Can a trapezoid ever have a right angle?

4. Can an isosceles trapezoid every have a right angle?

5. The *Honors Problem of the Century*: Prove that in any triangle, if the angle bisectors of two angles are congruent, then the triangle is isosceles.

Some notes about the *Honors Problem of the Century*:

♪#1: This problem is what you might call hard. It took Fred five minutes to solve it. When I took geometry at George Washington High School, my teacher, Mr. Kenneth

Irwin, wrote this problem on the board. I worked on it that evening. The next morning before school started, I brought my proof to Mr. Irwin. He pointed out an error that I had made.

I worked on it the next evening. And the day after that. I even looked ahead in the book to the topic of similar triangles and tried to use them. Still no luck. I worked on the problem, off and on over the years. Became a math major at U.C. Berkeley. Got married. Had my first daughter. Became a college math teacher. Finally, about 14 years after Mr. Irwin had given me the problem, I solved it. (I'm a lot slower than Fred.)

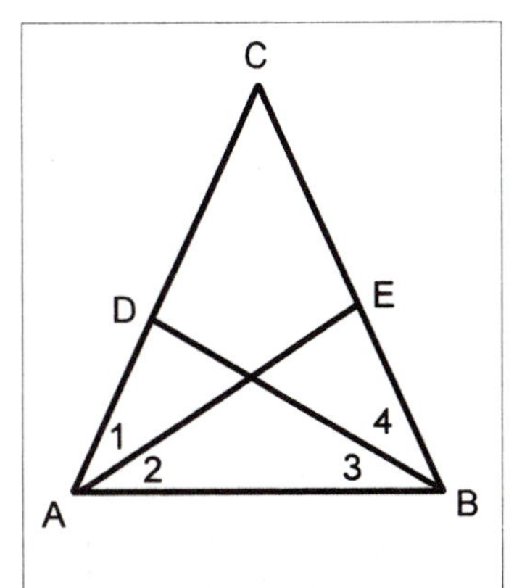

Honors Problem of the Century

♪#2: Since we won't be introducing any new postulates in this chapter, you *right now* have everything necessary to find a proof.

♪#3: This is not a problem intended for D, C, B, or even A students. You may consider yourself an A+++ student if you solve it.

♪#4: The solution will be given at the end of this chapter with three hints supplied along the way.

♪#5: There are two main converses to this *Honors Problem of the Century*:

Converse #1: AC = BC and AE = BD imply that $\overline{AE}$ and $\overline{BD}$ are angle bisectors.

Converse #2: AC = BC and $\overline{AE}$ and $\overline{BD}$ are angle bisectors imply AE = BD.

One of these is fairly easy to prove (using overlapping triangles which are shown to be congruent by ASA) and the other converse is false.

1. Every square is a rectangle.
Every square is a rhombus.
Every square is a parallelogram.
Every square is a quadrilateral.
Every rectangle is a rhombus.
Every rectangle is a parallelogram.
Every rectangle is a quadrilateral.
Every trapezoid is a quadrilateral.
Every isosceles trapezoid is a trapezoid.
Every isosceles trapezoid is a quadrilateral.
Every rhombus is a parallelogram.
Every rhombus is a quadrilateral.
Every parallelogram is a quadrilateral.

2. No trapezoid is a parallelogram.

3. Sure. In fact, if it had one right angle, it would have to have two of them. Otherwise none of the sides would be parallel.

4. Here we get into a bit of difficulty. I can't see how to draw that. When I try, I obtain a rectangle. Then *both* pairs of opposite sides are parallel and I no longer have a trapezoid.

5. No hints yet. Stay tuned.

Parallelograms are Fred's favorite quadrilaterals. Fred once mentioned in class that you see parallelograms much more frequently than squares or rectangles. The students were quiet. They couldn't imagine how that could be true.

Finally, one student responded, "What about clipboards and computer labels and suitcases and milk cartons? Those are all rectangles. Where are the parallelograms? Are you going to say that just because all squares and rectangles are parallelograms, that parallelograms are more common?

Fred smiled. "No. You really do see a lot more non-square, non-rectangular parallelograms than you see squares and rectangles in your everyday life. It's very rare that you see a rectangle."

They didn't believe him. "Show us!" they demanded.

Fred held up a clipboard, some computer labels, a suitcase, and a milk carton.

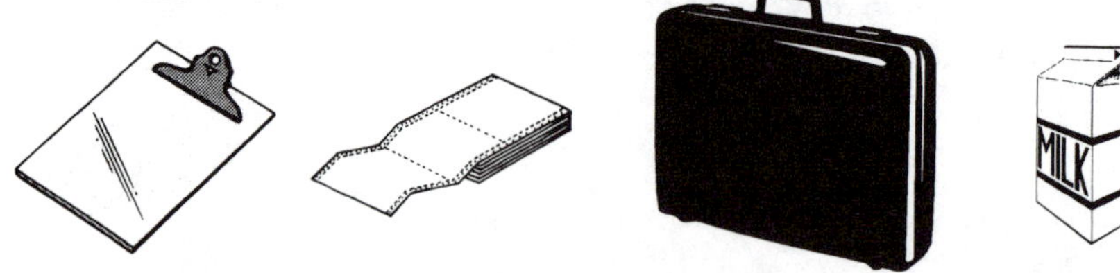

"It's only when you look at these from a very special angle," Fred explained, "that you really see a rectangle. Almost all the time you happen to see parallelograms that aren't rectangles."

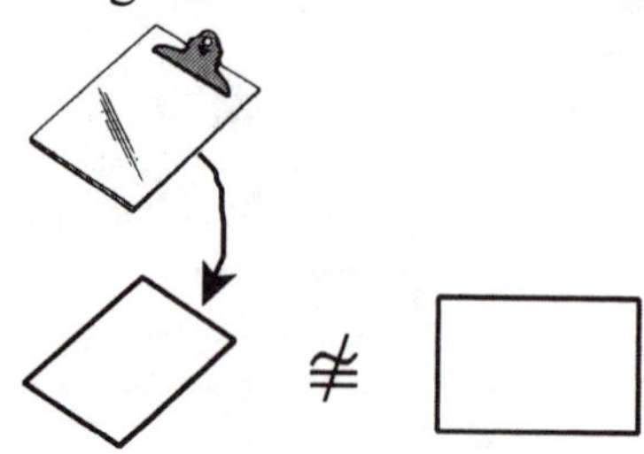

It was a little harder to find everyday illustrations of Isosceles trapezoids. Fred usually drew a picture of the U.S.S. Rubber Duck on the blackboard when he introduced trapezoids.*

Fred finished walking across the quad.

Often when crossing the quad, Fred would think of mathematics, but today was different. His thoughts had little hearts all around them.

The least that he could do for her was to send her a love letter. In every romantical (as he called them) movie he had ever seen, the fellow always sent his true love some kind of note.

When he got to the campus store, he headed to the stationery department. He figured that Ralph's Mouth Dust Powder could wait. First he had to get a letter written and mailed. There were so many different kinds of letterheads to choose among:

* U.S.S. = United States Ship

But none of them had the "feel" that he wished to express to P. They were all so r-e-c-t-a-n-g-u-l-a-r. He wanted her to know that he was a parallelogram type of guy.

He selected the ice cream stationery (since it was sweet), borrowed a pair of scissors from the clerk, and made his own special stationery—a parallelogram. This was the first love letter Fred had ever written.

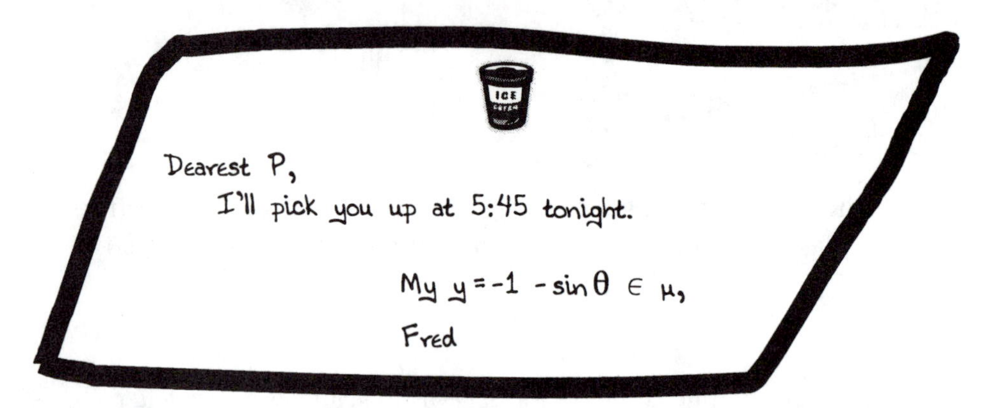

ICE
Dearest P,
I'll pick you up at 5:45 tonight.
My y = -1 - sin θ ∈ μ,
Fred

Fred's closing salutation may need a little explanation. He thought it would be a little corny just to write, "Yours truly," so he opted for something a little more mathematical.

The graph of $y = -1 - \sin\theta$ is a cardioid (which you will graph in trigonometry). It looks like a ♡, except that the bottom isn't as pointy. The symbol $\in$ is used in set theory and means "is an element of" or "belongs to." For example, ✈ $\in$ {☎, ✈, ✏}. The Greek letter μ is pronounced *mu*. Putting that all together, My $y = -1 - \sin\theta \in \mu$ becomes "My heart belongs to you."

He enclosed a monograph on parallelograms that he had recently written so that she would understand his attachment to them. If his love of parallelograms was a flaw in his character that she couldn't stand, he wanted to get it out into the open as soon as possible. If she preferred a rhombus type of guy, he wanted to know that before he obtained a marriage license.

He addressed the envelope and dropped it into a campus mailbox. He didn't know her exact address, but he was certain that there couldn't be two people with the name P Commonsizerinski on campus.

Monograph

Why Parallelograms Are the Neatest

by Prof. Fred Gauss
KITTENS University

Notation: ▱ means parallelogram.

People often disparage parallelograms. They will say things like, "It's only a parallelogram," or even worse, they will completely ignore them.

Think of the number of times you have engaged in conversations in which parallelograms are not even mentioned! How rude!

But as everybody knows from the title of this monograph, parallelograms are cool. Even just to say that quadrilateral ABCD is a parallelogram is to say a lot.

☞ ▱ ABCD implies that the opposite sides are congruent.

☞ ▱ ABCD implies that $\triangle ABC \cong \triangle CDA$, namely that the diagonal $\overline{AC}$ divides the parallelogram into two congruent triangles.

☞ ▱ ABCD implies that the opposite angles are congruent.

Here's the big picture:

▱ ABCD implies

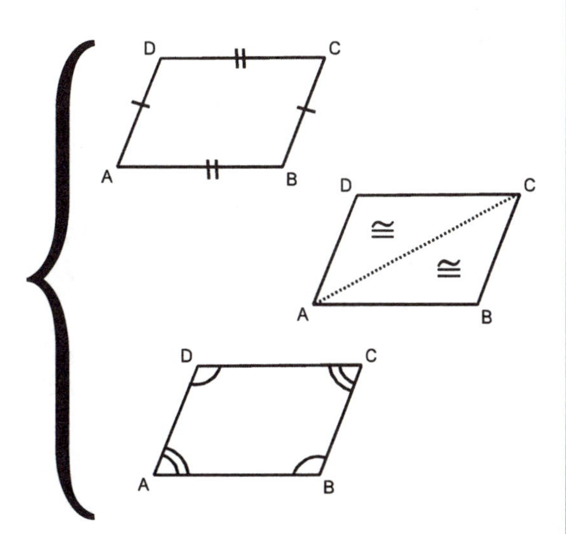

Your Turn to Play

1. (For English majors and others who speak the language) Fred had three theorems in his monograph. Combine them all into one big theorem.

2. Prove that ▱ABCD implies that $\triangle ABC \cong \triangle CDA$, namely that a diagonal of a parallelogram divides it into two congruent triangles.

3. Prove that the opposite sides of a parallelogram are congruent.

4. Prove that the opposite angles of a parallelogram are congruent.

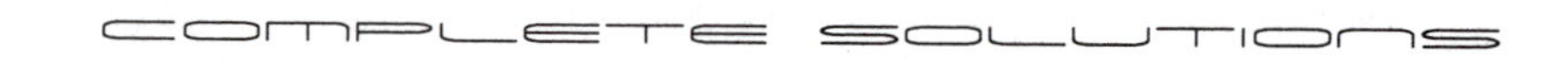

1. In ▱ABCD, the opposite sides are congruent, the opposite angles are congruent, and $\triangle ABC \cong \triangle CDA$.

2.

Statement	Reason
1. ▱ABCD	1. Given
2. $\overline{AD} \parallel \overline{BC}$ and $\overline{AB} \parallel \overline{DC}$.	2. Def of ▱
3. $\angle 1 \cong \angle 2$, $\angle 3 \cong \angle 4$	3. PAI Theorem
4. AC = AC	4. Algebra
5. $\overline{AC} \cong \overline{AC}$	5. Def of $\cong$ segments
6. $\triangle ABC \cong \triangle CDA$	6. ASA

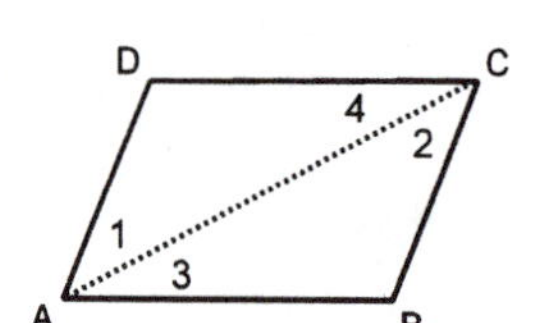

3.

Statement	Reason
1. ▱ABCD	1. Given
2. $\triangle ABC \cong \triangle CDA$	2. Theorem proven in problem 2, above
3. $\overline{AD} \cong \overline{BC}$	3. Def of $\cong$ triangles

4. This is proved almost exactly like problem 3. The only difference is that in the last line, the statement would be $\angle B \cong \angle D$. (If you wanted to establish that $\angle A \cong \angle D$, you would just draw the other diagonal.)

P received Fred's letter twenty minutes after he mailed it. (The campus mail system operates a little differently than the USPS.) She thought that the ice cream icon on the stationery was very sweet. Here is her reply:

Thursday

Dear Fred,

My favorite shape is the four-leaved rose that I learned about when I studied trig last fall. I tried cutting out a piece of paper to make some stationery in the shape of $r = \cos 2\theta$, but it fell into four pieces. (Parallelograms are a much better shape for stationery.)

Instead, I'll draw you one:

Four-leaved rose
$r = \cos 2\theta$

Enclosed is my essay that I wrote on parallelograms when I was studying geometry two years ago. I think it complements your monograph. I compliment your work in this important area of life.

Toowoomba till 5:45,

P Commonsizerinski

"Toowoomba" really threw Fred for a loop. He looked it up in his dictionary. (It's a city in the eastern part of Australia.) But the phrase did have a nice ring to it: *Toowoomba till five forty-five*. It would echo through his noggin all afternoon.

He read her essay.

My Essay

by P Commonsizerinski, age 4

Have You Got What It Takes?

Once upon a time, a long time ago, there lived a poor but honest quadrilateral. She wanted to grow up to be a brave, strong ▱.

She went to the frog and asked him, "What should I do to be a brave, strong parallelogram?"

"If both pairs of your opposite sides are congruent, you will be a parallelogram," the frog told her.

She went to the mouse and asked him, "What should I do to be a brave, strong parallelogram?"

"If you take one pair of opposite sides and make them both parallel and congruent, then you will be a parallelogram," the mouse told her.

She went to the spider and asked the same question.

Sammy Spider told her, "Just make both pairs of your opposite angles congruent, and you will become a lovely parallelogram."

Here is the big picture:

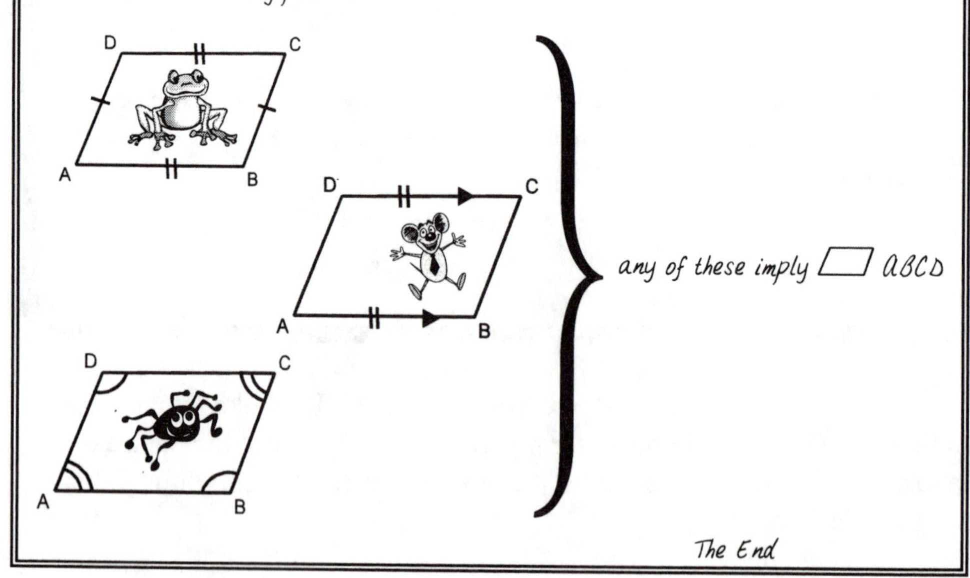

The End

Your Turn to Play

1. (For English majors again) Combine P's three theorems in one big theorem.

2. Prove the Frog Theorem: If the opposite sides of a quadrilateral are congruent, then it is a ▱.

3. Prove the Mouse Theorem: If one pair of opposite sides of a quadrilateral is both parallel and congruent, then it is a ▱.

4. The Sammy Spider Theorem is much more tricky to prove. (If both pairs of opposite angles of a quadrilateral are congruent, then it is a ▱.) I'll supply the proof and you supply the missing reasons.

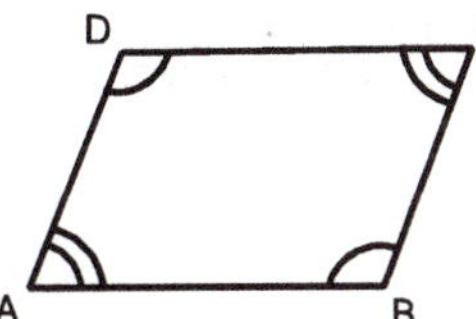

Statement	***Reason***
1. quadrilateral ABCD. $\angle A \cong \angle C$ and $\angle B \cong \angle D$.	1. ?
2. $m\angle A = m\angle C$ $m\angle B = m\angle D$	2. ?
3. Draw $\overline{AC}$.	3. ?
4. $m\angle DAB = m\angle 1 + m\angle 2$ $m\angle DCB = m\angle 3 + m\angle 4$	4. ?
5. $m\angle 1 + m\angle D + m\angle 3 = 180°$ $m\angle 2 + m\angle B + m\angle 4 = 180°$	5. ?
6. $m\angle DAB + m\angle B = 180°$	6. A lot of Algebra (steps 2, 4 and 5)
7. $\angle A$ and $\angle 5$ are supplementary. (We can't write $\angle B$ now since there is more than one angle at B.)	7. ?
8. $\angle 5$ and $\angle 6$ form a linear pair.	8. ?
9. $\angle 5$ and $\angle 6$ are supplementary.	9. ?
10. $\angle A \cong \angle 6$	10. Two angles that are supplementary to the same angle are congruent. This theorem was problem 2 in Hamlet on p. 85.
11. $\overline{AD} \parallel \overline{BC}$	11. ?

And then drawing the diagonal $\overline{BD}$ and going through the same steps, you can establish $\overline{AB} \parallel \overline{DC}$ and hence we have a parallelogram. ⊠

COMPLETE SOLUTIONS

1. A quadrilateral with either (1) both pairs of opposite sides congruent; (2) one pair of opposite sides both congruent and parallel; or (3) both pairs of opposite angles congruent is a parallelogram.

2. If the opposite sides of a quadrilateral are congruent, then it is a ▱.

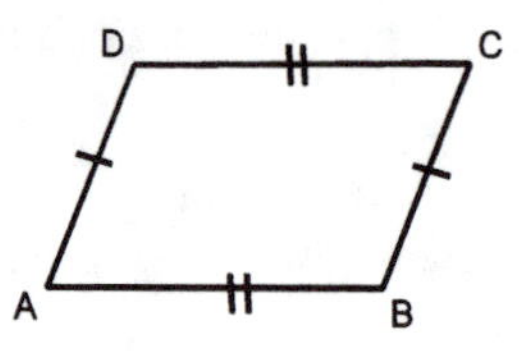

Statement	Reason
1. Quadrilateral ABCD. $\overline{AB} \cong \overline{CD}$ and $\overline{AD} \cong \overline{BC}$	1. Given
2. Draw $\overline{AC}$.	2. Postulate: Two points determine a line.
3. AC = AC	3. Algebra
4. $\overline{AC} \cong \overline{AC}$	4. Def of ≅ segments
5. $\triangle ACD \cong \triangle CAB$	5. SSS
6. $\angle 1 \cong \angle 2$, $\angle 3 \cong \angle 4$	6. Def of ≅ ▲
7. $\overline{AB} \parallel \overline{CD}$ and $\overline{AD} \parallel \overline{CB}$	7. AIP Thm (If the alternate interior angles are congruent, then the lines are parallel.)
8. ▱ ABCD	8. Def of ▱

3. To prove: If one pair of opposite sides of a quadrilateral are both parallel and congruent, then it is a ▱.

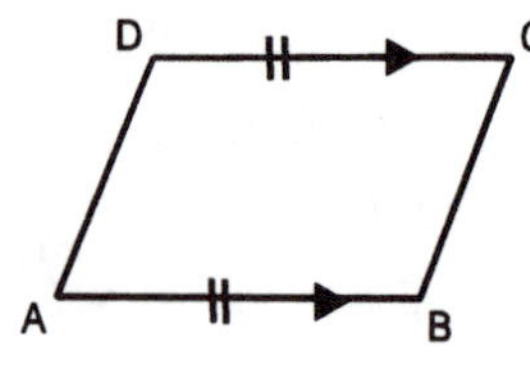

Statement	Reason
1. Quadrilateral ABCD. $\overline{AB} \cong \overline{CD}$ and $\overline{AB} \parallel \overline{CD}$	1. Given
2. Draw $\overline{AC}$.	2. Postulate: Two points determine a line.
3. AC = AC	3. Algebra
4. $\overline{AC} \cong \overline{AC}$	4. Def of ≅ segments
5. $\angle 3 \cong \angle 4$	5. PAI
6. $\triangle ACD \cong \triangle CAB$	6. SAS
7. $\overline{AD} \cong \overline{CB}$	7. Def of ≅ ▲

8. ▱ ABCD	8. Theorem (If the opposite sides of a quadrilateral are congruent, then it is a ▱.)

4. Here's the proof with all the missing steps supplied. To prove: If both pairs of opposite angles of a quadrilateral are congruent, then it is a parallelogram.

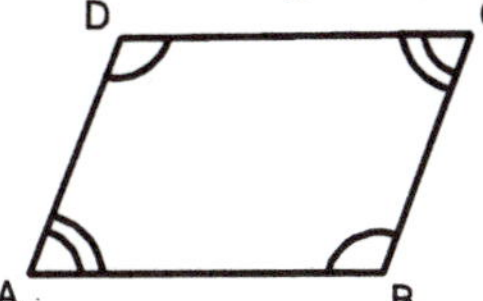

Statement	*Reason*
1. quadrilateral ABCD. $\angle A \cong \angle C$ and $\angle B \cong \angle D$.	1. Given
2. $m\angle A = m\angle C$ $m\angle B = m\angle D$	2. Def of $\cong$ $\angle$s
3. Draw $\overline{AC}$.	3. Postulate: Two points determine a line
4. $m\angle DAB = m\angle 1 + m\angle 2$ $m\angle DCB = m\angle 3 + m\angle 4$	4. Angle Addition Post.
5. $m\angle 1 + m\angle D + m\angle 3 = 180°$ $m\angle 2 + m\angle B + m\angle 4 = 180°$	5. Thm: The sum of the $\angle$s of a triangle add to 180°.
6. $m\angle DAB + m\angle B = 180°$	6. A lot of Algebra (steps 2, 4 and 5)
7. $\angle A$ and $\angle 5$ are supplementary. (We can't write $\angle B$ now since there is more than one angle at B.)	7. def of supplementary
8. $\angle 5$ and $\angle 6$ form a linear pair.	8. Def of linear pair
9. $\angle 5$ and $\angle 6$ are supplementary.	9. Postulate: Linear pair implies supplementary
10. $\angle A \cong \angle 6$	10. Two angles that are supplementary to the same angle are congruent. This theorem was problem 2 in Hamlet on p. 85.
11. $\overline{AD} \parallel \overline{BC}$	11. AIP Theorem

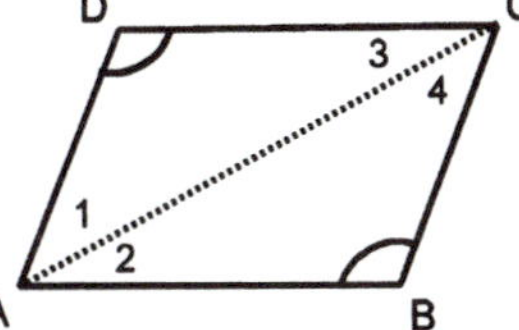

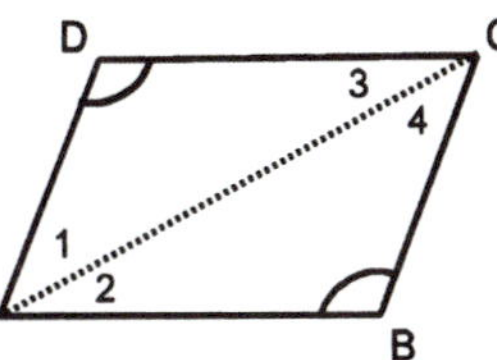

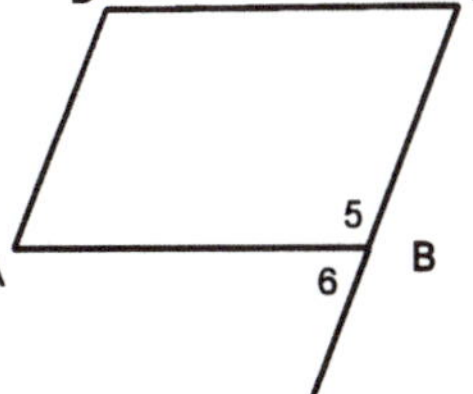

And then drawing the diagonal $\overline{BD}$ and going through the same steps, you can establish $\overline{AB} \parallel \overline{DC}$ and hence we have a parallelogram. ⊠

P became increasingly beautiful in Fred's mind. She seemed to be the perfect match for him. He noticed that her essay and his monograph were almost mirror reflections of each other. He talked about *necessary conditions* for parallelograms and she talked about *sufficient conditions* in order to have a parallelogram.

We need to take a little break.

Honors Problem of the Century

Hint # 1

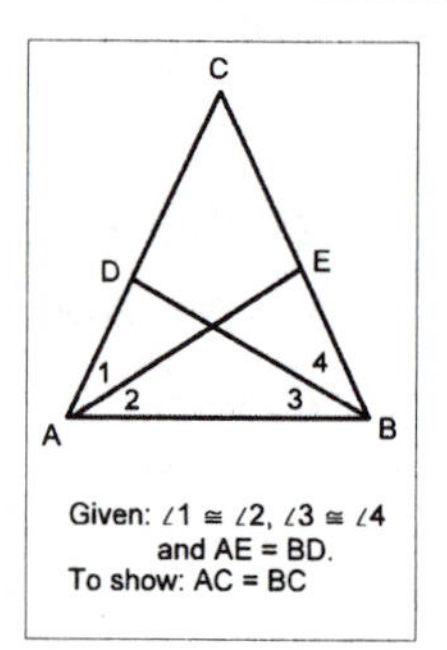

We are trying to prove that in any triangle, if the angle bisectors of two of the angles are congruent, then the triangle is isosceles.

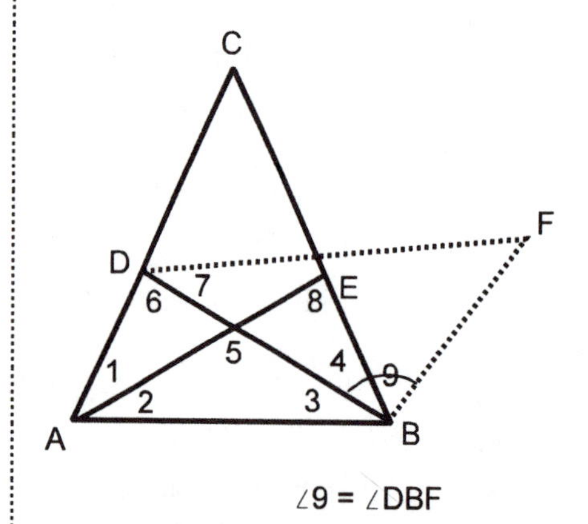

There are many ways to prove this theorem. Here is a diagram of the proof that I will furnish at the end of this chapter.

Toowoomba till five forty-five ran through his head. So few hours until I see her again. And so much to do!

First he dashed off another letter to P thanking her for her letter. He closed this short note with, "Looking forward to meeting you at 5:45," and signed his name. That closing wasn't half as snappy as her "Toowoomba till 5:45," but it would have to do. His mind wasn't firing on all its cylinders. After dropping his letter into the campus mail, he made his way to the department that sells Ralph's Mouth Dust Powder.*

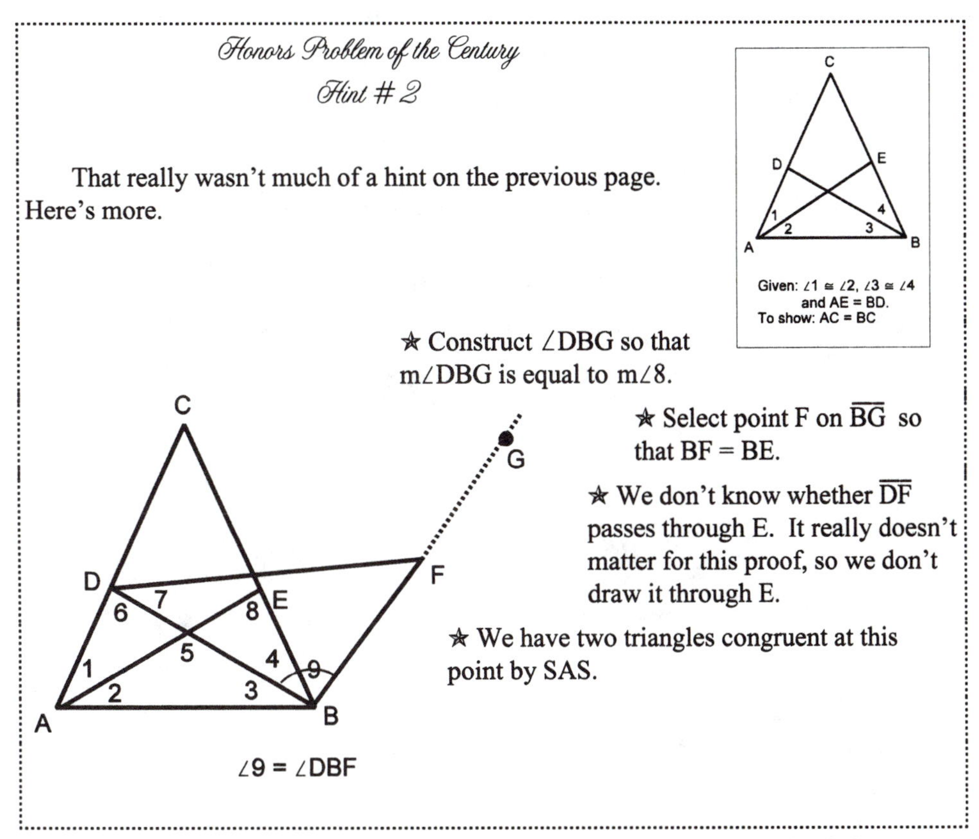

Honors Problem of the Century
Hint # 2
That really wasn't much of a hint on the previous page. Here's more.
C
D
E
1
2
3
4
A
B
Given: ∠1 ≅ ∠2, ∠3 ≅ ∠4 and AE = BD.
To show: AC = BC
★ Construct ∠DBG so that m∠DBG is equal to m∠8.
★ Select point F on $\overline{BG}$ so that BF = BE.
★ We don't know whether $\overline{DF}$ passes through E. It really doesn't matter for this proof, so we don't draw it through E.
★ We have two triangles congruent at this point by SAS.
C
G
D
F
E
6
7
8
5
1
4
9
2
3
A
B
∠9 = ∠DBF

Fred was almost breathless with anxiety when he reached the Soap & Stuff department. Will they be sold out? Do I have to be 18 in order to buy some? Will I have to tell them what I want to use it for?

He approached a clerk to ask her where they stocked Ralph's Mouth Dust Powder.

"Hi Professor Fred! How can I help you?" She was one of his students.

---

* Campus mailboxes are located on every pathway on campus. There are no more than a hundred yards between boxes. Messengers locate the recipients wherever they are on campus and personally hand letters to them. That is how Fred received P's letter when he was walking from the stationery department to the personal hygiene department.

Fred swallowed hard. He couldn't lie and say that he was looking for some shaving cream. He wasn't the lying type. But he didn't exactly want to shout it out for everyone to hear. He said, "Excuse me. Do you carry Ralph's Mouth Dust Powder?"

She made him repeat his whispered request, smiled, and then pointed to a shelf across the room. She told him, "I won't tell anyone."

Fred sneaked, slithered, and sidled over to the display. He hoped that it would be like displays of toothpaste: quiet, unobtrusive, and restrained.

Funny looking containers, Fred thought to himself. They look like they are only about half full. But the bottles are sure pretty. Clear crystal with a black stone knob on top.

He measured the left side and found that the powder came up to the midpoint of the left side of the triangle.

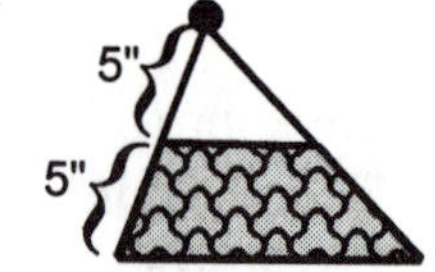

The same was true of the right side of the triangle.

The top of the powder was a **midsegment** (a segment joining the midpoints of two sides of a triangle).

<u>The Midsegment Theorem</u>: In any triangle, a midsegment is parallel to the third side and is equal to half of the length of the third side.

Your Turn to Play

1. Prove the Midsegment Theorem in statement–reason form. Here is a brief outline of the proof.

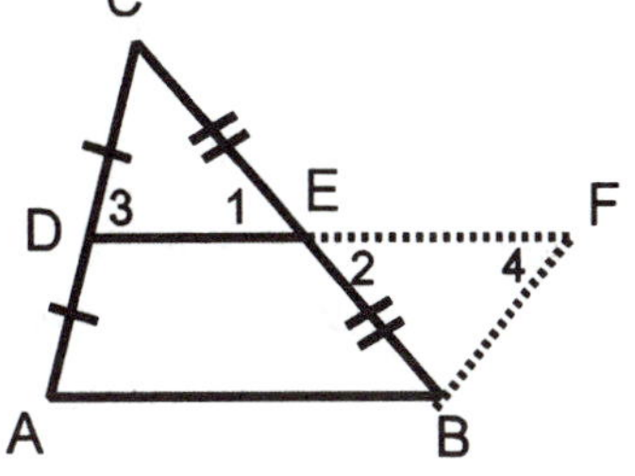

Locate F so that EF = DE. With vertical angles we have the triangles congruent.

Then $\angle 3 \cong \angle 4$ and CD = BF.

The $\cong$ $\angle$s give us $\parallel$ lines.

▱ ABFD.

Since DE is half of DF, it is half of AB.

2.

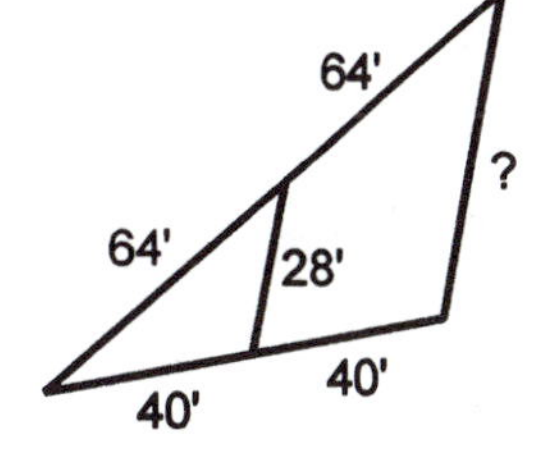

3. The <u>Converse of the Midsegment Theorem</u>: In any triangle, a line which passes through the midpoint of one side and is parallel to the third side must pass through the midpoint of the second side.

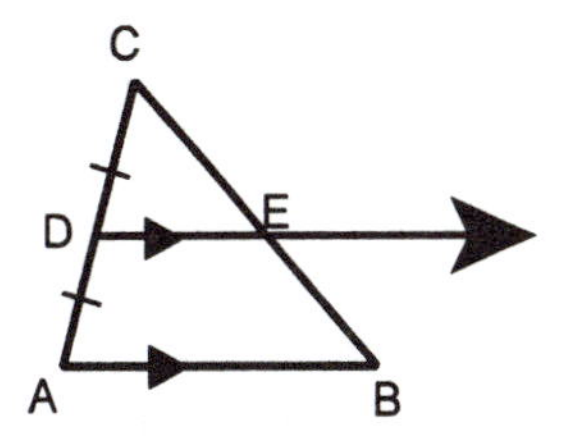

Prove the Converse of the Midsegment Theorem in statement–reason form. Here is a brief outline of the proof.

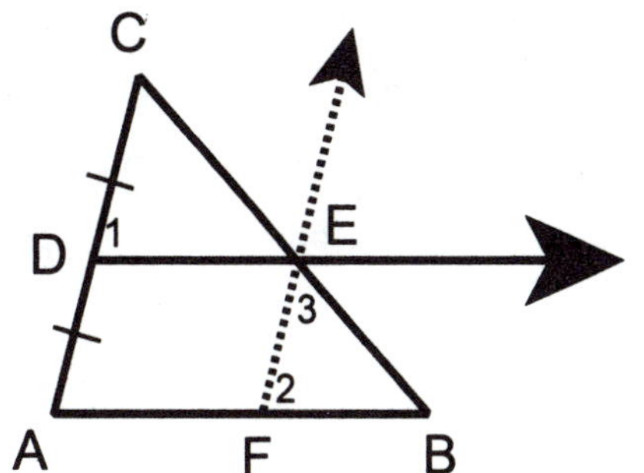

Draw a line parallel to $\overline{AC}$ through E.

▱ AFED

$\angle 1 \cong \angle A$, $\angle A \cong \angle 2$, and $\angle C \cong \angle 3$

ASA

$\overline{CE} \cong \overline{EB}$

COMPLETE SOLUTIONS

1.

Statement	*Reason*
1. $\triangle ABC$ with D the midpoint of $\overline{AC}$ and E the midpoint of $\overline{BC}$	1. Given
2. Draw the line through D and E.	2. Postulate: Two points determine a line.
3. Locate point F on $\overline{DE}$ so that EF = DE.	3. Ruler Postulate
4. $\overline{AD} \cong \overline{DC}$ and $\overline{CE} \cong \overline{EB}$	4. Def of midpoint
5. AD = DC, CE = EB, and EF = DE	5. Def of ≅ segments
6. $\angle 1$ and $\angle 2$ are vertical angles.	6. Def of vertical angles
7. $\angle 1 \cong \angle 2$	7. Vertical Angle Theorem
8. $\triangle DEC \cong \triangle FEB$	8. SAS
9. $\overline{DC} \cong \overline{BF}$ and $\angle 3 \cong \angle 4$	9. Def of ≅ $\triangle$s
10. DC = BF	10. Def of ≅ segments
11. $\overline{AD} \parallel \overline{BF}$	11. AIP
12. AD = BF	12. Algebra (steps 5 and 10)
13. $\overline{AD} \cong \overline{BF}$	13. Def of ≅ segments
14. ▱ ABFD	14. Thm: If two sides are both ≅ and ∥, then it's a ▱
15. $\overline{DF} \parallel \overline{AB}$ (We're half done. This is one of the two things we needed to prove.)	15. Def of ▱
16. $\overline{AB} \cong \overline{DF}$	16. Thm: Opposite sides of a parallelogram are congruent.
17. AB = DF	17. Def of ≅ segments
18. DE + EF = DF	18. Definition of D–E–F
19. AB = ½ DE (This is the other thing we needed to prove.)	19. Algebra (steps 3, 17, 18)

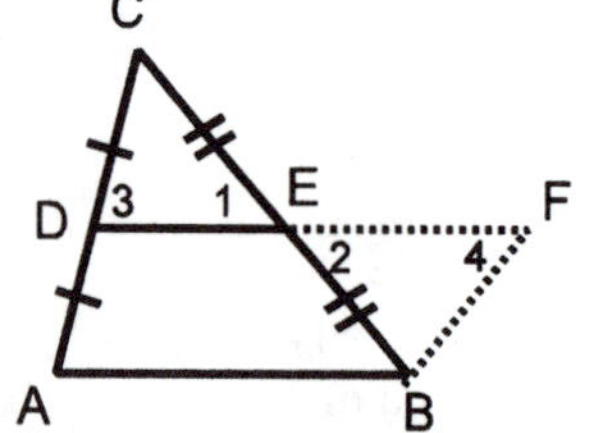

⊠

2. 56'

3.

Statement	*Reason*
1. $\triangle ABC$ with D the midpoint of $\overline{AC}$ and $\overrightarrow{DE} \parallel \overline{AB}$.	1. Given 
2. Draw a line through E which is parallel to $\overline{AC}$.	2. Thm: There exists at least one line through P which is parallel to ℓ.
3. ▱ AFED	3. Def of ▱
4. $\angle 1 \cong \angle A$, $\angle A \cong \angle 2$, $\angle C \cong \angle 3$	4. PCA Theorem: If parallel lines are cut by a transversal, then corresponding ∠s are ≅.
5. $\angle 1 \cong \angle 2$	5. Transitive Property of Congruent Angles Theorem
6. AD = DC	6. Def of midpoint (step 1)
7. $\overline{AD} \cong \overline{DC}$	7. Def of ≅ segments
8. $\overline{AD} \cong \overline{EF}$	8. Thm: Opposite sides of a parallelogram are congruent.
9. $\overline{DC} \cong \overline{EF}$	9. Transitive Property of Congruent Segments Theorem*
10. $\triangle CDE \cong \triangle EFB$	10. ASA
11. $\overline{CE} \cong \overline{EB}$	11. Def of ≅ ▵s

⊠

Part of the excitement of geometry proofs is that there is an element of creativity in constructing them. It is not like algebra where everyone pretty much does the same thing as they crunch $2x + 7 = 13$ to $2x = 6$ and then to $x = 3$.

On the next page is a totally different proof of the above theorem. It doesn't use any congruent triangles or any parallelograms.

---

* Which, as any sharp-eyed reader will note, is a theorem which we haven't proven. It states that if $\overline{PQ} \cong \overline{RS}$ and $\overline{RS} \cong \overline{TU}$, then $\overline{PQ} \cong \overline{TU}$. We should have proved this as a lemma before we did this theorem. Please put it on your page of theorems that I asked you to keep. Its proof is virtually identical to the proof of the Transitive Property of Congruent Angles Theorem which we proved on p. 140.

An alternative proof of the Converse of the Midsegment Theorem: In any triangle, a line which passes through the midpoint of one side and which is parallel to the third side must pass through the midpoint of the second side.

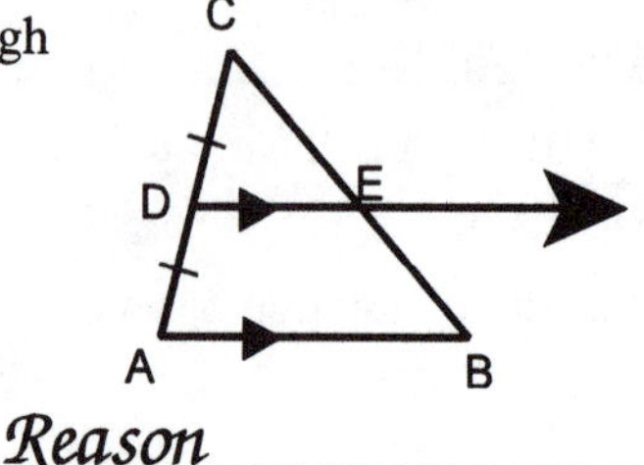

Statement	*Reason*
1. ΔABC with D the midpoint of $\overline{AC}$ and $\overrightarrow{DE} \parallel \overline{AB}$.	1. Given
2. Assume that E is not the midpoint of $\overline{BC}$.	2. Beginning of an indirect proof.
3. Let F be the midpoint of $\overline{BC}$.	3. Midpoint Theorem: Every segment has a midpoint.
4. $\overline{DF} \parallel \overline{AB}$	4. Midsegment Theorem
5. There is at most one line through D that is parallel to $\overline{AB}$.	5. Parallel Postulate
6. E is the midpoint of $\overline{BC}$.	6. Contradiction in steps 1, 4 and 5 and therefore the assumption in step 2 is false.

⊠

And this second proof is much shorter!

We are getting to the point in geometry where there is some room to play.

In arithmetic it was almost 100% rote learning. You memorized the addition table, the multiplication table, how to add decimals, how to multiply fractions, etc. It was "cookbook" mathematics: To make low-rise bread, take two cups of flour and one cup of water, mix, and bake at 300° for 40 minutes. To divide fractions, flip the one on the right upside down and then multiply.

The good news is that as you progress in mathematics there is less to memorize and more to understand and play with.

After the lockstep sequence of ARITHMETIC → BEGINNING ALGEBRA → GEOMETRY → ADVANCED ALGEBRA → TRIG → CALCULUS, the world of mathematics opens up like a field of wild flowers in spring.

Here are some areas of math that will be open to you after calculus. Most of these can be taken the minute that you finish calculus. They all contain surprises and delights (if they are taught right).

LOGIC In which you can prove "If the moon is made of green cheese, then mice made the holes." (An untrue hypothesis implies any conclusion.)

PROJECTIVE GEOMETRY In which every pair of lines in a plane intersect. There are no parallel lines.

ABSTRACT ALGEBRA In which groups, rings and fields are not associations, things for your fingers or places to plow. We'll discover patterns that you may never have noticed. For example, working with addition in the real numbers is virtually the same as ("isomorphic" is the fancy word) working with multiplication in all the non-zero real numbers. They are mirror images of each other.

ANALYSIS In which we go back and prove stuff in calculus that you just had to take on faith when you were a calculus student.

METAMATHEMATICS In which we look at mathematics as a whole from a mathematical point of view. We talk about mathematical things that are true, but which can never be proven.

SET THEORY In which we prove that there is no such thing as the set of all sets.

GRAPH THEORY In which we decide whether we can start at one pizza place and visit all the other pizza places on the map exactly once and come back to the original pizza place without ever walking down the same street twice.

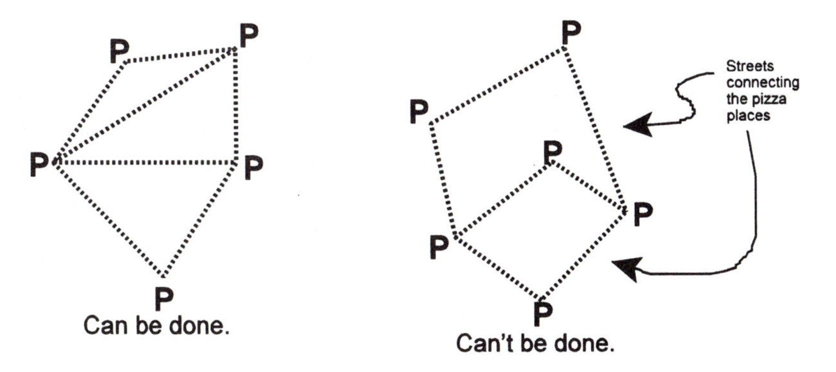

TOPOLOGY In which geometry is drawn on a rubber sheet which can be stretched, so that a circle is considered the same thing as a square.

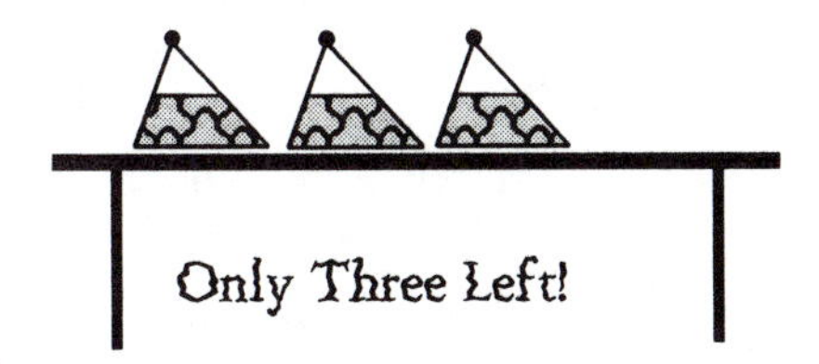

Fred selected a bottle and was surprised when the sign underneath the table changed.

After purchasing it, he took a little swig just "to get his mouth warmed up." It tasted like talcum powder.

He emptied the rest of the power into a box marked REFUSE.*

✶ When Fred was first learning to read, he thought those boxes were political signs, but he couldn't figure out what they were asking their readers to refuse.

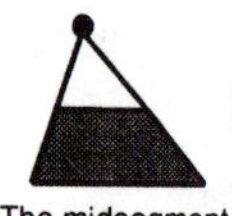

When he got back to his office, he planned to refill the bottle with raspberry syrup and then use it as a visual aid in his classroom teaching.

Honors Problem of the Century

Hint # 3

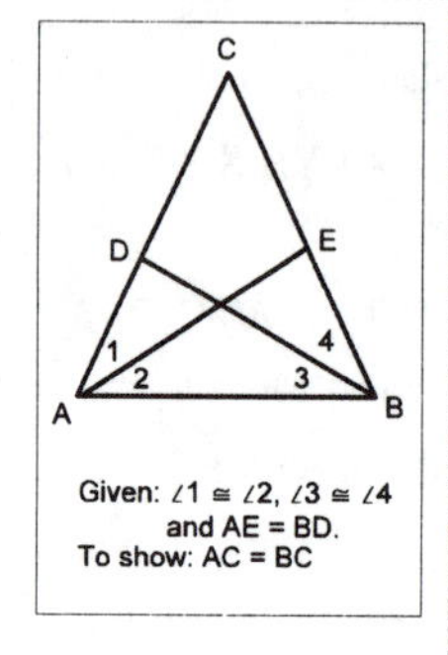

This is a lemma that you may find handy.

<u>Lemma</u>: 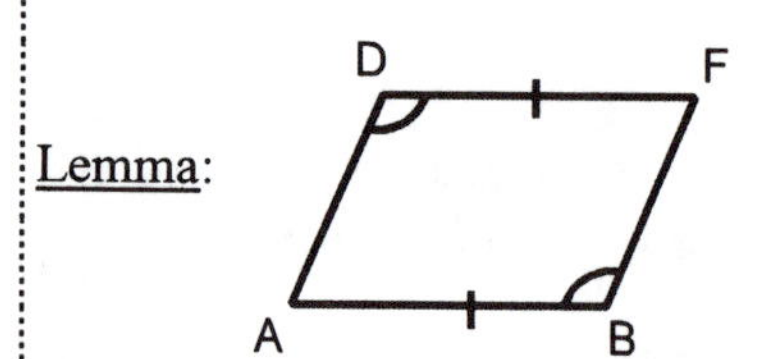 implies ▱ ABFD.

This is a seemingly simple little theorem, but finding its proof would stump many geometry teachers. If I could ask a favor of you, please shut this book and spend a couple of minutes trying to prove it. Even if you don't get the solution, your brain will be stronger because of your effort. Also, you will enjoy my solution a bit more after you have tried on your own.

Don't look below this line until you have given it your best two-minute try.
Don't look below this line until you have given it your best two-minute try.
Don't look below this line until you have given it your best two-minute try.
Don't look below this line until you have given it your best two-minute try.
Don't look below this line until you have given it your best two-minute try.
Don't look below this line until you have given it your best two-minute try.
Don't look below this line until you have given it your best two-minute try.
Don't look below this line until you have given it your best two-minute try.
Don't look below this line until you have given it your best two-minute try.
Don't look below this line until you have given it your best two-minute try.
Don't look below this line until you have given it your best two-minute try.
Don't look below this line until you have given it your best two-minute try.

Statement	Reason
1. Quadrilateral ABFD. $\overline{AB} \cong \overline{DF}$ and $\angle B \cong \angle D$	1. Given
2. Drop the ⊥ from F to $\overline{AD}$ and from A to $\overline{BF}$.	2. Theorem: There exists at least one ⊥ from a point to a line.
3. ∠1 and ∠2 are a linear pair. ∠3 and ∠4 are a linear pair. (See next page for the diagram.)	3. Def of linear pair.

Statement	Reason
4. $\angle 1$ and $\angle 2$ are supplementary. $\angle 3$ and $\angle 4$ are supplementary.	4. Linear Pair Postulate
5. $\angle 2 \cong \angle 4$	5. Thm: Supplements of congruent $\angle$s are $\cong$.
6. $\Delta DGF \cong \Delta BHA$	6. AAS
7. $\overline{DG} \cong \overline{BH}$, $\overline{GF} \cong \overline{HA}$	7. Def of $\cong$ $\triangle$s
8. Draw $\overline{AF}$.	8. Postulate: Two points determine a line.
9. $\overline{AF} \cong \overline{AF}$	9. Thm: Every segment is congruent to itself.
10. $\Delta AFG \cong \Delta FAH$	10. Hypotenuse-Leg
11. $\overline{AG} \cong \overline{FH}$	11. Def of $\cong$ $\triangle$s
12. AG = FH, DG = BH	12. Def of $\cong$ segments
13. AD + DG = AG FB + BH = FH	13. Def of betweenness
14. AD = BF	14. Algebra
15. $\overline{AD} \cong \overline{BF}$	15. Def of $\cong$ segments
16. ▱ ABFD	16. Thm: If the opposite sides of a quadrilateral are congruent, then it is a parallelogram.

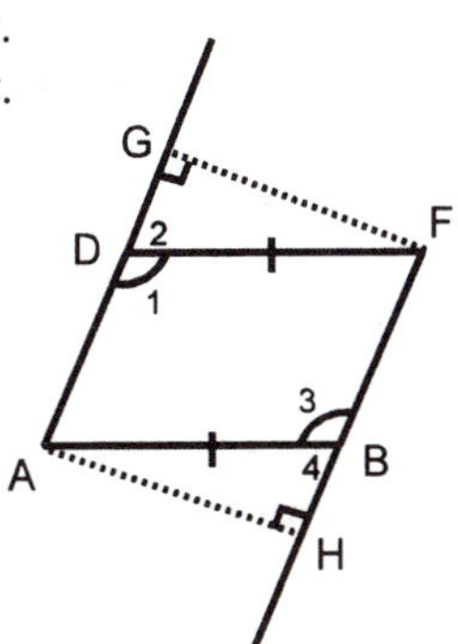

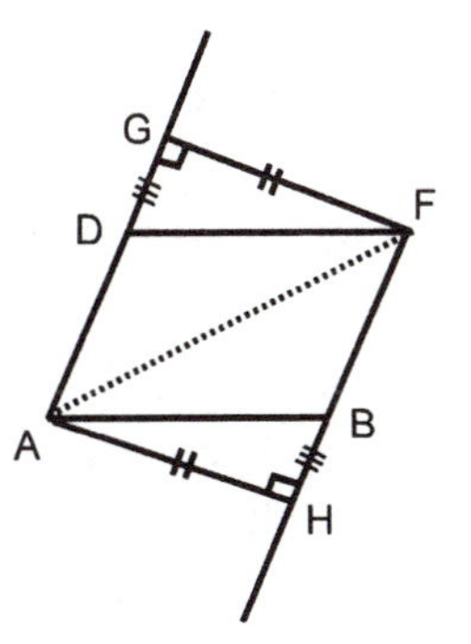

⊠

As Fred exited from the personal hygiene department, he noticed a department that made him think of P. (Everything make him think of her.)

Fred had instant fantasies of going on a picnic with his sweetheart. We could take along weenies and a couple of strawberry milkshakes. I could be the barbecue chef and she could watch me cook. I would feel so manful.

These were new thoughts for this 37-pound six-year-old. Before today, he had always enjoyed being a boy. Despite his years of being an emancipated minor, being employed as a university professor, and living on his own, there were many steps in the transition from boyhood to manhood that he had not yet taken. Learning to barbecue hot dogs was one of them.

With $29 in hand, he approached the BBQ clerk.

"Hi, Professor Fred! How can I help you?" It seemed that everyone on campus knew him.

Fred was trying, as best as he could, to act a little more manly. At least, he figured, no one in this department will ask me embarrassing questions. I didn't like buying that mouth dust powder because of that. We'll just keep this purchase on a business level. "I am thinking of purchasing the one on the sign. That should cook hot dogs just fine."*

"Oh yes. Our bottom-of-the-line electric model. That should be perfect in your office for you and her on these cold February evenings."

"Her?" Fred sputtered. How did the clerk know about her?

THE KITTEN Caboodle

The Official Campus Newspaper of KITTENS University

Thursday 10:10 a.m. Edition 10¢

exclusive

Fred Has Girlfriend!

KANSAS: It couldn't happen to a nicer guy. All of the campus is atwitter over the romance that has blossomed in the math department. According to our reporters, Fred has purchased a two-quart bottle of Ralph's. You all know what that means.

In addition, the campus mailman reports that Fred has sent two letters to a "P Commonsizerinski."

file photo

* Fred was going to say, "That should cook *our* hot dogs . . ." but quickly changed the wording to avoid letting anyone know he had a girlfriend.

Life in a small town is a lot different than living in Los Angeles or New York.

"Now if you're going to do some serious barbecuing, you'll need some accessories," the clerk announced. "The BBQ spice shaker ($6), a bottle of Filet Sauce ($8.49), a Filet Sauce dauber ($5), and a chef's hat ($18). Oh, and a solid oak cutting board ($36) to cut your hot dogs on. And you can't use that electric model for picnics unless you plan to haul a car battery along to power it. Here's the model you need to do it right."

The clerk pointed to the top-of-the-line:

Hecks Kitchen
Model No. 9097933395-290-2880-B
Surface Grill Area = 25" × 48"
Stainless Steel Grill Bars
—Guaranteed Absolutely Parallel
Shipping weight 638 lbs.
Price $1800.

"She deserves the best, doesn't she?"

Fred blanched. His face became the color of paper. That's three months' salary.

"And the stainless steel grill bars are absolutely parallel. She will be really impressed."

All I was thinking of was a little picnicking.

"Let me fire it up for you. I think you'll like the way this baby cooks."

The clerk turned a knob which opened the gas line to the 500-gallon propane tank—an optional accessory, $350—and pressed the automatic ignition button. Flames leaped through the grill bars three feet into the air. "That's if you really want to cook a moose in a hurry. Let me turn it down a tad."

Fred noticed that the clerk's eyebrows were missing.

Fred had one question. "You know the old expression P.I.E.?* I know these grill bars are guaranteed to be parallel, but they might be like this:

"I sure you wouldn't want uneven marks on her hot dog. Of course not, sir." (People who may be buying an $1800 hot-dog cooker, get called sir a lot.)

"Permit me to demonstrate the evenness of these grills." He dropped a piece of uncooked spaghetti onto the red hot grills. "Please note how the grill makes very even marks."

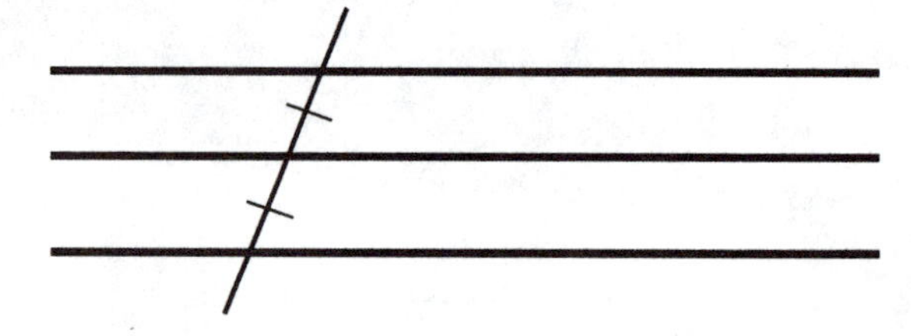

Fred mentally rephrased the clerk's spaghetti talk in the language he understood best: The parallel lines **intercept** (that means "cuts off") congruent segments on the transversal.

* Parallel Isn't Everything.

"If you wish, sir, I could throw a second strand of spaghetti on the grill at a different angle and it also would experience the same fine evenness of the grill bars."*

"No need to waste the pasta," Fred answered. "If it works for one, it'll work for any other."

Your Turn to Play

1. Theorem (Hecks Kitchen Theorem a.k.a. Spaghetti Theorem): If three parallel lines intercept congruent segments on one transversal, they will intercept congruent segments on any transversal.

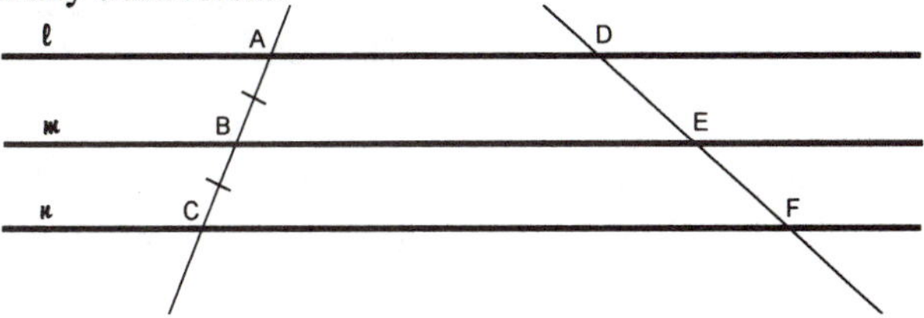

Prove this theorem. Here is a brief outline of the proof. Draw a line through D which is parallel to $\overleftrightarrow{AB}$. You have two parallelograms. The right sides of those two parallelograms are congruent. Then use the Midsegment Theorem.

COMPLETE SOLUTIONS

1.

Statement	*Reason*
1. ℓ, m and n are parallel. $\overline{AB} \cong \overline{BC}$.	1. Given
2. Draw a line through D which is parallel to $\overleftrightarrow{AB}$.	2. Theorem: Given a point P not on ℓ, there exists at least one line through P which is ∥ ℓ.

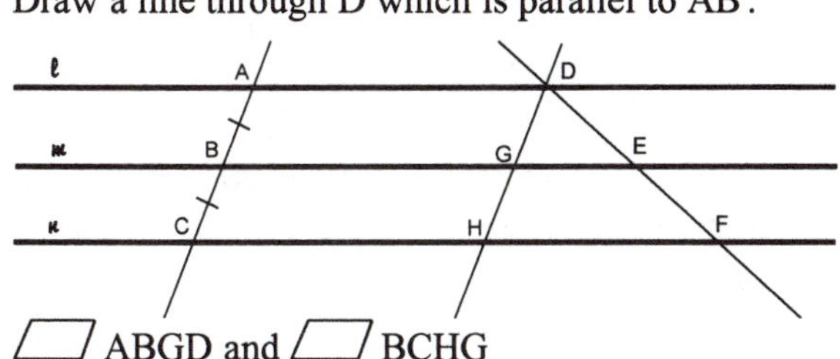

3. ▱ ABGD and ▱ BCHG	3. Def of ▱

✶ The clerk was starting to sound haute. The original "Hi, Professor Fred!" has changed into the high-toned ". . . strand of spaghetti . . . experiences the same fine evenness. . . ." If the clerk knew that Fred only made $600 a month, he might have said, "Lemme toss another stick of spaghetti on."

4. $\overline{AB} \cong \overline{DG}$ and $\overline{BC} \cong \overline{GH}$. — 4. Thm: Opposite sides of a ▱ are congruent.

5. $\overline{DG} \cong \overline{GH}$ — 5. Transitive Property for Congruent Segments Theorem

6. $\overline{DE} \cong \overline{EF}$ — 6. Converse of the Midsegment Theorem

"Of course," Fred continued, "you would need to use two pieces of spaghetti to test whether the grills are parallel as they are guaranteed to be."

Your Turn to Play

1. <u>Theorem</u> (Testing the Guarantee Theorem): If $\ell \parallel n$, $\overline{AB} \cong \overline{BC}$, and $\overline{DE} \cong \overline{EF}$, then $\ell \parallel m \parallel n$. If three lines intercept congruent segments on two transversals and two of the lines are parallel, then all three lines are parallel.

Is this a converse of the Hecks Kitchen Theorem?

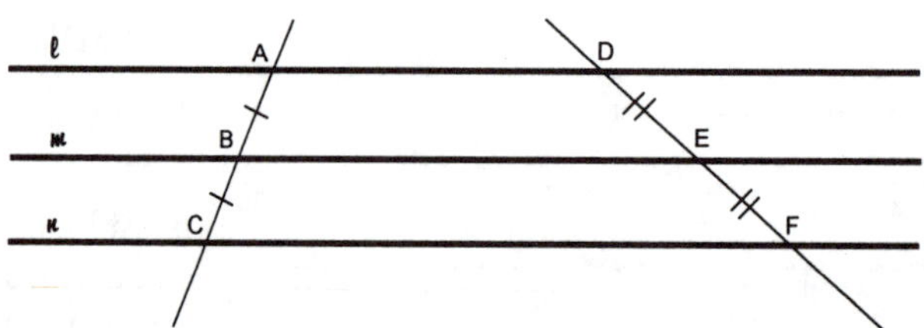

2. Prove the theorem. Here is a brief outline of the proof. Mentally erase line m for a moment. It will make the diagram less cluttered. Draw a line through D which is parallel to $\overleftrightarrow{AB}$.

Let M be the midpoint of $\overline{DG}$.
After several steps, $\overline{MG} \cong \overline{BC}$.
▱BCGM. $\overline{BM} \parallel n$.
On the other hand, $\overline{ME} \parallel n$.
B, M and E are collinear.
(collinear = they lie on the same line.)
$\overline{BE} \parallel n$.

COMPLETE SOLUTIONS

1. A converse of a theorem interchanges some (or all) of the hypothesis with some (or all) of the conclusion of the theorem. If we compare the two theorems:

Hecks Kitchen Theorem	
Hypothesis	Conclusion
i) $\ell \parallel m$	i) $\overline{DE} \cong \overline{EF}$
ii) $\ell \parallel n$	
iii) $\overline{AB} \cong \overline{BC}$	

Testing the Guarantee Theorem	
Hypothesis	Conclusion
i) $\ell \parallel n$	i) $\ell \parallel m$
ii) $\overline{AB} \cong \overline{BC}$	
iii) $\overline{DE} \cong \overline{EF}$	

They are converses of each other.

2.

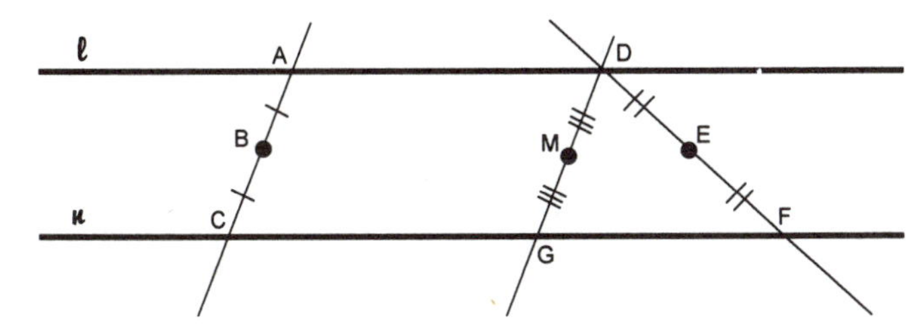

Statement	*Reason*
1. $\ell \parallel n$, $\overline{AB} \cong \overline{BC}$, and $\overline{DE} \cong \overline{EF}$	1. Given
2. Draw a line through D which is parallel to $\overleftrightarrow{AC}$.	2. Theorem: Given a point P not on ℓ, there exists at least one line through P which is $\parallel \ell$.
3. Let M be the midpoint of $\overline{DG}$.	3. Thm: Every segment has a midpoint.
4. $\overline{DM} \cong \overline{MG}$.	4. Def of midpoint
5. ▱ ACGD	5. Def of ▱
6. $\overline{AC} \cong \overline{DG}$	6. Thm: Opposite sides of a parallelogram are $\cong$.
7. AB + BC = AC DM + MG = DG	7. Def of betweenness
8. AB = BC, DM = MG, AC = DG	8. Def of $\cong$ segments
9. BC = MG	9. Algebra
10. $\overline{BC} \cong \overline{MG}$	10. Def of $\cong$ segments

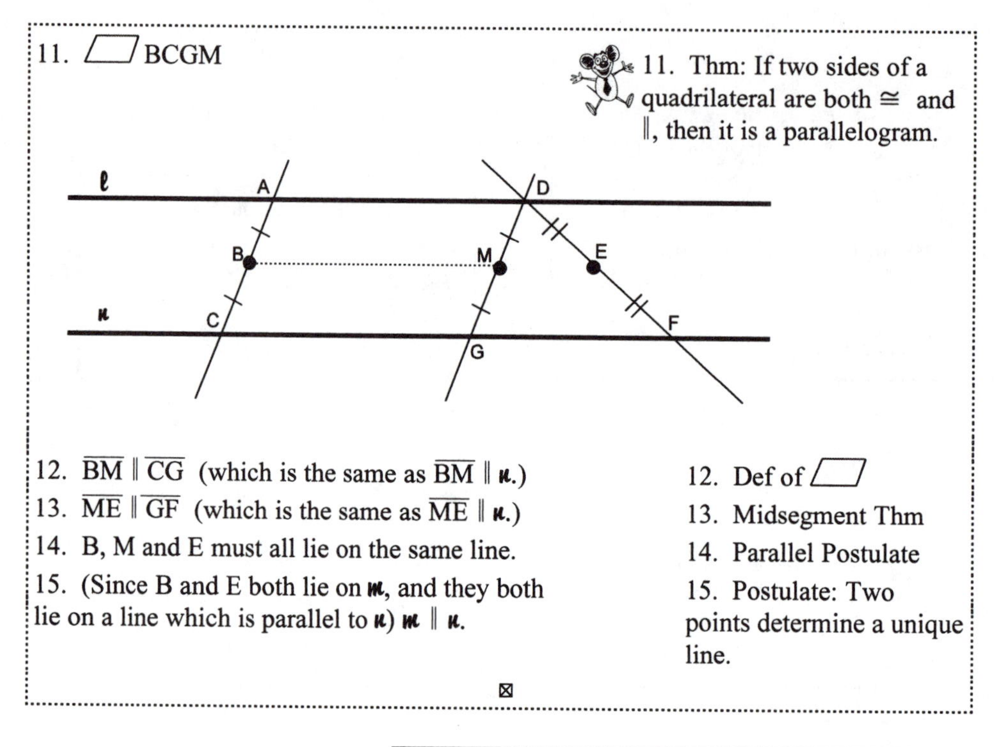

11. ▱ BCGM	11. Thm: If two sides of a quadrilateral are both ≅ and ∥, then it is a parallelogram.
12. $\overline{BM} \parallel \overline{CG}$ (which is the same as $\overline{BM} \parallel n$.)	12. Def of ▱
13. $\overline{ME} \parallel \overline{GF}$ (which is the same as $\overline{ME} \parallel n$.)	13. Midsegment Thm
14. B, M and E must all lie on the same line.	14. Parallel Postulate
15. (Since B and E both lie on m, and they both lie on a line which is parallel to n) $m \parallel n$.	15. Postulate: Two points determine a unique line.

⊠

"Okay," said Fred.

The clerk totaled up the bill and Fred wrote a check out to Barbecue Basics.

B-B-Q spice shaker	6.00
bottle of Filet Sauce	8.49
Filet Sauce dauber	5.00
chef's hat	18.00
oak cutting board	36.00
Hecks Kitchen	1800.00
500 gallon propane tank . . .	350.00
subtotal	2223.49
tax & gratuity . .	276.51
Total	2500.00

Fred decided not to carry his new Hecks Kitchen grill home since his hands were full with the Ralph's bottle. The clerk pasted a campus mail stamp on the barbecue grill. It would be delivered to Fred's office before Fred even left the store.

There are two promises to be kept before we end this chapter. First is the solution to the *Honors Problem of the Century* and second is the solution to the unlocking-any-two-locks-will-open-the-gate problem.

The Solution to the Honors Problem of the Century

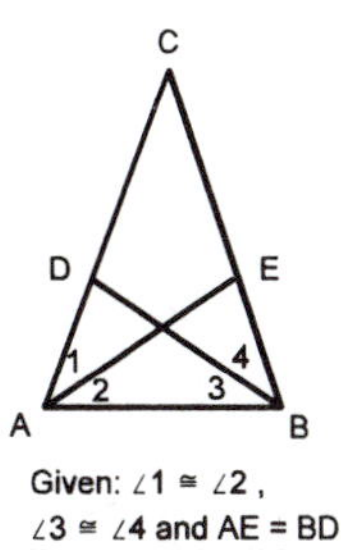

Theorem: In any triangle, if the angle bisectors of the two angles are congruent, then the triangle is isosceles.

With Hint #2 we had the diagram on the right with BF = BE and $m\angle 9 = m\angle 8$. (For reference, we'll call this line ✯)

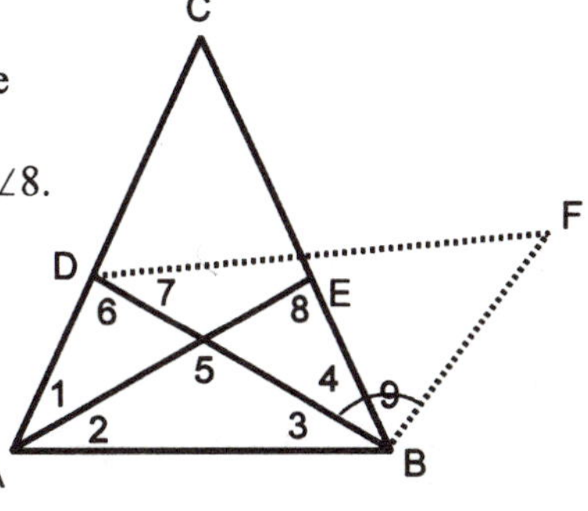

Since the given is AE = BD, we have by SAS that $\Delta DBF \cong \Delta AEB$. This is not especially easy to see since the triangles are overlapping and flipped.

$$\left.\begin{array}{ll} & \overline{DF} \cong \overline{AB} \\ (✯✯) & \angle ABE \cong \angle F \\ (✯✯✯) & \angle 2 \cong \angle 7 \end{array}\right\} \text{ by def of } \cong \triangle\text{s}$$

Now we need to "chase" a lot of angles:

$$\left.\begin{array}{l} m\angle 5 = m\angle 1 + m\angle 6 \\ m\angle 5 = m\angle 4 + m\angle 8 \end{array}\right\} \text{ by Ext. Angle Thm}$$

By algebra	$m\angle 1 + m\angle 6 = m\angle 4 + m\angle 8.$
By given	$m\angle 2 + m\angle 6 = m\angle 4 + m\angle 8$
By (✯✯✯)	$m\angle 7 + m\angle 6 = m\angle 4 + m\angle 8$
By given	$m\angle 7 + m\angle 6 = m\angle 3 + m\angle 8$
By (✯)	$m\angle 7 + m\angle 6 = m\angle 3 + m\angle 9$

By the lemma of Hint #3, ▱ ABFD.
$\angle DAB \cong \angle F$ since the opposite angles of a ▱ are ≅.

$\angle DAB \cong \angle ABE$ by the previous line and (✯✯).
ΔABC is isosceles by the converse of the Isosceles Triangle Theorem.

Q.E.D.

One thing to learn is that you can't always tell how difficult a problem is by its appearance. This was a very simple looking theorem: *If the two angle bisectors are congruent, then the triangle is isosceles,* but its proof wasn't.

Another example is Goldbach's "Theorem." It states that every even number greater than two is the sum of two prime numbers. For example:

4 = 2 + 2
6 = 3 + 3
8 = 3 + 5
10 = 5 + 5
12 = 5 + 7
14 = 7 + 7
16 = 3 + 13
18 = 7 + 11
20 = 7 + 13
22 = 11 + 11
24 = 5 + 19
. . .

A prime number is a number that can't be factored. 15 is not prime since it equals 3 × 5. 51 is not prime since 51 = 3 × 17. The prime numbers are: 2, 3, 5, 7, 11, 13, . . . *with a more complete list on the next page.*

It's not really a theorem since no one has ever proved it! And no one has ever found a counterexample—an even number that can't be expressed as the sum of two primes. We just plain don't know whether it is true or not.

On the other hand, there are problems that look devilishly difficult, but which turn out to be fairly easy to work out. For example, in advanced algebra in the chapter on sequences, series, and matrices, problem 1 in one of the *Your Turn to Plays* is to find the sum of the first 100 terms of 3 +2 + 4/3 + 8/9 + 16/27 +. . . (each term being two-thirds as large as the previous term). In about two steps we find the answer is $91(1 - (2/3)^{100})$. Even your kid brother could do it (after he learned a bit of advanced algebra).

The solution to the ultimate chain-the-gate problem is one of those why-didn't-I-think-of-that ideas.

Answer to Puzzle Question #2

(from chapter 5½)

#2. How to chain the gate with three padlocks so that opening any two locks will open the gate.

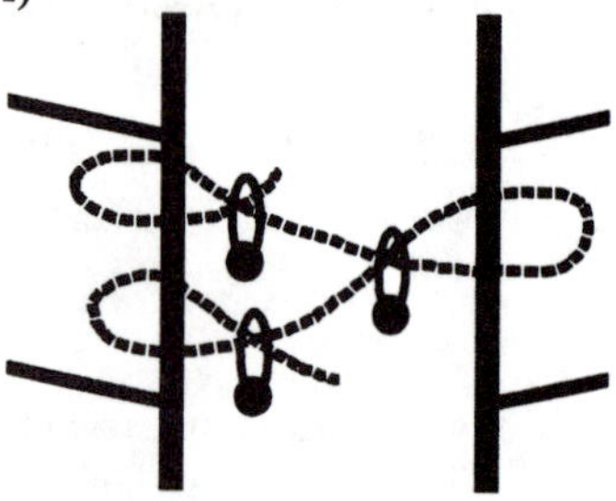

A List of the More Popular Primes

2, 3, 5, 7, 11, 13, 17, 19, 23, 29, 31, 37, 41, 43, 47, 53, 59, 61, 67, 71, 73, 79, 83, 89, 97, 101, 103, 107, 109, 113, 127, 131, 137, 139,149, 151, 157, 163, 167, 173, 179, 181, 191, 193, 197, 199, 211, 223, 227, 229, 233, 239, 241, 251, 257,263, 269, 271, 277, 281, 283, 293, 307, 311, 313, 317, 331, 337, 347, 349, 353, 359, 367, 373, 379, 383, 389,397, 401, 409, 419, 421, 431, 433, 439, 443, 449, 457, 461, 463, 467, 479, 487, 491, 499, 503, 509, 521, 523, 541, 547, 557, 563, 569, 571, 577, 587, 593, 599, 601, 607, 613, 617, 619, 631, 641, 643, 647, 653, 659, 661, 673, 677, 683, 691, 701, 709, 719, 727, 733, 739, 743, 751, 757, 761, 769, 773, 787, 797, 809, 811, 821, 823, 827, 829, 839, 853, 857, 859, 863, 877, 881, 883, 887, 907, 911, 919, 929, 937, 941, 947, 953, 967, 971, 977, 983, 991, 997, 1009, 1013, 1019, 1021, 1031, 1033, 1039, 1049, 1051, 1061, 1063, 1069, 1087, 1091, 1093, 1097, 1103, 1109, 1117, 1123, 1129, 1151, 1153, 1163, 1171, 1181, 1187, 1193, 1201, 1213, 1217, 1223, 1229, 1231, 1237, 1249, 1259, 1277, 1279, 1283, 1289, 1291, 1297, 1301, 1303, 1307, 1319, 1321, 1327, 1361, 1367, 1373, 1381, 1399, 1409, 1423, 1427, 1429, 1433, 1439, 1447, 1451, 1453, 1459, 1471, 1481, 1483, 1487, 1489, 1493, 1499, 1511, 1523, 1531, 1543, 1549, 1553, 1559, 1567, 1571, 1579, 1583, 1597, 1601, 1607, 1609, 1613, 1619, 1621, 1627, 1637, 1657, 1663, 1667, 1669, 1693, 1697, 1699, 1709, 1721, 1723, 1733, 1741, 1747, 1753, 1759, 1777, 1783, 1787, 1789, 1801, 1811, 1823, 1831, 1847, 1861, 1867, 1871, 1873, 1877, 1879, 1889, 1901, 1907, 1913, 1931, 1933, 1949, 1951, 1973, 1979, 1987, 1993, 1997, 1999, 2003, 2011, 2017, 2027, 2029, 2039, 2053, 2063, 2069, 2081, 2083, 2087, 2089, 2099, 2111, 2113, 2129, 2131, 2137, 2141, 2143, 2153, 2161, 2179, 2203, 2207, 2213, 2221, 2237, 2239, 2243, 2251, 2267, 2269, 2273, 2281, 2287, 2293, 2297, 2309, 2311, 2333, 2339, 2341, 2347, 2351, 2357, 2371, 2377, 2381, 2383, 2389, 2393, 2399, 2411, 2417, 2423, 2437, 2441, 2447, 2459, 2467, 2473, 2477, 2503, 2521, 2531, 2539, 2543, 2549, 2551, 2557, 2579, 2591, 2593, 2609, 2617, 2621, 2633, 2647, 2657, 2659, 2663, 2671, 2677, 2683, 2687, 2689, 2693, 2699, 2707, 2711, 2713, 2719, 2729, 2731, 2741, 2749, 2753, 2767, 2777, 2789, 2791, 2797, 2801, 2803, 2819, 2833, 2837, 2843, 2851, 2857, 2861, 2879, 2887, 2897, 2903, 2909, 2917, 2927, 2939, 2953, 2957, 2963, 2969, 2971, 2999, 3001, 3011, 3019, 3023, 3037, 3041, 3049, 3061, 3067, 3079, 3083, 3089, 3109, 3119, 3121, 3137, 3163, 3167, 3169, 3181, 3187, 3191, 3203, 3209, 3217, 3221, 3229, 3251, 3253, 3257, 3259, 3271, 3299, 3301, 3307, 3313, 3319, 3323, 3329, 3331, 3343, 3347, 3359, 3361, 3371, 3373, 3389, 3391, 3407, 3413, 3433, 3449, 3457, 3461, 3463, 3467, 3469, 3491, 3499, 3511, 3517, 3527, 3529, 3533, 3539, 3541, 3547, 3557, 3559, 3571, 3581, 3583, 3593, 3607, 3613, 3617, 3623, 3631, 3637, 3643, 3659, 3671, 3673, 3677, 3691, 3697, 3701, 3709, 3719, 3727, 3733, 3739, 3761, 3767, 3769, 3779, 3793, 3797, 3803, 3821, 3823, 3833, 3847, 3851, 3853, 3863, 3877, 3881, 3889, 3907, 3911, 3917, 3919, 3923, 3929, 3931, 3943, 3947, 3967, 3989, 4001, 4003, 4007, 4013, 4019, 4021, 4027, 4049, 4051, 4057, 4073, 4079, 4091, 4093, 4099, 4111, 4127, 4129, 4133, 4139, 4153, 4157, 4159, 4177, 4201, 4211, 4217, 4219, 4229, 4231, 4241, 4243, 4253, 4259, 4261, 4271, 4273, 4283, 4289, 4297, 4327, 4337, 4339, 4349, 4357, 4363, 4373, 4391, 4397, 4409, 4421, 4423, 4441, 4447, 4451, 4457, 4463, 4481, 4483, 4493, 4507, 4513, 4517, 4519, 4523, 4547, 4549, 4561, 4567, 4583, 4591, 4597, 4603, 4621, 4637, 4639, 4643, 4649, 4651, 4657, 4663, 4673, 4679, 4691, 4703, 4721, 4723, 4729, 4733, 4751, 4759, 4783, 4787, 4789, 4793, 4799, 4801, 4813, 4817, 4831, 4861, 4871, 4877, 4889, 4903, 4909, 4919, 4931, 4933, 4937, 4943, 4951, 4957, 4967, 4969, 4973, 4987, 4993, 4999, 5003, 5009, 5011, 5021, 5023, 5039, 5051, 5059, 5077, 5081, 5087, 5099, 5101, 5107, 5113, 5119, 5147, 5153, 5167, 5171, 5179, 5189, 5197, 5209, 5227, 5231, 5233, 5237, 5261, 5273, 5279, 5281, 5297, 5303, 5309, 5323, 5333, 5347, 5351, 5381, 5387, 5393, 5399, 5407, 5413, 5417, 5419, 5431, 5437, 5441, 5443, 5449, 5471, 5477, 5479, 5483, 5501, 5503, 5507, 5519, 5521, 5527, 5531, 5557, 5563, 5569, 5573, 5581, 5591, 5623, 5639, 5641, 5647, 5651, 5653, 5657, 5659, 5669, 5683, 5689, 5693, 5701, 5711, 5717, 5737, 5741, 5743, 5749, 5779, 5783, 5791, 5801, 5807, 5813, 5821, 5827, 5839, 5843, 5849, 5851, 5857, 5861, 5867, 5869, 5879, 5881, 5897, 5903, 5923, 5927, 5939, 5953, 5981, 5987, 6007, 6011, 6029, 6037, 6043, 6047, 6053, 6067, 6073, 6079, 6089, 6091, 6101, 6113, 6121, 6131, 6133, 6143, 6151, 6163, 6173, 6197, 6199, 6203, 6211, 6217, 6221, 6229, 6247, 6257, 6263, 6269, 6271, 6277, 6287, 6299, 6301, 6311, 6317, 6323, 6329, 6337, 6343, 6353, 6359, 6361, 6367, 6373, 6379, 6389, 6397, 6421, 6427, 6449, 6451, 6469, 6473, 6481, 6491, 6521, 6529, 6547, 6551, 6553, 6563, 6569, 6571, 6577, 6581, 6599, 6607, 6619, 6637, 6653, 6659, 6661, 6673, 6679, 6689, 6691, 6701, 6703, 6709, 6719, 6733, 6737, 6761, 6763, 6779, 6781, 6791, 6793, 6803, 6823, 6827, 6829, 6833, 6841, 6857, 6863, 6869, 6871, 6883, 6899, 6907, 6911, 6917, 6947, 6949, 6959, 6961, 6967, 6971, 6977, 6983, 6991, 6997, 7001, 7013, 7019, 7027, 7039, 7043, 7057, 7069, 7079, 7103, 7109, 7121, 7127, 7129, 7151, 7159, 7177, 7187, 7193, 7207, 7211, 7213, 7219, 7229, 7237, 7243, 7247, 7253, 7283, 7297, 7307, 7309, 7321, 7331, 7333, 7349, 7351, 7369, 7393, 7411, 7417, 7433, 7451, 7457, 7459, 7477, 7481, 7487, 7489, 7499, 7507, 7517, 7523, 7529, 7537, 7541, 7547, 7549, 7559, 7561, 7573, 7577, 7583, 7589, 7591, 7603, 7607, 7621, 7639, 7643, 7649, 7669, 7673, 7681, 7687, 7691, 7699, 7703, 7717, 7723, 7727, 7741, 7753, 7757, 7759, 7789, 7793, 7817, 7823, 7829, 7841, 7853, 7867, 7873, 7877, 7879, 7883, 7901, 7907, 7919, 7927, 7933, 7937, 7949, 7951, 7963, 7993, 8009, 8011, 8017, 8039, 8053, 8059, 8069, 8081, 8087, 8089, 8093, 8101, 8111, 8117, 8123, 8147, 8161, 8167, 8171, 8179, 8191, 8209, 8219, 8221, 8231, 8233, 8237, 8243, 8263, 8269, 8273, 8287, 8291, 8293, 8297, 8311, 8317, 8329, 8353, 8363, 8369, 8377, 8387, 8389, 8419, 8423, 8429, 8431, 8443, 8447, 8461, 8467, 8501, 8513, 8521, 8527, 8537, 8539, 8543, 8563, 8573, 8581, 8597, 8599, 8609, 8623, 8627, 8629, 8641, 8647, 8663, 8669, 8677, 8681, 8689, 8693, 8699, 8707, 8713, 8719, 8731, 8737, 8741, 8747, 8753, 8761, 8779, 8783, 8803, 8807, 8819, 8821, 8831, 8837, 8839, 8849, 8861, 8863, 8867, 8887, 8893, 8923, 8929, 8933, 8941, 8951, 8963, 8969, 8971, 8999, 9001, 9007, 9011, 9013, 9029, 9041, 9043, 9049, 9059, 9067, 9091, 9103, 9109, 9127, 9133, 9137, 9151, 9157, 9161, 9173, 9181, 9187, 9199, 9203, 9209, 9221, 9227, 9239, 9241, 9257, 9277, 9281, 9283, 9293, 9311, 9319, 9323, 9337, 9341, 9343, 9349, 9371, 9377, 9391, 9397, 9403, 9413, 9419, 9421, 9431, 9433, 9437, 9439, 9461, 9463, 9467, 9473, 9479, 9491, 9497, 9511, 9521, 9533, 9539, 9547, 9551, 9587, 9601, 9613, 9619, 9623, 9629, 9631, 9643, 9649, 9661, 9677, 9679, 9689, 9697, 9719, 9721, 9733, 9739, 9743, 9749, 9767, 9769, 9781, 9787, 9791, 9803, 9811, 9817, 9829, 9833, 9839, 9851, 9857, 9859, 9871, 9883, 9887, 9901, 9907, 9923, 9929, 9931, 9941, 9949, 9967, 9973.

There are more primes, but most of them are not as interesting as these.

Camden

1. Prove that the diagonals of a kite are perpendicular.

To show:
$\overline{AC} \perp \overline{BD}$

2. In Fred's monograph in which he listed things that must be true for every parallelogram, he didn't mention that the diagonals of every parallelogram bisect each other. Prove this theorem (and put it in your list of theorems).

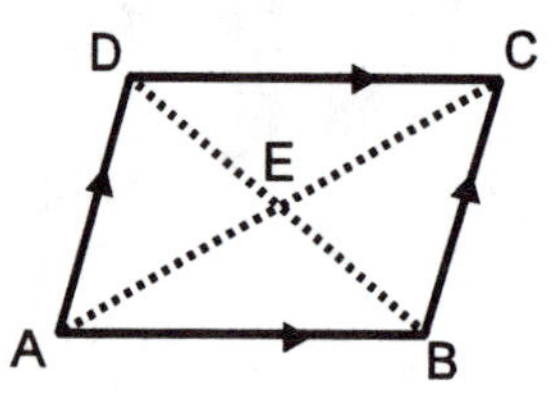

3. In P's essay in which she listed things that are sufficient conditions for having a parallelogram, she didn't mention that if the diagonals of a quadrilateral bisect each other, then the quadrilateral is a parallelogram. Prove this theorem.

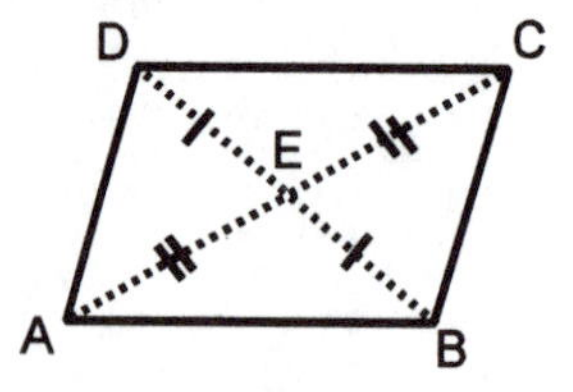

4. How would you show that if *four* parallel lines intercept congruent segments on one transversal, that they will intercept congruent segments on any other transversal? This can be done without resorting to parallelograms, the Midsegment Theorem, or congruent triangles.

In the diagram below, let a, b, c, x, y and z be the lengths of the segments.

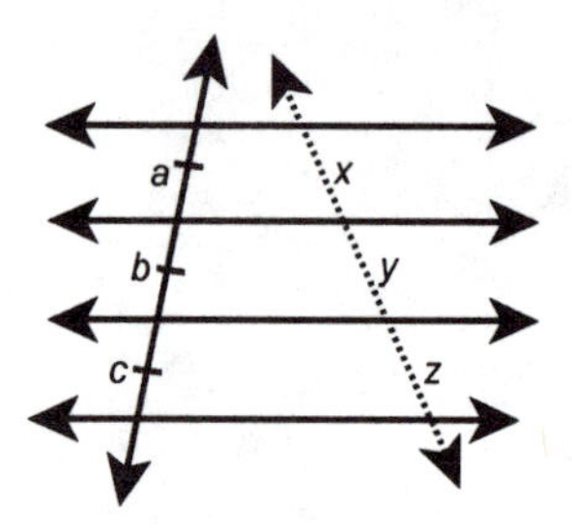

5. Explain in a sentence or two why a quadrilateral in which the diagonals bisected each other, couldn't be a trapezoid.

answers

1.

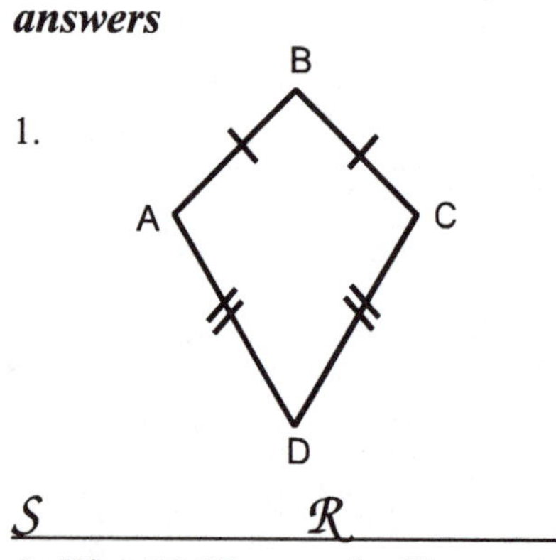

S	*R*
1. Kite ABCD	*1.* Given
2. AB = BC, AD = CD	*2.* Def of kite
3. B and D are on the ⊥ bisector of $\overline{AC}$.	*3.* Converse of the Perpendicular Bisector Theorem (see problems 2 and 3 on p. 174)
4. $\overleftrightarrow{BD} \perp \overline{AC}$	*4.* Two points determine a line.

2.

S	R
1. ▱ABCD	1. Given
2. $\angle 1 \cong \angle 2$, $\angle 3 \cong \angle 4$	2. PAI Theorem

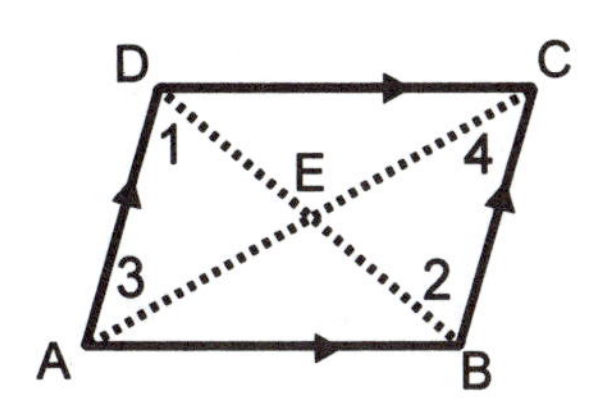

S	R
3. $\overline{AD} \cong \overline{CB}$	3. The opposite sides of a ▱ are congruent.
4. $\triangle AED \cong \triangle CEB$	4. ASA
5. $\overline{AE} \cong \overline{CE}$ and $\overline{DE} \cong \overline{BE}$	5. Def of $\cong$ ▵s
6. The diagonals bisect each other.	6. Def of bisect

3.

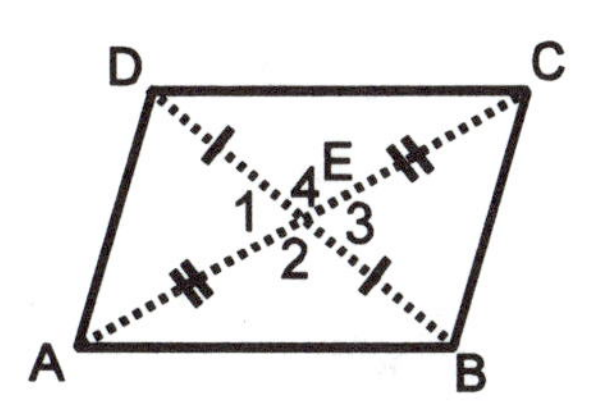

S	R
1. $\overline{DE} \cong \overline{BE}$ and $\overline{AE} \cong \overline{CE}$	1. Given
2. $\angle 1$ and $\angle 3$ are vertical angles. $\angle 2$ and $\angle 4$ are vertical angles.	2. Def of vertical $\angle$s
3. $\angle 1 \cong \angle 3$ and $\angle 2 \cong \angle 4$	3. Vertical $\angle$ Thm
4. $\triangle AED \cong \triangle CEB$ $\triangle DEC \cong \triangle BEA$	4. SAS
5. $\overline{AD} \cong \overline{CB}$ and $\overline{DC} \cong \overline{BA}$	5. Def of $\cong$ ▵s
6. ▱ABCD	6. Thm: If the opposite sides are congruent, then it is a ▱

4. We are given the four parallel lines and a = b = c.

Using a = b, we know that x = y by the Hecks Kitchen Theorem (If three parallel lines intercept congruent segments on one transversal, they will intercept congruent segments on any transversal).

Using b = c, we know that y = z by the same reason.

By algebra we have x = y = z.

5. If the diagonals bisect each other, then it is a parallelogram by the theorem we just proved in problem 3. Therefore it can't be a trapezoid since no parallelogram can be a trapezoid (parallelograms, by definition, have both pairs of opposite sides parallel, and trapezoids, by definition, have exactly one pair of opposite sides parallel).

Garfield

1. If AG = GF and DG = GE, show that $\angle C \cong \angle EFB$. This will either drive you crazy or you will have a flash of insight and write a proof that is only about five steps long.

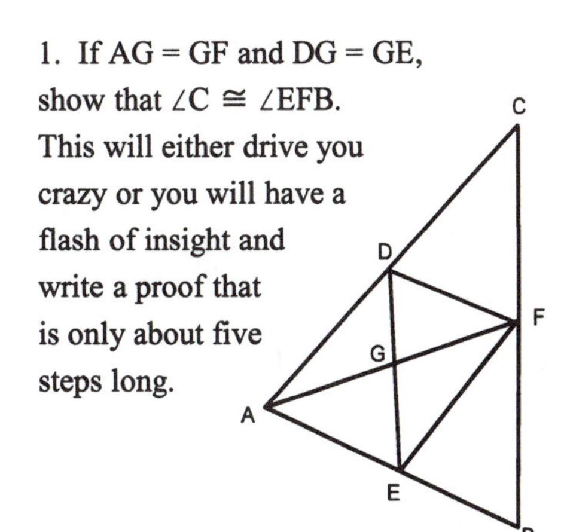

2. When Fred had mentioned in class that you see parallelograms much more frequently than either squares or rectangles, Joe raised his hand and asked, "You say that you like parallelograms because they are so common. Why then don't you like quadrilaterals more than parallelograms?"

Fred smiled. He had hoped he would get that question. He answered, "There are just as many parallelograms as there are quadrilaterals."

The class was quiet. That didn't seem true to them. Parallelograms seem so special in comparison to common old quadrilaterals.

Then Fred explained. "Take any quadrilateral. Inside it, just dying to get out, is a parallelogram. Just connect the midpoints of the sides of any quadrilateral and you get a parallelogram."

Joe headed to the blackboard and drew a bunch of quadrilaterals and Fred filled in the parallelograms.

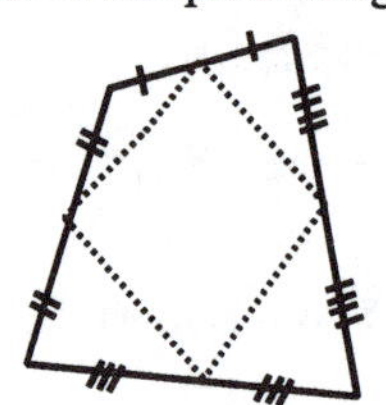

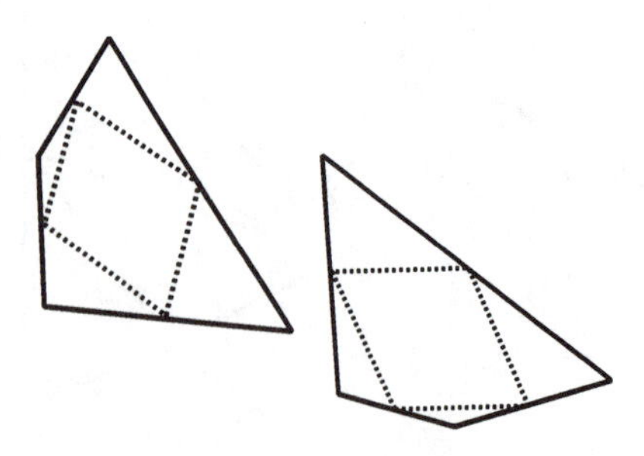

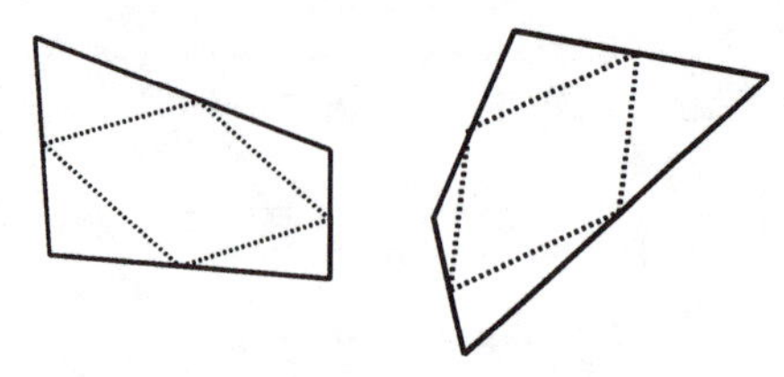

One of the hallmarks of Fred's special way of teaching was the number of surprises that he presented to his students.

"But how can that be true?" Joe asked.

Fill in the missing parts of Fred's proof.

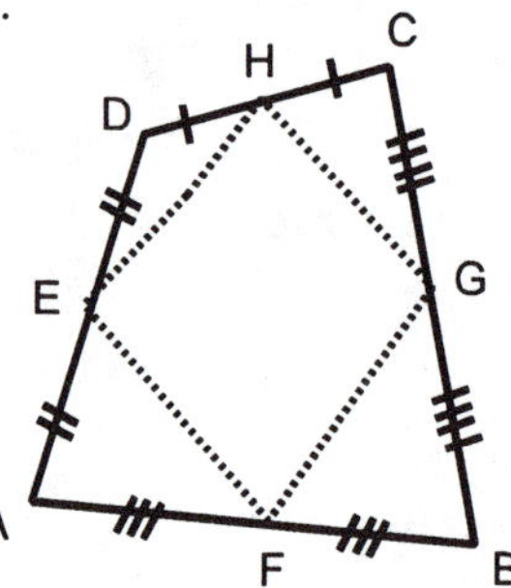

Given the figure as marked. Then draw $\overline{AC}$ by the postulate that two points determine a line.

$\overline{EH} \parallel \overline{AC}$ by __?__

$\overline{FG} \parallel \overline{AC}$ for the same reason.

$\overline{EH} \parallel \overline{FG}$ by the Transitive Property of Parallel Lines Theorem (If two lines are parallel to the same line, they are parallel to each other.)

Repeat the process again drawing $\overline{BD}$ and establishing that $\overline{EF} \parallel \overline{GH}$.

Thus the opposite sides of quadrilateral EFGH are parallel and EFGH must be a ▱ by __?__.

⊠

3. Prove that the base angles of an isosceles trapezoid are congruent.

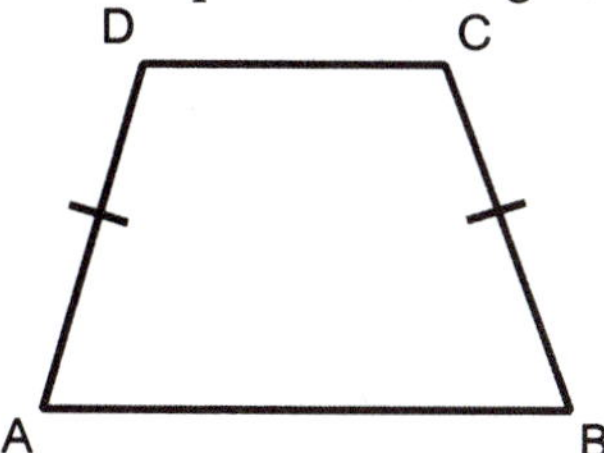

To show $\angle A \cong \angle B$

(Hint: Draw the line through C that is parallel to $\overline{AD}$.)

4. On the U.S.S. Rubber Duck (an isosceles trapezoid), prove that the diagonals are congruent.

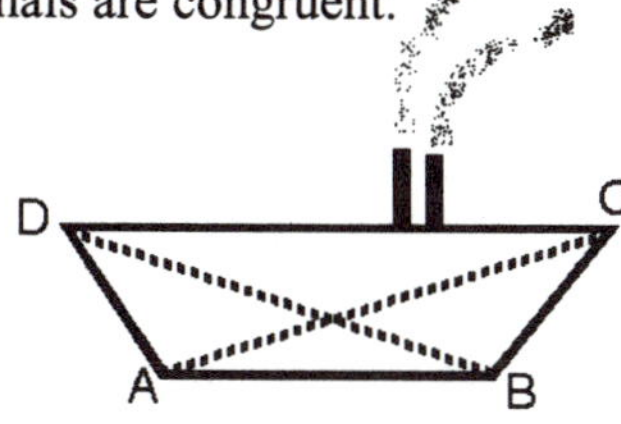

Hint: Use the theorem just proved in the previous problem and ΔDAB will be congruent to ΔCBA by SAS.

5. Draw a diagram to illustrate the proposition: *Through the point of intersection of the angle bisectors of two angles of a triangle, draw a line parallel to the base of the triangle (the side included by the two angles which are bisected). The length of the parallel line intercepted (cut off) by the sides of the triangle is equal to the sum of the two segments of the triangle which are intercepted by the parallel lines.*

answers

1.

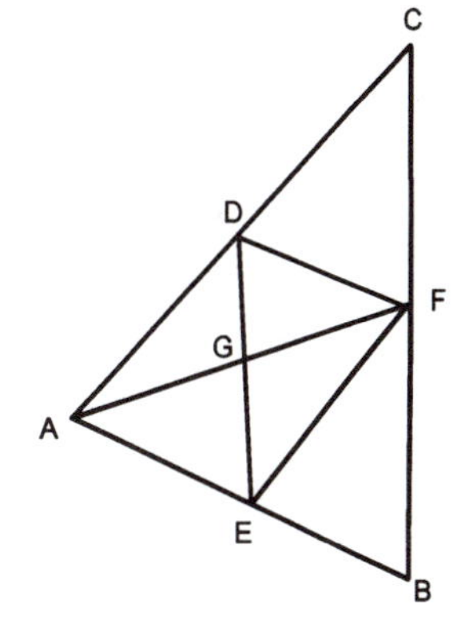

S	*R*
1. AG = GF and DG = GE	*1.* Given
2. ▱ AEFD	*2.* Thm: If the diagonals of a quadrilateral bisect each other, then it is a ▱. (Proved in problem 3 on p. 222)
3. $\overleftrightarrow{AD} \parallel \overleftrightarrow{EF}$	*3.* Def of ▱
4. $\angle C \cong \angle EFB$	*4.* P→CA Thm

2. $\overline{EH} \parallel \overline{AC}$ by Midsegment Thm

Thus the opposite sides of quadrilateral EFGH are parallel and EFGH must be a ▱ by definition of parallelogram.

3.

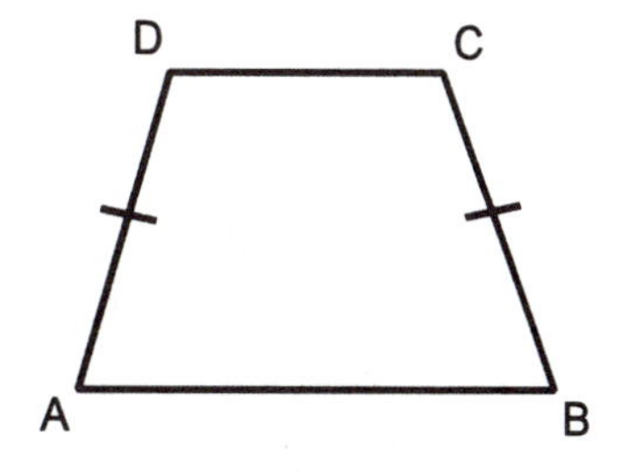

S	*R*
1. trapezoid ABCD with $\overline{AB} \parallel \overline{CD}$ and $\overline{AD} \cong \overline{BC}$	*1.* Given

2. Draw the line through C that is parallel to $\overline{AD}$.

2. Thm: There exists at least one line through a point parallel to a given line.

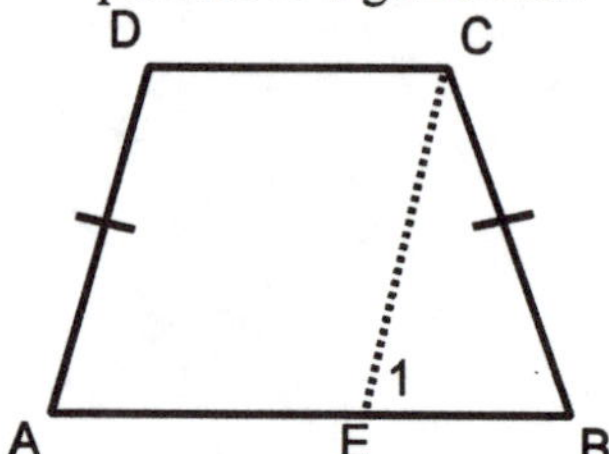

3. $\angle A \cong \angle 1$	*3.* PCA Theorem
4. ▱ AECD	*4.* Def of ▱
5. $\overline{AD} \cong \overline{EC}$	*5.* Thm: Opposite sides of a ▱ are congruent.
6. $\overline{BC} \cong \overline{EC}$	*6.* Transitive Property of ≅ Segments Thm (steps 1 and 5)
7. $\angle 1 \cong \angle B$	*7.* Isosceles △ Thm
8. $\angle A \cong \angle B$	*8.* Transitive Property of ≅ ∠s Thm (steps 3 and 7)

4.

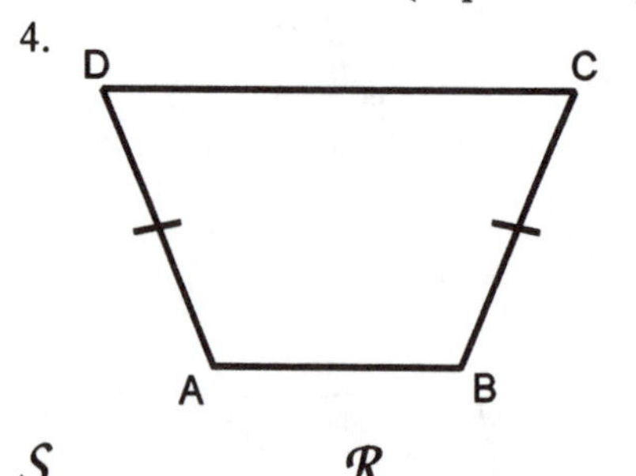

S *R*

1. trapezoid ABCD with $\overline{AB} \parallel \overline{CD}$ and $\overline{AD} \cong \overline{BC}$ *1.* Given

(Note: This is the same given as in the previous problem.)

2. $\angle A \cong \angle B$	*2.* By the theorem just proven in the previous problem.
3. AB = AB	*3.* Algebra
4. $\overline{AB} \cong \overline{AB}$	*4.* Def of ≅ segments
5. △DAB ≅ △CBA	*5.* SAS
6. $\overline{DB} \cong \overline{CA}$	*6.* Def of ≅ △s

5.

To show: DE = AD + BE

Tasco

1. Prove: In any quadrilateral, the segments joining the midpoints of the opposite sides bisect each other. This is a corollary of two previous theorems in Cities. The proof is quite short.

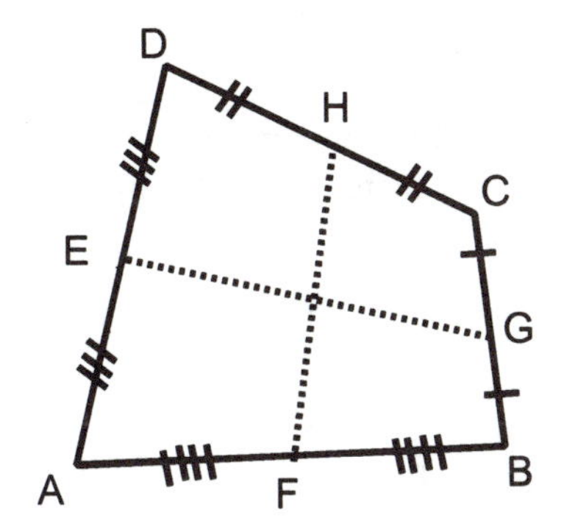

2. Given any $\triangle ABC$ where E is any point on $\overline{BC}$ and D is any point on $\overline{AC}$. Prove that $\overline{AE}$ and $\overline{BD}$ don't bisect each other.

An indirect proof is the easiest approach.

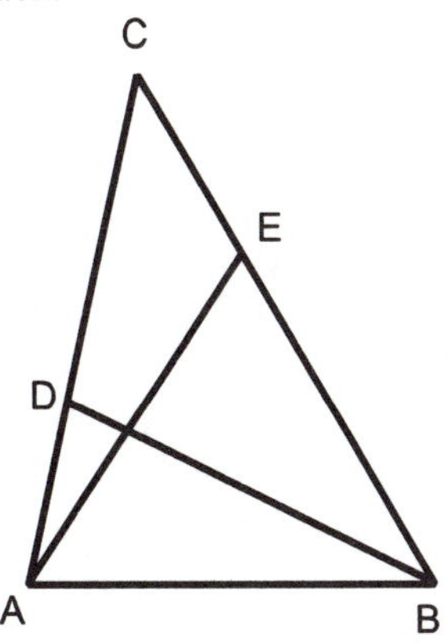

3. Given the figure as marked. Prove that ▱ ABCD.

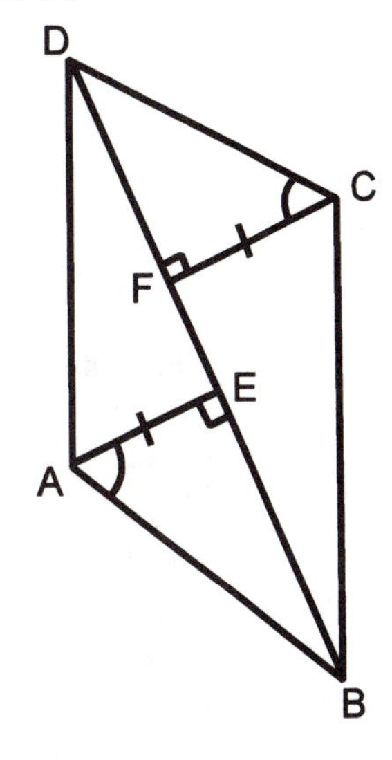

4. Can a trapezoid with two right angles ever be an isosceles trapezoid? Explain your answer.

5. Explain how you would prove that the segment which joins the midpoints of two opposite sides of a parallelogram is parallel to the other two sides. You need not give a full statement–reason proof.

Your explanation can be quite short (2–4 sentences) if you happen to notice something.

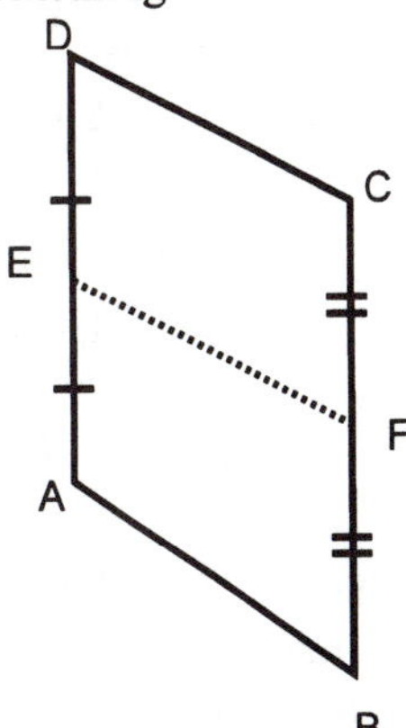

odd answers

1.

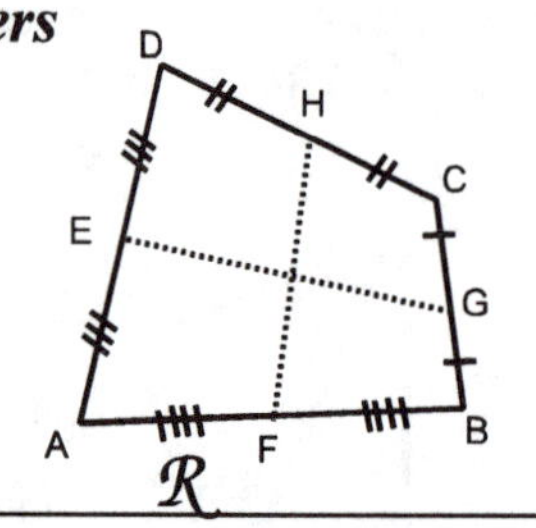

S	*R*
1. The diagram as marked.	*1.* Given
2. ▱ EFGH	*2.* Joining the midpoints of the sides of any quadrilateral gives you a ▱. (Problem 2 of Garfield)
3. $\overline{EG}$ and $\overline{FH}$ bisect each other.	*3.* Thm: The diagonals of a ▱ bisect each other. (Problem 2 of Camden)

3.

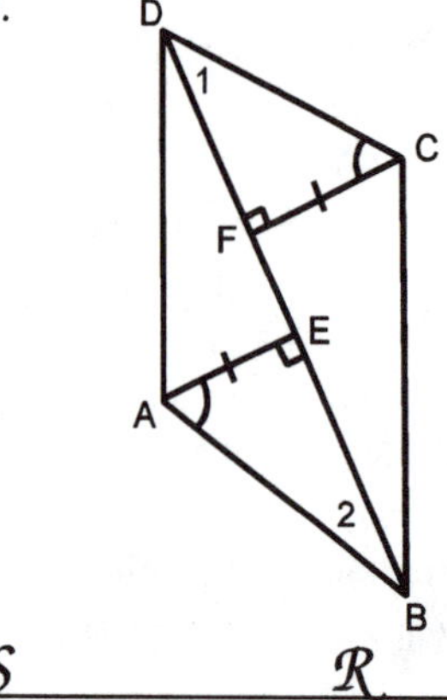

S	*R*
1. The diagram as marked.	*1.* Given
2. $\triangle ABE \cong \triangle CDF$	*2.* ASA
3. $\angle 1 \cong \angle 2$ and $\overline{AB} \cong \overline{CD}$	*3.* Def of ≅ ⚠
4. $\overline{AB} \parallel \overline{CD}$	*4.* AIP Thm
5. ▱ ABCD	*5.* Thm: If two sides are both parallel and ≅, then it is a parallelogram.

5. That's just the Testing the Guarantee Theorem from p. 216.

Garwood

1. Prove the proposition: Through the point of intersection of the angle bisectors of two angles of a triangle, draw a line parallel to the base of the triangle (the side included by the two angles which are bisected). The length of the parallel line intercepted (cut off) by the sides of the triangle is equal to the sum of the two segments of the triangle which are intercepted by the parallel lines.

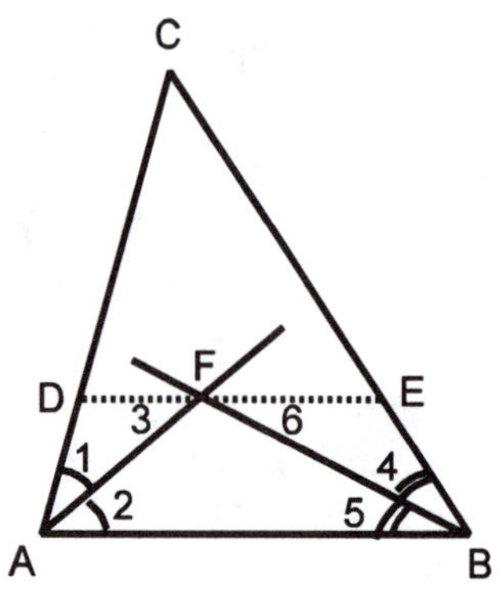

To show: DE = AD + BE

Hint: $\angle 1 \cong \angle 2 \cong \angle 3$.

2. If the diagonals of a quadrilateral are perpendicular to each other and bisect each other, prove that it is a rhombus.

3. Fill in the missing word: If the diagonals of a quadrilateral bisect each other and are perpendicular to each other, then the quadrilateral is a ___?___.

4. The four angle bisectors of a rhombus enclose some kind of quadrilateral inside the rhombus. Make several drawings and furnish

your best guess as what kind of quadrilateral it must be.

5. If $\overline{DE}$ and $\overline{DF}$ are midsegments of ΔABC, prove that $\angle 1 \cong \angle B$.

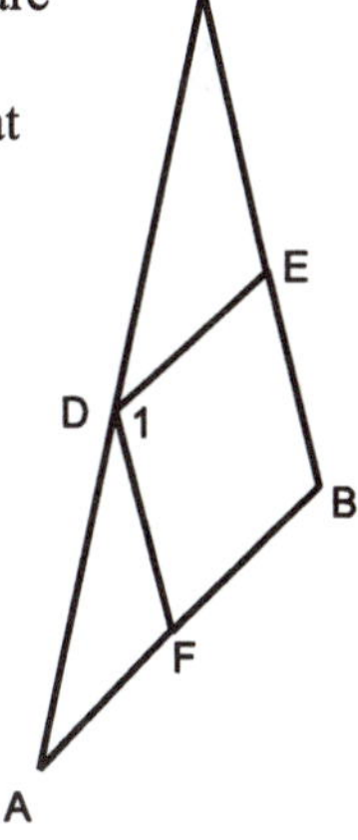

odd answers

1.

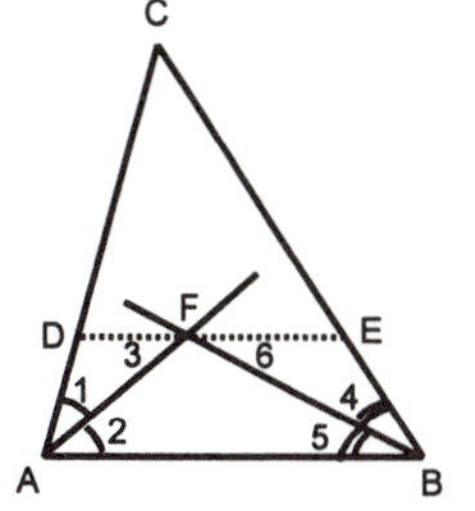

S *R*

1. The diagram as marked and the line through DFE is parallel to $\overline{AB}$.
1. Given

2. $\angle 2 \cong \angle 3$ *2.* PAI Theorem

3. $\angle 1 \cong \angle 3$ *3.* Thm: Transitive Property of $\cong \angle s$

4. $\overline{AD} \cong \overline{FD}$ *4.* Converse of the Isosceles Δ Theorem

(And the same would be done on the right side of the diagram to show that $\overline{BE} \cong \overline{FE}$.)

5. AD = FD and BE = FE
5. Def of $\cong$ segments

6. FD + FE = DE *6.* Def of D–F–E

7. AD + BE = DE *7.* Algebra

3. Rhombus

5.

S *R*

1. $\overline{DE}$ and $\overline{DF}$ are midsegments of ΔABC. *1.* Given

2. $\overline{DE} \parallel \overline{FB}$ and $\overline{DF} \parallel \overline{EB}$
2. Midsegment Thm

3. ▱DFBE *3.* Def of ▱

4. $\angle 1 \cong \angle B$ *4.* Thm: Opposite $\angle s$ of a ▱ are $\cong$.

Ingalls

1. Prove that the diagonals of a rectangle are congruent.

2. Do the diagonals of a kite bisect each other? If your answer is yes, then prove it. If your answer is no, draw a counterexample.

3. Does either diagonal of a kite bisect the angles? If your answer is yes, then prove it. If your answer is that neither diagonal bisects the angles of the kite, then draw a counterexample.

4. Given $\overline{DF} \parallel \overline{AB}$, $\overline{DE} \parallel \overline{BC}$, AD = 16, DC = 16, BF = 18, and AB = 40. Find AE and CF.

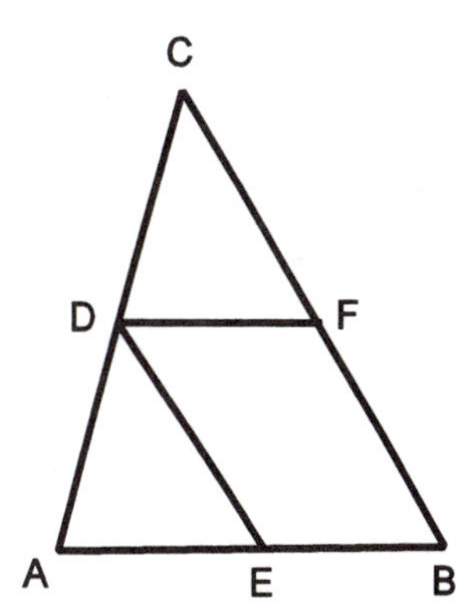

5. Given the diagram as marked and $\ell \parallel m \parallel n$ and BC = 60 and EG = 36, find FG and AD.

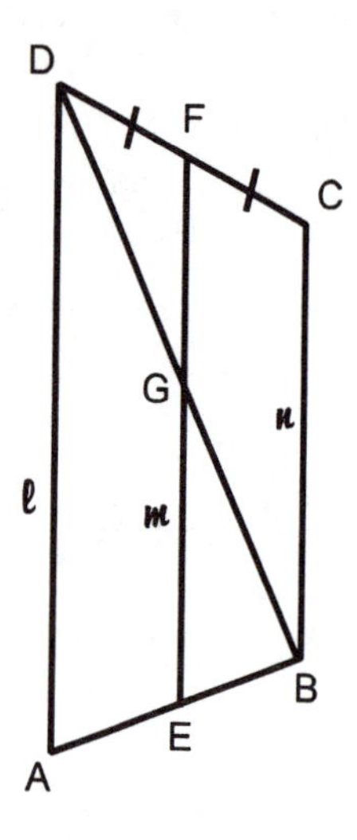

La Mesa

1. Fill in the missing word: If the diagonals of a parallelogram are congruent, then the quadrilateral is a __?__.

2. The four angle bisectors of a rhombus create four triangles. Prove that they are all congruent.

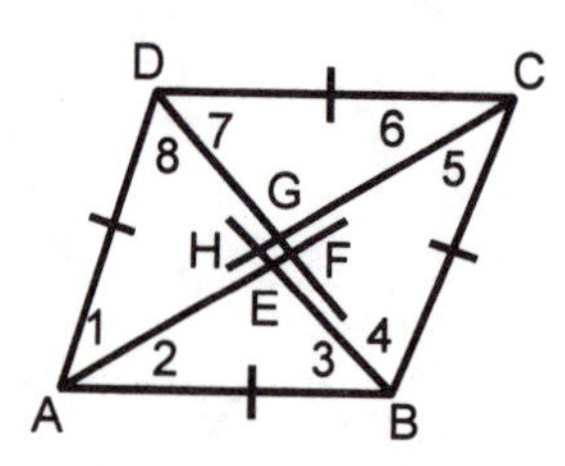

To prove:

$\triangle AEB \cong \triangle CHB \cong \triangle CGD \cong \triangle AFD$

3. Explain how you would prove that the segment which joins the midpoints of the two non-parallel sides of a trapezoid is parallel to the parallel sides. In exchange for not having to give a full statement–reason proof, please give your answer in complete English sentences.

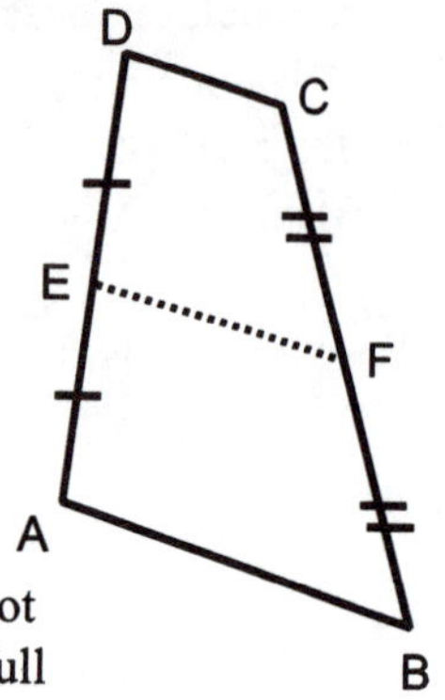

4. If a quadrilateral has one pair of opposite angles congruent and one pair of opposite sides parallel, prove it is a parallelogram.

5. Given the diagram as marked and given $\ell \parallel m \parallel n$. Find x and y.

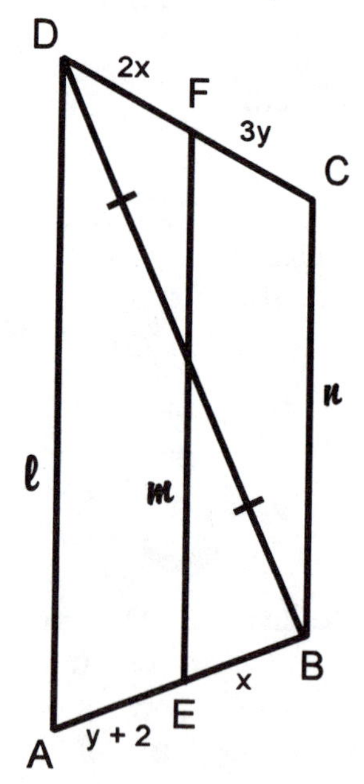

Chapter Seven

Area

Fred was lost in reverie standing outside of Barbecue Basics at the campus store. He stuck a campus mail stamp on the Ralph's bottle and mailed it to his office so that he didn't have to carry it to the awards ceremony which was scheduled to start in an hour.

Do I need a new bow tie? Maybe that would impress her. I could get one with polka dots. That's my favorite color. (Fred wasn't thinking too clearly.)

It's been a long time since I've had a haircut. I wouldn't want to look too shaggy for her. Or maybe a new hairstyle. I wonder which way she would like it best?

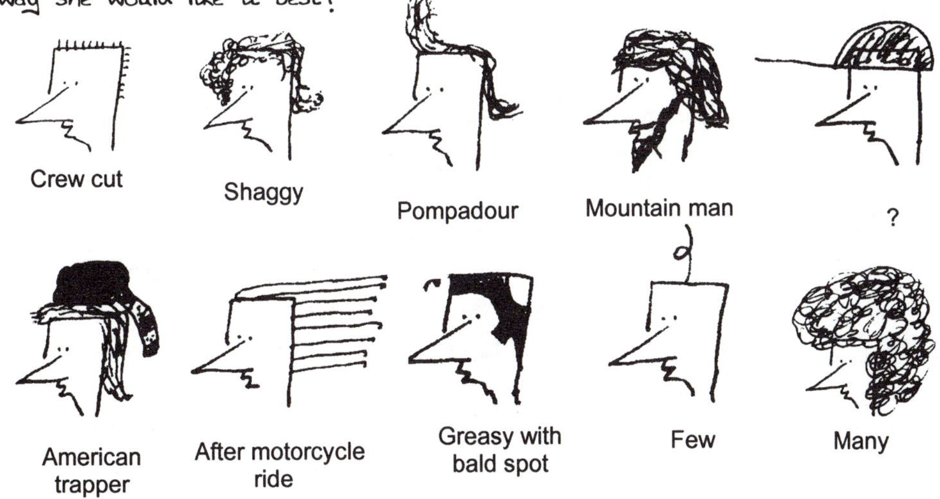

He thought about taking P to meet Betty and Alexander at the pizza place tonight.

I wonder what kind of pizza she likes. If I order a combination, will she want anchovies? Or maybe extra anchovies? There is so much that I want to learn about her.

He remembered the time Betty gave Alexander a rose.

Roses! Yes, roses. I'll have to stop by the campus florist at 4:45 and pick out a bouquet and a corsage and maybe a flower for her hair. And a miniature rose bush for her room.

He never saw them coming.

In the movies they always give chocolates with roses. Does she like milk chocolate or dark chocolate or mint chocolate? Maybe one of each to be sure.

They were walking in his direction.

I remember seeing Nelson Eddy singing a love song to Jeanette MacDonald in the movie "Rose Marie." I wonder if I could write a love song before I see P tonight. Then I could hire a violinist to come to our table and I could sing it to her. It would be called the The Toowoomba Troubadour. It would have a calypso beat.

♪♬ Toowoomba troubadours do tell a tale
Of two who met in Archimedes Hall.
A pizza tryst, perhaps to dance the night
Away. Conviviality! Cha-cha-cha. ♪♬

The last line's cha-cha-cha didn't quite scan correctly, but the rest was in perfect iambic pentameter. This was the first song in recorded human history to include the city of Toowoomba, the Greek mathematician Archimedes, pizza, troubadours, and conviviality.

Wait! Stop! Enough! you the reader exclaim. ***We get the picture that Fred is totally nutso about this six-year-old girl named P. Next thing you'll be telling me is that he plans to propose to her tonight.***

No I wasn't. Fred was thinking a little beyond that.

With the Commonsizerinskis all named with single letters for their first names—I wonder if the Commonsizerinski clan would allow us to name our kids with Greek letters. Our first four children could be named with the first four letters of that alphabet: α(alpha), β(beta), γ(gamma), and δ(delta).

They spotted him.

"Hi Fred!" said Betty.

It was Betty and Alexander holding hands with a little kid between them. The child was crying.

"I wanna go home! (sob!) I wanna see my Daddy and my Mommy! (sniff!)"

Alexander offered the little girl his handkerchief.

(honk!)

It's P!!!!!!

He explained to Fred, "When Betty and I went into the chapel to see the pastor for our premarital counseling session, we found her there in the pews. She told us that she was a new student on campus. We think she's got a terrible case of homesickness."

"Hello Freddy," P said in a sad little voice.

"Oh! You know each other," Betty observed with a smile on her face.

Fred now faced one of the toughest decisions of his life. He could respond in four different ways.

Response #1	Response #2	Response #3	Response #4
The Say-Everything-That's-On-Your-Mind Response ↓	*The Disavowal* ↓	*The Gloss-Over Approach* ↓	*The Minimalist Gambit* ↓
"Hello, my love." Ask about a wedding date and discuss the names of your future children. Then break out in a humongous love song.	"Golly. Have we met before? You must have seen my picture in one of the national magazines. Nice to make your acquaintance."	Turn toward P and say, "Hi. How are you?"	Answer Betty and say, "Yes. We met about an hour ago in Archimedes Hall."
This is often the response of very young children when asked any question. There is (continued) ↓	*Peter did this to Jesus.*	*This is the favorite approach of politicians. The trick is to not pay attention to the* (continued) ↓	*Tact, decorum, truth . . . need we say more?*

no conversational filter installed in their brains yet and they let it all hang out.

question. If you're asked if you ever turned over military secrets to the Chinese communists, you answer, "I really do enjoy Chinese food."

Betty continued, "We were going to drive her over to her parents' house. They just live over on Tangent Road. Would you like to come along?"

Fred was overjoyed. "Yes, I think it's about time I meet her parents."

Neither Betty nor Alexander (nor P) understood Fred's response.

They had no trouble finding the right house.

(Their sign required an easement onto their neighbor's property.)

I jogged by here this morning with Lambda and I never realized that this was the house of my future in-laws.

Mrs. Commonsizerinski met them at the door. P ran by them all and headed upstairs to her bedroom and shut the door.

"Won't you come in?" Mrs. Commonsizerinski asked her visitors.

Betty and Alexander offered their regrets since their pastor was waiting for them and they had to leave.

Fred walked in.

He had his speech all prepared. First he was going to introduce himself. Then tell P's parents how they had met, although he intended to leave out the part about P finding D Commonsizerinski's letter between his toes. Then he would segue (pronounced SEG-way) to the fact that they were both exactly six years and six days old. After establishing that they were ideally suited for each other age-wise, he would ask Mr. T Commonsizerinski for the hand of his daughter.

Before he could get a word out, P's mom asked, "Would you care to join us for lunch? It's a shame that your parents had to leave."

Fred blushed. Betty and Alexander are old friends. They used to be my students. What can I say that won't get me into trouble? Fred took a wild stab at it, but forgot to express aloud the part he had been thinking. He just began with, "They had to leave. They are going to premarital counseling. They're getting married this summer."

Now it was her turn to blush. She quickly changed the subject, "How do you like our square house?"

Personally, I prefer parallelograms. "Oh, that's quite unique. Why did you come to choose a square house?" Fred was learning to be a proper conversationalist.

"It was the math," she answered. "In order to save money when this house was built for us, we had to paint the front of the house ourselves. So we had to figure the area of the house so we could buy the right amount of paint. The area of a square is the only formula that we could remember, so we ordered a square house."

Something seemed strange to Fred. They had a daughter who finished high school at the age of four and yet her mother knew only one of the area formulas. It's time we take a break.

Your Turn to Play

Here are the area formulas. The letter *h* stands for height. (Pronunciation note: height doesn't rhyme with depth and width. It is pronounced "hite." Here's four lines in anapestic monometer to help you remember: *It is might*

That makes right

Is the height

Of brain blight.

Area of a triangle = ½ bh

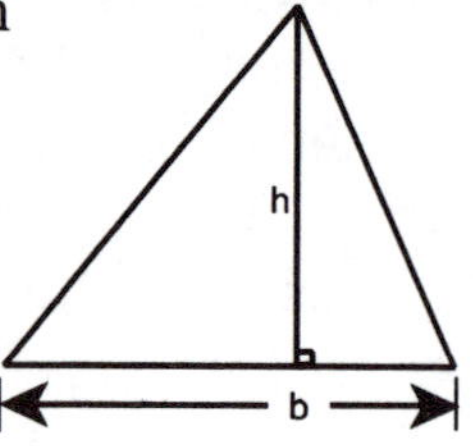

Area of a parallelogram = bh

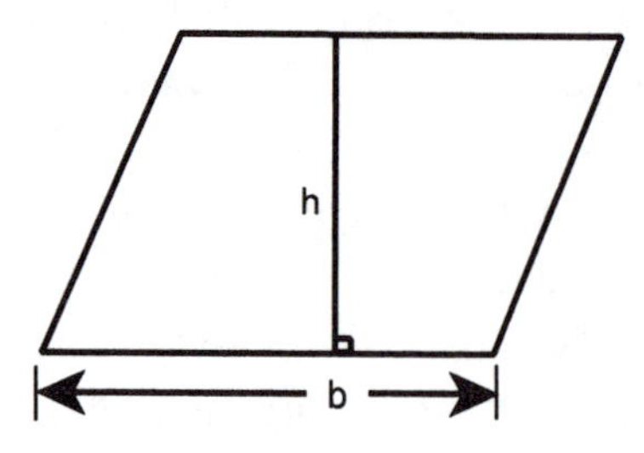

Area of a rectangle, rhombus, and square are all equal to bh since they are all parallelograms.

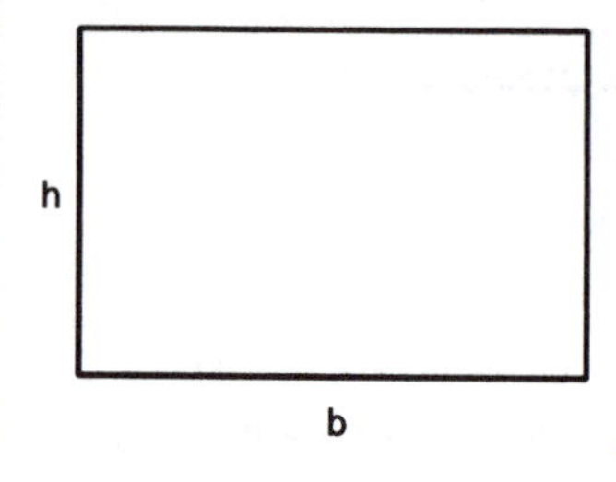

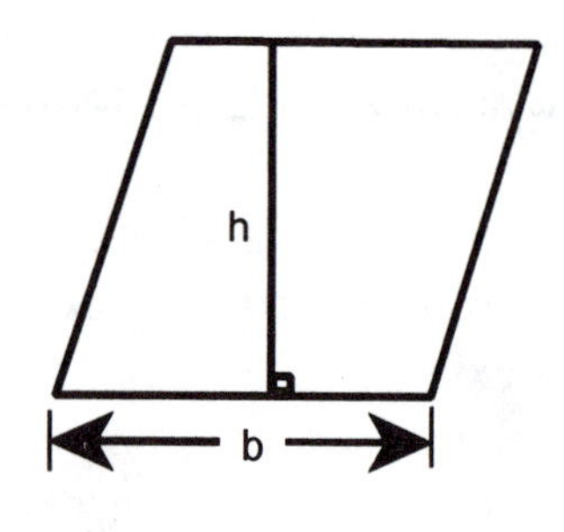

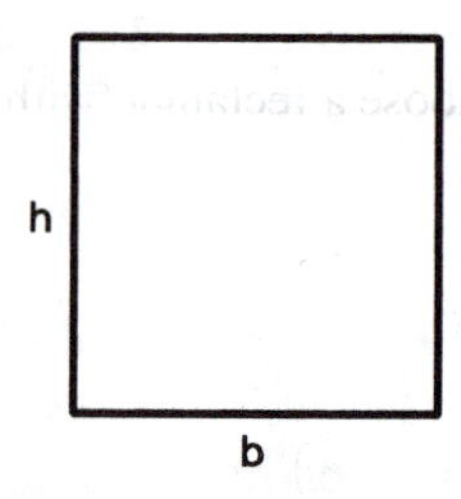

Area of a trapezoid = ½(a + b)h.

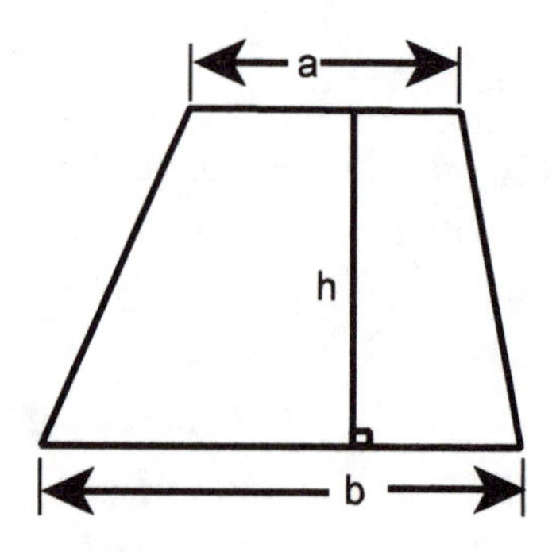

1. If the Commonsizerinski dinning room table is a rectangle that is 2 feet by 4 feet, what is its area?

2. If the front of their house is 25 feet wide and 25 feet tall, what is its area?

3. If you take a square that is 6 feet on each side, and you sit on the top of it and squish it slightly, can you determine its area?

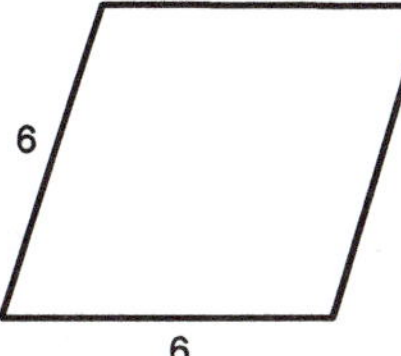

4. The **perimeter** of a triangle or a quadrilateral is the distance around the outside. It is the sum of the length of the sides. If a square has sides of length *s*, what is its perimeter and its area?

5. If a rectangle has a short side equal to *w*, and a long side equal to *l*, what is its perimeter and its area?

6. Your rich uncle had a million acres of flat land on the Great Plains. When he died, his will stated that you would receive a "rectangle of land" (those were his words), that had a perimeter of 36 miles. You would be permitted to choose the exact shape.

You went to work immediately making a chart to figure out the possibilities. Your first thought was to choose a rectangle with a width of 2 miles. That would make a nice long bowling alley. You filled in the first line of the chart with a width of 2. In order that the perimeter would equal 36 miles, you reasoned that if the width of one side of the rectangle were 2 miles, then the other width would also have to be two miles. That leaves 36 – 2 – 2 = 32 miles for the combined lengths. That would make each length equal to 16 miles. Here's the first line of the chart as you filled it out

This Land Will Be My Land

Proposed width	Resulting length	How much area will I get?
2	16	2 × 16 = 32 square miles of glorious land

Then you wondered if you might get even more area using a larger width. Fill in the blank spaces on this new expanded chart:

This Land Will Be My Land

Proposed width	Resulting length	How much area will I get?
2	16	2 × 16 = 32 square miles of glorious land
3	?	?
4	?	?
1	?	?
5	?	?
6	?	?

7. Fill out the blank spaces in this chart:

This Land Will Be My Land

Proposed width	Resulting length	How much area will I get?
8	?	?
8.3	?	?
x	?	?

(The answers in the last row will be expressions containing x.)

8. What is the best choice of width so that you will receive the largest area of Great Plains land?

9. Everyone knows that 5 inches plus 7 inches equals 12 inches which is 1 square foot. Does 5 square inches plus 7 square inches equal 1 square foot?

10. How many square feet are in a square yard?

11. How many square feet are in a square mile? (1 mile = 5280 feet)

1. Since the area of a rectangle is base times height, the area of their dinning room table is 2 feet × 4 feet = 8 square feet.

2. The area of a square is base times height.

Sometimes this is written as $A = s^2$.

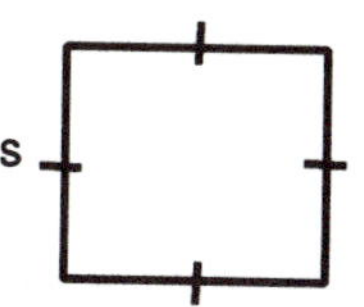

Area of their square house = 25^2

= 625 square feet.

3. If you squish is just a little bit, its area doesn't decrease very much. When the rhombus is almost a square, the area might be 35 square feet.

But if you take that square and collapse it almost to the ground, its area might be only one square foot.

4. For a square: $p = 4s$ and $A = s^2$

5. For a rectangle: $p = 2w + 2l$ and $A = lw$.

6.

2	16	32 square miles of glorious land
3	15	45 square miles of glorious land
4	14	56 square miles of glorious land
1	17	17 square miles of glorious land
5	13	65 square miles of glorious land
6	12	72 square miles of glorious land

7.

8	10	80 square miles of glorious land
8.3	9.7	80.51 square miles of glorious land
x	½(36 − 2x)	x(½)(36 − 2x) square miles of glorious land

8. If you play around with the above chart and keep increasing the proposed width, you'll find that a width of 9 miles, which gives a resulting length of 9 miles will give you 81 square miles of glorious land. This is the largest area you can inherit.

9. Certainly, 5 square inches plus 7 square inches equals 12 square inches. The question is whether 12 square inches equals 1 square foot. In the diagram below, let each square represent 1 square inch. The whole diagram is 1 square foot. Five square

inches and 7 square inches have been darkened in. There are 12^2 *(twelve squared)* square inches in a square foot.

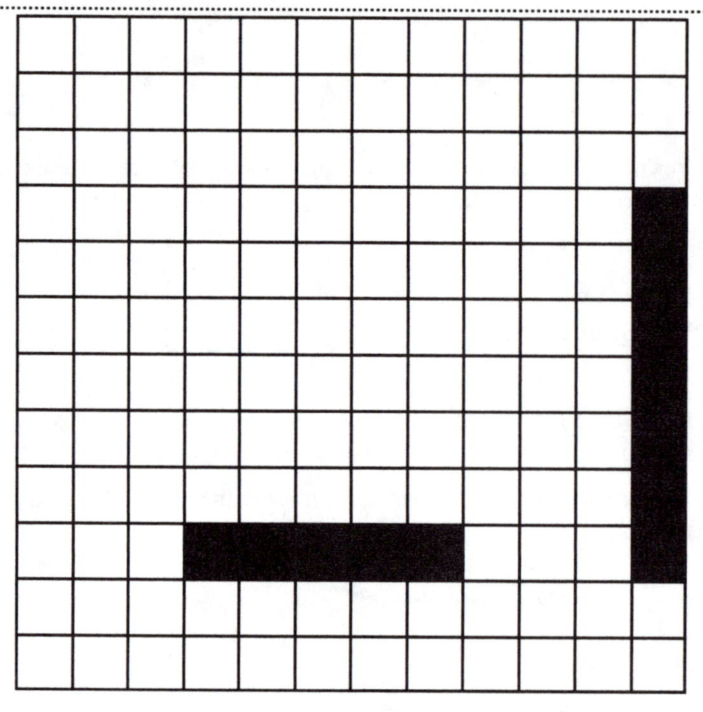

10. There are 9 square feet in a square yard.
11. There are 5280^2 square feet in a square mile. (= 27,878,400)

"Would hot dogs be okay for you for lunch?" Mrs. Commonsizerinski asked. "That's what we usually have on Thursdays."

Wonderful! Fred thought. *Then the $2500 I spent on my new Hecks kitchen grill wasn't a waste.* "Oh yes. That would be great."

She pulled a little grill out of the closet and poured in a cup of charcoal. "How many hot dogs are you good for? You're a growing boy. Will it be three or four? Don't be bashful."

Fred, who had never finished even one in his life, couldn't get the words out. He just held up his index finger. He needed to change the subject before she asked him about some potato salad, an apple, a large size (44 oz.) glass of prune juice, some homemade chocolate chip cookies, and a piece or two of her pumpkin pie. "I have a question, Mrs. Commonsizerinski. Your daughter's first name is P. And you and your husband's first names are L and T. What does P stand for? What's her full name?"

L was used to being asked that question. "Since we have such a long last name, it's become a family tradition to have very short first names. Do you ever remember hearing about a U.S. president named Truman?"

History was one of Fred's strong points. He had read a lot of United States history on his own since he had had very little formal schooling before he began teaching at KITTENS years ago. "Sure, I remember reading about Harry Truman. He was president from 1945 to 1953. And I know exactly what you're thinking. His middle name was S and that wasn't an initial but his whole middle name."

"Well," said L, "that's the idea, except that we Commonsizerinskis have been doing that for generations. Have you ever seen the cemetery on the south side of town? The Commonsizerinski section is something to behold."

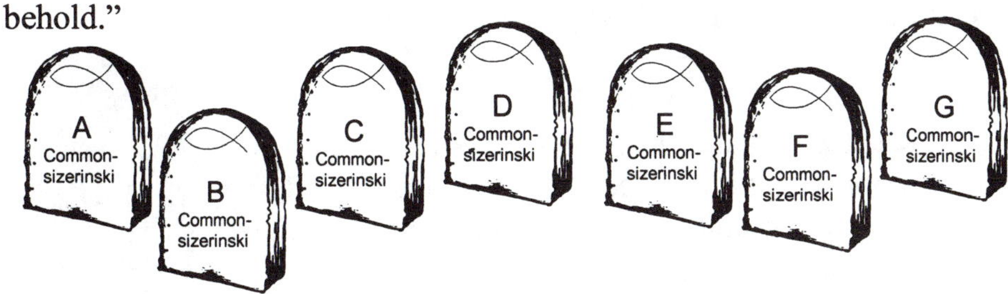

Fred was wondering if this was the right time to introduce to L his idea of naming Fred and P's children with Greek letters. He also thought about his own name: Fred Gauss. He had never been given a middle name. He imagined: Fred A Gauss; Fred B Gauss, and Fred ABCDEFGHIJKLMNOPQRSTUVWXYZ Gauss. *If I change my name, maybe I would be more accepted by the Commonsizerinski clan.*

Mrs. Commonsizerinski threw six hot dogs on the grill (one for herself, one for her husband, one for P and three for Fred). "Would you please go and tell T and P that lunch is almost ready. He's in the library and she's probably upstairs in her room."

The library! On a Thursday morning? I hope I'm not disturbing his studies. I don't want to start off on the wrong foot. I'm only six years old. If he's been studying all his life like this, he'll be a lot more educated than I am. I know about Harry S. Truman, but what

if he wants to talk about Ur
(an ancient Sumerian city where Abraham
came from) or maybe ogee arches
which are arches that look like →

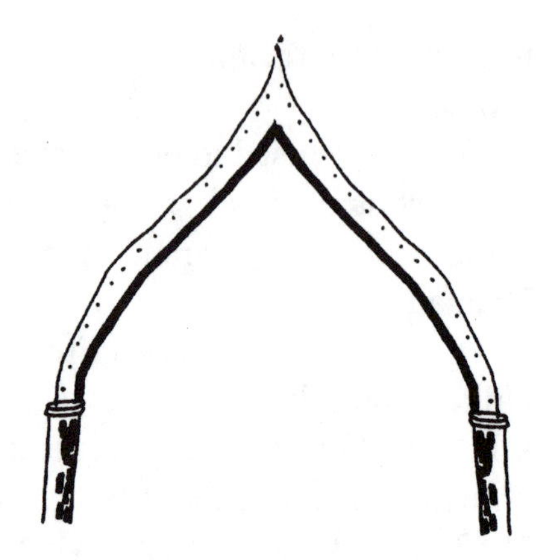

Wait! Stop! Forget Ur and ogee arches. You stuck a period after the "S" in Harry S Truman's name! What's going on here?

Could we handle this in a footnote? We're at an important part of the story, and we don't want to interrupt things. Fred is approaching his potential future father-in-law and this is an important moment in his life.

Sure. Give me a footnote.*

Fred stood outside the door and hesitated. The top of the door looked a bit crooked. The door was shaped like a trapezoid. Fred could hear muffled sounds coming from within the library.

Maybe this isn't a good time to disturb him.

Door to
The Library

Your Turn to Play

1. The iibrary door measured 28" across the bottom edge. The left edge was 78" and the right side was 80". What is the area? (You may assume that the bottom of the door is not crooked.)

2. Make a guess: What is the length of the segment which joins the midpoint of the top edge of the door with the midpoint of the bottom edge?

✶ This is weird. Even though Mr. Truman's middle name is just S, the *U.S. Government Printing Office Style Manual* declares that his name will be spelled Harry S. Truman. So when you visit the museum and library dedicated to him, you will find it named "The Harry S. Truman Museum and Library."

3. In the diagram, let m = the length of the midsegment of the triangle. What does m equal in terms of b and h?

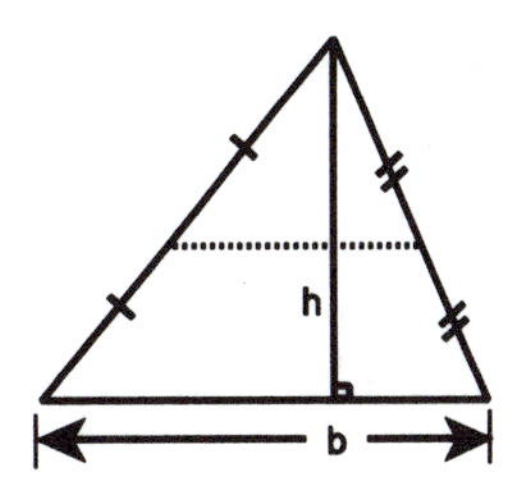

4. What is the area of the above triangle in terms of m and h?

5. In the diagram, let m = the length of the midsegment of the trapezoid. What does m equal in terms of a and b?

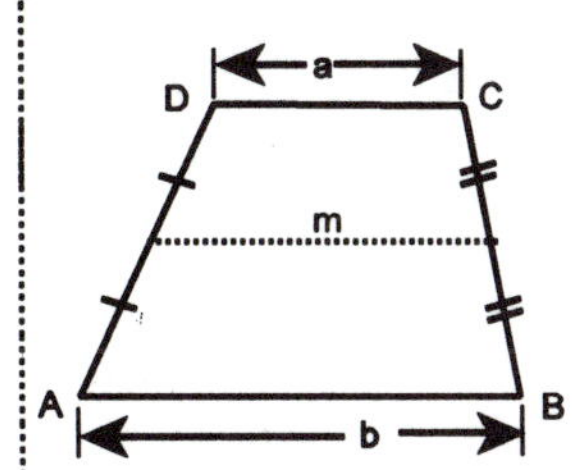

Hint: Draw $\overline{AC}$. That forms $\triangle ACD$ and $\triangle ACB$. Use the Midsegment Theorem in each triangle.

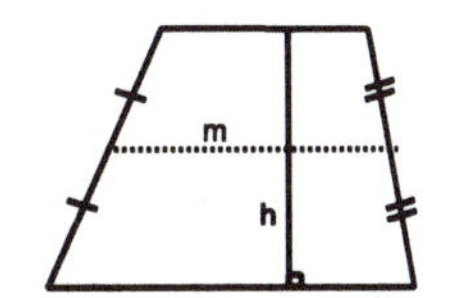

6. What is the area of the trapezoid in terms of m and h?

7. What is the area of a rectangle in terms of m and h?

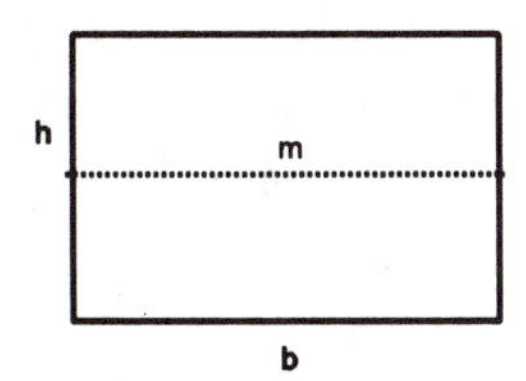

8. Is A = *mh* the correct area formula for triangles, trapezoids, rectangles, rhombuses, and squares? *Great leaping ogee arches!*—a common exclamation among some people—could it all be *that* simple: A = *mh*?

COMPLETE SOLUTIONS

1. Since the bottom edge of the door isn't crooked, the 28" will be the height of the trapezoid. The two parallel edges are 78" and 80". A = ½(a + b)h = ½(78 + 80)28 = 2212 square inches.

2. If you guessed that the length of that segment was 79", you could not have had a better guess.

3. The Midsegment Theorem states that in any triangle, a midsegment is parallel to the third side and is equal to half of the length of the third side. Therefore, $m = \frac{1}{2}b$.

4. $A = mh$

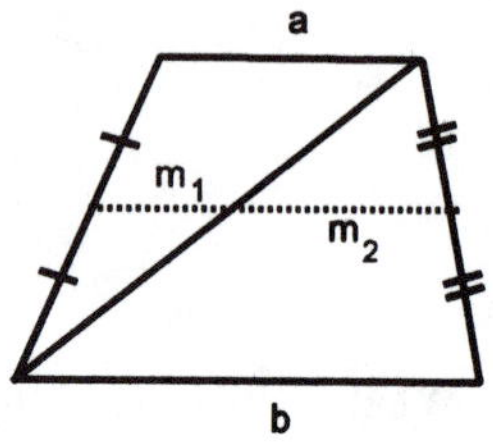

5. After drawing $\overline{AC}$ the midsegment is divided into two smaller lengths, m_1 and m_2.

$$m = m_1 + m_2$$

m_1 is half of the length of a by the Midsegment Theorem.

$$m_1 = \frac{1}{2}a$$

Similarly, $m_2 = \frac{1}{2}b$

Adding these equations: $m_1 + m_2 = \frac{1}{2}a + \frac{1}{2}b$

$$m = \frac{1}{2}a + \frac{1}{2}b$$

Using the distributive law: $m = \frac{1}{2}(a + b)$

6. The usual formula for the area of a trapezoid is $A = \frac{1}{2}(a + b)h$.

By the formula we have $A = mh$.

7. $A = mh$

8. yes

Taking a big swallow, Fred knocked and entered the library. T was sitting in front of a television set watching reruns of a roller derby. The room was filled with stacks of old newspapers and magazines.

"Sit down boy and enjoy the show!" Mr. Commonsizerinski said.

There was no other chair in the room so Fred moved a stack of magazines ("The Weekend Miniature Golfer") near T's chair and sat down. Fred had never seen roller derby before. They must not be very good skaters since they keep bumping into each other.

Fred waited until a commercial break and then told T that lunch was almost ready. When T got up, Fred knew that this was his opportunity to see P. It seemed like it had been forever since he had spoken to her. He was also curious to see what her room looked like. He wanted to know everything he could about her.

When they left the library, Fred saw that P was already down in the kitchen helping her mother set the table. L got an extra chair from the living room so that Fred would have a place to sit. She got three phone books to put on Fred's chair. P's chair already had three phone books on hers.

They all washed up and came to the table. When they said grace, P prayed that she wouldn't be so scared being at the University.

When L passed Fred the apple-cinnamon biscuits he took the smallest one he could find and put it on his plate next to the three hot dogs. "Well, son," Mr. Commonsizerinski began. "Do you live around here?"

"Over in room 314," Fred answered. "I used to live with Lambda, but she died this morning."

L, having seen what she thought were Fred's parents, assumed that Lambda was probably Fred's aunt.

Mr. Commonsizerinski had a different understanding of Fred's answer. Since Fred had given a room number, T figured that Fred and his parents were so poor that they lived in one of the rooms at the local temporary housing for the homeless. He responded, "Sometimes you just have to tough it out."

Fred thought T was talking about Lambda's death.

In short, no one was communicating very effectively.

Fred played with his food. He was never one to gobble down his food. He cut the hot dogs lengthwise and arranged them in a large polygon on his plate.

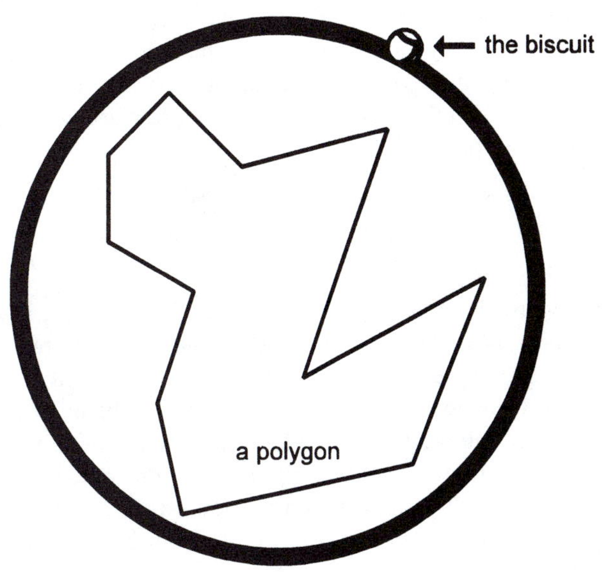

A **polygon** can be either a triangle, or a quadrilateral, or a five-sided figure (a pentagon), or a six-sided figure (a hexagon), or even more sides. Fred had created an 11-gon with the hot dogs on his plate.

The official definition of a polygon is fairly difficult to express. For most geometry students, they should put on their definitions page: "Definition: A **polygon** is a triangle, quadrilateral, pentagon, hexagon, etc." We'll put the official definition in a footnote.*

Fred looked over at P. Apparently she also wasn't a big eater. She had sliced up her hot dog lengthwise and took the pieces and transferred them over to Fred's plate. She turned his polygon into a whole bunch of triangles. (Her hot dogs are the dotted segments.)

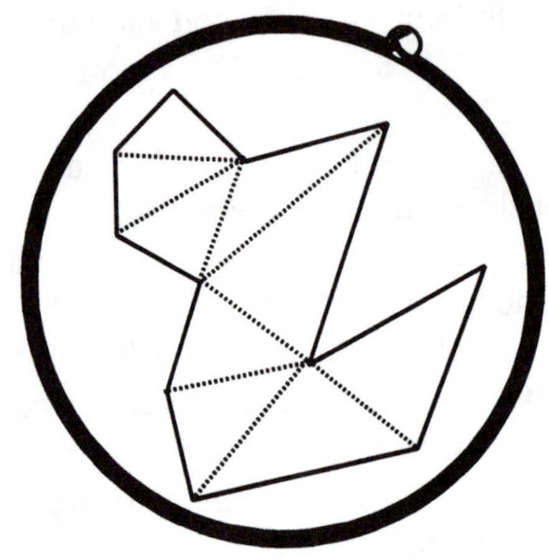

"You know that if we add up the areas of the triangles, we could have the area of the polygon," she said to Fred.

---

* The definition of a polygon with n sides: Take n points, $A_1, A_2, A_3, \ldots, A_{n-1}, A_n$ that are all distinct (different from each other). They will be the vertices (singular: vertex) of the polygon. A polygon is the union of the segments $\overline{A_1A_2}$, $\overline{A_2A_3}$, $\overline{A_3A_4}$, . . . , $\overline{A_{n-1}A_n}$ and $\overline{A_nA_1}$ provided that none of the segments intersect except at their endpoints. We ask that the segments only intersect at their endpoints so that we don't get polygons that look like:

A word about the subscripts on the letter A. We could have defined a polygon with vertices A, B, C, D, . . . , but we would have run into real trouble if the polygon had more than 26 sides. A_1, A_2 and A_3 are three different points like A, B and C. Now, if we want to name a million points, that's no problem: $A_1, A_2, A_3, \ldots, A_{999999}, A_{1000000}$. If someday you want to have a lot of children but don't want to be bothered with trying to figure out names for all of them, you could call them: kid_1, kid_2, kid_3, kid_4, kid_5 . . . . Actually, there are quite a few advantages to this naming system. You could announce, "Okay, odd numbers, go do the dishes. Even numbers, go clean up your rooms."

Fred rearranged her hot dogs.

"Of course," said Fred, "it wouldn't matter how you triangulated my polygon. The sum of the areas of the triangles will always equal the area of the polygon." Fred was in his element now. He was glad he wasn't being quizzed about ogee arches. "But since we officially don't know anything about area, we'll need some beginning assumptions in order that our geometry theory may include area."

On a paper napkin he wrote:

Postulate 10: (Area Postulate) Every polygon has a number attached to it which we will call its area. These areas will be positive numbers.

Postulate 11: If two triangles are congruent, then their areas are equal. (In symbols: If $\triangle ABC \cong \triangle DEF$, then $a\triangle ABC = a\triangle DEF$. "$a\triangle$" is the area associated with the triangle. We did the same thing with angles: $m\angle$ is the angle measurement associated with the angle.)

Postulate 12: (Area Addition Postulate) The area of a polygon is the sum of the areas of the (non-overlapping) triangles inside of it, and it doesn't matter which way you cut up the polygon into triangles.

Ha! I caught you! I know that P is too polite to object, but I, your ever vigilant reader, won't stand for this nonsense. You talked about the triangles that were <u>inside</u> of the polygon. Do you remember what the undefined terms of geometry are?

Sure. They are point, line, and plane.

Well. Did you ever define what it means for a point to be <u>inside</u> of a polygon?

Hey. This is a beginning geometry book. If I were to give an official definition of what it means for a point to be inside of a polygon, I would blow away most of my readers. After all, isn't it obvious which points are inside of a polygon? What could be more simple?

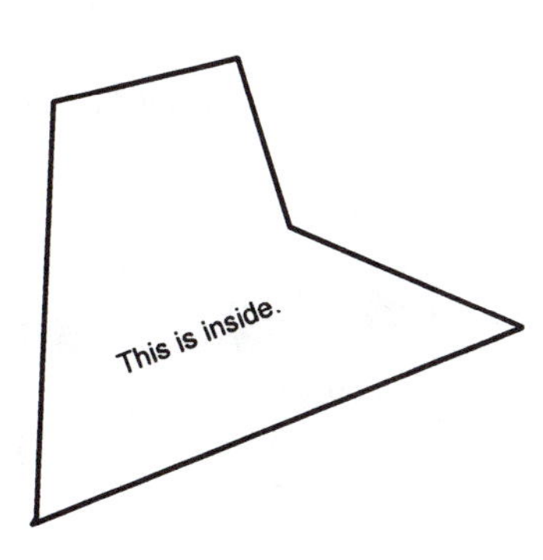

Okay, bright guy. Here's a polygon. Is it so simple to tell me whether the point I've marked with a star ☆ is inside or outside this polygon?

Okay, cough up the definition and only use either words we've already defined or the undefined terms. Don't say something like, "A point is inside a polygon if you can't wander around and find your way to the outside."

I only ask one thing. Let me put the definition in a box and leave instructions on the outside to warn normal readers away from it.

What!? Are you calling me abnormal?

No. Not at all. You just happen to be a very bright and picky reader who won't let me get away mentioning the **interior of a polygon** without demanding that I define it.

Okay. Make your box and issue your warnings.

You need not put this definition on your definitions page.

How to Tell If You're Inside of a Polygon

Definition: Let $A_1A_2A_3...A_n$ be a polygon and let ℓ be any line that does not intersect the polygon $A_1A_2A_3...A_n$. A point P is said to be in the **interior of the polygon** iff there exists a segment $\overline{PQ}$ where:

i) Q is on ℓ,

ii) $\overline{PQ}$ does not intersect any of the vertices of $A_1A_2A_3...A_n$, and

iii) $\overline{PQ}$ intersects the sides of $A_1A_2A_3...A_n$ an odd number of times.

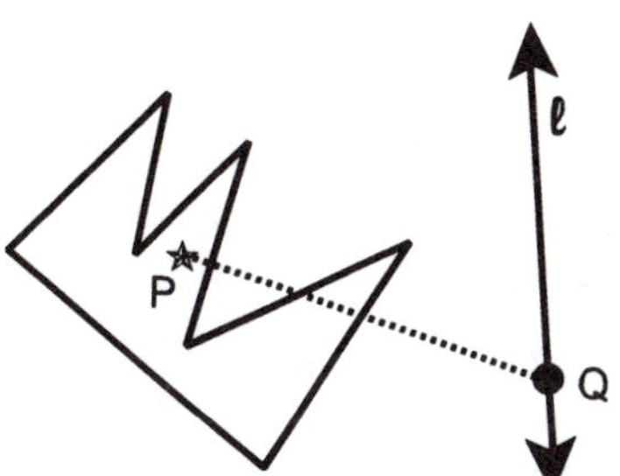

Some notes:

♪#1: Since ℓ doesn't intersect $A_1A_2A_3...A_n$, it is definitely outside of the polygon.

♪#2: In the diagram $\overline{PQ}$ intersects $A_1A_2A_3...A_n$ three times. If Q were located "higher" on ℓ, then $\overline{PQ}$ might intersect $A_1A_2A_3...A_n$ only once. But no matter where Q is located on ℓ, (as long as $\overline{PQ}$ doesn't intersect any of the vertices of the polygon) $\overline{PQ}$ will intersect $A_1A_2A_3...A_n$ an odd number of times.

♪#3: This is a definition, not a theorem. It doesn't have to be proved.

♪#4: We could now define the **exterior of a polygon** to be any points that are neither in the interior nor on $A_1A_2A_3...A_n$.

Hey! Wouldn't it be simpler to first define the points that are on the exterior of a polygon to be any points that lie on any line that doesn't intersect the polygon? Then you could define the points on the interior to be any points that are not on the exterior or on the polygon.

No.

What do you mean, "No"?

I mean that your definition of the exterior points of a polygon won't always work.

Show me.

Here's a point P that is clearly on the exterior of the polygon and yet you can't draw a line ℓ containing P that doesn't intersect the polygon.

Well, thanks for humoring my request for a definition of the inside of a polygon. I'm glad to know that you were up to the task. Now let's get back to the regular geometry.

You're welcome.

P took the biscuit off of Fred's plate and put it on the table. After pounding it flat, she took her knife and cut it into a parallelogram.

Fred's biscuit-less plate

Fred wasn't the only one who was a light eater.

She pushed the hot dog slices off his plate and plopped her parallelogram in front of him. "Okay, what's the area?"

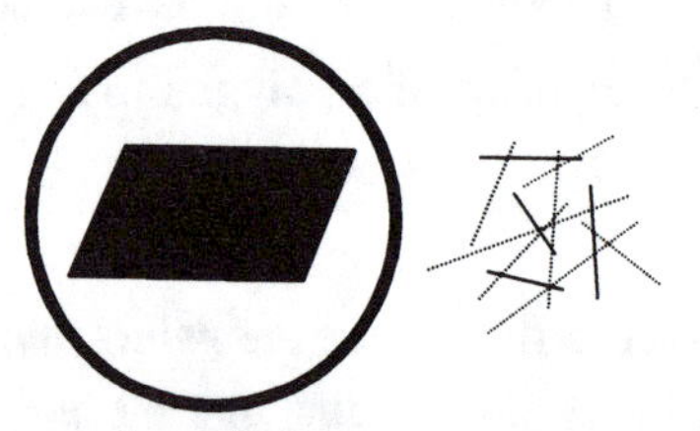
A parallelogram on a plate

Fred estimated that the height was 2" and the base was 3" and said that the area (A = bh) was 6 square inches.

"Okay. But you can't prove that using your three area postulates," she asserted.

Her parents were completely lost at this point. L thought that her daughter had just made a very lousy rectangle, whereas Fred thought it was a perfectly delightful parallelogram.

At this point T would usually say something like, "P! Don't play with your food. Eat!" but he couldn't this time since there was a guest in the house who was also playing with his food.

"I can't," he confessed. "All I can say is that your biscuit-parallelogram has some unique positive number associated with it which is its area, but I can't prove what that area is. I need a fourth area postulate."

"Who gets to name the postulate?" she asked.

Fred nodded toward her, and P began writing out:

P's Area Postulate✶*: We hereby assume all of the area formulas. Namely, the area of a triangle = ½ bh. The area of a parallelogram (which includes all rectangles, all rhombuses and all squares) = bh. The area of a trapezoid = ½(a + b)h.*

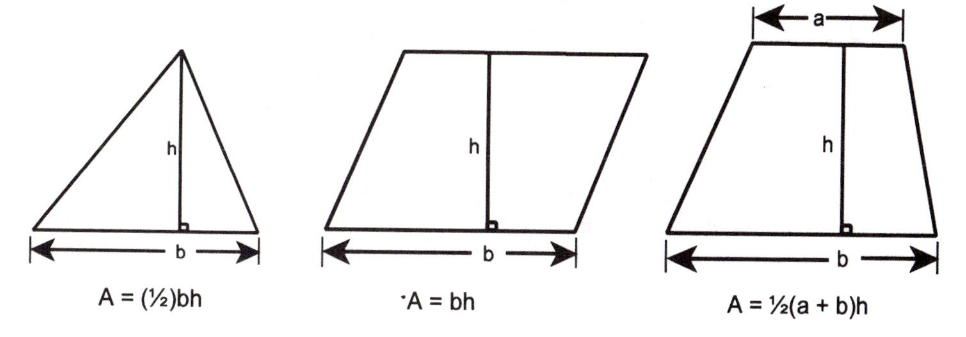

Your Turn to Play

1. What we do in mathematics is build theories. We start with some undefined terms and some beginning assumptions (postulates). From there we grow a tree (a theory) by adding on definitions and theorems. A theory is considered beautiful if it is tall and slender.

Tall = lots of theorems can be proved from the base of the undefined terms and postulates.

Slender = the base isn't too fat. The less you assume in your postulates, the more elegant is the theory you create.

What would be the objection to P's Area Postulate?

2. Suppose P's Area Postulate just assumed that we only knew the area formula for a triangle. $A = (\frac{1}{2})bh$. Describe (you need not use the statement–reason form) how you would prove that the area of a parallelogram is $A = bh$.

✶ Don't put this on your postulates page. It's not a very good postulate.

3. Suppose P's Area Postulate just assumed that we only knew the area formula for a triangle. A = (½)bh. Describe (you need not use the statement–reason form) how you would prove that the area of a trapezoid is A = ½(a + b)h.

COMPLETE SOLUTIONS

1. If you assume e-v-e-r-y-t-h-i-n-g, then what is there to prove? There is no theory, just a pile of "facts" at the base. The theory would be neither tall nor slender. When you first studied arithmetic in elementary school, it was probably taught as a pile of facts and procedures.

For example, when you were taught how to divide.................... 2/3 ÷ 5/8 you were told,

Turn the one on the right upside down and then multiply............ 2/3 × 8/5

There were no proofs, just stuff to stick in your head. No creativity. Everything the teacher said was to be taken as the truth. When she spoke, it was a postulate. "Why?" would have been considered a four-letter word.*

There was nothing creative in dividing fractions. After three days of learning to turn-the-right-one-upside-down-and-multiply, you were given zillions of them to grind out for homework. No wonder many people think they hate math! What they really hate is the uncreative drudgery like . . .

YOUR FOURTH GRADE HOMEWORK FOR TONIGHT:

69/43 ÷ 12/75
48/2 ÷ 95/9
80/39 ÷ 20/58
82/35 ÷ 29/41
17/65 ÷ 69/60
78/42 ÷ 15/68
73/90 ÷ 24/36
12/13 ÷ 15/18
27/91 ÷ 71/31
99/85 ÷ 9/58
54/48 ÷ 86/11

60/59 ÷ 57/55
48/22 ÷ 30/18
56/52 ÷ 37/94
29/84 ÷ 96/24
70/73 ÷ 68/56
52/37 ÷ 14/62
64/62 ÷ 64/61
44/27 ÷ 33/13
58/96 ÷ 71/21
21/20 ÷ 22/12
32/42 ÷ 78/50

7/93 ÷ 70/17
29/48 ÷ 66/25
56/93 ÷ 70/34
76/43 ÷ 3/22
36/64 ÷ 57/18
41/99 ÷ 89/7
19/41 ÷ 20/28
73/37 ÷ 65/47
85/62 ÷ 15/59
94/5 ÷ 29/76
69/8 ÷ 27/64

92/46 ÷ 54/26
69/7 ÷ 32/82
78/74 ÷ 10/38
45/52 ÷ 17/97
62/58 ÷ 60/92
32/28 ÷ 10/72
98/44 ÷ 52/63
92/33 ÷ 27/14
89/38 ÷ 41/33
24/94 ÷ 55/58
76/13 ÷ 29/71

✶ Famous four-letter words: *love, math, rose, sing, pray, grin, work, Fred,* and *food.*

64/68 ÷ 73/83	65/90 ÷ 44/33	22/9 ÷ 34/83	84/23 ÷ 96/50
56/99 ÷ 37/15	58/6 ÷ 9/3	8/73 ÷ 7/20	67/83 ÷ 27/14
67/75 ÷ 40/16	15/94 ÷ 35/53	95/51 ÷ 33/75	89/39 ÷ 39/39
70/80 ÷ 47/32	16/88 ÷ 15/41	84/73 ÷ 7/21	39/38 ÷ 40/38
51/93 ÷ 27/41	3/27 ÷ 73/40	4/39 ÷ 57/26	42/33 ÷ 78/28
20/81 ÷ 52/76	99/93 ÷ 82/61	83/75 ÷ 84/78	64/77 ÷ 52/3
91/17 ÷ 98/59	16/52 ÷ 25/70	91/64 ÷ 12/9	93/38 ÷ 25/51
32/79 ÷ 70/4	36/67 ÷ 53/28	73/67 ÷ 92/58	99/32 ÷ 66/68
21/6 ÷ 25/64	75/72 ÷ 39/96	1/89 ÷ 63/12	88/23 ÷ 79/42
91/44 ÷ 53/19	88/94 ÷ 37/79	96/77 ÷ 40/54	17/71 ÷ 89/28
70/14 ÷ 90/44	22/9 ÷ 38/54	82/49 ÷ 72/17	44/37 ÷ 26/33
51/16 ÷ 95/52	23/10 ÷ 36/80	5/84 ÷ 42/55	bonus problem:
55/32 ÷ 20/81	70/12 ÷ 3/21	82/50 ÷ 85/43	99792/723 ÷ 23/177
59/3 ÷ 15/94	9/34 ÷ 84/83	58/18 ÷ 87/75	
36/50 ÷ 22/78	86/83 ÷ 87/53	16/84 ÷ 25/16	

2. First draw a diagonal. Then note that the two triangles are congruent and each has an area equal to (½)bh.

Then by the Area Addition Postulate, the area of the parallelogram is equal to (½)bh + (½)bh which is bh by algebra.

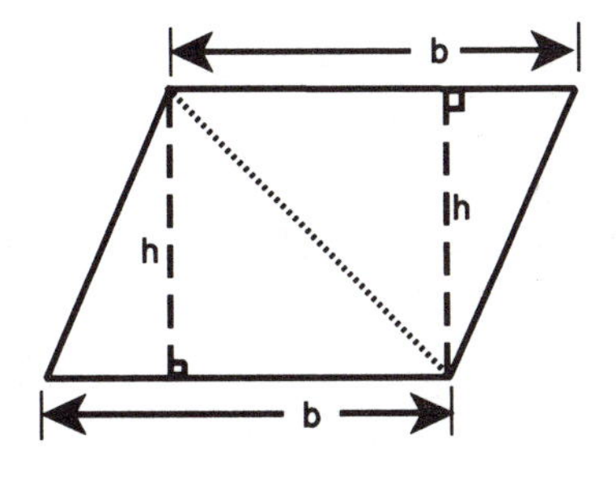

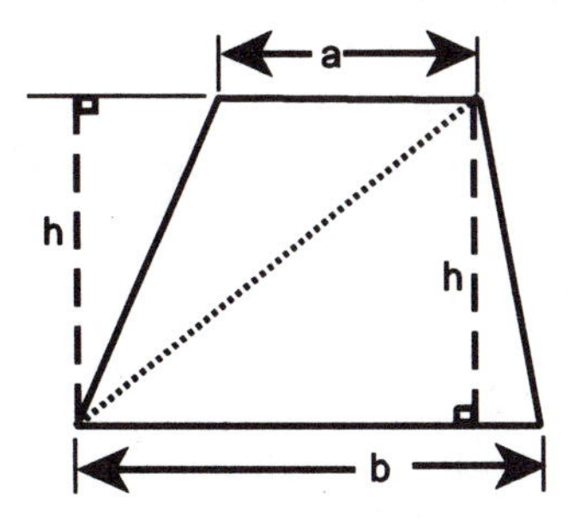

3. Draw the diagonal again. The area of the lower triangle is (½)bh. The area of the upper triangle is (½)ah. (In obtuse triangles, the segment representing the height can fall outside the triangle.)

Using the Area Addition Postulate, the area of the trapezoid is equal to (½)bh + (½)ah. By algebra, this equals (½)(a + b)h.

Thus, starting with a postulate that gives us the area of a triangle, gives us the area formulas for parallelograms and trapezoids without too much difficulty.

Without thinking, Fred had frowned when P wrote her Area Postulate that assumed all the area formulas. He imagined a geometry book written in that style. Page one would state: "Assume everything," and all the rest of the pages would just be number problems—elementary school stuff. Ultra boring.

L saw Fred's face and asked, "You don't like my hot dogs? I could whip up some of my fried eggs with strudel. Would you like that? I know that growing boys need a lot of calories."

Fred quickly took one of the hot dog slices in his hand and stuck the end in his mouth. "Oh, you needn't go to such trouble. This is quite satisfactory." I wish that I were growing. It seems like I've been 37 pounds forever and haven't gained an ounce for months. (Every week Fred weighs himself on the postal scale in the math department office.)

If I could just ask P to use the formula for the area of a triangle, $A = \frac{1}{2}bh$, as the fourth area postulate,* then we could have at least a couple of little theorems: one for the area of a parallelogram and one for the area of a trapezoid.**

Of course, it would be a bit more exciting if our fourth area postulate were the area of a parallelogram, $A = bh$.

Your Turn to Play

1. Before Fred has a chance to suggest to P that Postulate 13 should read, "The area of a parallelogram is bh," let's see how hard it would be to work with that as a possible Postulate 13.

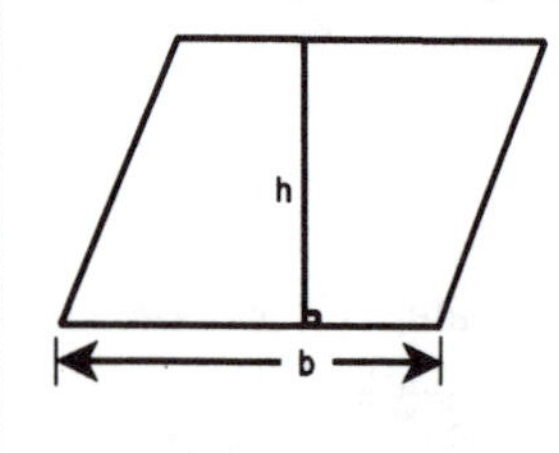

Since in the previous *Your Turn to Play* we showed that if you know the area formula for a triangle, you can find all the other formulas, all we need to do is show that if we know the area of a parallelogram, then we can find the area of a triangle. Do it. Describe how you would do it. You need not use statement-reason form. (Hint: Start with any $\triangle ABC$ as the given.)

✶ in addition to the first three area postulates, which are, roughly speaking:

Postulate 10: Every polygon has an area.

Postulate 11: $\cong$ $\triangle$s → equal areas

Postulate 12: You can add areas together if they don't overlap.

✶✶ which is what we did on the previous page in problems 2 and 3.

COMPLETE SOLUTION

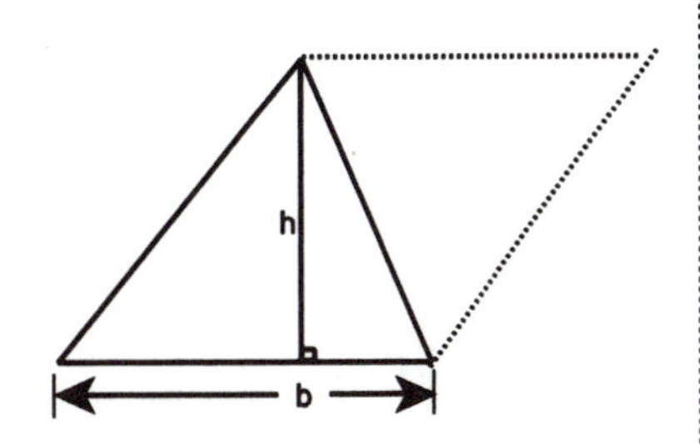

1. Start with any triangle. Draw the two (dotted) lines parallel to the sides of the triangle. (You can do this by the theorem which states: Given a point P not on ℓ, there is a line through P parallel to ℓ.) Then we have a parallelogram by definition and the diagonal divides it into two congruent triangles. By Postulate 11, the two triangles have equal areas. By the Area Addition Postulate, the sum of the areas of those triangles equals the area of the parallelogram (which we are assuming to be bh). Algebra gives the area of the original triangle = (½)bh.

But, I've got to remember that P is a pretty smart cookie. She finished her high school studies when she was four. If I suggest that we use the area of a parallelogram as our fourth area postulate, she might sense some condescension—which I really don't feel. I think P is the coolest, neatest, most wonderful . . .

Your Turn to Play

1. Before Fred goes completely off the deep end in his puppy love, let's see what happens if our fourth area postulate is the area of a rectangle, A = ℓw, length times width. Is it possible to start with A = ℓw and derive all the other area formulas from that? The answer is yes.

All we really need to do is start with A = ℓw as the postulate and find the area formula for a parallelogram. Once we have the area formula for a parallelogram, we may use the three previous proofs in *Your Turn to Play* and that will give us all the area formulas.

Here is the proof that the area of a parallelogram is ½bh using as our fourth area postulate the area of a rectangle is A = ℓw. Fill in the missing reasons.

D C
A B 1

Statement	***Reason***
1. ▱ ABCD	1. ?
2. $\overline{AD} \parallel \overline{BC}$	2. ?
3. $\angle A \cong \angle 1$	3. ?

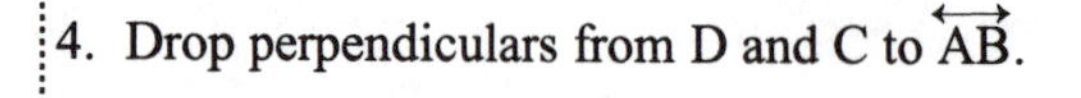
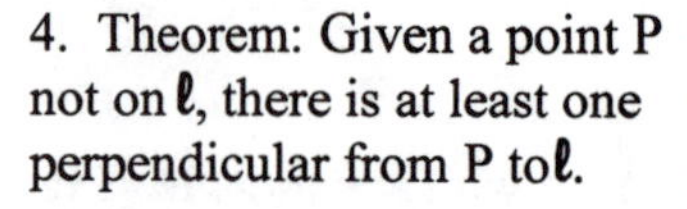
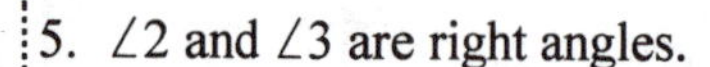
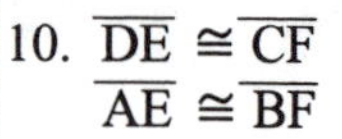

Statement	Reason
4. Drop perpendiculars from D and C to $\overleftrightarrow{AB}$.	4. Theorem: Given a point P not on ℓ, there is at least one perpendicular from P to ℓ.
5. $\angle 2$ and $\angle 3$ are right angles.	5. Def of $\perp$ lines
6. $m\angle 2 = m\angle 3 = 90°$	6. Def of right angles
7. $\angle 2 \cong \angle 3$	7. Def of $\cong$ angles
8. $\overline{AD} \cong \overline{BC}$	8. ?
9. $\Delta AED \cong \Delta BFC$	9. AAS
10. $\overline{DE} \cong \overline{CF}$ $\overline{AE} \cong \overline{BF}$	10. ?
11. $a\Delta AED = a\Delta BFC$	11. Postulate 11: If two Δs are $\cong$, their areas are equal.
12. a▱ABCD = $a\Delta AED$ + a(EBCD) a(EFCD) = $a\Delta BFC$ + a(EBCD)	12. Area Addition Postulate
13. a▱ABCD = a(EFCD)	13. Algebra
14. a(EFCD) = DE × EF	14. We have temporarily assumed that Postulate 13 is the area of a rectangle: $A = \ell w$
15. AE = BF	15. Def of $\cong$ segments (step 10)
16. AE + EB = AB and EB + BF = EF	16. Def of A–E–B and E–B–F
17. AB = EF (from steps 15 and 16) a(EFCD) = DE × AB (from above equation and step 14) a▱ABCD = DE × AB (from above equation and step 13)	17. Algebra

This is a very long proof by beginning geometry standards. It takes time and effort to read and understand each step. It can't be read like reading the comics in the Sunday newspaper.

Learning to follow (and create) logical arguments such as this one is at the heart of geometry. Many of the area formulas you may have learned in elementary school. But unless you are like P or Fred, it takes some more years until your mind is ready to follow the steps in logical reasoning. If you can remember back a hundred years ago to your elementary school days, most kids struggled at that time with the easiest logical deduction: If B then P. (B ➜ P) That was beyond many of your classmates. (Let B = you do something Bad and let P = you will be Punished.)

Reasoning is a lot harder than memorizing area formulas. *Expect* that the reading of proofs will take time and effort.

COMPLETE SOLUTION

1.

1. Given
2. Def of parallelogram
3. P ➜ CA Theorem
8. Theorem: Opposite sides of a ▱ are ≅.
10. Definition of congruent triangles

It's easy to get lost in the details of a proof like this one. Except for the algebra, none of the steps is especially hard to follow. There are just a lot of steps.

But if you step back from the details to look at the overall picture, the concept behind this proof isn't very deep.

We started with a parallelogram for which we wished to find the area.

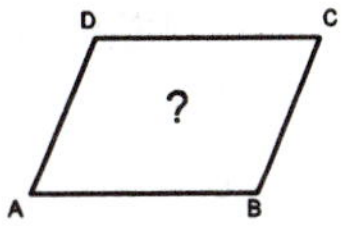

Then we chopped off a triangle on the left side.

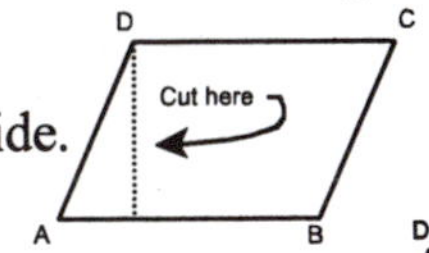

And we stuck it on the right side to create a rectangle.

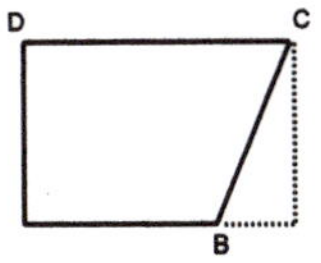

And since we knew the area of a rectangle, we were done.

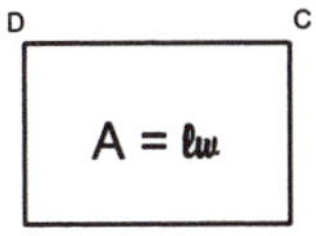

$A = \ell w$

P looked up from writing out her area postulate. She saw Fred with a slice of hot dog hanging from his mouth and she giggled. She took another piece and imitated him. She thought it looked funny.

The only difference is that she actually chewed on her piece.

Her dad frowned. This new kid was teaching his daughter bad table manners.

"You know," she began, "I was just kidding about making the fourth area postulate so dorky. If you assumed everything, of course, you couldn't have any fun doing any theorems."

She took a crayon and crossed out her writing.

~~*P's Area Postulate: We hereby assume all of the area formulas. Namely, the area of a triangle = ½ bh. The area of a parallelogram (which includes all rectangles, all rhombuses and all squares) = bh. The area of a trapezoid = ½(a + b)h.*~~

"Since you're going to let me name the postulate, I'm going to make it really exciting." She smiled and began to write.

<u>Postulate 13</u>: We hereby assume the area formula for. . . .

Fred held his breath. Is she going to use the area formula for a triangle and make everything super easy? Is she going to write the area formula for a parallelogram? That would make everything hard enough. Or is she going to go for the gold and say that the only area formula we are given is the one for rectangles? Then it's a 17-step proof to get from the area formula for a rectangle to the one for parallelograms.

<u>Postulate 13</u>: We hereby assume the area formula for a SQUARE! A = s^2.

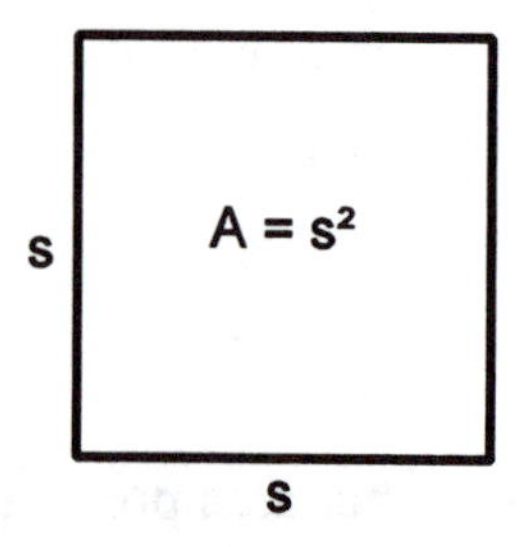

[It's official now. P has spoken. Place this postulate on your page of postulates.]

Fred's mouth dropped open. P had written the **ultimate** area postulate. Nothing could be tougher.

These four area postulates

> Postulate 10: Every polygon has an area.
> Postulate 11: ≅ ▲→ equal areas
> Postulate 12: You can add areas together if they don't overlap.
> Postulate 13: The area of a square is s^2.

are the most difficult axioms* imaginable. With these forming the base of

---

* Axiom is another word for postulate. AX-ee-um

our geometry theory, it will be hard to grow any theorem concerning area at all. The theory would have such a small base of undefined terms and postulates that we might not be able to grow any theorems at all.

The whole development of our geometry theory for area will depend on whether we can find all the area formulas (rectangle, parallelogram, triangle, trapezoid) from our given knowledge that the area of a square is $A = s^2$.

Since we can start from the area of a rectangle and prove the area of a parallelogram,

RECTANGLE ➜ PARALLELOGRAM

Done on p. 255 in that long 17-step proof in which we chopped a triangle off the left side of a parallelogram and stuck it on the right side.

Since we can start from the area of a parallelogram and prove the area of a triangle,

PARALLELOGRAM ➜ TRIANGLE

Done on p. 254 in which we took a triangle and built it up into a parallelogram.

Since we can start from the area of a triangle and prove the area of a trapezoid,

TRIANGLE ➜ TRAPEZOID

Done on p. 252 in problem 3.

. . . our whole job boils down to finding the area of a rectangle if we know the area of a square.

And what a job that is.

Now almost everyone who wants to try and prove that the area of a rectangle is A = ℓw, starts out the same way. They are given a rectangle ABCD with length ℓ and width w. They draw a picture.

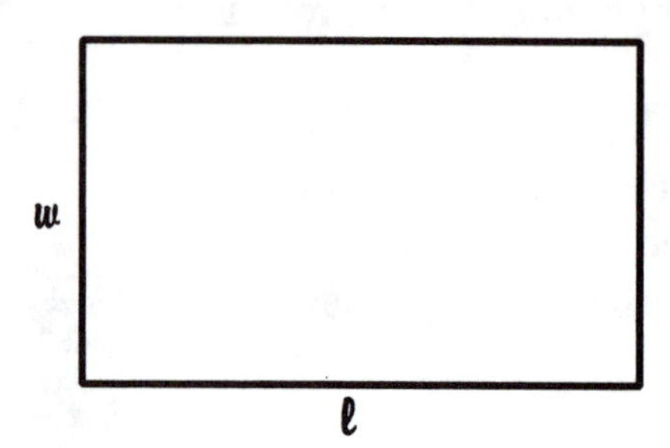

Then after a little head scratching, they say "Aha!" and subdivide the rectangle into little squares.

If, for example, the width were five feet and the length were nine feet, then you could subdivide the original rectangle into 45 little squares. Each square would be one square foot by Postulate 13 and by the Area Addition Postulate, the area of the rectangle would be 45 square feet.

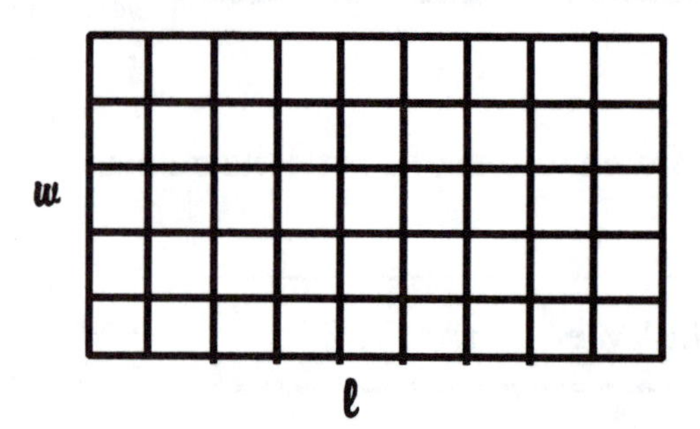

Very often, when mathematicians are trying to prove a general theorem, they will look at specific examples to try and get an idea of how to proceed in the general case.

Your Turn to Play

1. If w were 3 inches and ℓ were 20 inches, how might you subdivide the rectangle into little squares?

2. If w were 1/5 of a mile and ℓ were 3/5 of a mile, how might you subdivide the rectangle into little squares?

3. If w were 2.31 meters and ℓ were 88.7 meters, how might you subdivide the rectangle into little squares?

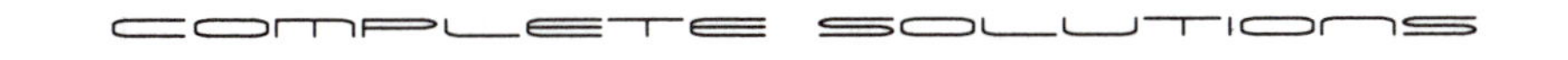

1. You would cut the width into three pieces and the length into 20 pieces and you would have 60 little squares.

2. You would cut the rectangle into little squares that were 1/5 of a mile on each side.

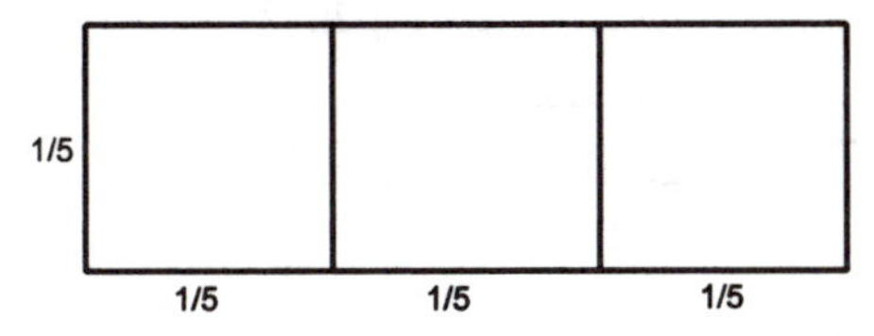

3. You would cut the rectangle into little squares that were 0.01 meters on each side. Then the width would be cut into 231 parts and the length into 887 parts.

The whole trick is to find a number that divides evenly into both the width and the length. If *w* were 2/3 and *ℓ* were 3/4, then we would divide everything into squares that were 1/12 on each side.*

If *w* were 1/7 and *ℓ* were 97883.039, then we would use squares that were 1/7000 on each side (since 1/7000 divides evenly into both numbers).**

What we don't want to happen is to chop up the rectangle and find out that all the little spaces on the inside weren't squares.

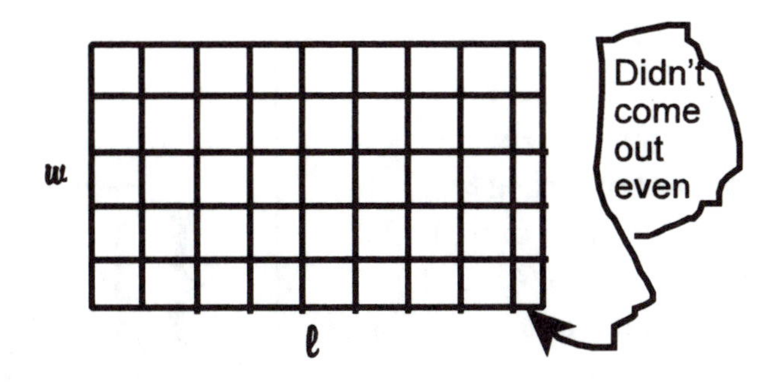

---

* Here's the arithmetic. 2/3 is equal to 8/12. 3/4 is equal to 9/12. (Twelve is the common denominator.) So the width would be cut into 8 parts and the length into 9 parts.

** Change 97883.039 into a fraction: 97883039/1000. Then the common denominator for 1/7 and 97883039/1000 is 7000. Squares with a sides of 1/7000 will work nicely.

Depending on the values for w and ℓ, we might have to chop things up pretty finely.

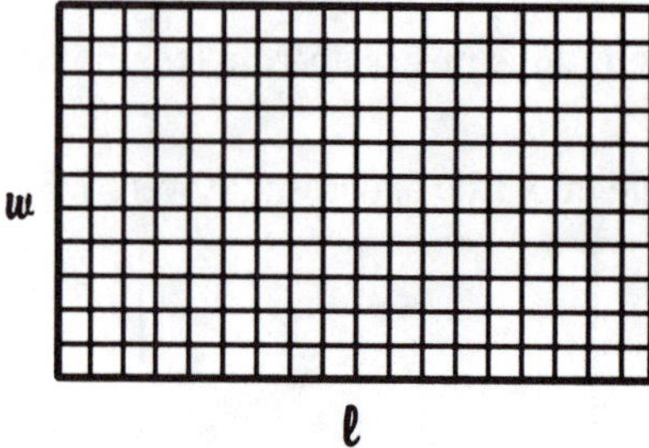

Let's start the proof of the theorem and see how it goes.

<u>Theorem</u>: The area of a rectangle of width w and length ℓ is given by the formula $A = \ell w$.

Our area postulates are:

Postulate 10: Every polygon has an area which is a positive number.

Postulate 11: Congruent triangles have equal areas.

Postulate 12: You can add areas together if they don't overlap.

Postulate 13: The area of a square is s^2.

Statement	*Reason*
1. Rectangle ABCD with width w and length ℓ.	1. Given
2. Find a positive number x which divides evenly into both w and ℓ. (This means that w/x and ℓ/x are both natural numbers = {1, 2, 3, 4, . . . }.) Let w/x be the natural number m and let ℓ/x be the natural number n.	2. Algebra 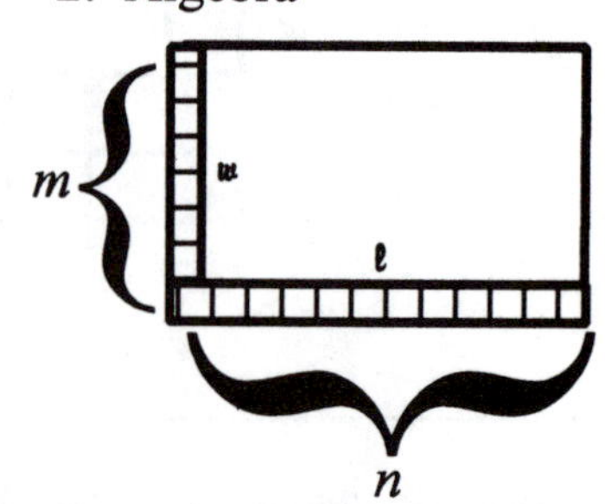
3. Locate the m evenly spaced points on the width and the n evenly spaced points on the length.	3. Ruler Postulate.
4. Consider the mn little squares, etc., etc., etc.	

Why'd you stop the proof? You were doing so well (for a change). This proof was going to be a lot shorter than that monster 17-step proof you gave on p. 255. I as your reader demand you finish the proof.

I can't.

What seems to be the difficulty?

It was back in step 2. I was given the width w and the length ℓ of the rectangle and we were to find a number x which divided evenly into both w and ℓ.

Yeah. So what's so hard about that? In the *Your Turn to Play* ***that you had me do, it was child's play. All you needed to do is turn both w and ℓ into fractions and find the common denominator. Anybody can do that. Watch. Let me write part of this book:***

My Turn to Play

1. If w is 3/4 and ℓ is 2/7, what number works? Ans. 1/28

2. If w is 1/10 and ℓ is 32/6, what number will divide evenly into both of them? Ans. 1/60

3. If w is 3989/5 and ℓ is 3.07? Ans. First convert 3.07 to a fraction (=307/100) and then the number that divides into both of them is 1/500. In fact, even 1/100 will work.

That's nice, but it won't always work.

Why not? You can always find a common denominator. You can always add any two fractions.

That's true, but what if you can't turn w into a fraction?

I'm listening.

Back in algebra we talked about numbers which can't be expressed as fractions—numbers which can't be written as *a*/*b* where *a* and *b* are natural numbers {1, 2, 3, 4 . . .}. They were numbers like π and $\sqrt{2}$. They were called irrational numbers and we'll talk about them a lot more in advanced algebra. Any unending, non-repeating decimal is also an irrational number, such as 0.937920377340822304296293648923492390423894239489034234929494949395794573495040359349054345 3450345435472394793479823494823979582759475983479587957234985793475923479349237925792375923794529429469236942984389274929492349239492892439696 9781121363693421496192369216319232998342489573892359823694632989789734929823949194924023474262346962612446246238948923498239849369842314893498 93812243943986o349234923489662348923806238642386482382334284280899876966666612234496239463947398234938279924191236982639841238912366329. . . .

So if the width of the rectangle happens to be an irrational number and the length is some nice number like 8, then there is no way that I can find an *x* that divides evenly into both of them. I can't chop the rectangle up into little squares. My proof fails at step 2.

That's too bad.

I know. It would have made things really easy. P's Postulate 13: *The area of a square is* s^2 wouldn't be that difficult to deal with.

So what's the proof for showing that A = s^2 for a square will imply that A = ℓw for a rectangle?

It's one of the most clever proofs in geometry. It's one of those how-did-they-ever-think-of-that kind of proof.

You start with a rectangle whose area is A. You are trying to show that A = ℓw.

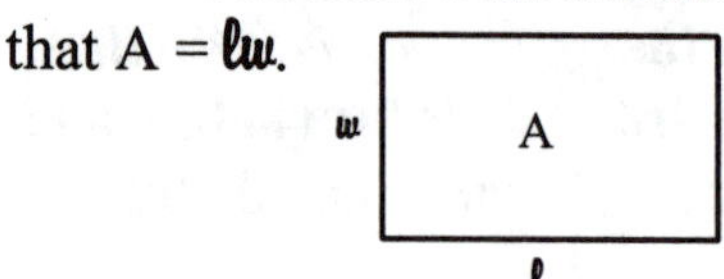

Then you add on a square on its right. (The Ruler Postulate helps here.)

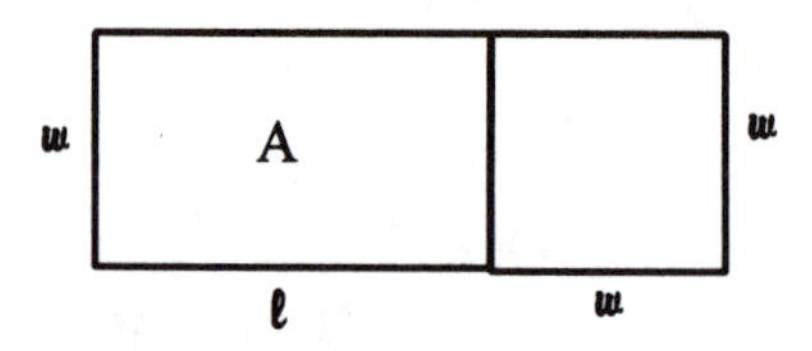

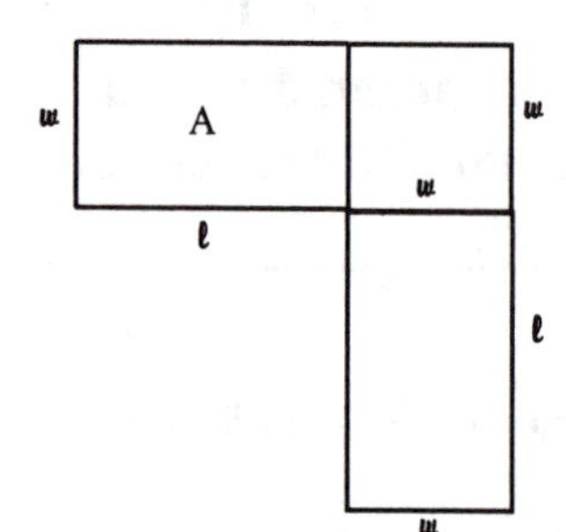

Then you add on a rectangle below the square with the same dimensions as the original rectangle. Its area will also be equal to A.

Finish the diagram with a square.

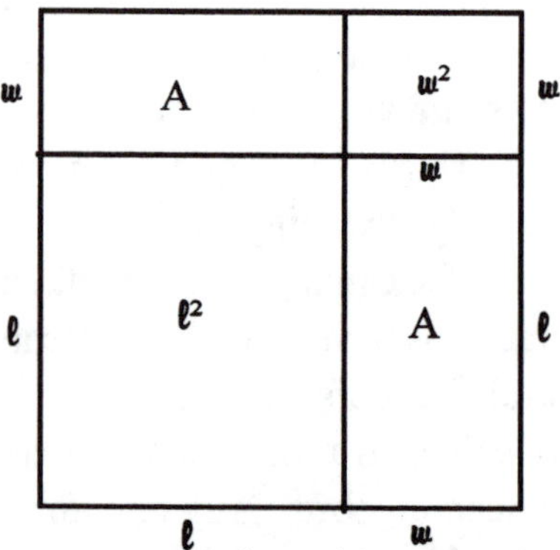

The area of the two squares and the two rectangles adds up to the area of the giant square that encloses them (by the Area Addition Postulate): $\ell^2 + w^2 + A + A = (w + \ell)^2$

Some algebra: $\ell^2 + w^2 + A + A = w^2 + 2w\ell + \ell^2$

More algebra: $A + A = 2w\ell$

More algebra: $A = w\ell$ which finishes the proof.

The bell rang from the bell tower on the KITTENS campus. "Bong!" It was starting to announce the noon hour. Fred was supposed to be at the awards ceremony at noon. The whole school would be waiting for him. He was the reason for this meeting.

Fred quickly excused himself from the table and dashed to the door of P's square house and then broke into a full run down Tangent Road. "Bong!"

He turned on Newton Street and headed onto the campus. "Bong!" He was so grateful for his years of faithful jogging every morning. They won't have to carry him off in a stretcher after he arrives at the ceremony grounds with a case of tachycardia. (*tachy* = swift; *-cardia* = suffix indicating something bad about the heart. *tachycardia* = excessively fast heartbeat.)

"Bong!" He raced to the math building and up the stairs to his office. He put on his mortarboard and straightened his bow tie. Then down the stairs. "Bong!"

Formal academic clothing

Now Fred faced a tough decision. "Bong!" The ceremony was diagonally across the brick courtyard. It would be tough to run on those uneven bricks, but the alternative was to head east 528 feet and then north 201 feet, which was longer than

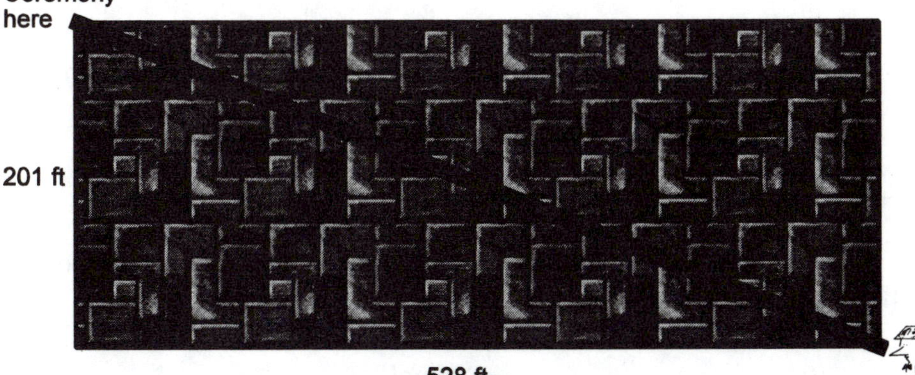

just cutting across the courtyard. "Bong!" If he ran along the edge of the courtyard, it would be 528 + 201 = 729 feet. What would be the distance he would have to run if he headed straight across the courtyard?

To find the length of the hypotenuse of a right triangle, knowing the length of the two legs—that was Fred's challenge.

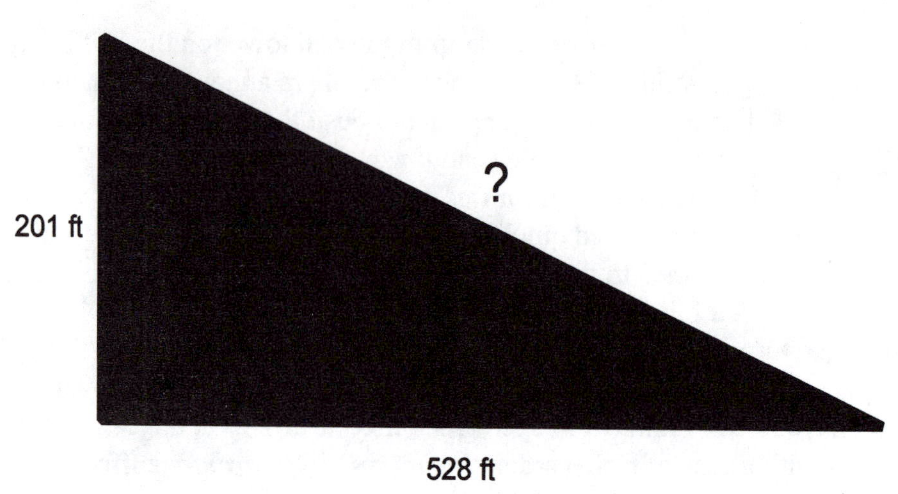

It's not obvious what the length of the hypotenuse is. Really not obvious at all. You might look at this triangle all day long and during commercials on television tonight and still not arrive at the fact that the hypotenuse is exactly $\sqrt{319185}$ feet long.*

We need a moment of silence right now. Fred can stop running for a second.

"Bong!"

And that bell tower. Could someone check the wiring on that tower and figure out a way that we can shut that thing down for a moment?

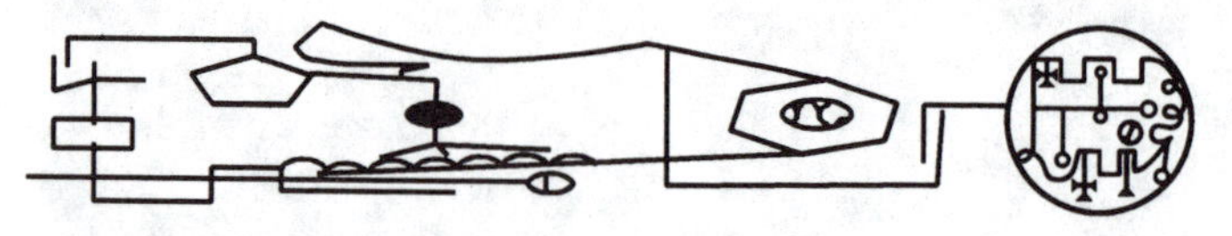

This is *the* most important moment in geometry and it would be nice not to have that silly bell disturbing us.

"Bong!"

May I as your ever helpful reader suggest . . .

Thank you.

---

* That's approximately 564.965 feet.

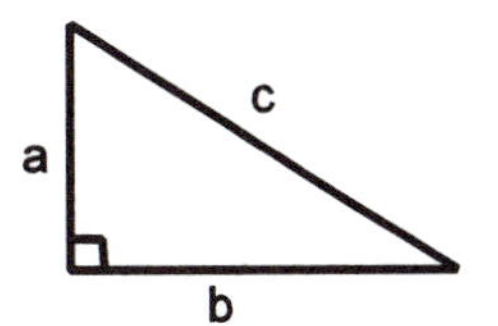

Finding the length of the hypotenuse, given the lengths of the legs,

- ✻ is the most famous theorem of geometry,
- ✻ is one of the handiest theorems of geometry,
- ✻ is one of the easiest to memorize: $a^2 + b^2 = c^2$
- ✻ is probably the first great theorem in all of mathematics, and
- ✻ is called the Pythagorean Theorem.

<u>The Pythagorean Theorem</u>: In any right triangle, $a^2 + b^2 = c^2$.*

Some notes about this super-famous theorem:

♪#1: It's a surprise. Why in the world should $a^2 + b^2 = c^2$? When you look at an isosceles triangle, for example, it seems pretty clear that the base angles will be congruent, but when you gaze at a right triangle, nothing seems to shout $a^2 + b^2 = c^2$.

♪#2: This theorem has been around for a long, long time. Your grandmother knew about it. George Washington knew about it. (He was a surveyor before he became more famous.) It was known at the time of Christ. As far as we can tell, it was first proved by Pythagoras (born about 2500 years ago) or by one of the members of his brotherhood. Since proving a statement makes it a theorem, Pythagoras established $a^2 + b^2 = c^2$ as a theorem about 530 B.C. And over a thousand years before Pythagoras, the Babylonians marked in clay tablets $a^2 + b^2 = c^2$, but we have no evidence that they ever figured out how to prove it.

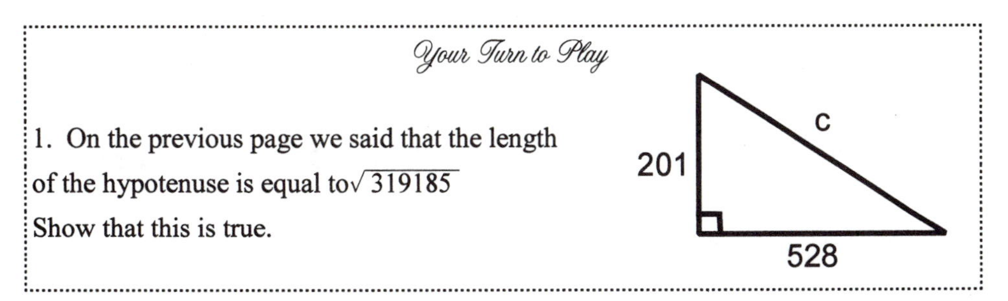

Your Turn to Play

1. On the previous page we said that the length of the hypotenuse is equal to $\sqrt{319185}$ Show that this is true.

✶ In words: In any right triangle, the sum of the squares of the legs is equal to the square of the hypotenuse.

With even more words and more accuracy: In any right triangle, the sum of the squares of the lengths of the legs is equal to the square of the length of the hypotenuse.

COMPLETE SOLUTIONS

1. By the Pythagorean Theorem, $a^2 + b^2 = c^2$.

So $(201)^2 + (528)^2 = c^2$

$40401 + 278784 = c^2$

$319185 = c^2$

And then taking the square root of both sides of the equation $\sqrt{319185} = c$

♪#3: Since Pythagoras, we have figured out *several* more proofs of the Pythagorean Theorem.

♪#4: The previous note is not quite true. We have constructed *a lot* of proofs for the fact that $a^2 + b^2 = c^2$ is true in any right triangle.

♪#5: The previous note isn't the whole truth. This theorem has probably been proven in more ways than any other theorem in all of mathematics.

Some people collect postage stamps. Some collect paintings. E. S. Loomis collected proofs of the Pythagorean Theorem. In 1940 he wrote a book containing 370 different proofs of this theorem. Of course, more proofs have been found since then.

Even Garfield has written a proof! Honest. And that's the one we will look at now.

Garfield's Proof of the Pythagorean Theorem

Statement	*Reason*
1. right $\triangle ABC$	1. Given
2. On $\overleftrightarrow{AC}$, locate the point D so that AD = a.	2. Ruler Postulate
3. Erect a perpendicular to $\overleftrightarrow{AD}$ through point D. 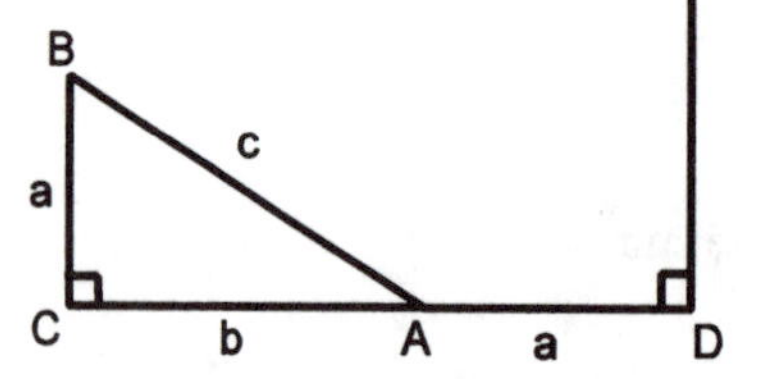	3. Theorem: From every point on a line, there is a perpendicular to that line through that point. (Problem 3 on p. 162)

4. On that perpendicular, locate the point E so that DE = b.	4. Ruler Postulate

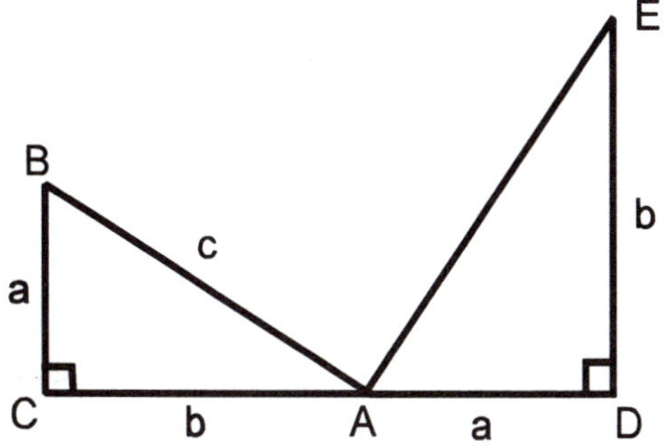

5. $\triangle ABC \cong \triangle EAD$	5. SAS

E
B
c
a
b
2
1
3
C
b
A
a
D

6. $\angle 1$ and $\angle B$ are complementary. $\angle 3$ and $\angle E$ are complementary.	6. Theorem: The acute angles of a right triangle are complementary.
7. $m\angle 1 + m\angle B = 90°$	7. Def of complementary
8. $m\angle 1 + m\angle 2 + m\angle 3 = 180°$	8. Def of the measure of a straight angle (See p. 59)
9. $\angle B \cong \angle 3$	9. Def of $\cong$ ⨺
10. $m\angle B = m\angle 3$	10. Def of $\cong$ ∠s
11. $m\angle 2 = 90°$	11. Algebra
12. $\angle 2$ is a right angle.	12. Def of right $\angle$
13. $\overline{AB} \cong \overline{EA}$	13. Def of $\cong$ ⨺

14. $\overline{BC} \parallel \overline{DE}$ 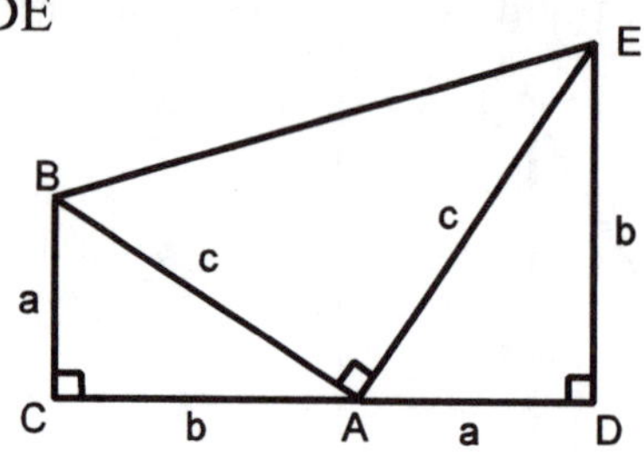	14. Thm: Two lines ⊥ to the same line are parallel.
15. trapezoid BCDE	15. Def of trapezoid
16. (The height, h, of trapezoid BCDE is b + a. The lengths of the parallel sides are a and b.) The area = ½(b + a)(a + b).	16. Thm: The area of a trapezoid = ½ h(a + b)
17. a∆ABC = ½ab a∆ADE = ½ab a∆BAE = $½c^2$	17. Thm: The area of a triangle = ½bh
18. ½ab + ½ab + $½c^2$ = ½(b + a)(a + b)	18. Area Addition Postulate (The area of the three triangles add up to the area of the trapezoid.)
19. $c^2 = a^2 + b^2$	19. Algebra (step 18)

⊠

Wait a minute! That's a nice proof (except for the algebra in going from step 18 to step 19 which I had to work out on scratch paper). But you have got to be kidding. Garfield wrote this proof? I bet you dollars to doughnuts that no comic strip cat ever constructed a proof of the Pythagorean Theorem!

Cat? Oh. You're thinking of *that* Garfield. I'm thinking of the real Garfield—James Abram Garfield—President of the United States of America. About five years before he became our 20th president, he figured out this proof. It's a lot nicer proof than the one that Euclid (300 B.C.) wrote in *The Elements*.

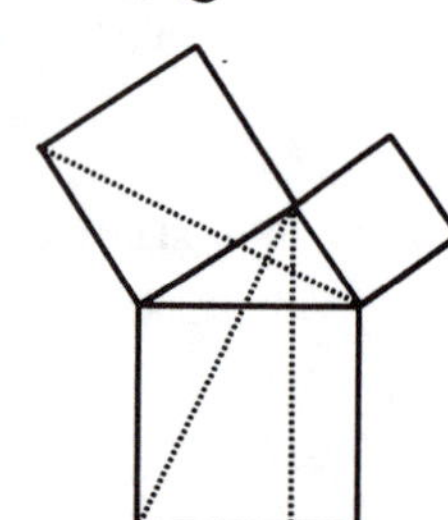
Euclid's Diagram

Elbert

1. Find the areas:

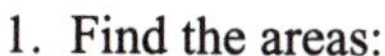

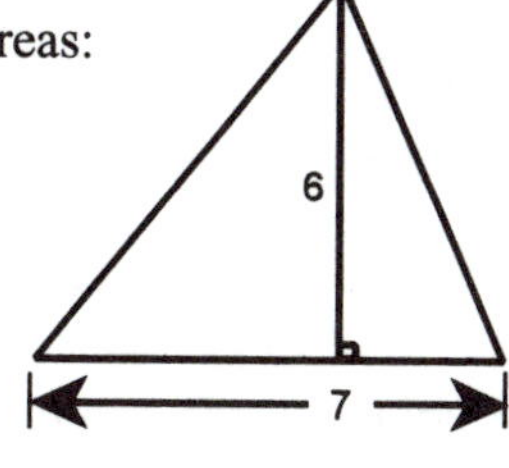

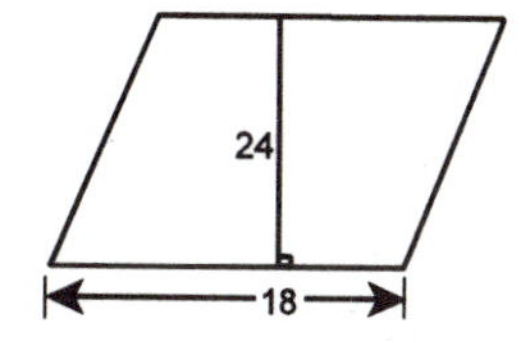

2. This is an unfair question, but take a stab at it anyway. Find the area of:

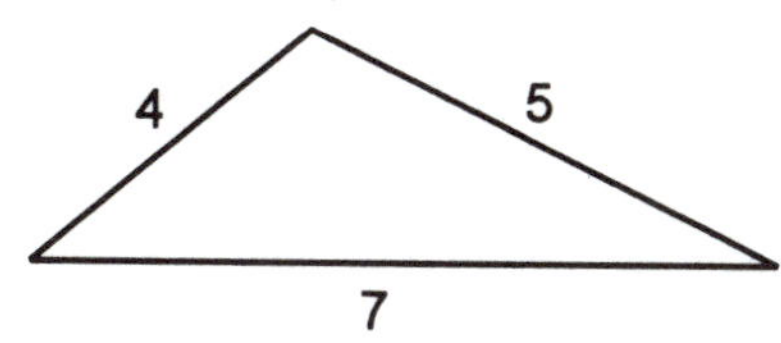

If turns out to be too difficult, then as a consolation prize, just find the perimeter of the triangle instead.

3. One of the corollaries of the Pythagorean Theorem is called **Heron's formula**. It is a formula for finding the area of any triangle if you know the lengths of the three sides. Heron lived at the time of St. Paul (first century A.D.), but there is no record of their having met each other since Heron lived in the city of Alexandria which Paul never visited.

Heron's formula is the easiest way to find the area of the triangle when you know the lengths of the three sides. The only problem is that it isn't as easy as the traditional formula $A = \frac{1}{2}bh$. Of course, $A = \frac{1}{2}bh$ doesn't work when all you know is the lengths of the three sides.

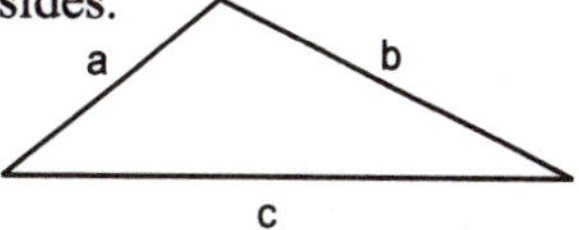

Using Heron's formula, the area of a triangle whose sides are a, b and c is equal to

$$A = \sqrt{s(s-a)(s-b)(s-c)}$$

where s is the **semiperimeter**

which means that s is half of the perimeter of the triangle,

which means that $s = \frac{1}{2}(a + b + c)$.

Using Heron's formula, find the area of the triangle in the previous problem.

While you are working on that, here is an outline of the proof of Heron's formula which all but the most intrepid readers may ignore.

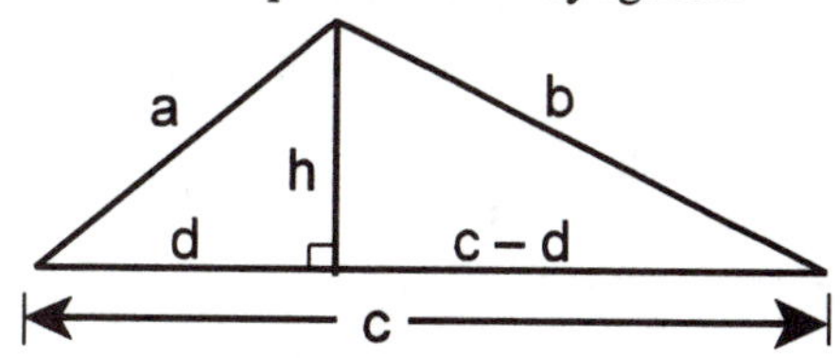

First drop the ⊥ and call its length h. That splits c into two parts d and c – d. That gives us two right triangles and we use the Pythagorean Theorem on each of them:

$d^2 + h^2 = a^2$ and $h^2 + (c-d)^2 = b^2$

Solving each of these equations for h^2,

$h^2 = a^2 - d^2$ and $h^2 = b^2 - (c-d)^2$

and equating the two equations,

$$a^2 - d^2 = b^2 - (c-d)^2$$

and solving for d,

$$d = \frac{a^2 - b^2 + c^2}{2c}$$

(We skipped about three steps which you can do on scratch paper if you wish.)

Now starting again with $h^2 = a^2 - d^2$ (look upward about eight lines), we factor, using difference of squares:

$h^2 = (a - d)(a + d)$.

Then we get rid of the d in the above equation by using

$d = \dfrac{a^2 - b^2 - c^2}{2c}$

(look upward eleven lines).

This yields an intimidating $h^2 =$

$(a - \dfrac{a^2 - b^2 - c^2}{2c})(a + \dfrac{a^2 - b^2 - c^2}{2c})$

Multiplying through by $4c^2$ will eliminate the fractions, giving

$4c^2h^2$ = a lot of stuff, which by some fancy factoring equals a nicely symmetrical

$(a + b + c)(a - b + c)(-a + b + c)(a + b - c)$

That cleans up using $s = \frac{1}{2}(a + b + c)$ into $4c^2h^2 = 16s(s - a)(s - b)(s - c)$.

Take the square root of both sides,

$2ch$ = a lot of stuff.

And divide by 4,

$\frac{1}{2}ch = \sqrt{s(s - a)(s - b)(s - c)}$

which finishes the proof since $\frac{1}{2}ch$ is our traditional formula for the area of a triangle.

4. Given $a\Delta ABC = 12$ and $a\Delta ACD = 7$, must it be true that the area of the quadrilateral ABCD = 19?

5. The area of a rectangle is 7.5 square feet and its perimeter is 11 feet. Showing all your algebra, find its length ℓ and its width w.

answers

1. 21 and 432

2. Right now you don't have enough information to find the area of the triangle. Its perimeter is 16.

3. $\sqrt{96}$ (This may be simplified to $\sqrt{16}\,\sqrt{6}$ which is equal to $4\sqrt{6}$.)

4. The Area Addition Postulate applies only when the areas do not overlap. Here is an example when $a\Delta ABC = 12$ and $a\Delta ACD = 7$, but the area of the quadrilateral ABCD $\neq 19$.

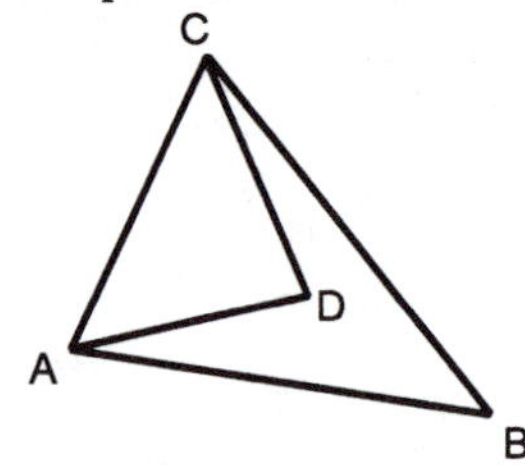

5. perimeter: $2\ell + 2w = 11$
area: $\ell w = 7.5$

Solving the first equation for w,

$w = \frac{1}{2}(11 - 2\ell)$

Substituting that into the second equation:

$\ell(\frac{1}{2}(11 - 2\ell)) = 7.5$

Multiplying out the left side:

$(11/2)\ell - \ell^2 = 7.5$

Doubling the equation to eliminate the fractions and decimals:

$11\ell - 2\ell^2 = 15$

Transposing everything to the right side of the equation:

$0 = 2\ell^2 - 11\ell + 15$

Factoring:

$0 = (2\ell - 5)(\ell - 3)$

Setting each factor equal to zero:

$2\ell - 5 = 0$ OR $\ell - 3 = 0$

$\ell = 5/2$ OR $\ell = 3$

Putting $\ell = 5/2$ into either of the two original equations gives $w = 3$. Putting $\ell = 3$ into those equations gives $w = 5/2$.

Rampart

1. How is the area of a rhombus related to the product of the lengths of its diagonals?

In order to figure this out, you may need to draw a diagram and play with the areas of the four triangles.

2. What is the semiperimeter of a triangle whose sides measure 20, 30 and 40?

3. What is wrong with this question:

What is the semiperimeter of a triangle whose sides measure 2.03, 3¾ and 8?

4. Find the area of a triangle whose sides measure 4, 7 and 9.

5. Is it possible to find the area of a rhombus if all you are given is that one of its sides is 100 miles long?

answers

1. The four triangles are each right triangles since we know that the diagonals of a rhombus are perpendicular bisectors of each other. And those triangles are all congruent by SAS and hence all have the same area. If the lengths of the diagonals were x and y, then each of the right triangles would have legs of lengths $x/2$ and $y/2$ and areas of $xy/8$. Adding together the areas of the four triangles, we would have the area of the rhombus equal to $xy/8 + xy/8 + xy/8 + xy/8$ which is $xy/2$. Hence the area of a rhombus is equal to one-half the product of the diagonals.

2. 45

3. If you tried to draw that triangle to scale, you would find that the two shorter sides were so short that you couldn't make a triangle. In any triangle whose sides are a, b and c, the lengths of any two sides must always be greater than the length of the third side: $a + b > c$. This is called the **triangle inequality**.

You may have seen the triangle inequality expressed as, "The shortest distance between two points is a straight line."

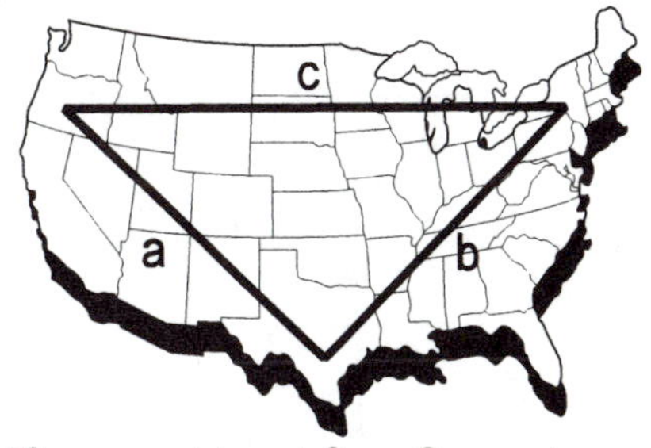

If you want to get from Oregon to New York, the shortest route isn't by way of Texas. $a + b > c$

4. $\sqrt{180}$ or $6\sqrt{5}$

5. If one of the sides is equal to 100 miles, then all four sides are equal to 100 miles. But this still isn't enough information. The rhombus might be a square (which would have an area of 10000 square miles) or it might have an area of one square inch if the rhombus were really squished.

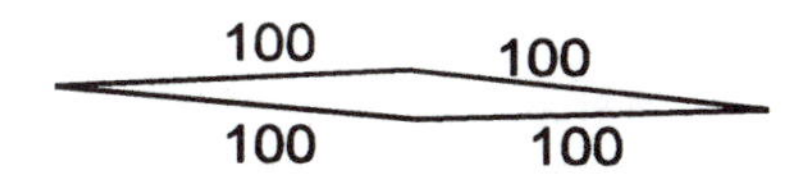

Upton

1. Find the area of a rhombus whose sides are each 6 inches long and where one of the diagonals is 8 inches.

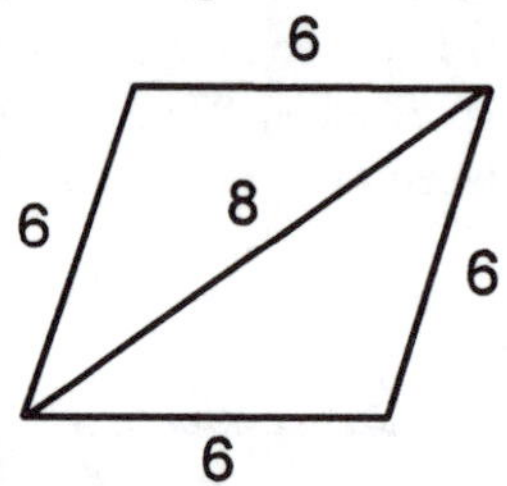

2. Find the area, if possible, of the trapezoid:

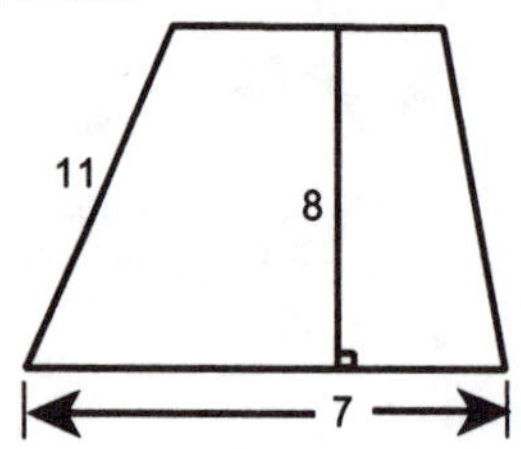

3. State the converse of the Pythagorean Theorem. Does it seem to be true? (Sometimes converses of true statements are true and sometimes they are false. *If I stand naked in the snow, then I will be cold* is a true statement, but its converse, *If I am cold, then I am standing naked in the snow* is false.)

4. Find the area, if possible, of the trapezoid pictured in problem 2 if you also know that its perimeter is equal to 32.

5. Fill in the missing reasons in this proof of the converse of the Pythagorean Theorem. (Also place this theorem on your list of theorems that you are keeping.)

You are given a ΔABC in which $a^2 + b^2 = c^2$.

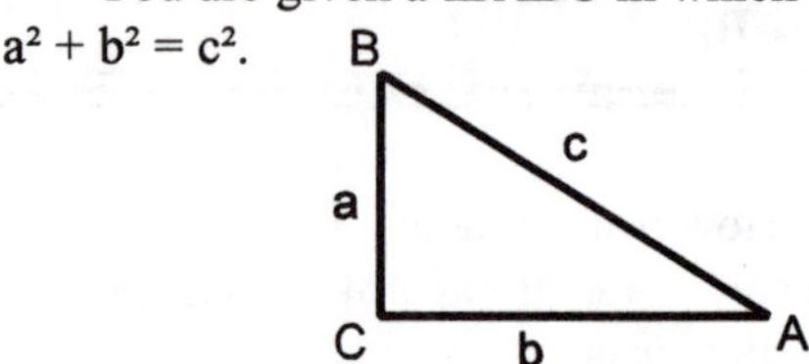

Create a second triangle: (by the Ruler Postulate) draw $\overline{DF}$ so that DF = b, and then erect a $\perp$ on $\overline{DF}$ at F by (?).

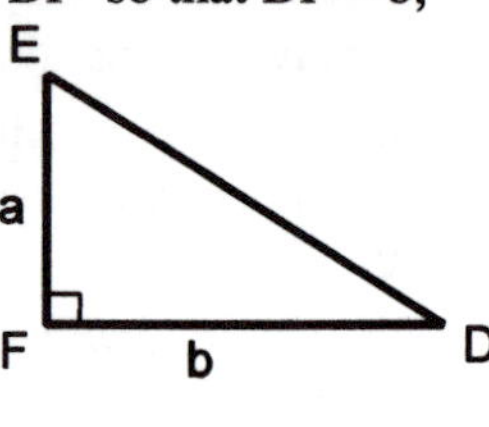

By (?), locate a point E on the $\perp$ so that FE = a.

By Pythagorean Thm, $DE = a^2 + b^2$

$\Delta ABC \cong \Delta DEF$ by SSS.

By definition of $\cong$ ▲s,

$\angle C \cong \angle(?)$

Since $\angle F$ is a right angle by definition of $\perp$ lines, then $\angle C$ must also be a right angle and finally, ΔABC must be a right triangle.

odd answers

1. By Heron's formula, the area of one of the triangles must be $\sqrt{320}$ and since the triangles are congruent, the area of the rhombus is $2\sqrt{320}$ which can be simplified to $16\sqrt{5}$.

3. In any triangle, if $a^2 + b^2 = c^2$, then it must be a right triangle.

5. Erect a $\perp$ on $\overline{DF}$ at F by the theorem which states that given a point P on ℓ, there exists a line $\perp$ ℓ through P. (This theorem was on p. 162.)

By the Ruler Postulate, locate a point E on the $\perp$ so that DE = a.

By definition of $\cong$ ▲s,
$\angle C \cong \angle F$.

Dania

1. In trigonometry (which is usually studied right after advanced algebra) there are two right triangles that are frequently referred to. One of them is the 30°–60°–90° triangle.

If the shortest side is equal to one, find the lengths of the other two sides.

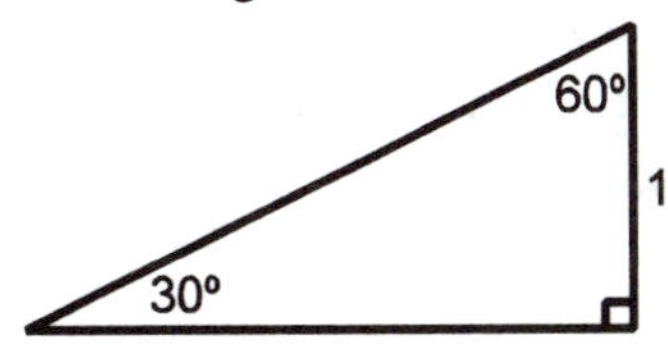

In chapter 9, (p. 330, problem 6) we will prove that in a 30°–60°–90° triangle, the side opposite the 30° (which is equal to 1 in this problem) is half the length of the hypotenuse.

2. Find the areas:

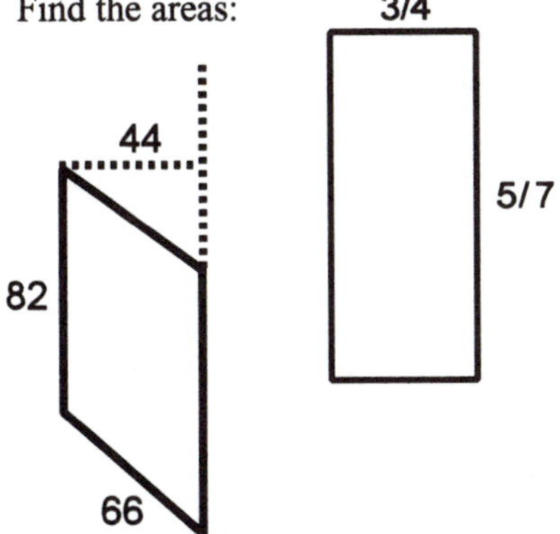

3. In trigonometry there are two right triangles that are frequently referred to. One of them is the 45°–45°–90° triangle.

If the length of each of the legs is each equal to 1, find the length of the hypotenuse.

4. Find the area of this trapezoid if you know that the perimeter equals 32.

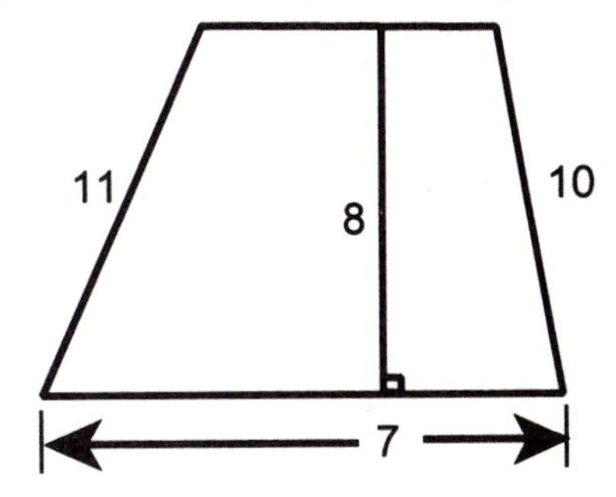

5. The measure of one angle of a parallelogram is 30°. Its sides measure 10 and 16. What is its area?

odd answers

1. The hypotenuse is equal to 2. If we call the length of the third side b, then by the Pythagorean Theorem,

$$1^2 + b^2 = 2^2$$

$$b^2 = 3$$

$$b = \sqrt{3}.$$

3. $\sqrt{2}$

5.

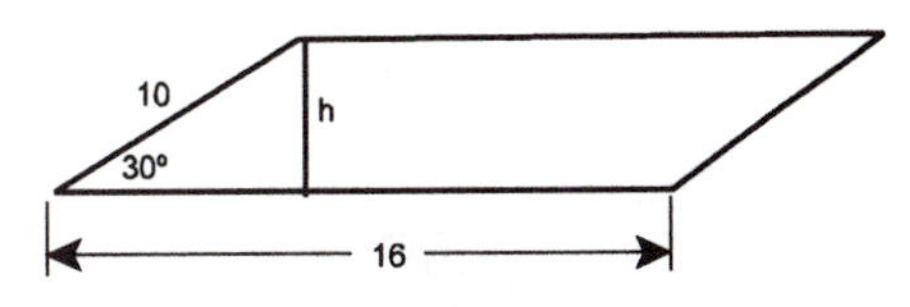

By the theorem mentioned in problem 1, we know that h = 5. The area would be 80.

Jay

1. Find the areas:

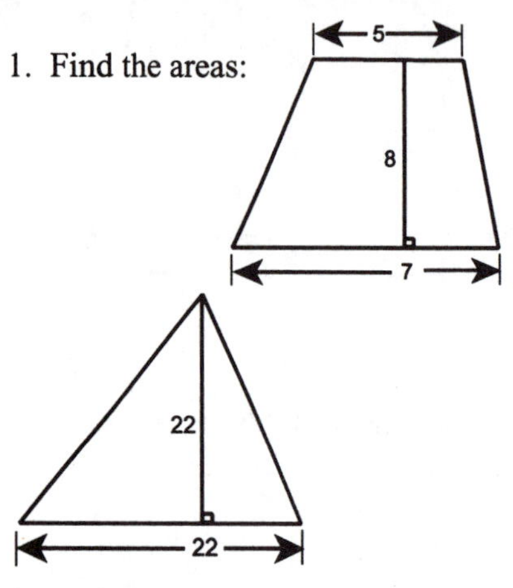

2. A right triangle has a hypotenuse with a length of 13 and one leg with a length of 5. What is the length of the other leg?

3. You are standing one yard north of point C.

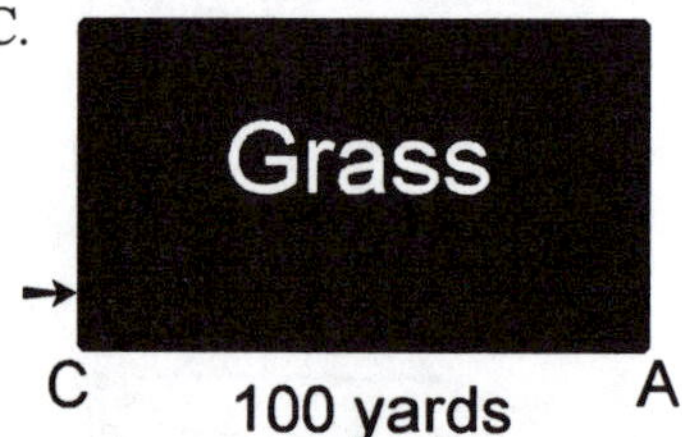

The sign says,

Approximately how much distance would you save by cutting across the lawn to point A, rather than going down to C and then over to A?

4. Find the exact area of

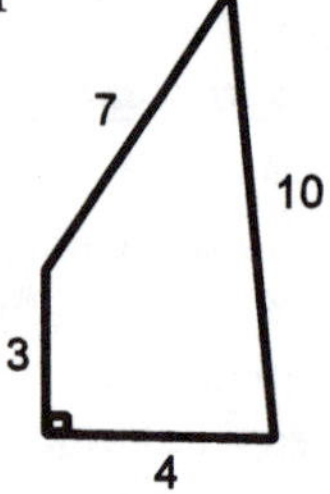

(Hint: If you were to round off your answer to the nearest tenth, it would be 22.2.)

5. How many square inches are there in a square yard?

Kanopolis

1. You are given the lengths of the three sides of each triangle. Which of these are right triangles?

Δ#1: 36 inches, 4 ft. , 5 ft.

Δ#2: 4 ft., 5 ft., 8 ft.

Δ#3: 6 meters, 8 meters, 10 meters

Δ#4: 30 miles, 40 miles, 50 miles

2. Find the area of

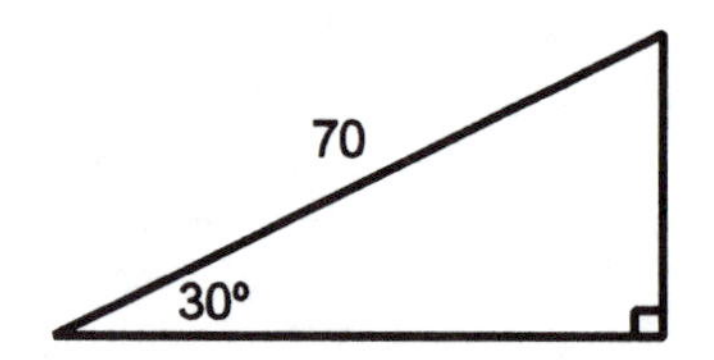

(Hint: Problem 1 in Dania)

3. $\overline{PQ}$ is the midsegment of the trapezoid in the diagram. Find the area of the trapezoid.

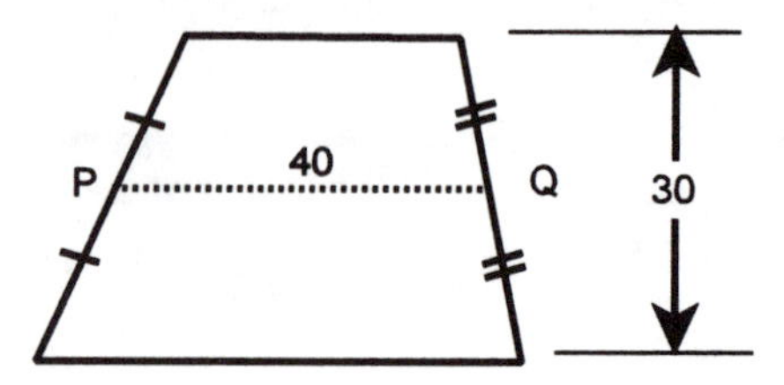

4. Find the exact value of h.

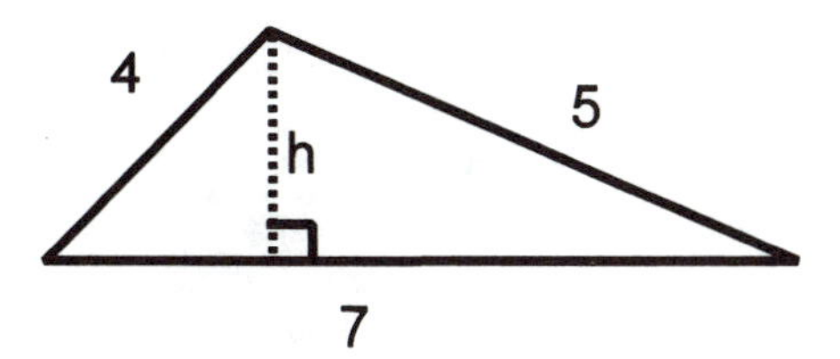

(Hint: The exact value is very close to 2.8.)

5. How would you show that the large triangle in the previous problem is *not* a right triangle?

Solution to the last chain-the-gate problem in chapter 5½. Unlocking any one of the locks opens the gate. $P \vee Q \vee R$

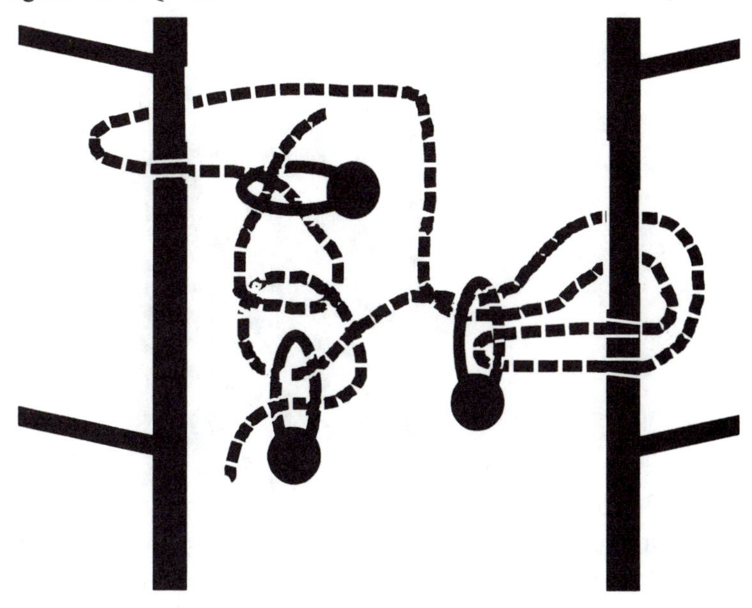

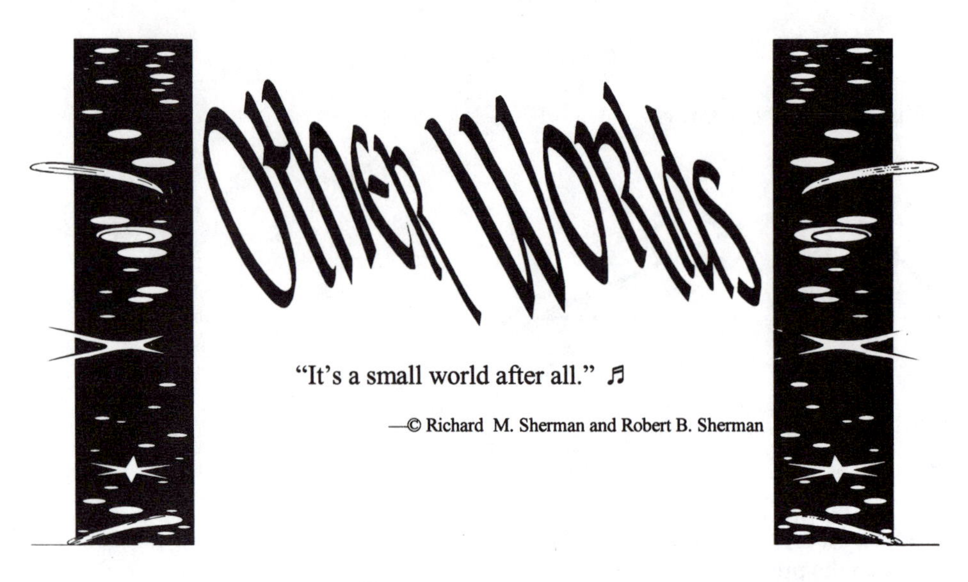

"It's a small world after all." ♬

—© Richard M. Sherman and Robert B. Sherman

Chapter 7½
Junior Geometry and Other Little Tiny Theories

There are so many points in "regular" geometry. It's hard to keep track of all of them. P, Q, R, S . . .

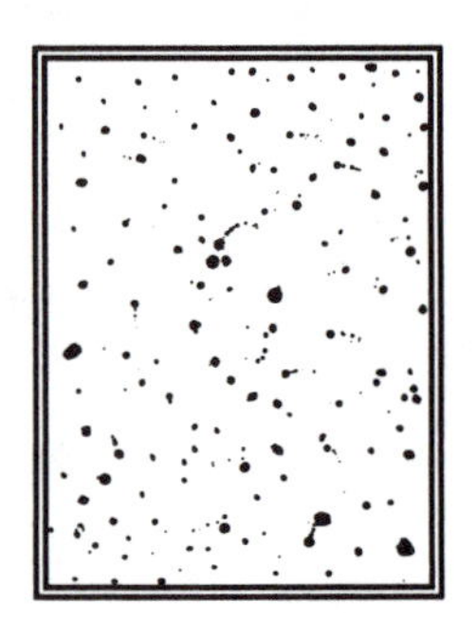

And so many lines through all those points.
ℓ, m, n . . .

And all those postulates relating the undefined terms of *point* and *line.*

Postulate 1: One and only one line can be drawn through any two points.

Postulate 2: There is a one-to-one correspondence between the points on a line and the real numbers so that every point matches up with a single real number and every real number matches up with a single point.

Postulate 3: (Angle Measurement Postulate) You can match up every angle with a number between 0 and 180.

Postulate 4: If two angles form a linear pair, then they are supplementary.

Postulate 5: For any two angles, ∠AOB and ∠BOC, if A–B–C, then m∠AOB + m∠BOC = m∠AOC.

Postulate 6: (SSS) If AB = DE, BC = EF, and AC = DF, then ΔABC ≅ ΔDEF.

Postulate 7: (SAS) If AB = DE, AC = DF, and ∠A ≅ ∠D, then ΔABC ≅ ΔDEF.

Postulate 8: (ASA) If ∠A ≅ ∠D, ∠C ≅ ∠F, and AC = DF, then ΔABC ≅ ΔDEF.

Postulate 9: (The Parallel Postulate) If you have a line ℓ and a point P not on ℓ, then there is at most one line through P that is parallel to ℓ.

Postulate 10: (Area Postulate) Every polygon has a number attached to it which we will call its area.

Postulate 11: If two triangles are congruent, then their areas are equal.

Postulate 12: (Area Addition Postulate) The area of a polygon is the sum of the areas of the (non-overlapping) triangles inside of it, and it doesn't matter which way you cut up the polygon into triangles.

Postulate 13: The area of a square is equal to one of its sides squared.

Wouldn't it be nice if life (and geometry) were just a bit simpler? (Multiple choice: ☐ yes; ☐ no)

Let's start with three undefined terms: *point*, *line*, and *on*.

We will say that points can be *on* lines and also that lines can be *on* points. (Rather than say that lines *contain* points.) That will cut down on the number of different words we need to use.

In this chapter we are going to create a whole new mathematical theory.* It may turn out to be the World's Smallest Geometry.

Now we need our axioms (postulates) to spell out the relationship between the undefined terms. Since we don't want a million points in our little geometry theory, let's start with:

Axiom 1: There exist exactly three points.

We are off to a great start. It is time for our first definition:

Definition Call those three points P, Q and R.

✶ Recall that a mathematical theory starts with undefined terms and assumptions about those undefined terms (postulates) and then builds up definitions and theorems on that base.

This illustration keeps getting tinier

Now we need a postulate that talks about lines:
Axiom 2: Each two points are on exactly one line.

Now if we wished to, we could start cranking out theorems.

Theorem P and Q are on a line.

Statement	*Reason*
1. Points P and Q.	1. Given
2. P and Q are on a line.	2. Axiom 2

Now to prevent 

which would be a pretty boring geometry, we add another postulate:

Axiom 3: Not all the points are on the same line.

Your Turn to Play

1. How many lines are there in this geometry?

COMPLETE SOLUTION

1. Your reasoning might have been:

a) P and Q are on a line (by the theorem).

b) Q and R are on a line, but it can't be the same line as the line on P and Q since by axiom 3 not all points are on the same line. So the line through Q and R is different than the line through P and Q. That gives us two lines so far.

c) Then there has to be a line through P and R and that can't be the same as the two lines we've discovered so far.

d) That covers all the possibilities that I can think of. There has to be three lines in this geometry.

And your reasoning, if it was like that outlined in this *Your Turn to Play* would be—how shall we say it?—wrong. Certainly, you have shown that there are *at least* three lines.

But there could be a fourth line. Nothing in our axioms prevents this.

Axiom 1: There exist exactly three points.
Axiom 2: Each two points are on exactly one line.
Axiom 3: Not all the points are on the same line.

There could be a line which wasn't on any point. It's important to remember that when we build a theory, the only things we are allowed to know are the things that our postulates tell us. Right now we could have:

line ℓ which is on P and Q;
line m which is on Q and R;
line n which is on P and R;
line a which isn't on any point;
line b which isn't on any point;
line c which isn't on any point;
line d which isn't on any point.

There could be an infinite number of lines in our little geometry.

Our last axiom (which will limit the number of possible lines):

Axiom 4: Every two lines are on *at least* one point.

Roughly speaking, this means that every two lines intersect, although we haven't defined *intersect*. This will prevent lines a, b, c and d as described above.

Theorem Every two lines are on *at most* one point.

Statement	*Reason*
1. Two different lines. Call them ℓ and m.	1. Given
2. Assume that ℓ and m are on more than one point.	2. Beginning of an indirect proof
3. ℓ and m are the same line.	3. Axiom 2
4. ℓ and m are not on more than one point.	4. Contradiction in steps 1 and 3 and therefore the assumption in step 2 is false.

Your Turn to Play

1. How many lines are there in this geometry? (This turns out to be the hardest theorem of this junior geometry.)

COMPLETE SOLUTIONS

1. With the addition of our fourth axiom, we could now prove that exactly three lines exist. First, you would demonstrate that there were at least three lines, as you did in the previous *Your Turn to Play*. Then you would begin an indirect proof by assuming that there existed a fourth line. By axiom 4 it would have to share a point in common with each of the other three lines. That would mean that at least two of the points would have to be on this fourth line. Then this fourth line would share two points in common with one of the three original lines. This contradicts axiom 2.

Models

We began with three undefined terms: *point*, *line*, and *on* and four axioms:

<u>Axiom 1</u>: There exist exactly three points.
<u>Axiom 2</u>: Each two points are on exactly one line.
<u>Axiom 3:</u> Not all the points are on the same line.
<u>Axiom 4</u>: Every two lines are on *at least* one point.

We officially know only what the axioms tell us about point, line, and on.

When I wrote *point*, you were probably thinking of ●. When I wrote *line*, you were probably thinking of ←——→. When I wrote *on*, you were probably thinking ←●——→.

I was thinking of *point* = ■, and *line* = ˋ- - -ˋ-ˎ_ - -ˋ , and *on* = ˋ- -■ˎ_ - -ˋ

We had different **models** for the same theory. After all, the undefined terms are *undefined*, aren't they?

Or, as a different model for the theory, I might have been thinking of the points P, Q and R as three friends of mine: Paul, Quincy and Rose. And *lines* are the school clubs. Paul and Quincy belong to the Dance Club. Quincy and Rose belong to the Circus Club and Paul and Rose belong to the Magic Club. If you replace *point* by friend of mine, and *line* by school club, and *on* by belongs to, then all of the axioms are true:

<u>Axiom 1</u>: There are exactly three friends.
<u>Axiom 2</u>: Each two friends belong to exactly one club.
<u>Axiom 3</u>: Not all friends belong to the same club.
<u>Axiom 4</u>: Every two clubs have at least one of my friends who belongs to it.

Paul & Quincy

Paul & Rose

Quincy & Rose

Another model for these four axioms might be bees (points), hives (lines), and visits (on).

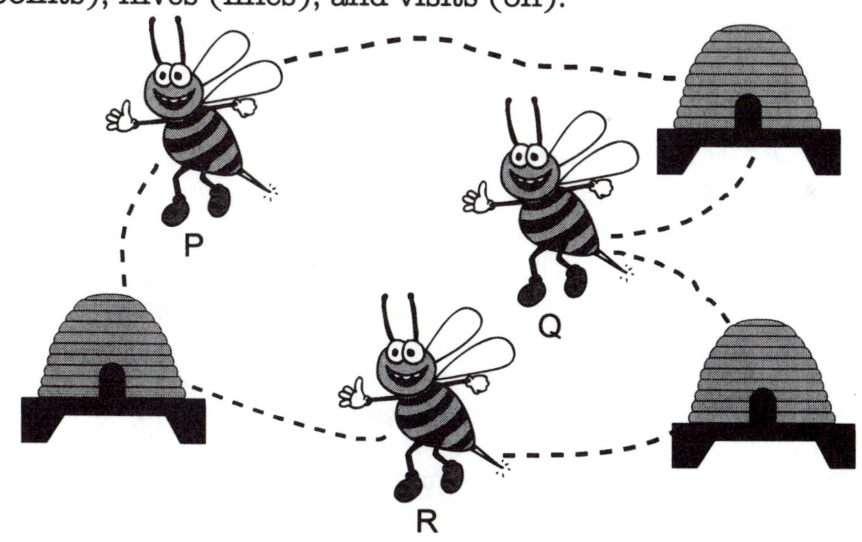

Or maybe just nails and boards.

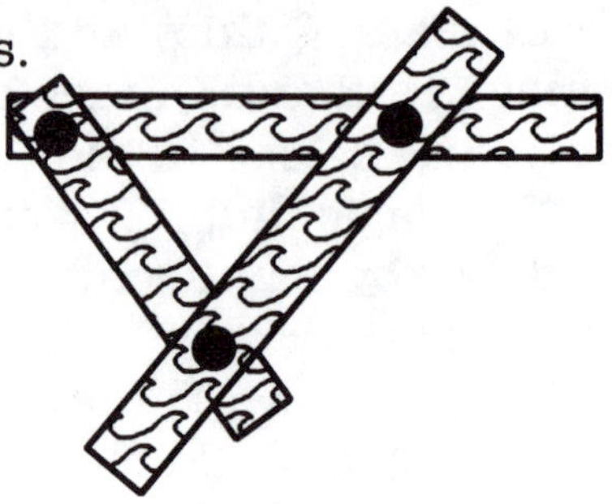

The important thing about theories and their models is that once you've proved a theorem, it will hold true for all the models of the theory.

Three pages ago we proved "Every two lines are on at most one point."

✻ Since the Paul/Quincy/Rose/school clubs is a model for the axioms and undefined terms, we can then say that "Every two school clubs have at most one friend of mine that belongs to both of them."

✻ Since the bees/hives/visits is a model, we can say that any two hives are visited by at most one bee.

✻ Since the nails and boards are a model, we can say that any two boards share at most one nail.

Create an axiom system with 50 models and every time you prove a theorem, you are establishing 50 different facts about the world.

In abstract algebra, which is a math course that is often taken by juniors in college after finishing calculus, they create a little theory called a **group**.*

A group consists of undefined elements a, b, c, d. . . . and an undefined operation ☼ so that if you take two

---

✶ Please notice the phrases "*abstract* algebra," "taken by juniors in college," and "after finishing calculus." These may indicate that the discussion we are about to have about groups may be a tad more recondite (= difficult/arcane/obscure) than "Two points determine a line."

elements and combine them with the operation, you get another element. For example, a ☼ d might equal c. Or b ☼ c might equal a.

There are three axioms for a group:

Axiom 1: (Associative Law) For any three elements, a ☼ (b ☼ c) = (a ☼ b) ☼ c.

Axiom 2: (Identity) There is an element (call it i) such that for every element, a, of the group, a ☼ i = a.

Axiom 3: (Inverse) For every element, a, of the group, there is an element, a', such that a ☼ a' = i.

It takes about a week of teaching for college juniors to start to feel comfortable with these three axioms. Then we prove theorems such as "A group can have only one identity element."

Groups have many important models. Several years ago when nuclear physicists were discovering lots of new subatomic particles, they noticed that a bunch of them formed a group. Or, rather, they almost formed a group. One element was missing. Because of group theory, they were able to predict properties of this undiscovered particle and knew where to look in order to find it.

Here are three models of this theory:

Model #1: Let the elements a, b, c, d . . . be the real numbers (which are all the numbers on the number line). Let ☼ be addition. Then all the axioms are satisfied.

Axiom 1: (Associative) For all real numbers, $a + (b + c) = (a + b) + c$.

Axiom 2: (Identity) Let the identity be zero. For all real numbers, a, it is true that $a + 0 = a$.

Axiom 3: (Inverse) For every real number a, there is a real number $-a$ such that $a + (-a) = 0$.

Model #2: Let the elements a, b, c, d . . . be the positive numbers and let ☼ be multiplication.

Associative: $a(bc) = (ab)c$

Identity: For every positive number a, $a1 = a$.

Inverse: For every positive number a, there is a number $1/a$ such that $a(1/a) = 1$.

And my favorite model of a group . . .

Model #3: (The Spin-A-Duck Model) Start with your average duck: 

The elements of the group are the different ways that you can spin this duck around.

For example, Q will be "rotate 90°":

R means rotate 180°: 

S means rotate 270°: 

And P will mean, "leave the poor duck alone." P is the identity element for this group. 

The group operation ☼ will be, "is followed by."

In this model, Q ☼ Q means rotate 90° followed by another rotation of 90°. So Q ☼ Q has the same effect as R. This may be written: Q ☼ $Q = R$.

In the same way, Q ☼ S is interpreted as rotate 90° followed by a rotation of 270°. So Q ☼ S equals P.

Since Q ☼ $S = P$, and since P is the identity element, we may say that S is the inverse of the element Q.

All this is sooooo much simpler than what you had to endure in elementary school. Back then you were given the MULTIPLICATION TABLE to memorize. Since there is an infinite number of natural numbers, {1, 2, 3, 4, 5 . . . }, you had a lot of memorizing to do.

THE MULTIPLICATION TABLE

you had to memorize
when you were a kid

x	1	2	3	4	5	6	7	8	9	10	11	12	13	14	15	16	17
1	1	2	3	4	5	6	7	8	9	10	11	12	13	14	15	16	17
2	2	4	6	8	10	12	14	16	18	20	22	24	26	28	30	32	34
3	3	6	9	12	15	18	21	24	27	30	33	36	39	42	45	48	51
4	4	8	12	16	20	24	28	32	36	40	44	48	52	56	60	64	68
5	5	10	15	20	25	30	35	40	45	50	55	60	65	70	75	80	85
6	6	12	18	24	30	36	42	48	54	60	66	72	78	84	90	96	102
7	7	14	21	28	35	42	49	56	63	70	77	84	91	98	105	112	119
8	8	16	24	32	40	48	56	64	72	80	88	96	104	112	120	128	136
9	9	18	27	36	45	54	63	72	81	90	99	108	117	126	135	144	153
10	10	20	30	40	50	60	70	80	90	100	110	120	130	140	150	160	170
11	11	22	33	44	55	66	77	88	99	110	121	132	143	154	165	176	187
12	12	24	36	48	60	72	84	96	108	120	132	144	156	168	180	192	204
13	13	26	39	52	65	78	91	104	117	130	143	156	169	182	195	208	221
14	14	28	42	56	70	84	98	112	126	140	154	168	182	196	210	224	238
15	15	30	45	60	75	90	105	120	135	150	165	180	195	210	225	240	255
16	16	32	48	64	80	96	112	128	144	160	176	192	208	224	240	256	272
17	17	34	51	68	85	102	119	136	153	170	187	204	221	238	255	272	289
18	18	36	54	72	90	108	126	144	162	180	198	216	234	252	270	288	306
19	19	38	57	76	95	114	133	152	171	190	209	228	247	266	285	304	323
20	20	40	60	80	100	120	140	160	180	200	220	240	260	280	300	320	340
21	21	42	63	84	105	126	147	168	189	210	231	252	273	294	315	336	357
22	22	44	66	88	110	132	154	176	198	220	242	264	286	308	330	352	374
23	23	46	69	92	115	138	161	184	207	230	253	276	299	322	345	368	391

The duck rotation table
is quite small in comparison.

☼	P	Q	R	S
P	P	Q	R	S
Q	Q	R	S	P
R	R	S	P	Q
S	S	P	Q	R

Chapter Eight

Similar Triangles

Bong! Bong! Bong! As the twelfth tone sounded from the bell tower, Fred was sitting in his assigned seat next to the university president. Fred was an on-time kind of person.

The president arose to begin the first of the Special Awards Ceremonies. They were scheduled today at noon, 1 P.M., 2 P.M., 3 P.M., 4 P.M., and 5 P.M. Tomorrow the hourly ceremonies would begin at 8 A.M. and continue all day.

Fred was impressed with the 6000 chairs that the Wistrom brothers had set up. There was a chair for every student on campus, every administrator, and, of course, chairs for faculty, families, friends, alumni, and the media.

The first six rows were occupied by the administrators. (They were required to attend.)

The rest of the seats were empty.

The president opened the binder he was carrying and began to read his 57-minute speech, which traced the history of KITTENS University. It mentioned all the former deans who had retired. It described each building on campus—when it was built, who was the architect, and what its square footage was. It mentioned the great advances that the engineering department had made—such as the last day that a student ever carried a slide rule on campus.

On and on he droned oblivious to his audience, many of whom were talking on their cell phones during his speech. He would use the same speech at all the ceremonies. The only variation would come in the last three minutes of each hour when he would present a trophy. His speech notes at this point read: `Pick up trophy. Read inscription. Hand to recipient.`

As the president read, "To Professor Fred Gauss for his bravery in solving the Bocci Ball Lawn Vandalism case," an assistant held up a sign which read, "PREPARE TO APPLAUD."

When Fred was handed his trophy, the assistant held up a second sign, "APPLAUD NOW." Three administrators who happened to notice the sign clapped perfunctorily. When they stopped after two seconds, Fred noticed a tiny, but much more vigorous clapping coming from the back row. A six-year-old girl yelled, "Way to go!"

After the ceremony was over, Fred hopped off the dais and headed back to see his sweetheart.

"I'm sorry I had to dash from your house so quickly," he began. "I hadn't been watching the clock."

"Oh, don't worry about that. You did swell up there. Everyone is so proud of you."

Fred thought to himself, Swell? All I did was avoid falling asleep. This speech was the same one that the president gave last year at the Special Awards Ceremonies.

"Well," she smiled. "I gotta get going now. See you at a quarter to six."

Fred had half expected her to say, "Toowoomba till 5:45."

P skipped off. She didn't look back.

Suddenly Fred was alone. He could hear the president in the distance beginning his speech again. With classes canceled today and tomorrow he didn't have teaching to do for the rest of the week. It was too soon to write P another love letter. He had done his shopping for the day. The Ralph's Mouth Dust Powder was a disappointment, but the Hecks Kitchen grill should be a delight.

He began walking.

Not for the exercise. Not to get anywhere in particular. Just to experience a spring-like afternoon. He thought about the two thousand books he'd read and enjoyed thus far in his life. They had done much to enlarge his view of life. When Fred was young—a relative term for a six-year-old—he headed to the library and asked the librarian what was good to read. He was directed to a collection of books* which pointed to the best books that have ever been written. There are a hundred or so books that seem to pop up on almost everyone's list of most significant books ever written: from earliest times (Homer's *Odyssey,* Sophocles' *Antigone*, and the Bible) through the plays of Shakespeare and Lewis Carroll's *Alice's Adventures in Wonderland*, to J. R. R. Tolkien's *The Lord of the Rings*. In his reading Fred had been in "conversation" with some of the greatest visionaries our world has ever known.

✶ Such as Philip Ward's *A Lifetime's Reading*, Clifton Fadiman's *The Lifetime Reading Plan*, and David Denby's *Great Books*.

Fred thought about how his love for P compared with that of Tristan for Iseult. (Not very well, since Fred and P are only six years old and Fred has only known P for an hour or so.)

Fred walked past the tennis courts, the university chapel, and through the KITTENS arboretum with its stepping stone paths. He was lost in his thoughts, a sweet reverie. He was grateful to God for his days of his life.

Out on the Great Lawn (as listed on the KITTENS campus map) which surrounded the arboretum, he saw students lying on blankets enjoying the Kansas sun. Some were reading, some conversing with each other, and some just sunbathing. Beside one of the fellows who had just finished doing a bunch of pushups was a very familiar bottle.

He watched as the guy opened a bottle of Ralph's Mouth Dust Powder and poured some powder into his hands. Fred immediately flashed on the converse of the Midsegment Theorem: A line which passes through the midpoint of one side of a triangle and is parallel to the third side will pass through the midpoint of the second side.

Fred's mouth went dry, anticipating that the guy was going to stick the powder in his mouth. Instead he rubbed it all over his arms and chest! It smelled a lot better than it tasted.

"Hi, Professor Fred!" It was one of his calculus students. "Would you like to try some?" He handed Fred the bottle.

Fred took the bottle and turned it over and read the label on the back:

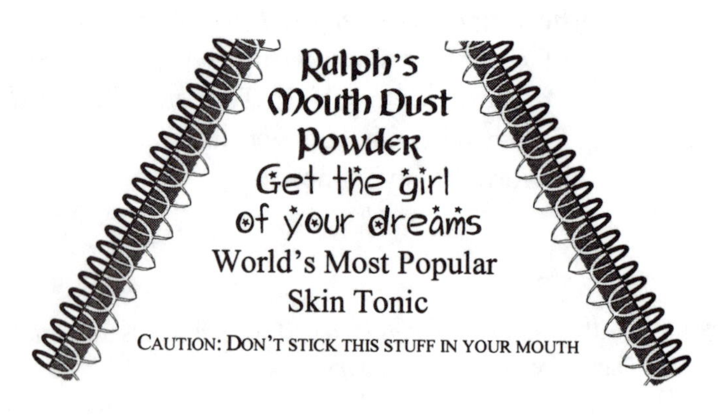

Ralph's
Mouth Dust
Powder
Get the girl
of your dreams
World's Most Popular
Skin Tonic
CAUTION: DON'T STICK THIS STUFF IN YOUR MOUTH

Son of a gun Fred thought. I should have read the label. It's a dusting powder. Ralph's Mouth is maybe the name of some small town somewhere. They should have called it Dusting Powder from Ralph's Mouth. Yuck! Even that sounds bad. Fred set the bottle down next to the student's blanket.

The bottle no longer qualified as a good illustration of the Midsegment Theorem. The line didn't pass through the midpoint of one side of the triangle. When Fred looked at the bottle now, he could see two triangles. They were the same shape, but not the same size. The two triangles were not congruent.

The smaller triangle

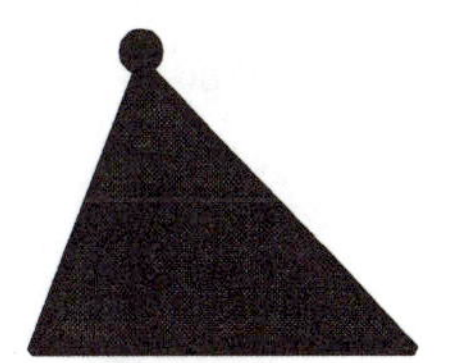

The larger triangle

These two triangles reminded him of a photo that he had taken of his beloved llama. He had two copies printed.

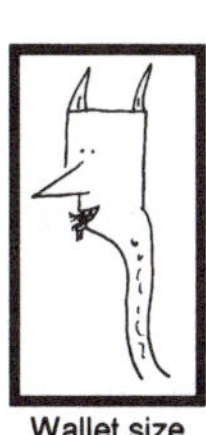

Wallet size

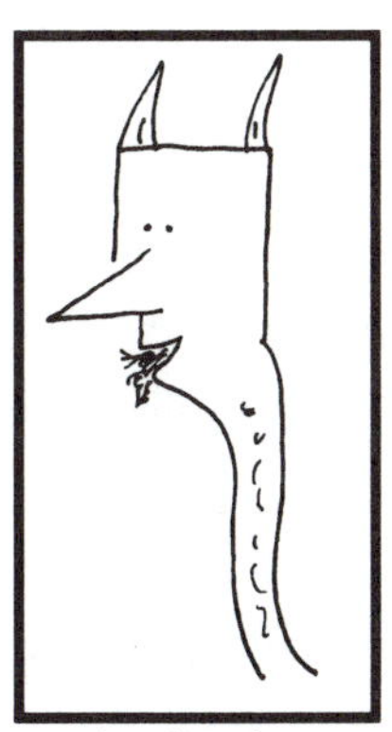

Jumbo size

The two photos were **similar**. They had the same shape, but not necessarily the same size. The two triangles in the Ralph's Mouth Dust Powder were also similar.

Your Turn to Play

1. Which pairs of illustrations are similar?

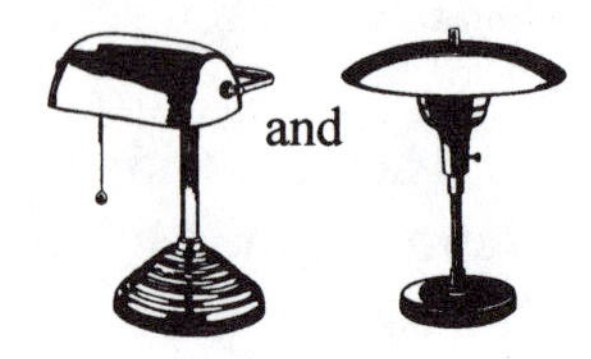

2. Are all squares similar?

3. Are all equilateral triangles similar?

4. Are all right triangles similar?

5. What would be wrong with the following definition? ΔABC is similar to ΔDEF iff they have the same shape.

6. Are all circles similar?

7. In advanced algebra we will study ellipses. (Some people call them ovals.)

If you take a coin and look at it from an angle, its edge will be an ellipse.

Are all ellipses similar?

1. The birds are similar. Rotation doesn't affect the shape.

The lamps are the same size, but they have different shapes. They aren't similar.

The snowmen are similar. Reflection noitɔelʇeЯ doesn't change the shape.

2. No matter what size you draw them or how you rotate them, any two squares will always be similar.

3. Any two equilateral triangles will always have the same shape. They will be similar.

4. 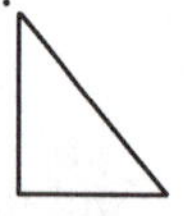 and 

don't have the same shape. No matter how you enlarge or reflect or rotate the first triangle, you will never make it look like the second triangle. They are not similar.

5. We have not defined what "shape" means in our geometry theory and "shape" is not one of the undefined terms. When it comes time to define similar triangles, we will need to do it *using words we have already defined.* If you ask what the German word Dunst means, would you be very happy if you were told it is a thin Nebel? Or would you rather have Dunst defined as a mist? (A Nebel, of course, being a thick mist.) (For the record, note that an "N" is an N, not an R. R's are written "R" in the old fashioned German script.)

6. Yes.

7. Ellipses can come in various shapes. In calculus, we will describe very flat ellipses by saying that their **eccentricity** is very close to one. Ellipses that are almost circles will have eccentricities close to zero.

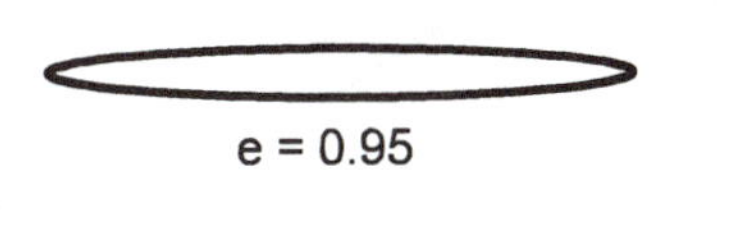

e = 0.95

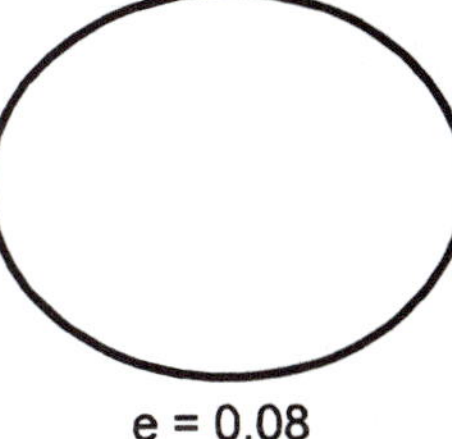

e = 0.08

When we defined congruent triangles back in chapter three, we said that the corresponding angles must be congruent and the corresponding sides must be congruent.

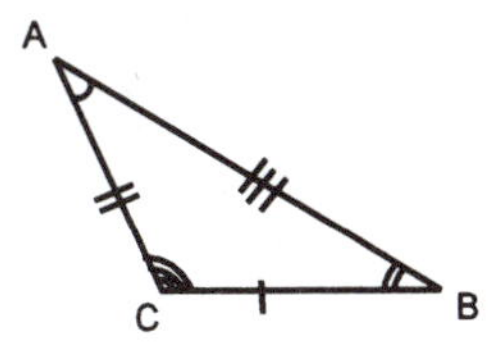

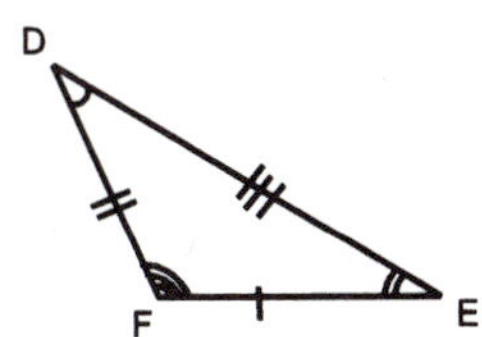

We now define similar triangles and say that the corresponding angles must be congruent and the corresponding sides must be *proportional.*

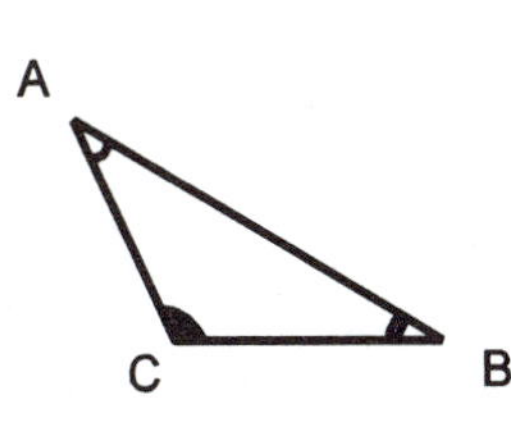

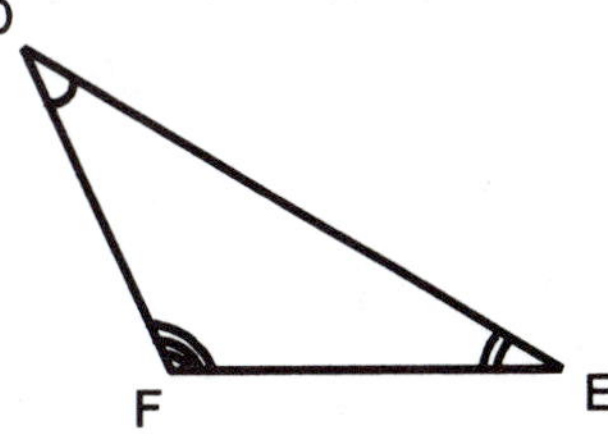

<u>Definition</u>: $\triangle ABC$ is **similar** to $\triangle DEF$ iff
$\angle A \cong \angle D$, $\angle B \cong \angle E$, $\angle C \cong \angle F$, and $\dfrac{AB}{DE} = \dfrac{BC}{EF} = \dfrac{CA}{FD}$

The equation $\frac{AB}{DE} = \frac{BC}{EF} = \frac{CA}{FD}$ is hard to read with all those letters in it.

It's much more pleasant to express this in terms of the lengths of the sides of the triangles:

$$\frac{a}{d} = \frac{b}{e} = \frac{c}{f}$$

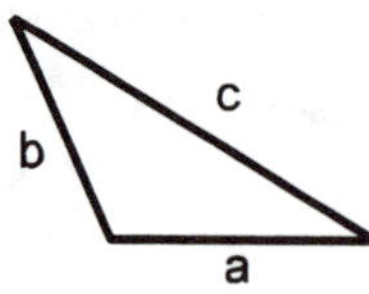

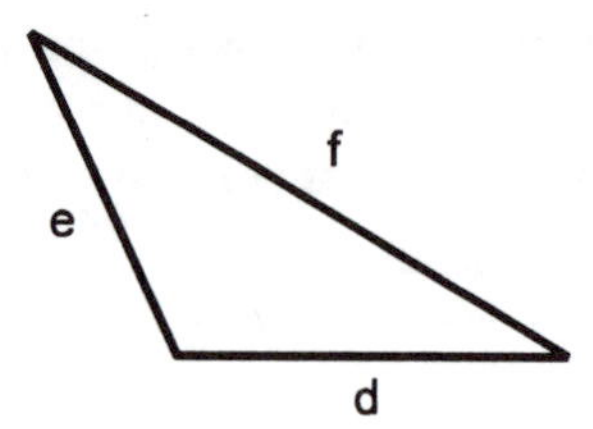

And if you really want to make it simple, you could use primes:

$$\frac{a}{a'} = \frac{b}{b'} = \frac{c}{c'}$$

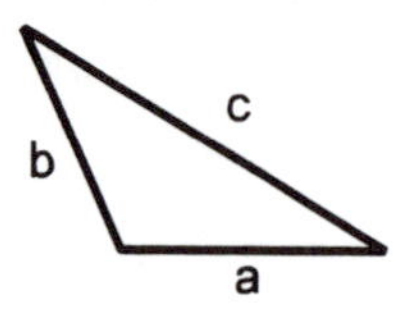

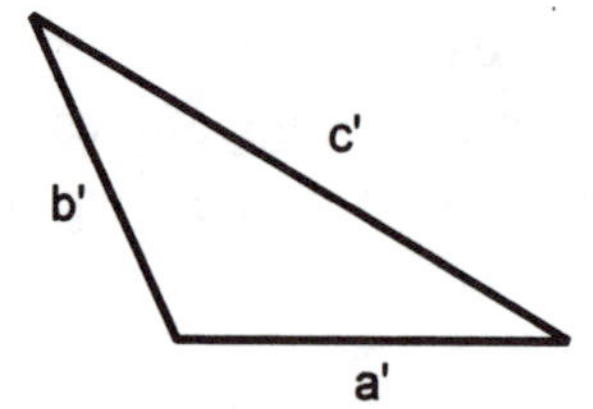

It's your choice.

In working with congruent triangles we had a postulate (SSS) that if the corresponding sides were congruent, then the triangles were congruent. We also had a postulate (SAS) that if two pairs of corresponding sides were congruent and the included angles were congruent, then the triangles were congruent. And we had the ASA postulate.

For similar triangles, things are simpler:

Postulate 14: (AA) $\triangle ABC$ is similar to $\triangle DEF$ if $\angle A \cong \angle D$ and $\angle B \cong \angle E$.

If you like abbreviations, Post 14: AA ➔ ~ ⩟.
("~" is the symbol for similar.)

Your Turn to Play

1. If two angles of one triangle are 30° and 40° and two angles of a second triangle are 30° and 40°, are the triangles always/sometimes/never similar? (Choose one alternative.)

2. If two angles of one triangle are 60° and 70° and two angles of a second triangle are 50° and 60°, are the triangles always/sometimes/never similar?

3. If two angles of one triangle are 10° and 50° and two angles of a second triangle are 20° and 50°, are the triangles always/sometimes/never similar?

4. If one angle of a rhombus is 60° and an angle of a second rhombus is 120°, are the rhombuses similar?

5. If the two triangles in the diagram are similar, find x and y.

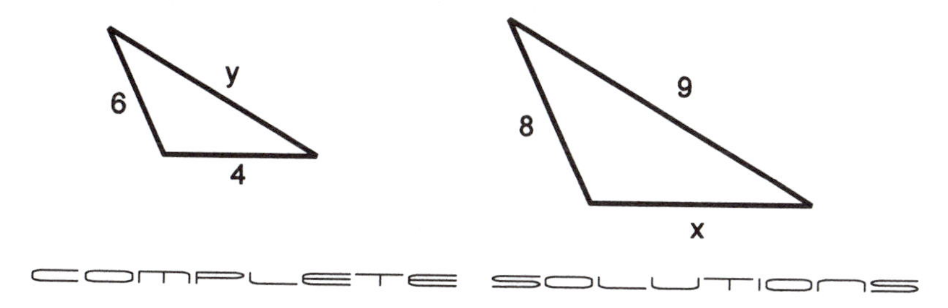

1. By the AA $\rightarrow \sim \triangle$ postulate, they must always be similar.

2. In the first triangle, if two of the angles are 60° and 70°, then the third angle must be 50° since the sum of the angles in any triangle equals 180°. By AA they must always be similar.

3. The first triangle's angles are 10°, 50° and 120°. The second triangle's angles are 20°, 50° and 110°. They are never similar.

4. The consecutive angles of a rhombus (or even a parallelogram) are supplementary (they add to 180°). Both rhombuses look like [rhombus: 60°, 120°] and will both have the same shape. They will always be similar.

5. Since the triangles are similar, the corresponding sides will be proportional. $\frac{6}{8} = \frac{4}{x}$ As we did in beginning algebra, to solve an equation containing fractions we multiply both sides by an expression that both denominators divide into evenly. In this case we multiply both sides by 8x. $\frac{6 \cdot \mathbf{8x}}{\mathbf{8}} = \frac{4 \cdot \mathbf{8x}}{\mathbf{x}}$

This eliminates the denominators $6x = 32$

Divide both sides by 6 $x = 32/6$ or $16/3$ or $5\ 1/3$.

Now to find y.

$$\frac{y}{9} = \frac{6}{8}$$

Multiplying both sides by 72

$$\frac{y \cdot 72}{9} = \frac{6 \cdot 72}{8}$$

$$8y = 54$$

$$y = 54/8 \text{ or } 27/4 \text{ or } 6\ 3/4 \text{ or } 6.75$$

On the blanket next to Fred were a couple who were playing cat's cradle with a loop of string. Next to them was a juggler. Near him was a woman who was busy converting part of her pet into a sweater. The fellow who had been doing pushups was now sorting his vitamin pills into little containers.

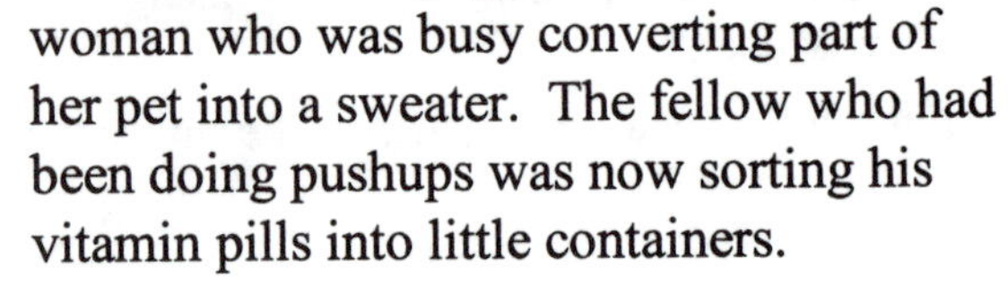

Fred felt the urge to **manipulate**. Everyone else was. He needed to flip, toss, squish, wiggle s-o-m-e-t-h-i-n-g. The herd instinct was starting to have an effect on him. Among Fred's favorite objects to manipulate are proportions, and he could do all kinds of tricks with them.

In beginning algebra, proportions were just defined as two ratios set equal to each other: $\frac{a}{b} = \frac{c}{d}$ which wasn't a tremendously exciting concept back in algebra since we didn't use them very much then. We'll have much more use for proportions now* in working with similar triangles.

---

* Dealing with "proportions now" is a lot easier than back in the old days when the notation was more difficult to deal with. The old way of writing $\frac{a}{b} = \frac{c}{d}$ was a:b::c:d which was introduced back in 1657. Colons are not the world's most popular punctuation; at least they weren't until William Oughtred invented a:b::c:d. His colon-filled way of writing proportions remained in common use even into the 1920s.
In case you have trouble picturing 1657, here it is on the time line:

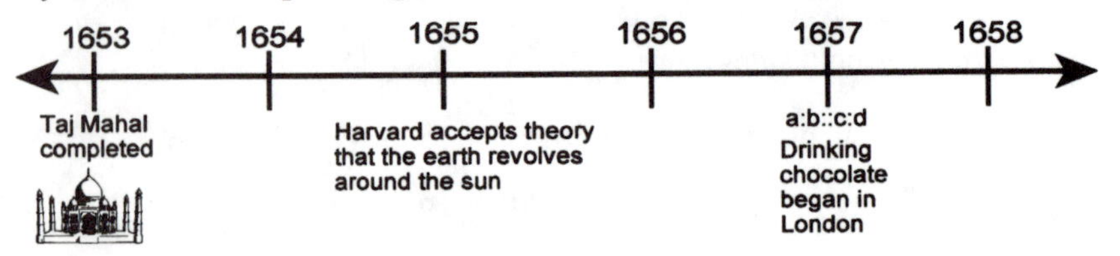

Here's a chart of . . .

Fred's Five Favorite Tricks With Proportions		
The Trick	**Why It Works**	**An Example**
Inverting $\frac{a}{b} = \frac{c}{d}$ becomes $\frac{b}{a} = \frac{d}{c}$	Multiply both sides of $\frac{a}{b} = \frac{c}{d}$ by $\frac{\mathbf{bd}}{\mathbf{ac}}$ and you get $\frac{\mathbf{abd}}{\mathbf{bac}} = \frac{\mathbf{cbd}}{\mathbf{dac}}$ which simplifies to $\frac{d}{c} = \frac{b}{a}$ Then switch the sides of the equation.	$\frac{3}{4} = \frac{6}{8}$ $\frac{4}{3} = \frac{8}{6}$
Cross multiplying $\frac{a}{b} = \frac{c}{d}$ becomes ad = bc	Multiply both sides of $\frac{a}{b} = \frac{c}{d}$ by **bd** and you get $\frac{\mathbf{abd}}{b} = \frac{\mathbf{cbd}}{d}$ which simplifies to ad = bc.	$\frac{2}{7} = \frac{6}{21}$ $2 \times 21 = 7 \times 6$
Interchange the means $\frac{a}{b} = \frac{c}{d}$ becomes $\frac{a}{c} = \frac{b}{d}$	Multiply both sides of $\frac{a}{b} = \frac{c}{d}$ by $\frac{\mathbf{b}}{\mathbf{c}}$ and you get $\frac{\mathbf{ab}}{\mathbf{bc}} = \frac{\mathbf{cb}}{\mathbf{dc}}$ which simplifies to $\frac{a}{c} = \frac{b}{d}$ In a:b::c:d, b and c are called the **means**.	$\frac{1}{3} = \frac{4}{12}$ $\frac{1}{4} = \frac{3}{12}$
Adding bottoms to tops $\frac{a}{b} = \frac{c}{d}$ becomes $\frac{a+b}{b} = \frac{c+d}{d}$	Add one to both sides of $\frac{a}{b} = \frac{c}{d}$ $\frac{a}{b} + 1 = \frac{c}{d} + 1$ $\frac{a}{b} + \frac{b}{b} = \frac{c}{d} + \frac{d}{d}$	$\frac{2}{3} = \frac{12}{18}$ $\frac{5}{3} = \frac{30}{18}$
Subtract bottoms from tops $\frac{a}{b} = \frac{c}{d}$ becomes $\frac{a-b}{b} = \frac{c-d}{d}$	Subtract one to both sides of $\frac{a}{b} = \frac{c}{d}$ $\frac{a}{b} - 1 = \frac{c}{d} - 1$ $\frac{a}{b} - \frac{b}{b} = \frac{c}{d} - \frac{d}{d}$	$\frac{6}{5} = \frac{42}{35}$ $\frac{1}{5} = \frac{7}{35}$

Your Turn to Play

Fred invented a proportion in his head: $\frac{32}{55} = \frac{\xi}{9}$

(ξ, xi, is my favorite Greek letter. Rhymes with "eye.") Then he mentally tossed the proportion in the air and bounced it around like skillful soccer players can do with a ball.

For each bounce, tell which of Fred's five favorite tricks he used to manipulate the proportion.

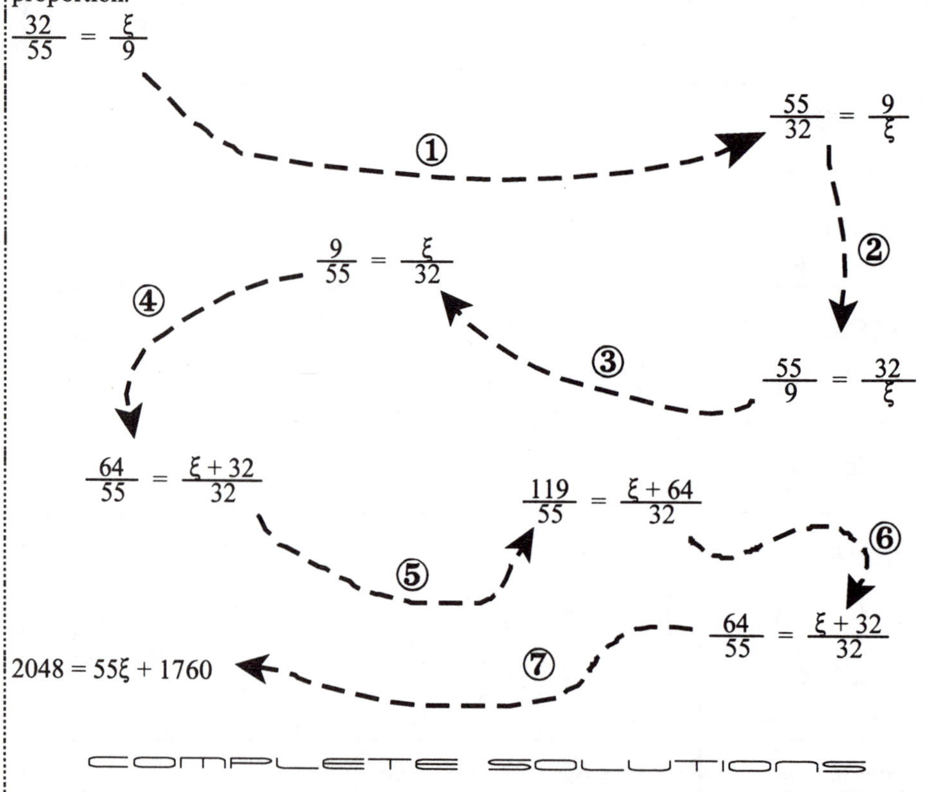

COMPLETE SOLUTIONS

1. invert 2. interchange the means 3. invert 4. add bottom to top
5. add bottom to top again 6. subtract bottom from top 7. cross multiply

The **means** of $\frac{a}{b} = \frac{c}{d}$ are b and c. The **extremes** are a and d. In the first three steps of Fred's bouncing around his proportion he transformed $\frac{32}{55} = \frac{\xi}{9}$ into $\frac{9}{55} = \frac{\xi}{32}$ which might be called **interchanging the extremes**.

Fred lay down on his tummy on the grass and looked at the bottle again. He could now easily prove the generalization of the converse of Midsegment Theorem.

(The converse of the Midsegment Theorem: A line which is parallel to one side of a triangle and which passes through the midpoint of one side will pass through the midpoint of the second side.)

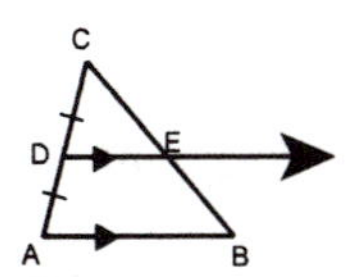

Theorem: (The Generalization of the Converse of the Midsegment Theorem*) A line parallel to one side of a triangle intercepts proportional segments on the other two sides.**

Translation: If $\overrightarrow{DE} \parallel \overline{AB}$, then $\frac{x}{w} = \frac{z}{y}$

It's easy to prove.

☞ Since the lines are parallel, we have $\angle CDE \cong \angle A$ by PCA.

☞ We note that $\angle C \cong \angle C$.

☞ Then our new postulate (AA ➜ ~ ⚠) gives us $\triangle CDE \sim \triangle CAB$.

☞ By definition of similar triangles, the corresponding sides are proportional: $\frac{x}{x+w} = \frac{z}{z+y}$

☞ And now anyone familiar with Fred's Favorite Five can invert: $\frac{x+w}{x} = \frac{z+y}{z}$

Subtract bottom from top: $\frac{w}{x} = \frac{y}{z}$

And invert again: $\frac{x}{w} = \frac{z}{y}$ ⊠

(The tricks with proportions *do* come in handy.)

✶ Yuck! That's an extremely long name. We might call it the G of the C of the M Theorem. But even that's too long. How about GCM Theorem?

✶✶ Assuming, of course, that it hits them.

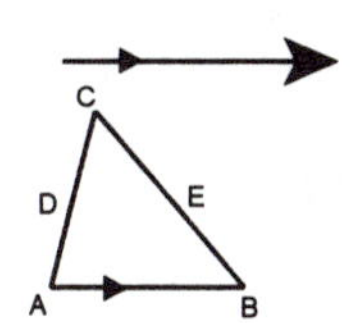

Fred rolled over on his back and looked at the clouds. He giggled. What about the Converse of the Generalization of the Converse of the Midsegment Theorem? If a line intercepts proportional segments on two sides of a triangle, then it's parallel to the third side.

<u>Theorem</u>: (The Converse of the Generalization of the Converse of the Midsegment Theorem)

If $\frac{x}{w} = \frac{z}{y}$ then $\overrightarrow{DE} \parallel \overline{AB}$.

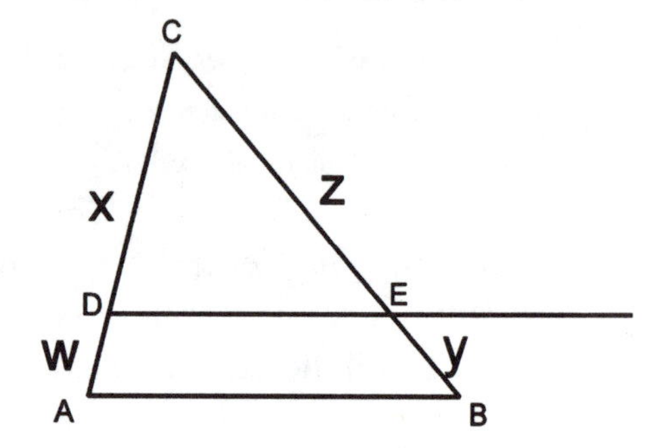

Now wait a minute! I as your reader can only take so much of this nonsense. Would you look at the length of the name of this theorem! It's twice as long as the theorem itself. Watch me. I'll measure them. I even have to put the word "theorem" in 3-point type to fit everything on one line.

The Converse of the Generalization of the Converse of the Midsegment Theorem

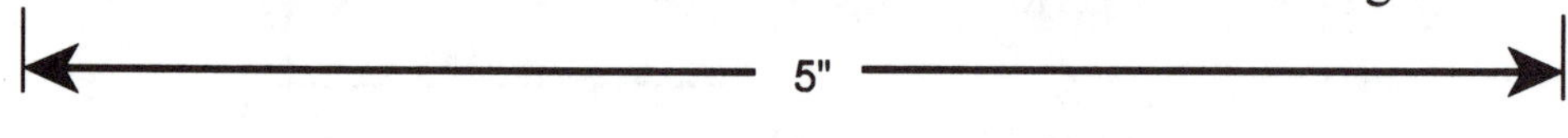

$\frac{x}{w} = \frac{z}{y}$ then $\overrightarrow{DE} \parallel \overline{AB}$

1.52"

And now you're probably going to tell me that we can abbreviate this theorem as the CGCM Theorem.

Hey! You left out the "If." With the "if" in there, it will be longer than 1.52".

If $\frac{x}{w} = \frac{z}{y}$ then $\overrightarrow{DE} \parallel \overline{AB}$

1.72"

It's still less than half as long as the silly name you gave for this theorem. I'd be happy calling it "If a line intercepts proportional segments on two sides of a triangle then it is parallel to the third side."

It's a deal.

I should point out that the reason Fred was giggling when he named this theorem "The Converse of the Generalization of the Converse of the Midsegment Theorem" is that it really is just the Generalization of the Midsegment Theorem (GM Thm). In the Midsegment Theorem, the line passes through the midpoints of two of the sides of the triangle, and in this generalization, it cuts the two sides proportionally.

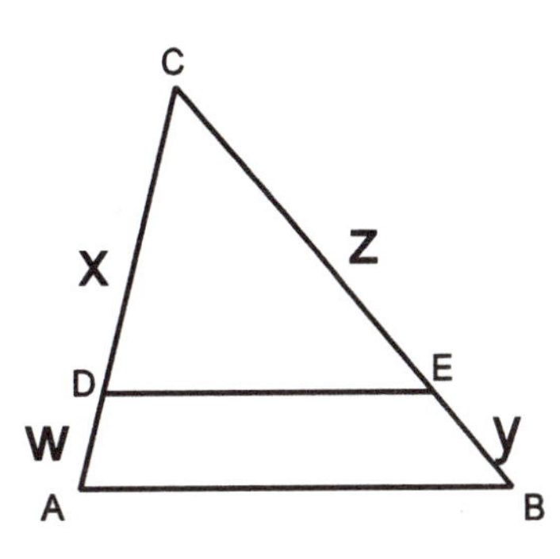

To prove the GM Theorem, we assume that $\overline{DE}$ is not parallel to $\overline{AB}$. (The beginning of an indirect proof.)

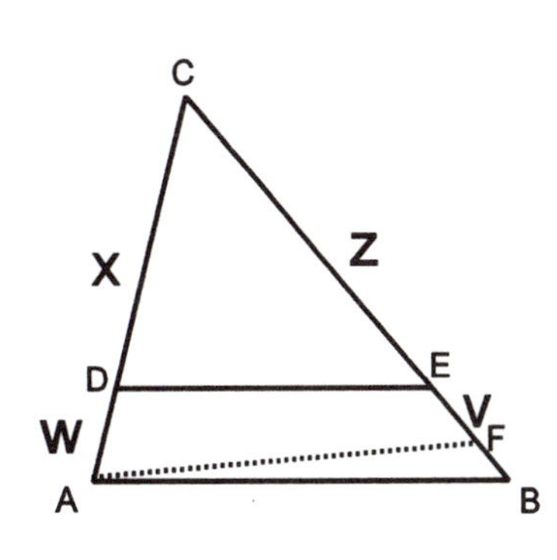

Then there must be a line $\overline{AF}$ through point A which is parallel to $\overline{DE}$.

(F isn't the same as B, otherwise $\overline{AB}$ and $\overline{AF}$ would be the same line and we're operating under the assumption that $\overline{DE} \nparallel \overline{AB}$.)

We let EF = v. Then by the GCM Theorem which we just proved $\frac{x}{w} = \frac{z}{v}$

Mix that in with the given equation $\frac{x}{w} = \frac{z}{y}$ and we have v = y. By the Ruler Postulate point F and point B must be the same point. ⊠

Combining these two theorems: $\overrightarrow{DE} \parallel \overline{AB}$ iff $\frac{x}{w} = \frac{z}{y}$

(Recall: "iff" means "if and only if.")

Your Turn to Play

1. Find x, y and z.

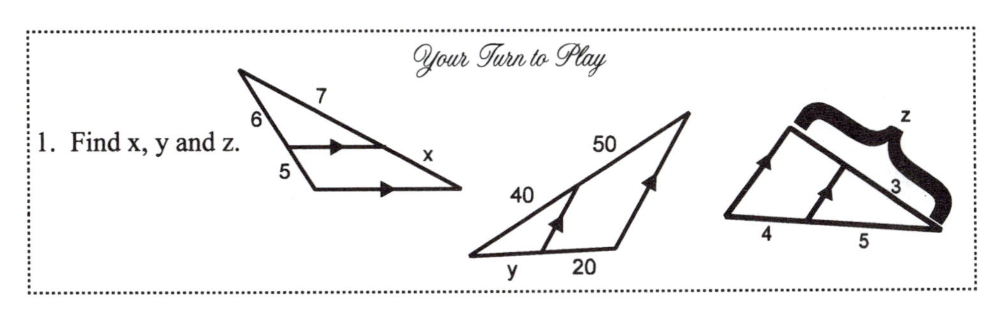

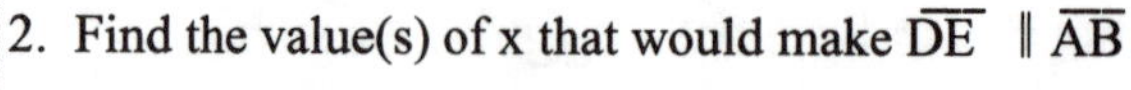
2. Find the value(s) of x that would make $\overline{DE} \parallel \overline{AB}$

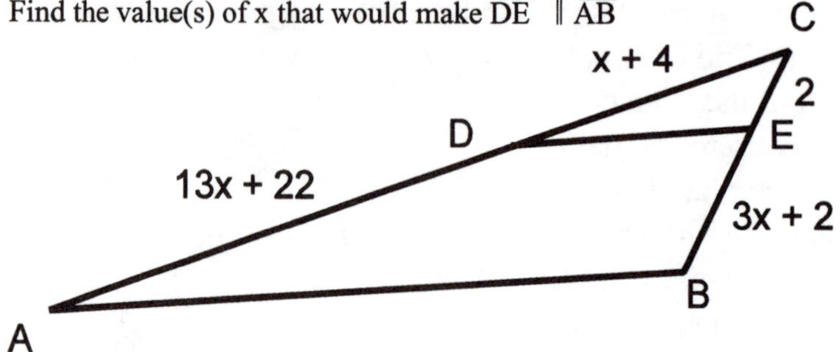

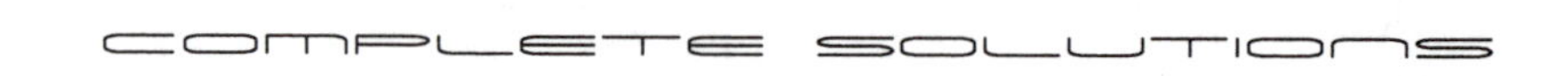

1. $6/5 = 7/x$ Cross multiplying: $6x = 35$ $x = 35/6$ or $5\ 5/6$

$y/20 = 40/50$ Cross multiplying: $50y = 800$ $y = 16$

Filling in the missing length:

$$\frac{5}{4} = \frac{3}{z-3}$$

$5(z - 3) = 12$

$5z - 15 = 12$

$5z = 27$

$z = 27/5$ or $5\ 2/5$ or 5.4

2. $$\frac{x+4}{13x+22} = \frac{2}{3x+2}$$

Cross multiply $(x + 4)(3x + 2) = 2(13x + 22)$

Multiply out $3x^2 + 2x + 12x + 8 = 26x + 44$

Transpose everything to the left side $3x^2 - 12x - 36 = 0$

Divide by 3 $x^2 - 4x - 12 = 0$

Factor $(x - 6)(x + 2) = 0$

Set each factor equal to zero $x - 6 = 0$ OR $x + 2 = 0$

Solve $x = 6$ OR $x = -2$

We can't use the $x = -2$ since that would make the $3x + 2$ length negative.

Using the $x = 6$ we have

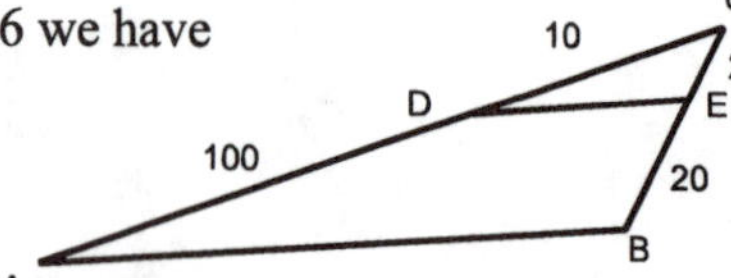

As Fred looked up at the clouds, he thought of all those artists who could paint wonderful cloud pictures. When he had tried to draw landscapes they always seemed to have a six-year-oldish feel to them.

What Fred Saw

What Fred Drew

Being a famous landscape painter wasn't going to be the road to fame and fortune for him. Maybe cartooning! Some comic strip artists draw like I draw. I could combine my art skills with my math teaching. He thought of the SAS Similarity Theorem which many geometry books never mention and even fewer prove:

Theorem: (SAS Similarity) Two triangles are similar if one pair of corresponding angles are congruent and the including sides are proportional.

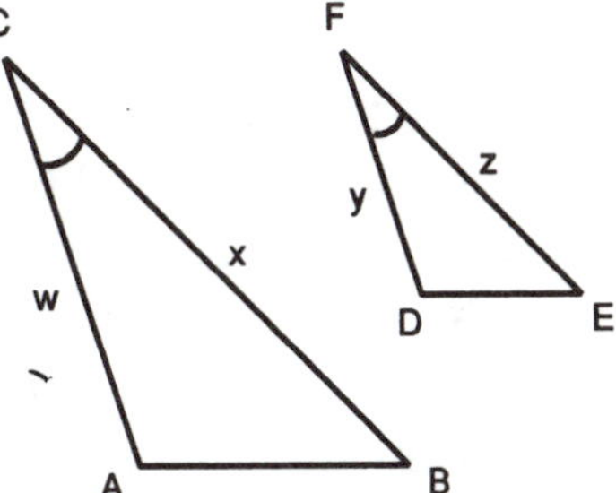

Given: $\angle C \cong \angle F$ and $\frac{w}{y} = \frac{x}{z}$

To show: $\Delta ABC \sim \Delta DEF$

Fred's cartoon proof

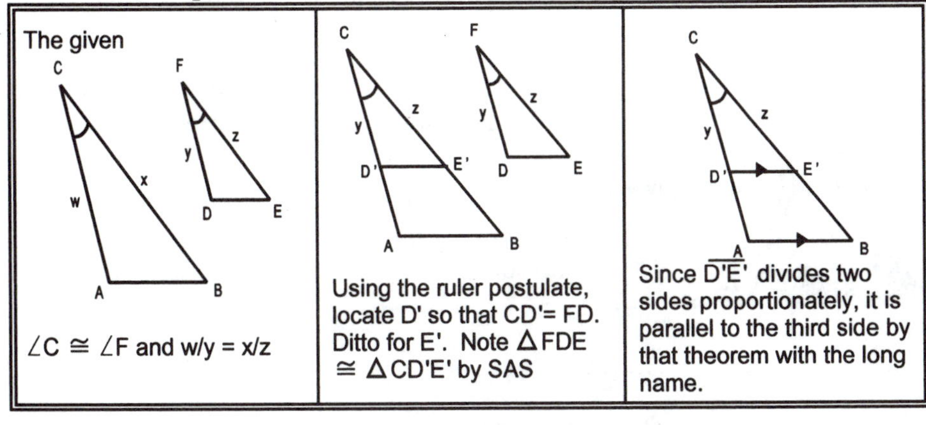

The given
$\angle C \cong \angle F$ and w/y = x/z
Using the ruler postulate, locate D' so that CD'= FD. Ditto for E'. Note ΔFDE $\cong$ ΔCD'E' by SAS
Since $\overline{D'E'}$ divides two sides proportionately, it is parallel to the third side by that theorem with the long name.

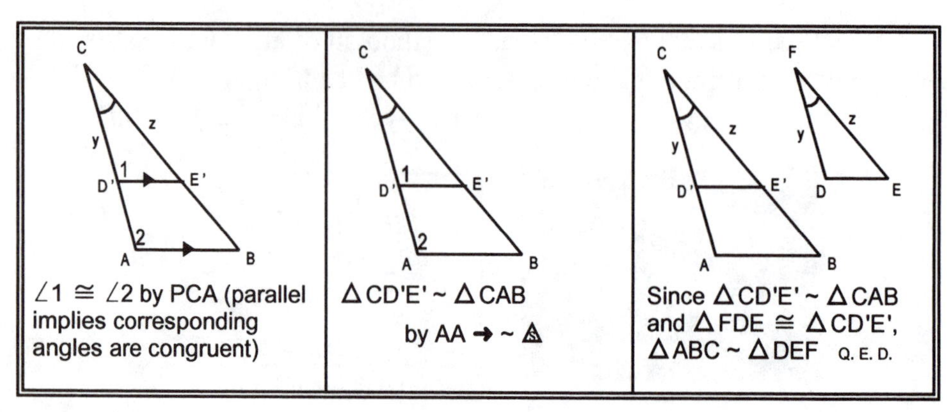

In that last frame Fred used a lemma that seemed so obvious to him that he forgot to prove it before drawing his comic strip. (Lemma: If the first triangle is similar to the second, and the second is congruent to the third, then the first must be similar to the third.) The proof uses just the definitions of similarity and congruence and algebra, and we will also skip it.

He thought of doing a similar cartoon strip for the proof of:
Theorem: (SSS Similarity) Two triangles are similar if their respective sides are proportional.

It would start in almost exactly the same way. After showing the given in the first frame, he would again create D' and E' in the second frame. Then in the third frame he'd establish $\triangle CAB \sim \triangle CD'E'$ by the SAS Similarity Theorem. etc. In the last frame he would again use that lemma that if $\triangle\#1 \sim \triangle\#2$, and if $\triangle\#2 \cong \triangle\#3$, then $\triangle\#1 \sim \triangle\#3$.

Instead he sat up and looked around. At the edge of the Great Lawn was a lake that all the students called the Great Lake. It really wasn't one of the Great Lakes,* but no one wanted to call it the Small Pond (which it was).

Students were out sailing. He could hear their laughter and it seemed to call to him.

The first lines of *Moby Dick* ran through his mind: Call me Ishmael. Some years ago--never mind how long precisely--having little or no money in my purse, and nothing particular to interest me on shore, I thought

---

* Erie, Huron, Michigan, Ontario and Superior.

I would sail about a little and see the watery part of the world. It is a way I have of driving off the spleen and regulating the circulation. Whenever I find myself growing grim about the mouth; whenever it is a damp, drizzly November in my soul; whenever I find myself involuntarily pausing before coffin warehouses, and bringing up the rear of every funeral I meet; and especially whenever my hypos get such an upper hand of me, that it requires a strong moral principle to prevent me from deliberately stepping into the street, and methodically knocking people's hats off--then, I account it high time to get to sea as soon as I can. This is my substitute for pistol and ball. With a philosophical flourish Cato throws himself upon his sword; I quietly take to the ship. There is nothing surprising in this. If they but knew it, almost all men in their degree, some time or other, cherish very nearly the same feelings towards the ocean with me.

Fred made minor changes: Call me Fred. And it wasn't a damp, drizzly November in his soul. He would be seeing P at 5:45 this afternoon. Years ago when Fred had first read this passage, he misunderstood ". . . whenever the hypos get such an upper hand of me."

Later he looked up the word in the dictionary. *Hypo* was short for hypodermic syringe, but that didn't make any sense either. Then he noticed the archaic meaning of the word. It was a shortening of hypochondria—worrying too much about your health.

In any event, Fred was ready to become a seafaring man. Or in his case, a pondfaring boy.

From the shoreline, he carefully studied the sailboats. Larger boats had larger sails.

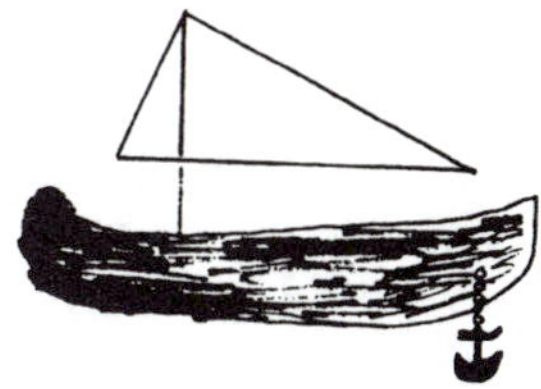

The triangles (the sails) were similar. He looked at the parts of the mast that were covered by the sails. Those were **altitudes**.* He knew the corresponding sides were proportional. He wondered whether the altitudes were also proportional to the corresponding sides.

✶ Definition An altitude of a triangle is a segment from the vertex drawn perpendicular to the opposite side.

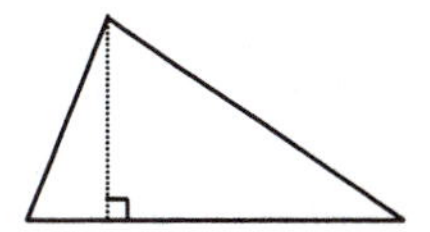

Your Turn to Play

1. Prove it.

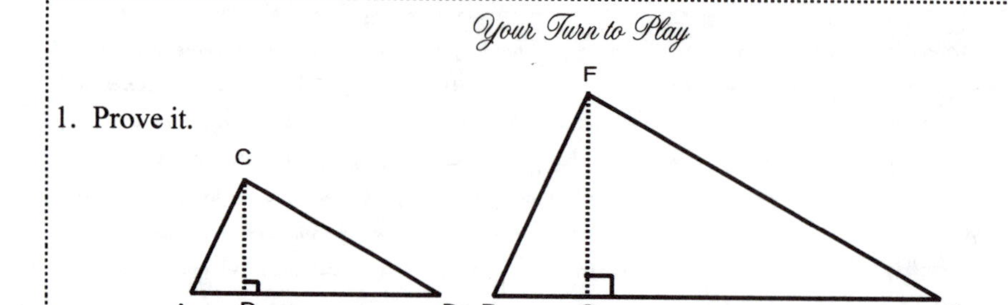

Given: $\Delta ABC \sim \Delta DEF$, altitudes $\overline{CP}$ and $\overline{FQ}$.

To prove: $\dfrac{AC}{DF} = \dfrac{CP}{FQ}$

COMPLETE SOLUTION

1.

Statement	Reason
1. $\Delta ABC \sim \Delta DEF$, altitudes $\overline{CP}$ and $\overline{FQ}$.	1. Given
2. $\angle A \cong \angle D$	2. Def of similar Δs
3. $\overline{CP} \perp \overline{AB}$ and $\overline{FQ} \perp \overline{DE}$	3. Def of altitude
4. $\angle APC$ and $\angle DQF$ are right angles.	4. Def of perpendicular
5. $\angle APC \cong \angle DQF$	5. All right $\angle$s are $\cong$*
6. $\Delta APC \sim \Delta DQF$	6. AA $\rightarrow \sim \Delta$s
7. $\dfrac{AC}{DF} = \dfrac{CP}{FQ}$	7. Def of $\sim \Delta$s

But how to obtain a boat? He noticed the sign at the pier:

★ If we haven't proven this proposition already, it is very easy to establish. If $\angle G$ and $\angle H$ are right angles, then $m\angle G = 90°$ and $m\angle H = 90°$ and by algebra, $m\angle G = m\angle H$. Then by definition of congruent angles, $\angle G \cong \angle H$.

Fred approached the man operating the boat-rental concession. "Excuse me sir. The sign says 25¢."

"That's right son."

"Is that 25¢ per minute or what?" (Fred had been taken in by scams too often in his life. He was wary of offers that looked too good to be true.)

"No, not per minute. That would be outrageously expensive. The charge is two bits to rent the boat. You can stay out there till dark if you want to. Just pick out your vessel and we'll make you captain of it."

Fred picked out a medium sized "ship" (as Fred called it) and handed the attendant a fifty-dollar bill. "I'm sorry. That's the smallest I have."

"That's okay little feller." He lifted Fred up and set him in the boat and handed him his change and a receipt. He untied the painter and with his foot shoved the boat away from the pier.

Fred looked at his change: $3.78. Then he looked at the receipt that the operator had given him.

C.C. Coalback
Boat Rentals

Amount tendered	50.00
Base rental price	.25
Docking fee	7.39
Launch fee	5.81
Administration fee	8.25
Collision insurance	9.00
County tax	4.24
State tax	3.33
Federal tax	7.95
YOUR CHANGE	3.78

no refunds once boat has left pier

Then a half dozen thoughts ran through his mind:

- This is a mighty expensive little boat ride. Since I only make $600 a month, it's going to take several days of work to earn this trip.
- It's 1:15 right now. Only four and half hours till I pick P up.
- P should see me now. I think she'd be proud of how I'm driving this ship.
- How DO you drive a boat?

I've never been . . . I mean I've never . . . This is water. It looks very wet. Swimming . . . I don't know how. I wonder if there is a book on swimming on board. To Fred's surprise, there was a copy of Professor Eldwood's *Treatise on Aquatic Endeavors for the Modern European Woman,* 1845. Fred flipped it open and read: The modern woman, if she is quite daring, may choose to employ the fashionable nine-piece bathing costume. This is a most avant garde ensemble in that it leaves the ankles and parts of the forearms entirely bare. Indeed, quite the thing on some of the more risqué beaches in the south of France. Fred wasn't interested in ancient women's clothes and turned to the chapter entitled "Actual Aquatic Maneuvers." The only thing Eldwood mentioned after several paragraphs describing how "the modern European woman should enter the waters gradually," was his observation: It should be noted that inhalation while one's mouth and nose are submerged is something to be avoided.

Fred tossed the book back into the bottom of the boat.

Even if I don't know how to steer this thing, I'm really very safe. There's hardly any breeze so I won't start going very fast. The sun is warm and I'm only feeling a little seasick right now. Not much really. I'll try to think of something else.

Those were his six thoughts. For those readers who have ever experienced seasickness, they know that it is difficult to direct one's thoughts to any other topic.

But one thought did come to his mind as he heard the sail ripping. It was tearing from the top. Neatly tearing. An angle bisector.

The Angle Bisector Theorem! A simple theorem with a very clever proof. Every educated person should know this, but lots of students, teachers, parents, school administrators, college presidents, and philatelists claim that they've never heard of it.

Before you look at the next page, can you make a guess about what proportion must be true in a triangle with an angle bisector? Draw some pictures on scratch paper. Active learning works much better than passive.

<u>Theorem</u>: (The Angle Bisector Theorem)

In any triangle, an angle bisector divides the opposite side in the same ratio as the other two sides.

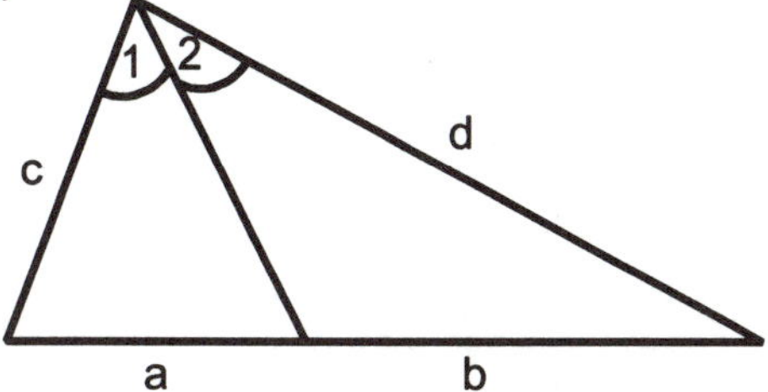

Given: $\angle 1 \cong \angle 2$

To prove: $\frac{a}{b} = \frac{c}{d}$

What is fun about this theorem is that there isn't an obvious way to prove it. The triangles share only one pair of congruent angles and they are clearly not similar to each other. (No obtuse triangle could be similar to an acute triangle.)

And yet we need similar triangles if we are ever going to establish the proportion $\frac{a}{b} = \frac{c}{d}$

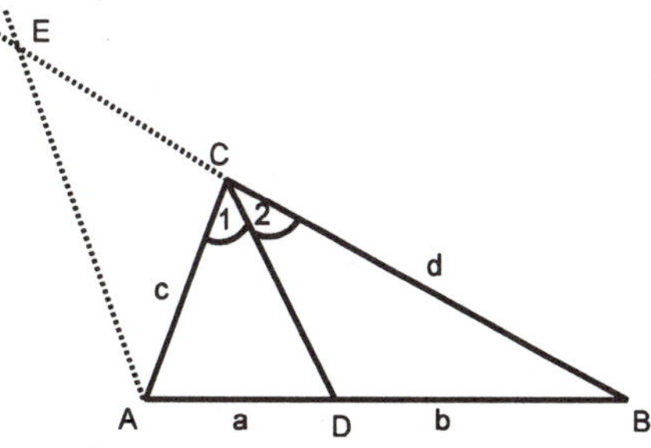

Here's the trick:
We make similar triangles.

Draw a line through point A which is parallel to $\overline{DC}$. Then extend $\overline{BC}$ (two points determine a line).

This new triangle (ΔACE) will turn out to always be isosceles.* We can then mark EC = c.

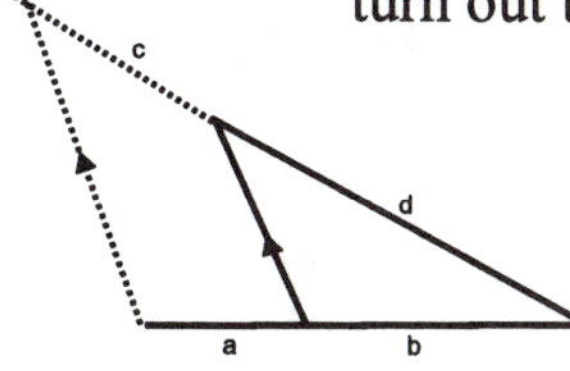

The last step is to use the Generalization of the Converse of the Midsegment Theorem (p. 299) to obtain $\frac{a}{b} = \frac{c}{d}$ ⊠

✶ $\angle 1 \cong \angle 3$ by PAI. $\angle 2 \cong \angle 4$ by PCA.

And since $\angle 1 \cong \angle 2$ (that was given), we have $\angle 3 \cong \angle 4$

Then the converse of the Isosceles Triangle Theorem gives us $\overline{AC} \cong \overline{EC}$.

Fairbanks

1. When your pizza has three times the diameter of Joe's pizza, you don't have three times the area of his pizza. Instead you have nine times the area. With similarly shaped polygons, the area changes as the square of the change in the lengths of corresponding sides.

For triangles, the theorem is: *In similar triangles the areas are proportional to the square of the ratios of corresponding sides.*

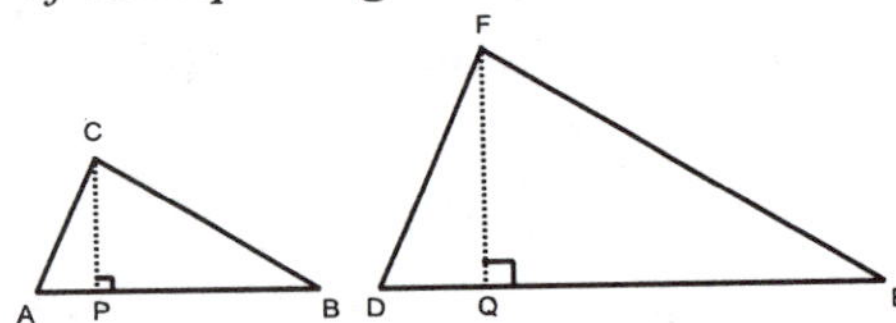

Fill in the missing parts of the proof of this theorem.

S	*R*
1. ?	*1.* Given
2. aΔABC = ½(AB)(CP) aΔDEF = ½(DE)(FQ)	*2.* Thm: Area of a triangle is ½ bh
3. $\frac{AB}{DE} = \frac{CP}{FQ}$	*3.* ?
4. $\frac{a\,\Delta ABC}{a\,\Delta DEF} = \left(\frac{AB}{DE}\right)^2$	*4.* Algebra

2. The algebra in the last step of the proof in the previous problem was a little tricky. Show how we get from steps 2 and 3 to step 4.

3. Solve $\frac{x}{8} = \frac{2x-9}{4}$

4. From the diagram and the GCM Theorem we know that $\frac{a}{b} = \frac{c}{d}$

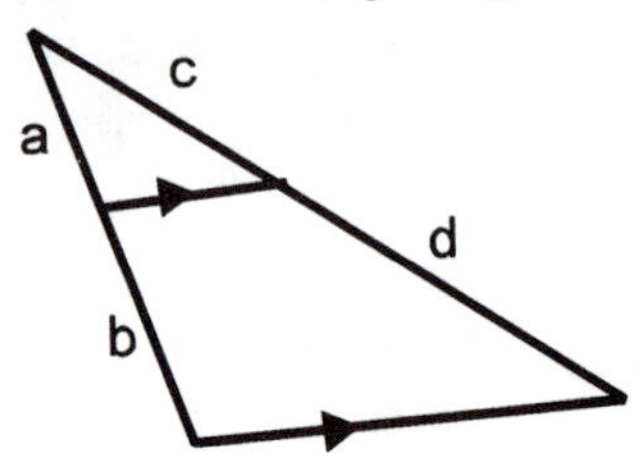

Using Fred's Five Favorite Tricks with Proportions, show that $\frac{a}{a+b} = \frac{c}{c+d}$

5. Find z.

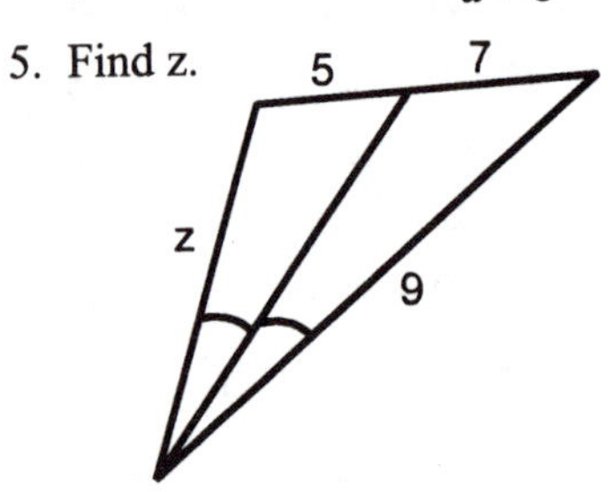

answers

1. *1.* $\Delta ABC \sim \Delta DEF$

3. Thm: The altitudes of similar triangles are proportional to corresponding sides of the triangles, which was proved on p. 306.

2. Dividing the two equations in step 2 we obtain

$$\frac{a\,\Delta ABC}{a\,\Delta DEF} = \frac{½(AB)(CP)}{½(DE)(FQ)}$$

Cancel the ½'s $= \frac{(AB)(CP)}{(DE)(FQ)}$

Substituting from step 3 $= \frac{(AB)(AB)}{(DE)(DE)}$

which is $= \left(\frac{AB}{DE}\right)^2$

3.

Cross multiplying	4x = 8(2x – 9)
Distributive law	4x = 16x – 72
Transpose	72 = 12x
Divide by 12	6 = x

4. We start with $a/b = c/d$

Invert $b/a = d/c$

Add bottom to top $(a + b)/a = (c + d)/c$

Invert $a/(a + b) = c/(c + d)$

5. 45/7

Santa Cruz

1. Prove the Transitive Property of Similar Triangles.

(If $\Delta ABC \sim \Delta DEF$ and if $\Delta DEF \sim \Delta GHI$, then $\Delta ABC \sim \Delta GHI$.)

2. Prove Theorem: *Three parallel lines intercept proportional segments on any two transversals.* (Place this theorem on your list-of-theorems page that you have been keeping.)

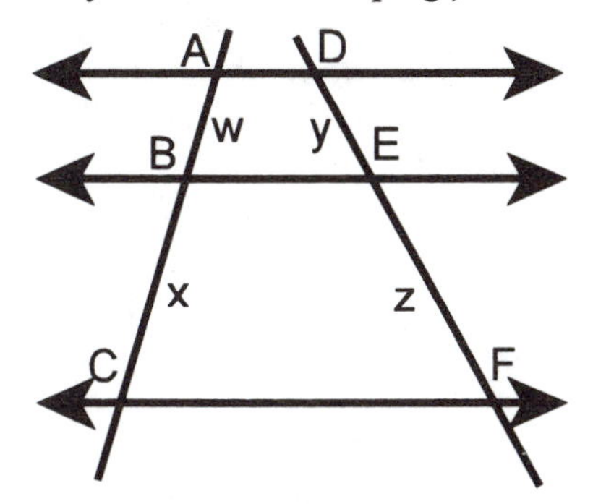

Given $\overleftrightarrow{AD} \parallel \overleftrightarrow{BE} \parallel \overleftrightarrow{CF}$

To prove: $w/x = y/z$

Hint: Students of geometry often attempt this proof by extending the transversals until they meet and then getting a diagram like:

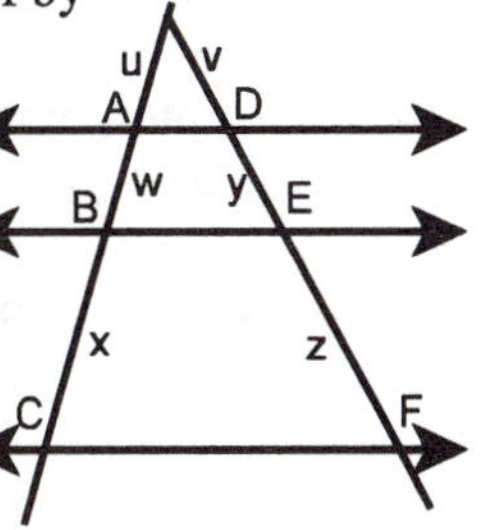

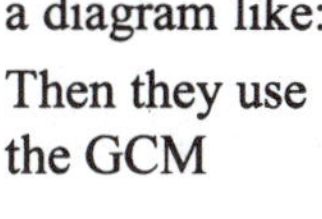

Then they use the GCM Theorem and get several proportions:

$u/w = v/y$, $(u + w)/x = (v + y)/z$ and $u/(w + x) = v/(y + z)$. Then after a lot of nasty algebra they eliminate the u and the v from the three equations and get $w/x = y/z$.

But there is a much easier way. Draw $\overline{CD}$ and use the GCM Theorem (The Generalization of the Converse of the Midsegment Theorem—We have to get a better name for that theorem!) and easy algebra gives you $w/x = y/z$. About a four-step proof.

3. Warning: this problem is very short (about three steps) if you have a good short-term memory, but slightly agonizing otherwise.

Given trapezoid ABCD.

To prove: $w/x = y/z$

(One student of mine once called this proposition: *The diagonals of a trapezoid do it to each other.*)

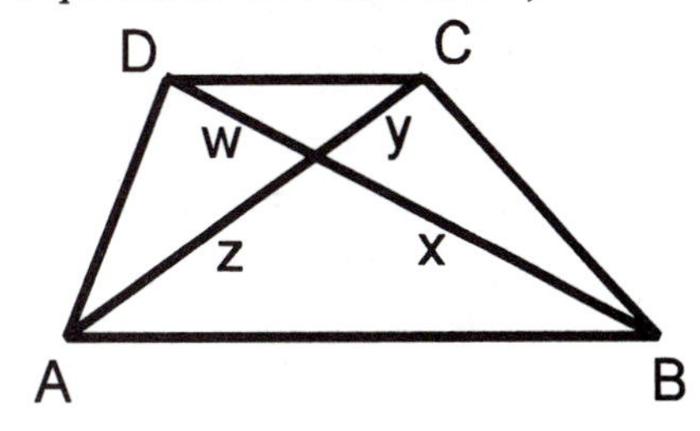

4. Find r.

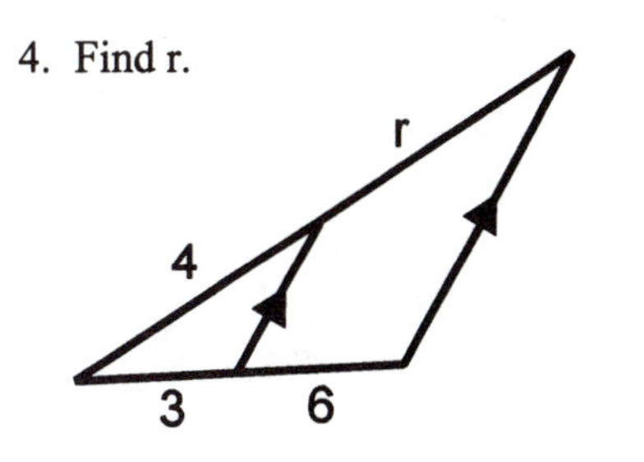

5. If $\Delta ABC \sim \Delta DEF$ and if $a\Delta ABC = 4$ and $a\Delta DEF = 16$ and if $AB = 7$, find DE.

answers

1.

$\mathcal{S}$	$\mathcal{R}$
1. $\triangle$ABC ~ $\triangle$DEF and $\triangle$DEF ~ $\triangle$GHI	*1.* Given
2. $\angle A \cong \angle D$, $\angle D \cong \angle G$, $\angle B \cong \angle E$, $\angle E \cong \angle H$	*2.* Def of similar $\triangle$s
3. $\angle A \cong \angle G$, $\angle B \cong \angle H$	*3.* Thm: Transitive Property of congruent $\angle$s
4. $\triangle$ABC ~ $\triangle$GHI	*4.* Post 14 (AA → ~ $\triangle$s)

2.

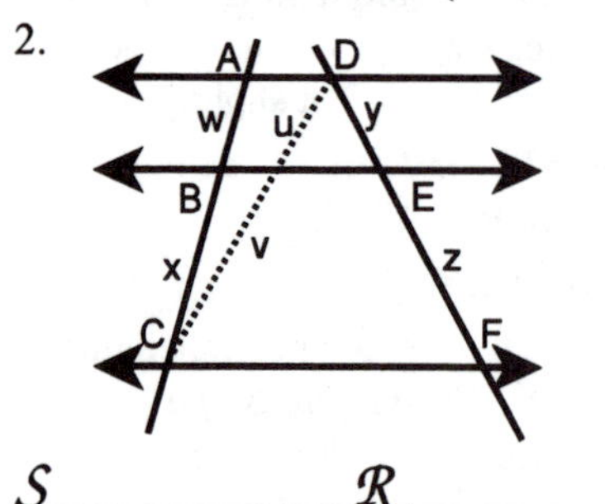

$\mathcal{S}$	$\mathcal{R}$
1. $\overleftrightarrow{AD} \parallel \overleftrightarrow{BE} \parallel \overleftrightarrow{CF}$	*1.* Given
2. Draw $\overline{CD}$	*2.* Postulate: Two points determine a line
3. w/x = u/v and u/v = y/z	*3.* GCM Theorem
4. w/x = y/z	*4.* Algebra

3.

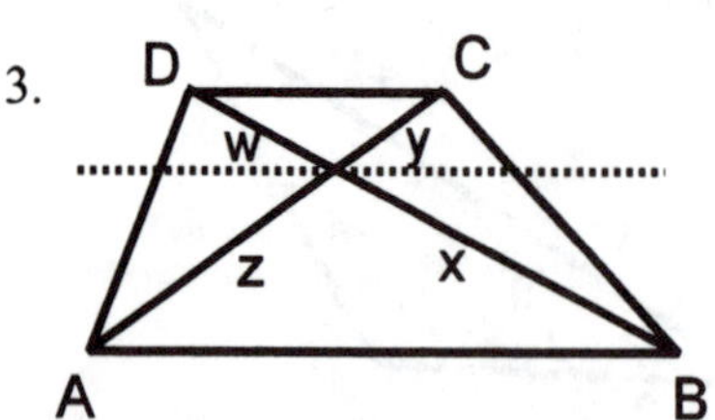

The first step is stating the given. The second step is to draw a line through the point of intersection of the diagonals. (We have a theorem: *If point P is not on ℓ, then there exists at least one line through P that is parallel to ℓ.*) The third step is to see that this is just a direct application of the theorem we just proved in the previous problem. w/x = y/z.

4. r = 8

5. Using the theorem proved in the first problem of Fairbanks, we note that since the areas are in the ratio of 1/4, the ratio of corresponding sides will be 1/2. So DE will be twice as large as AB. DE = 14.

Tate

1. Prove the converse of the Angle Bisector Theorem: *If a segment from the vertex of a triangle to the opposite side divides the side in the same ratio as the other two sides, then it is an angle bisector.*

Given: a/b = c/d To prove: $\angle 1 \cong \angle 2$

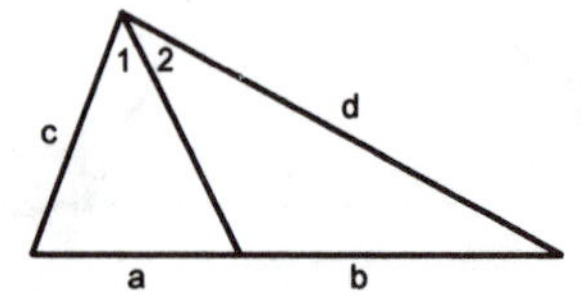

(Its proof is very similar to the proof of the Angle Bisector Theorem.)

2. If two quadrilaterals (ABCD and EFGH) have corresponding angles congruent ($\angle A \cong \angle E$, $\angle B \cong \angle F$, $\angle C \cong \angle G$, and $\angle D \cong \angle H$), are they always/sometimes/never similar?

3. If two quadrilaterals (ABCD and EFGH) have corresponding sides that are proportional (AB/EF = BC/FG = CD/GH = DA/HE), are they always/sometimes/never similar?

4. Find m∠x

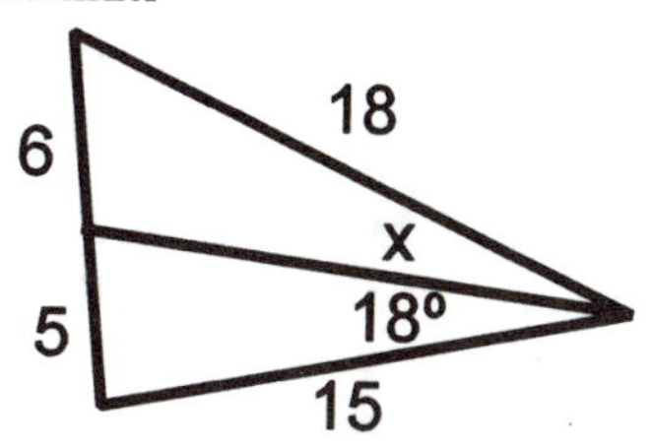

5. What theorem/postulate/definition justifies your answer to the previous problem?

odd answers

1.

S	*R*
1. a/b = c/d	*1.* Given

2. Draw a line through point A which is parallel to $\overline{DC}$.

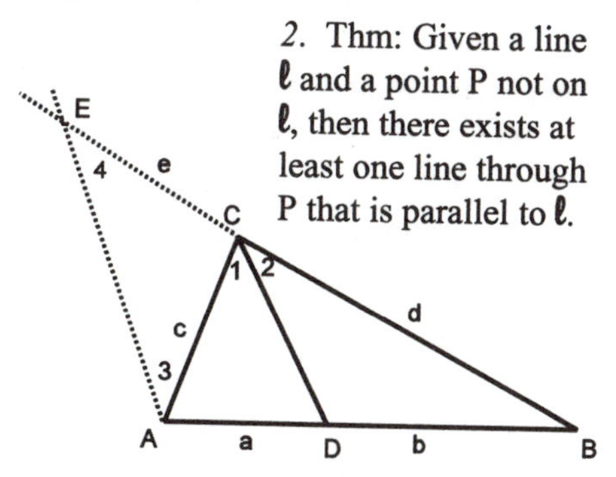

2. Thm: Given a line ℓ and a point P not on ℓ, then there exists at least one line through P that is parallel to ℓ.

3. Draw $\overleftrightarrow{BC}$.	*3.* Two points determine a line.
4. a/b = e/d	*4.* Generalization of the Midsegment Thm
5. c = e	*5.* Algebra (steps 1 and 4)
6. ∠3 ≅ ∠4	*6.* Isosceles Δ Thm
7. ∠1 ≅ ∠3	*7.* PAI Theorem
8. ∠2 ≅ ∠4	*8.* PCA Theorem
9. ∠1 ≅ ∠2	*9.* Thm: Transitive Property of ≅ ∠s

(Note: You also could have changed steps 6, 7 and 8 into m∠3 = m∠4, etc. and then used algebra, but that would have taken longer.)

3. They are sometimes similar. If the two quadrilaterals were, for example, both squares, then their corresponding sides would be proportional and they would be similar. Here is an example of two quadrilaterals with proportional corresponding sides which are not similar: ▭ ▱

5. The theorem we proved in the first problem of this city.

Vanderbilt

1. Find BC.

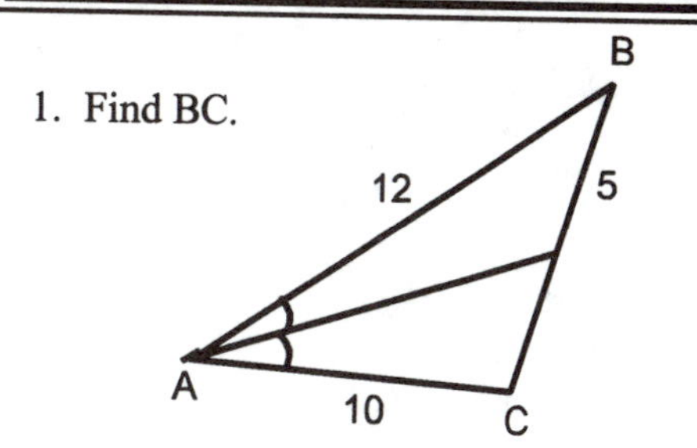

2. Using Fred's Five Favorite Tricks with Proportions, go from

$$\frac{8}{3} = \frac{w}{7}$$

to

$$\frac{11}{w+7} = \frac{3}{7}$$

(It takes two steps.)

3. If it takes 600 lbs. of grass seed to cover the triangular lawn in front of the president's mansion, how many pounds would be needed to cover the triangular lawn in front of the dean's house. The two triangles are similar, and the longest side of the president's lawn is five times the length of the longest side of the dean's lawn.

4. A student wrote: *If two triangles that are not isosceles are similar, then the longest side of the first triangle divided by the longest side of the second triangle must be equal to the shortest side of the first triangle divided by the shortest side of the second triangle.*

Is this true for:

A) all triangles
B) only obtuse triangles
C) only right triangles
D) only acute triangles
E) no triangles

5. Fill in the missing parts of the proof of: *In similar triangles, corresponding angle bisectors are proportional to corresponding sides.*

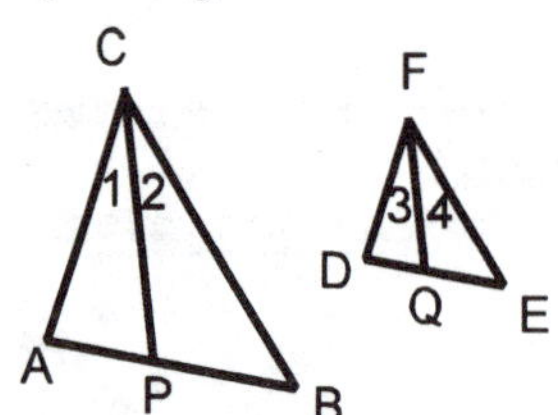

S	R
1. $\triangle ABC \sim \triangle DEF$, $\overline{CP}$ and $\overline{FQ}$ are angle bisectors.	*1.* ?
2. $m\angle 1 = m\angle 2$; $m\angle 3 = m\angle 4$	*2.* ?
3. $\angle ACB \cong \angle DFE$ and $\angle A \cong \angle D$	*3.* ?
4. ?	*4.* Angle Addition Postulate
5. $m\angle 1 = m\angle 3$	*5.* Algebra
6. ?	*6.* Def of $\cong$ $\angle$s
7. $\triangle ACP \sim \triangle DFQ$	*7.* ?
8. $\dfrac{AC}{DF} = \dfrac{CP}{FQ}$	*8.* ?

odd answers

1. 9 1/6
3. 24 lbs.
5.
1. Given
2. Def of angle bisector
3. Def of similar $\triangle$s
4. $m\angle 1 + m\angle 2 = m\angle ACB$ and $m\angle 3 + m\angle 4 = m\angle DFE$
6. $\angle 1 \cong \angle 3$
7. AA → ~ $\triangle$s
8. Def of similar $\triangle$s

Jennings

1. Prove that if $\triangle ABC \sim \triangle DEF$ and if $\triangle DEF \cong \triangle GHI$, then $\triangle ABC \sim \triangle GHI$.

2. Explain why an obtuse triangle may never be similar to an acute triangle.

3. If the corresponding sides of two triangles are parallel, must the triangles be similar? If it is true, prove it. If not, then draw a counterexample.

4. Using Fred's Five Favorite Tricks with Proportions, go from

$$\frac{z+4}{z} = \frac{32}{10}$$

to

$$\frac{4}{22} = \frac{z}{10}$$

5. Find x.

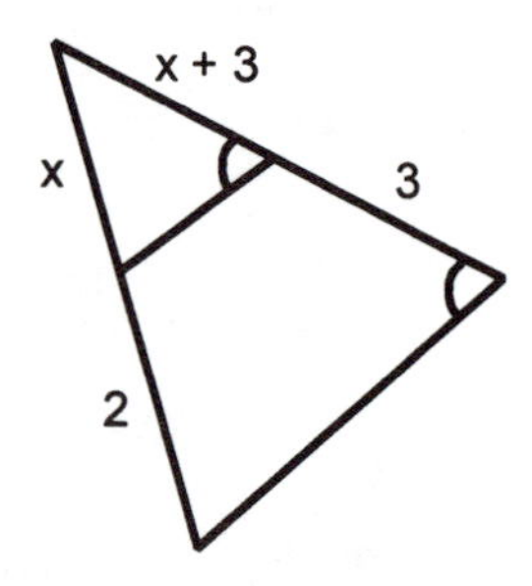

Landa

1. Find x.

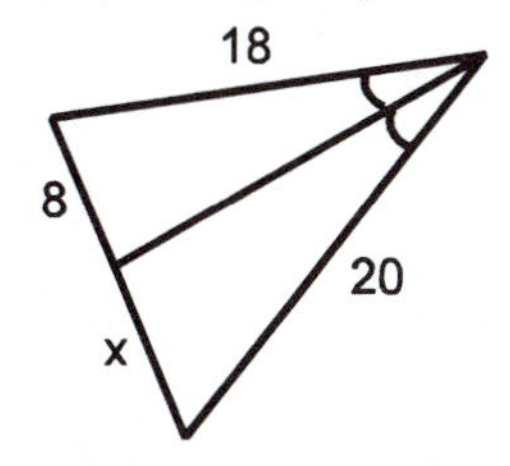

2. True or False: If two triangles are congruent, then they must be similar. If it is true, prove it. If it is false, find a counterexample.

3. If ΔABC has sides with lengths 4, 6, 7 and ΔDEF has sides with lengths 8, 12, 14, must the two triangles be similar? If yes, explain why. If no, then give a counterexample.

4. Fill in the missing parts of the proof of: *In similar triangles, corresponding medians are proportional to corresponding sides.*

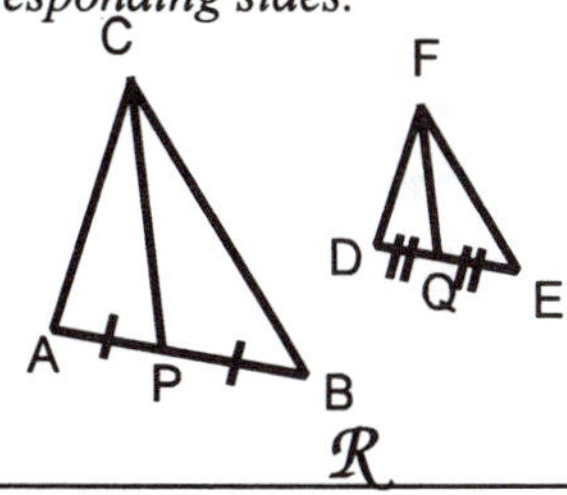

S	R
1. $\Delta ABC \sim \Delta DEF$, $\overline{CP}$ and $\overline{FQ}$ are medians.	*1.* ?
2. AP = PB and DQ = QE	*2.* Def of median
3. $\frac{CA}{FD} = \frac{AB}{DE}$ and $\angle A \cong \angle D$	*3.* ?
4. AP + PB = AB and DQ + QE = DE	*4.* Def of between
5. $\frac{CA}{FD} = \frac{AP}{DQ}$	*5.* ?

5. Given the figure as marked. Find the value, if possible, of y.

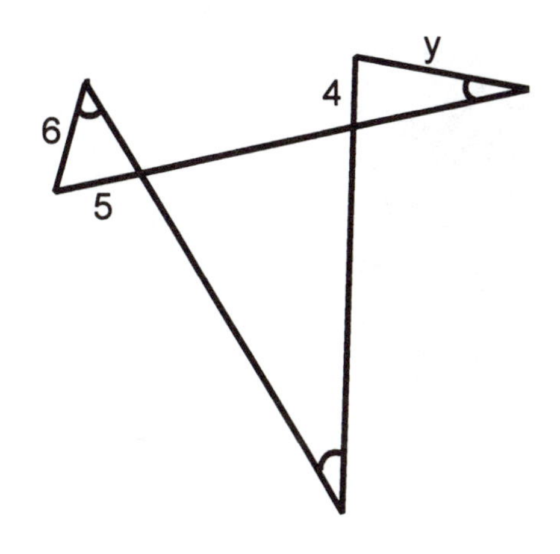

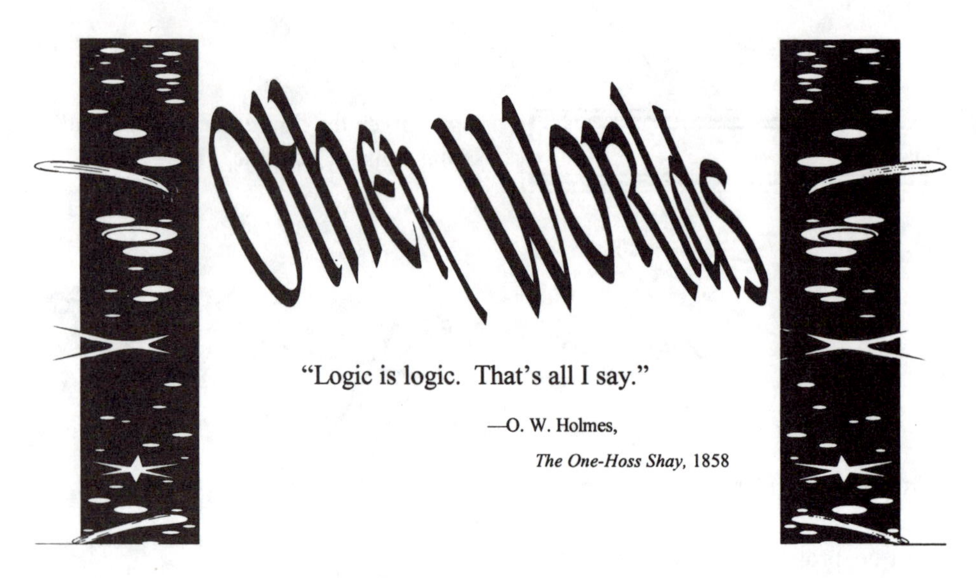

"Logic is logic. That's all I say."

—O. W. Holmes,

The One-Hoss Shay, 1858

Chapter 8½
Symbolic Logic

Seasickness is not generally considered a lot of fun. Fred had heard that eating something very mild might help settle his stomach. Next to Eldwood's book was a box marked "Naval Chow."

The box was sealed with a paper tape which read, "Those who open this box will be charged 10¢ plus applicable fees and taxes." "Applicable fees and taxes" meant lots more money, but he was desperate and broke the seal.

+ A can of anchovy paste with an easy-open lid.
+ A jar of chocolate-flavored mayonnaise.
+ A large yellow squash, slightly soft because of its age.
+ A carton of Coalback-brand® fortune cookies.

Fred opened the carton and dumped out the cookies. He broke one open and read the fortune: *If a boat is not a Coalback Rental boat, you can be sure that it is steady.* *That seems pretty true,* Fred thought to himself. He cracked a second cookie open and was informed: *If the boat passengers are not doomed, you can be sure that the sails are not ripped.* Fred looked up at his ripped sail.

A third fortune: *If it doesn't have moldy sails, then it isn't a Coalback Rental boat.* A fourth: *Steady boats don't have blue bottoms.* And the last one Fred opened informed him: *A sail will rip if it's moldy.*

The cookies were stale and he tossed them overboard. The seagulls came and ate them before they had a chance to sink. Fred looked at the slips of paper and realized that there was just one logical conclusion he could draw from those five fortunes. It's not especially easy to see.

First, he let S stand for "a boat is steady."

He let B stand for "the boat has a blue bottom."

Then the fourth fortune (*Steady boats don't have blue bottoms*) became $S \rightarrow \neg B$.*

In chapter two (p. 73) we first encountered the contrapositive. The contrapositive of $P \rightarrow Q$ is $\neg Q \rightarrow \neg P$. The contrapositive of true statements are true, and the contrapositive of false statements are false.

Your Turn to Play

1. This statement is true: *If it is raining, then the streets are wet.* What is its contrapositive?

2. To find the contrapositive of an *If ••• then •••* statement, you interchange the two parts of the sentence and negate each part. We know that the contrapositive of $P \rightarrow Q$ is $\neg Q \rightarrow \neg P$.

   What is the contrapositive of $\neg Q \rightarrow \neg P$?

3. If Q stood for "this is quince pie," then $\neg\neg Q$ would stand for "it is not the case that this is not quince pie." How could you express $\neg\neg Q$ more simply in English?

4. What is the contrapositive of the fourth fortune ($S \rightarrow \neg B$)?

---

* The symbol "→" you have seen before in chapter 5½. It stands for "implies." → is used in *If ••• then •••* situations.

The symbol "¬" is new to you. It is the logical symbol for "not." If B stands for "the boat has a blue bottom," then $\neg B$ would stand for "the boat does not have a blue bottom."

COMPLETE SOLUTIONS

1. If the streets are not wet, then it isn't raining.

2. ¬¬P ➜ ¬¬Q

3. "It is the not case that this is not quince pie" could be expressed more simply as "This is quince pie." ¬¬Q can be replaced by Q.

4. The contrapositive of S ➜ ¬B is ¬¬B ➜ ¬S. Or, using the results of the previous problem, we can say that the contrapositive of S ➜ ¬B is B ➜ ¬S.

Fred noticed some seagulls floating in the water near the boat. They were motionless.

Fred converted all five fortunes into symbols. He let:

S stand for "a boat is steady."
B stand for "the boat has a blue bottom."
C stand for "the boat is a Coalback rental."
D stand for "the boat passengers are doomed."
R stand for "the sails are ripped."
M stand for "the sails are moldy."

Fortune Number	In English	In Symbols
1	If a boat is not a Coalback Rental boat, you can be sure that it is steady.	¬C ➜ S
2	If the boat passengers are not doomed, you can be sure that the sails are not ripped.	¬D ➜ ¬R
3	If it doesn't have moldy sails, then it isn't a Coalback Rental boat.	¬M ➜ ¬C
4	Steady boats don't have blue bottoms.	S ➜ ¬B
5	A sail will rip if it's moldy.	M ➜ R

Now take the contrapositive of the 4th:	B ➜ ¬S
followed by the contrapositive of the 1st:	¬S ➜ C
followed by the contrapositive of the 3rd:	C ➜ M
followed by the 5th:	M ➜ R
followed by the contrapositive of the 2nd:	R ➜ D

Now it is easy to see the chain of reasoning: B implies ¬S, which implies C, which implies M, which implies R, which implies D.

So B implies D.* If the boat has a blue bottom, then its passengers are doomed. I need not tell you what color the bottom of Fred's boat was.

We now have four logical symbols: & (and), ∨ (or), ➜ (implies), and ¬ (not). With these we can build **truth tables.**

The easiest truth table to build is the one for the ¬ symbol. We know that if some statement P is true, then ¬P must be false. If P is false, then ¬P is true. The truth table looks like:

P	¬P
T	**F**
F	**T**

This truth table will serve as our definition of ¬.

When we construct the truth table for P & Q, we first have to think of all the possible combinations of T and F for both P and Q. They could both be true; or just P could be true; or just Q could be true; or both P and Q could be false.

✶ We have used the Transitive Property of ➜, which we will be able to prove by the end of this chapter. Formally stated: (The Transitive Property of ➜) If you know that P ➜ Q and if you know that Q ➜ R, then you know that P ➜ R.

If you *really* like symbols, the Transitive Property of ➜ could be written as ((P ➜ Q) & (Q ➜ R)) ➜ (P ➜ R). We need all those parentheses to keep everything straight.

Here is the truth table for P & Q which will serve as our definition of &:

P	Q	P & Q
T	T	T
T	F	F
F	T	F
F	F	F

We wanted P & Q to be true only in the case that both P and Q were true.

The symbol ∨ stands for the non-exclusive "or." We want P ∨ Q to be true if either one or both of them are true:

P	Q	P ∨ Q
T	T	T
T	F	T
F	T	T
F	F	F

The truth table for → may not be what you expect.

P	Q	P → Q
T	T	T
T	F	F
F	T	T
F	F	T

The only time that P → Q is false is when a true statement implies a false conclusion. If the "if" part of an *If ••• then •••* statement is false, we don't care what the conclusion is.

When you assert P → Q, you are saying that if P is true, then Q has to also be true.

Take an everyday example. Suppose you tell your kid sister, "If you eat a whole jar of peanut butter, then your stomach will feel queasy." That is a true P → Q statement. Nobody could argue with that. Nobody except your kid sister. She ate a whole box of chocolates and her stomach was queasy. She came to you and shouted, "Liar! Liar! Pants on fire. I didn't eat any peanut butter and my stomach is queasy." You try to explain to her that what you said was that IF you ate the jar of PEANUT BUTTER, then your stomach would be queasy. You weren't talking about other idiocies that she might perform.

In short, if she didn't consume a jar of peanut butter, there was no way that she could logically prove that you were wrong.

These are all true statements in logic:

✳ If 2 + 2 = 5, then I'm a monkey's uncle.

✳ If the moon is made of cheese, then mice made the craters.

✳ If there was not God, there would be no atheists. (G.K. Chesterton said that.)

With these truth tables you can check the validity of any argument that is built up from &, $\vee$, ➔ and ¬. A valid argument is one that is true for every line on the truth table.

Suppose your kid brother makes the argument that starting with assuming P, you can conclude that P & Q. In the words of geometry, this would be:

Given P;

To prove P & Q.

Your kid brother's argument is equivalent to P ➔ (P & Q). Is this *always* true? We set up a truth table where the first two columns are every possible combination of T's and F's. Then the next column is P & Q and the final column is P ➔ (P & Q).

P	Q	P & Q	P ➔ (P & Q)
T	**T**	**T**	**T**
T	**F**	**F**	**F**
F	**T**	**F**	**T**
F	**F**	**F**	**T**

Since P ➔ (P & Q) is not always true, your kid brother's argument is not valid.

We can now prove the validity of the Transitive Property of → by using truth tables. The Transitive Property of → states that if we know P → Q and if we know that Q → R, then we may say that P → R. We want to check the validity of ((P → Q) & (Q → R)) → (P → R).

Since this truth table has three different letters, there will be eight possible combinations of **T**'s and **F**'s.

P	Q	R	P → Q	Q → R	(P → Q) & (Q → R)	P → R	((P → Q) & (Q → R)) → (P → R)
T	**T**	**T**	**T**	**T**	**T**	**T**	**T**
T	**T**	**F**	**T**	**F**	**F**	**F**	**T**
T	**F**	**T**	**F**	**T**	**F**	**T**	**T**
T	**F**	**F**	**F**				**T**
F	**T**	**T**					**T**
F	**T**	**F**					**T**
F	**F**	**T**					**T**
F	**F**	**F**	**T**	**T**	**T**	**T**	**T**

Your Turn to Play

Try your hand at filling in the rest of this chart. The answer is on the next page.

An argument which is always true, such as the one above, is called a **tautology**. It doesn't matter what statements the P, Q and R stand for. P might stand for "Plutonium is toxic," or it might stand for "Pillows are more comfortable than pillories."

Tautologies are arguments that are true because of their *form* and not their *content*. The simplest tautology (taut-TAUL-oh-gee) is A → A. Its proof is a very small truth table:

A	A → A
T	**T**
F	**T**

We can now establish that if an implication is true, then its contrapositive must also be true. We can show that $(P \to Q) \to (\neg Q \to \neg P)$ is a tautology.

P	Q	P → Q	¬Q	¬P	¬Q → ¬P	(P → Q) → (¬Q → ¬P)
T	T	T	F	F	T	T
T	F	F	T	F	F	T
F	T	T	F	T	T	T
F	F	T	T	T	T	T

Your friend informs you, "You know that if the plumbing is done right, then Quincy did the work [$P \to Q$]. And I would like to think that if Quincy didn't do the work, then the plumbing was probably screwed up [$\neg Q \to \neg P$]."

When your friend tells you that, you may respond, "Of course, that's *tautologically correct*." Six-syllable words like *tautologically* are often great conversation starters.

Quincy at work

Here's the rest of the chart from the previous page:

P	Q	R	P → Q	Q → R	(P → Q) & (Q → R)	P → R	((P → Q) & (Q → R)) → (P → R)
T	T	T	T	T	T	T	T
T	T	F	T	F	F	F	T
T	F	T	F	T	F	T	T
T	F	F	F	T	F	F	T
F	T	T	T	T	T	T	T
F	T	F	T	F	F	T	T
F	F	T	T	T	T	T	T
F	F	F	T	T	T	T	T

Perhaps the most famous tautology is: *If you know that $P \to Q$ and if you know that P is true, then you know that Q is true.* $((P \to Q) \& P) \to Q$. In fancy logic discussions this is called Modus Ponens. Everyone else calls it common sense.

Chapter Nine
Right Triangles

The sail on Fred's blue-bottomed boat was ripped. The food was bad enough to kill seagulls. The expenses were mounting up.* And the new worry was how was he going to get back to the pier in time to meet P at 5:45?

He consoled himself with the thought, *I've been in much worse situations in my life. It's time I grow up a little bit and stop thinking and worrying like a five-year-old. After all, I'm six now. And my Sunday school teacher told me that if I'm one of His lambs, then it's okay to be concerned about things, but not to get all tied up in knots worrying about stuff. As the Shepherd expressed it hyperbolically, I should take no thought for tomorrow.***

Fred took a proportion $\frac{a}{b} = \frac{c}{d}$ and inverted it $\frac{b}{a} = \frac{d}{c}$ and interchanged the means $\frac{b}{d} = \frac{a}{c}$ but that didn't occupy his attention for long.

He looked out over the water. He saw a boat in the distance.

Fred could see that its sail wasn't ripped at all. From the third fortune (in chapter 8½) he knew that C ➜ M (where C stood for "the

---

* The nominal 25¢ boat rental had turned out to be $46.22. He wasn't sure how much the 10¢ Naval Chow was going to turn out to be after the "applicable fees and taxes," but with a food tax and an entertainment tax (the fortunes in the cookies), this little boat ride wasn't going to be cheap.

** The Sunday school teacher never used the word "hyperbolically." That's Fred's word. Even ministers would hesitate to use that word in the pulpit addressing adult congregations. Your ability to think is conditioned in part upon the vocabulary you have. Without knowing what *hyperbole* (hi-PER-bel-lee) means, it would be difficult to understand what was meant when He said that if you want to follow Him you have to hate your father and your mother. (Luke 14:26) And you have to hate your brothers and sisters. And you have to hate your children. And you have to hate your spouse.

boat is a Coalback rental, and M stood for "the sails are moldy) and from the fifth fortune M ➜ R (where R stood for "the sails are ripped).

C ➜ M and M ➜ R together yielded C ➜ R (by the Transitive Property of ➜). The contrapositive of C ➜ R is ¬R ➜ ¬C. Finally, since their sail was not ripped (¬R), Fred knew that it wasn't a Coalback rental boat. (If you didn't "do" chapter 8½, it's okay. Just take our word for it: the boat that the happy couple were in wasn't rented from C.C. Coalback.)

Fred admired their beautiful unripped sail. A lovely right triangle. And the mast was an altitude. (Altitude = passes through a vertex and is perpendicular to the opposite side.)

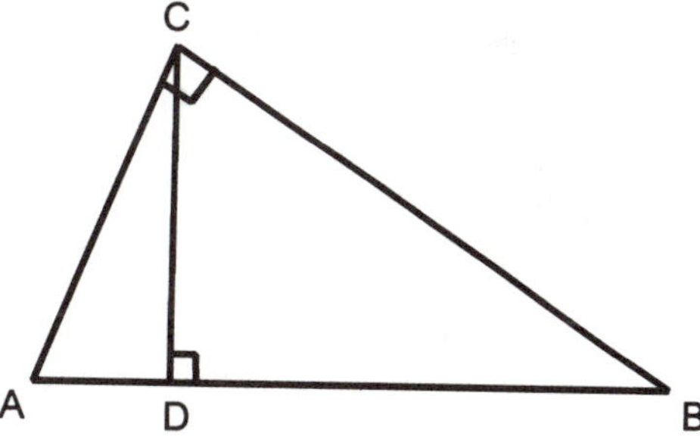

He could see three triangles: the big one ($\triangle ABC$), and the two little ones ($\triangle ACD$ and $\triangle CBD$).

And every pair of triangles shared two pairs of congruent angles. For example, the big triangle and the triangle on the left both shared $\angle A$ and both had right angles in them.

Your Turn to Play

1. Since the big triangle and the triangle on the left share two pair of congruent angles we may say the $\triangle ABC \sim \triangle ACD$. What reason may we use?

2. What two pairs of congruent angles do the big triangle and the triangle on the right have in common?

3. Since, by the previous problem, the big triangle and the triangle on the right share two pairs of congruent angles, we may say that $\triangle ABC \sim \triangle$??? (Be sure to place the B, C and D in the correct order. For example, since the C in $\triangle ABC$ is the right angle, it will have to correspond with the right angle in the triangle on the right.)

4. Since the big triangle is similar to the triangle on the left and since the big triangle is similar to the triangle on the right, we may say by the Transitive Property of Similar Triangles that ______________________. (Fill in the blank with some words.)

5. Translate the conclusion of the previous problem into symbols by completing:

   $\triangle ACD \sim \triangle$???

6. If $m\angle A = 60°$, what is $m\angle DCB$?

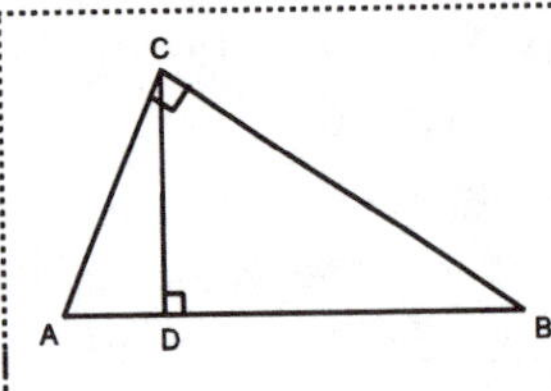

7. <u>Theorem</u>: In a right triangle, the altitude to the hypotenuse divides the triangle into three similar triangles. (From the diagram: $\triangle ABC \sim \triangle ACD \sim \triangle CBD$.) Using the similarity of the left and right triangles, complete the following proportion:

$$\frac{AD}{CD} = \frac{CD}{?}$$

8. In a proportion in which the means are equal (such as $\frac{x}{y} = \frac{y}{z}$) we say that y is the **mean proportional** to x and z. What is the mean proportional to 4 and 9?

As an easy consequence of the theorem in problem 7, we may write: <u>Corollary</u>: In a right triangle, the altitude to the hypotenuse is the mean proportional of the lengths of the two segments on the hypotenuse. (Please place this on your theorem page.)

9. (A fairly tough question) Complete the following: AC is the mean proportional to ____ and ____. (Look at the left triangle and the big triangle.) This is the second corollary to the theorem in problem 7.

10. If AD = 4 and AC = 8, what does AB equal?

11. Complete: BC is the mean proportional to ____ and ____ . This is the third corollary to the theorem in problem 7.

12. If AB = 54 and BC = 18, what does BD equal?

13. What is the name of the theorem proved below?

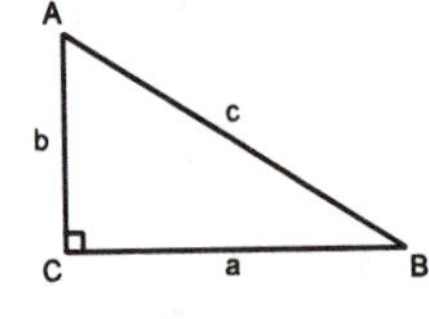

Statement	*Reason*
1. $\triangle ABC$ in which $\angle C$ is a right angle.	1. Given
2. Drop a $\perp$ from C to $\overline{AB}$.	2. Theorem: Given a point P not on ℓ, there is at least one $\perp$ from P to ℓ. (p. 162)
3. $x + y = c$	3. Definition of A–D–B
4. $\frac{x}{b} = \frac{b}{c}$ and $\frac{y}{a} = \frac{a}{c}$	4. The second and third corollaries to the theorem in problem 7. (See problems 9 and 11.)
5. $a^2 + b^2 = c^2$	5. Algebra

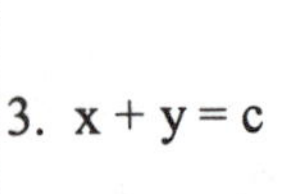

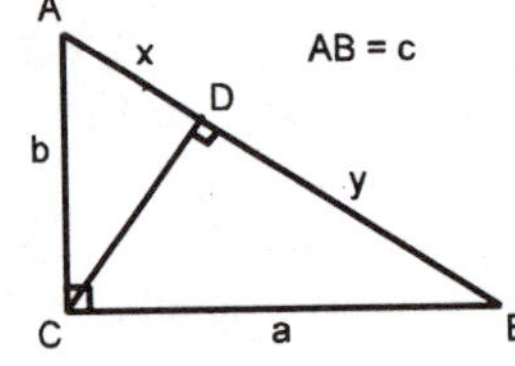

⊠

COMPLETE SOLUTIONS

1. Postulate 14: AA ➜ ~ $\triangle$s
2. They both contain $\angle B$ and they both contain right angles.
3. $\triangle ABC \sim \triangle CBD$
4. We may say that the two triangles are similar.
5. $\triangle ACD \sim \triangle CBD$
6. It is also equal to 60°, because they are corresponding angles of similar triangles.
7. BD. In order to picture the similarity of the left and right triangles, imagine that the left triangle were detached and spun 90° to the right and enlarged. Then it would fit over the triangle on the right.

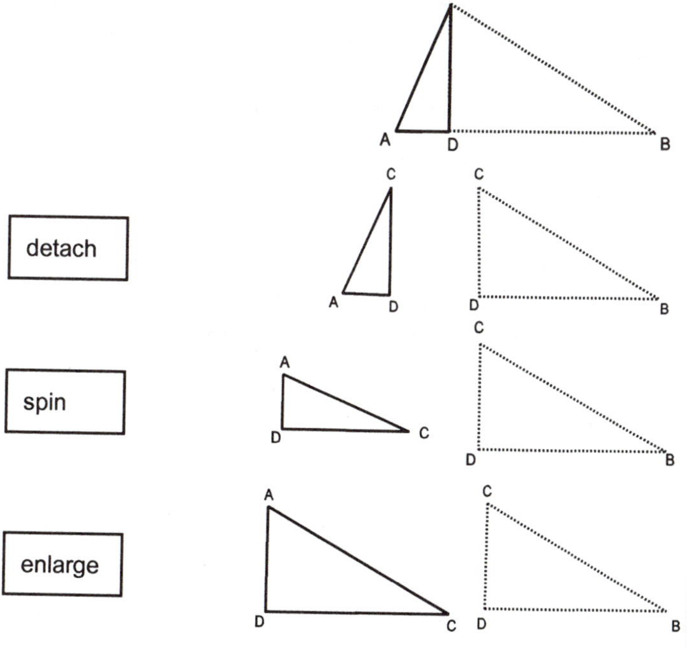

8. $$\frac{4}{m} = \frac{m}{9}$$

Cross multiplying $36 = m^2$

Taking the square root of both sides $\pm 6 = m$ (If m is a length, then we omit the -6)

9. AC is the mean proportional to AD and AB.

10. Using the results of problem 9, $\frac{4}{8} = \frac{8}{AB}$ and after some algebra, AB = 16.

(Actually the numbers are simple enough in this case that you could just "see" that AB had to be 16.)

11. BC is the mean proportional to BD and AB.

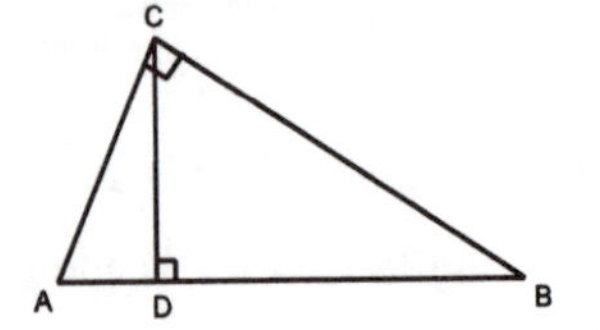

12. $\frac{BD}{18} = \frac{18}{54}$ and so BD = 6

13. We started out with a right triangle and in five steps proved that $a^2 + b^2 = c^2$. This is one of the shortest proofs of the Pythagorean Theorem.

Fred was always amazed at how many things you could do with a right triangle with an altitude to the hypotenuse. The couple on the boat with the unripped sail drifted into the distance and Fred was all alone on the "high seas." He wondered whether the Coast Guard would come and rescue him. It is not often you hear much about the Kansas Coast Guard.

Fred lay back in the boat and thought some more about right triangles. This will be a good way to pass the time until the KCG arrives. There are Three Famous Right Triangles. They are the ones that everyone talks about when the subject of right triangles comes up.

Imagine you and your sweetie are walking in the woods. It's a warm summer day and you hear, "Honey, I've been wanting to ask you." Instinctively your breath quickens and your adrenal glands (located above your kidneys) kick out a couple extra quarts (speaking hyperbolically) of adrenaline.*

"Ask me what?" you respond.

"Which of the three famous right triangles is your favorite? When we design our dream house, I think we should use lots of the kind of right triangles that you like best."

You don't want to answer, "I don't care," since that makes you look callous and insensitive.

You don't want to answer, "Oh really dear, what is *your* favorite?" because that makes you look like a total wimp.

You don't want to answer, "Three famous right triangles? I never heard of them," because that makes you look completely uneducated.

✶ Adrenaline is the old-fashioned name for epinephrine (EP-eh-NEF-rin). It's the juice (pardon me, the hormone) that your body gets bathed in whenever you get excited. It makes the heart beat faster and your blood pressure go up.

Before you get into that most difficult situation, here's a quick overview of the

Three Famous Right Triangles

Famous Triangle #1: The 3–4–5 right triangle.

In the early days surveyors would lay out on the ground a large rope with knots evenly spaced along the rope. (You don't need a tape measure in order to make knots that are evenly spaced.)

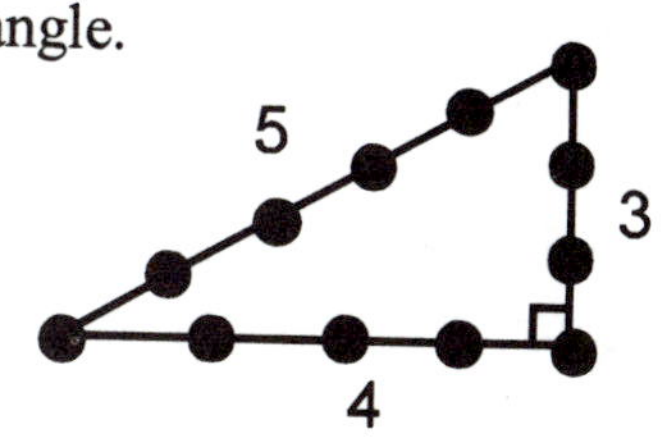

When they laid out a triangle so that the lengths of the three sides were 3, 4 and 5 they knew they had a right triangle.

Since $3^2 + 4^2 = 5^2$, the proof of the converse of the Pythagorean Theorem establishes that it really is a right triangle. The surveyors were doing this years before Pythagorus was born, and they apparently never went beyond, "Yup, it seems to work."

The 3–4–5 right triangle is the smallest triangle that has integral* sides. Any multiple of 3–4–5, such as 6–8–10 or 9–12–15, is also a right triangle.

Famous Triangle #2: The 45° right triangle.

If the length of the legs is equal to 1, then by the Pythagorean Theorem, the length of the hypotenuse must be equal to $\sqrt{2}$.

If you start with a different length for the legs, then the hypotenuse will be proportionally larger, since all 45° right triangles are similar.

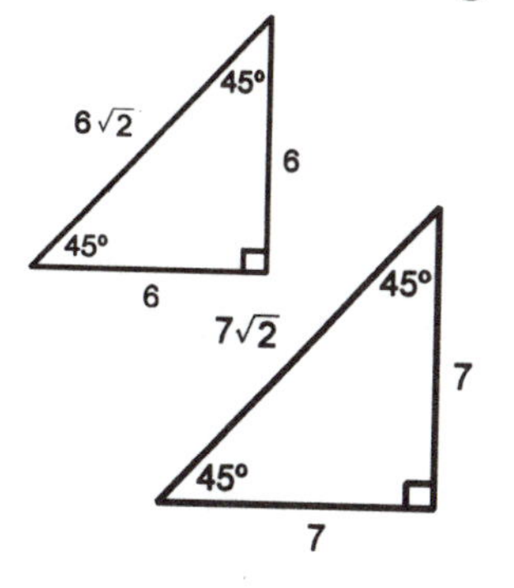

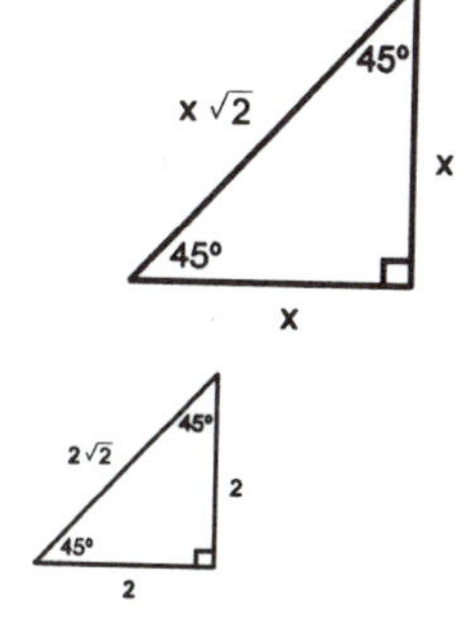

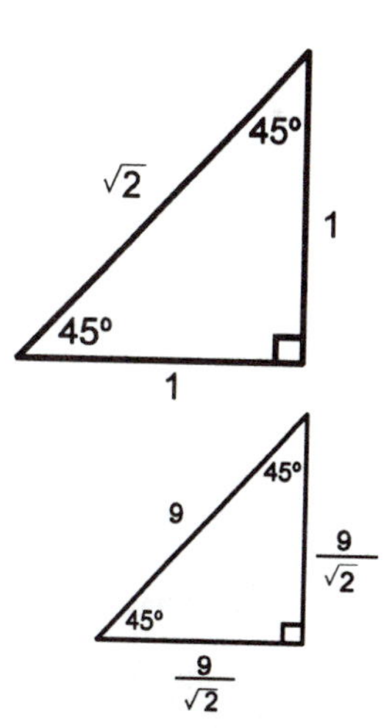

In this last triangle it is also true that if you multiply the length of the leg by $\sqrt{2}$ you will get the length of the hypotenuse.

★ Integer (noun) = {. . . –3, –2, –1, 0, 1, 2, 3 . . .}. Integral (adjective).

Famous Triangle #3: The 30°–60°–90° triangle.

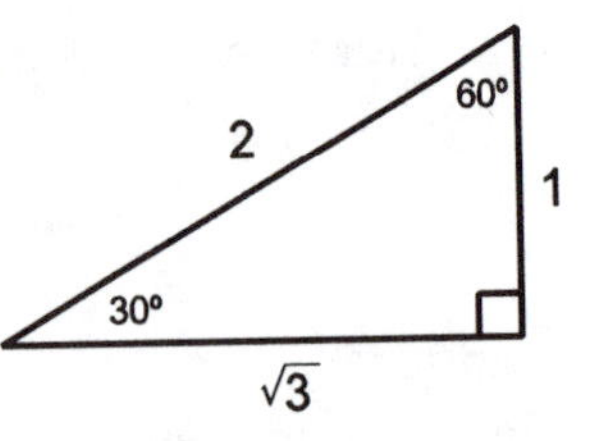

If the length of the shortest side is 1, then the hypotenuse will be 2 and the other leg will be $\sqrt{3}$. Kindergarten students count 1, 2, 3, but trig students who use the 30°–60°–90° triangle a lot learn to count 1, 2, $\sqrt{3}$.

And, just as in the case of the 45° right triangle, if you start out with a different length for the shortest side, then the other sides will be proportionally larger since all 30°–60°–90° triangles are similar.

Your Turn to Play

1. Why are all 30°–60°–90° triangles similar?
2. Find the lengths of the sides that are not given.

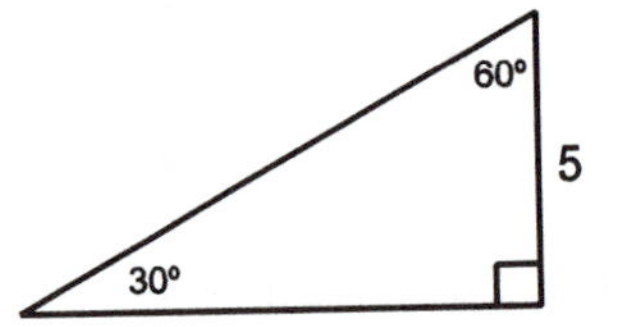

3.

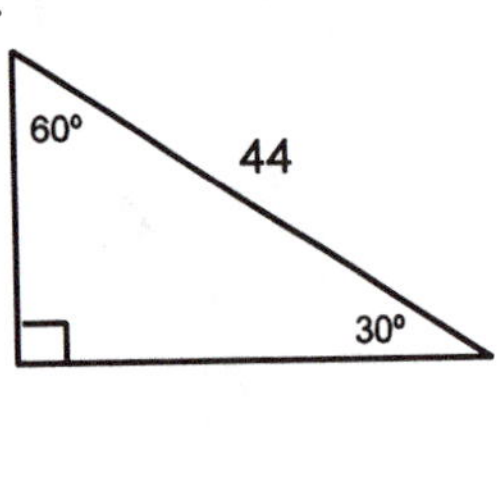

4.

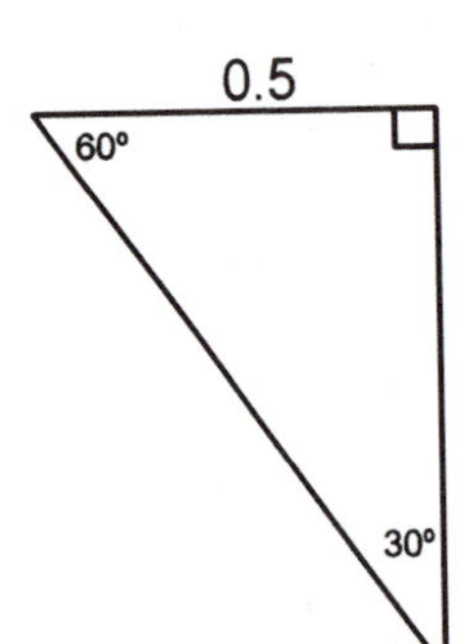

5.

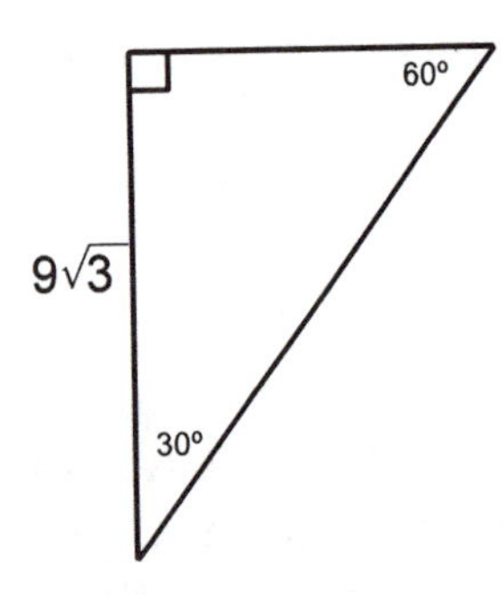

6. Here is a proof that in every 30°–60°–90° triangle, the sides are in the ratio 1, 2, $\sqrt{3}$. Fill in the missing parts.

Statement	*Reason*
1. $\triangle ABC$. $m\angle A = 30°$, $m\angle B = 60°$ and $m\angle C = 90°$	1. ?
2. On $\overleftrightarrow{BC}$, locate D so that CD = a.	2. Ruler Postulate
3. Draw $\overline{AD}$.	3. Two points determine a line.
4. $\angle 1$ and $\angle 2$ form a linear pair.	4. Def of linear pair
5. $\angle 1$ and $\angle 2$ are supplementary.	5. Linear Pair Postulate
6. $m\angle 1 + m\angle 2 = 180°$	6. Def of supplementary

7. $m\angle 2 = 90°$ and AC = AC	7. ?
8. $\overline{AC} \cong \overline{AC}$	8. ?
9. $\Delta ACB \cong \Delta ACD$	9. ?
10. $\angle D \cong \angle B$	10. ?

Then we establish that all three angles of ΔABD are 60° angles. It is an equiangular triangle and therefore, an equilateral triangle. Thus, $2a = c$.

Since the original ΔABC is a right triangle, we have $a^2 + b^2 = c^2$.

By algebra, these two equations imply that $b = a\sqrt{3}$. Relabeling the diagram using $2a = c$ and $b = a\sqrt{3}$ we have what we wanted to prove.

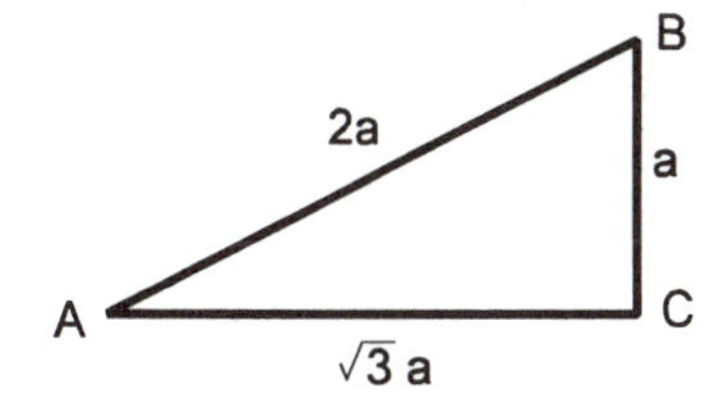

1. Postulate: AA ➜ ~ ⚠

2. 10 and $5\sqrt{3}$ 3. 22 and $22\sqrt{3}$ 4. 1 and $0.5\sqrt{3}$ 5. 9 and 18

6.
1. Given
7. Algebra
8. Definition of congruent segments
9. SAS
10. Definition of congruent triangles.

Fred imagined that P's parents would choose the 45° right triangle if they were building their dream house. Their current house is square and the 45° right triangle is great for making squares.

Fred, himself, liked the 30°–60°–90° triangle. He pictured that his dream house would have lawns, picture windows and roof lines in that pleasing shape. Perhaps it would be built on the side of a hill with a slope of 30° so that the trees would make 60° angles with the ground.

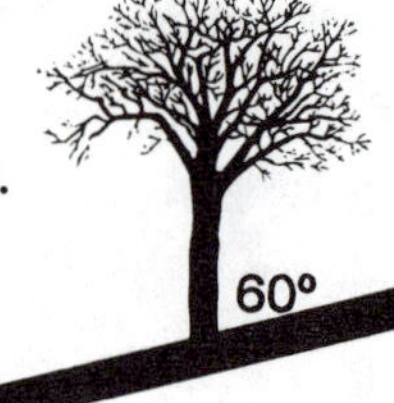

He then tried to imagine what kind of triangle that P would select for their dream house. She is such a free spirit, Fred thought. I wouldn't be surprised if she wanted the 3-4-5 right triangle for the architectural motif. I can just hear her saying, "Toowoomba, it's gotta be 3-4-5."

3-4-5 triangles have an air of mystery about them. With my old fuddy-duddy 30°-60°-90° triangles you know all the angles and you know all the sides: 1, 2, $\sqrt{3}$. But with P's 3-4-5 triangle you know all the sides, but you only know the 90° angle. Angle A is a great unknown—at least to most people.

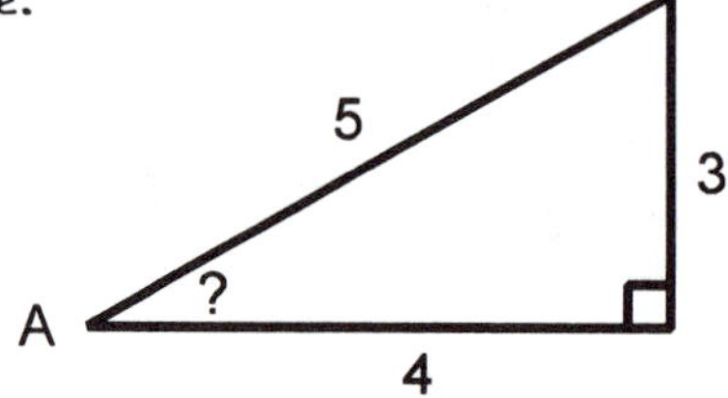

P's Poser
How big is A?

Fred took a scrap of paper out of his pocket and carefully drew a 3–4–5 right triangle and measured ∠A with a protractor that he always carried with him. He estimated the measure of ∠A to be approximately 36.8698976458°. Not bad for using a 99¢ protractor.

That truly was a fine bit of drawing and measuring. The last "8" in 36.8698976458° was a little shaky. The measure of A might have been as little as 36.8698976457° or as much as 36.8698976459°, but you have to give that kid a break since the boat was rocking a little.*

✶ Think of it this way. The Empire State Building weighs 365,000 tons, which is 11.68×10^9 ounces. If you were to be as accurate in dealing with the weight of the building as Fred was in drawing ∠A, you would be reporting the weight to within the weight of a single piece of paper.

Wait a minute! you the reader exclaim. ***Nobody, even on dry land with the finest drafting equipment, could possibly draw that 3–4–5 right triangle with that much accuracy. I know that Fred has those little beady eyeballs which give him 20/20 vision, but still—give me a break. Math books are supposed to contain as little fiction as possible. Admit it, Mr. Author, you just made up most of those digits in 36.8698976458°.***

Fred in sailor hat with his little beady eyes

This book does contain as little fiction as possible. Fred really is out on a boat on the Great Lake on the KITTENS campus. His sails really are torn. And all those digits, except maybe the last one, are true. The measure of angle A in a 3–4–5 right triangle is 36.8698976458°.

Okay. Are your eyes as beady as Fred's?

Heavens no.

Then how'd you get...

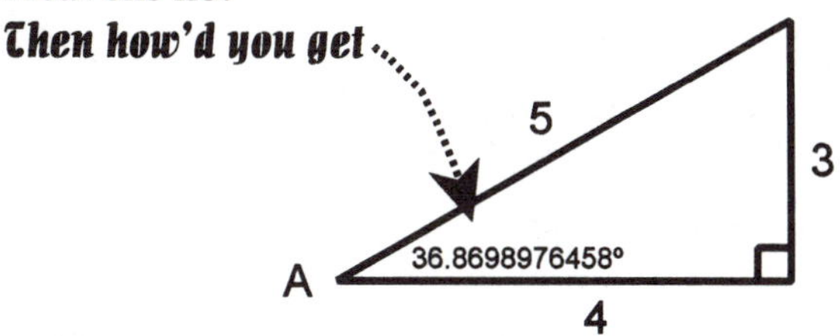

That's easy. I used trig.

Trig?

Trig—that's short for trigonometry.

I know that! But trig is the math course that comes after advanced algebra.

- ***arithmetic***
- ***beginning algebra***
- ***geometry***
- ***advanced algebra***
- ***trigonometry***
- ***calculus***

That's true, but everyone puts a little bit of trigonometry in their geometry books nowadays as a preview of coming attractions. They call it a "Touch o' Trig."

Lemme guess. You are about to give me a Touch o' Trig.

When you say it, you make it sound like a disease. It really isn't that difficult. And since trig deals mostly with right triangles, this is a great place to insert a tiny bit of trigonometry.

Okay. Fire away with your trigonometry, but remember, not too much. You have to keep something for the *Life of Fred: Trig* ***book.***

There are six trig functions: sine (pronounced "sign"), cosine, tangent, cotangent, secant, and cosecant.

Stop! Too much. Just pick one. In fact, I'll pick one for you. I pick the tangent function. That one sounds the most like geometry.

Excellent choice.

Now start with a right triangle and label one of the acute angles as angle A.

A

You haven't lost me so far.

Next label the three sides of the triangle as "hypotenuse," "opposite," and "adjacent." The hypotenuse is the longest side. The opposite side is that side of the triangle that isn't a part of angle A. The adjacent side is the third side. The adjacent side is one of the sides of angle A—the side that isn't the hypotenuse.

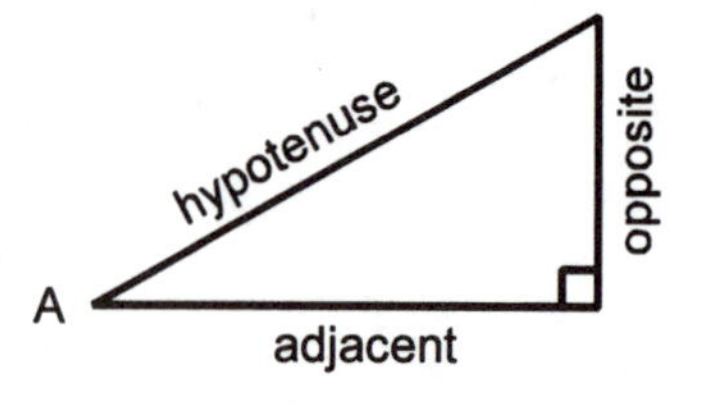

Your Turn to Play

1. Label the hypotenuse, opposite, and adjacent sides in each of the triangles.

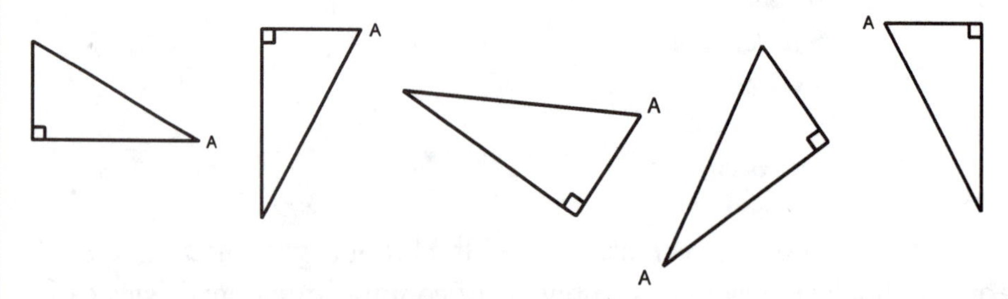

2. Now the only thing left to do is define tan A (which is short for "tangent of A"). We will define tan A as the length of the opposite side divided by the length of the adjacent side. This is often written: $\tan A = \frac{\text{opp.}}{\text{adj.}}$

In the 3–4–5 right triangle, what is tan A?

5 3 A 4

COMPLETE SOLUTIONS

1.

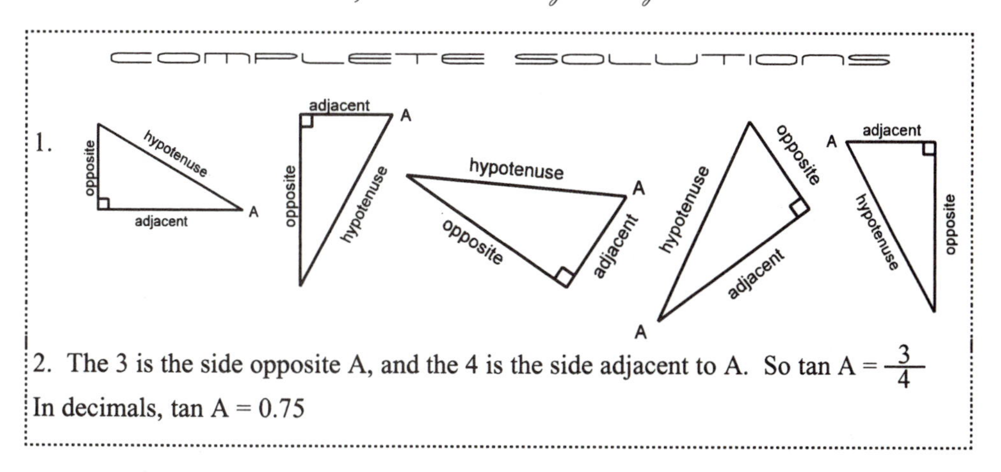

2. The 3 is the side opposite A, and the 4 is the side adjacent to A. So $\tan A = \frac{3}{4}$

In decimals, tan A = 0.75

Tangent is a function.* Tangent is a rule which assigns to each angle A, a fraction. You name an angle A between 0° and 90° and tan A will be some fraction which represents $\frac{\text{opposite}}{\text{adjacent}}$ for the right triangle in which A is an acute angle.

For example, if you are asked what tan 30° equals, you know the exact answer. You first draw the triangle.

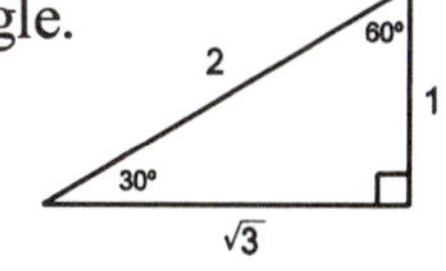

The side opposite 30° is 1.
The side adjacent 30° is $\sqrt{3}$.
Therefore, tan 30° = $1/\sqrt{3}$.

If you are asked to find tan 31°, then you would get out a protractor and draw a right triangle with an acute angle of 31° and measure the sides with a ruler and divide. You would get an answer of about 0.6008.

If you don't have a protractor or the drafting skills of Fred, on the next page I've done the work for you. I call it a table of tangents.

✶ In *Life of Fred: Beginning Algebra* there is a chapter ("Functions & Slope") in which we define what a function means and talk about the ideas of domain, codomain, and image. In *LOF: Advanced Algebra* we devote a whole chapter to functions. In *LOF: Trig* we also have a whole chapter on functions. Finally, in *LOF: Calculus*, the whole first chapter is about functions. In each book, we add a little more to the concept of functions. For example, in advanced algebra, we add the ideas of one-to-one functions, onto functions and inverse functions. In calculus, we use functions to count how many natural numbers {1, 2, 3, . . .} there are. (ans: there are aleph-null of them.) How many rational numbers there are. (ans: there are aleph-null of them.) And now many real numbers there are. (ans: there are aleph-one of them.)

A	tan A
1°	0.02
2°	0.03
3°	0.05
4°	0.07
5°	0.09
6°	0.11
7°	0.12
8°	0.14
9°	0.16
10°	0.18
11°	0.19
12°	0.21
13°	0.23
14°	0.25
15°	0.27
16°	0.29
17°	0.31
18°	0.32
19°	0.34
20°	0.36
21°	0.38
22°	0.40
23°	0.42
24°	0.45
25°	0.47
26°	0.49
27°	0.51
28°	0.53
29°	0.55
30°	0.58
31°	0.60
32°	0.62
33°	0.65
34°	0.67
35°	0.70
36°	0.73
37°	0.75
38°	0.78
39°	0.81
40°	0.84
41°	0.87
42°	0.90
43°	0.93
44°	0.97
45°	1.00
46°	1.04
47°	1.07
48°	1.11
49°	1.15
50°	1.19
51°	1.23
52°	1.28
53°	1.33
54°	1.38
55°	1.43
56°	1.48
57°	1.54
58°	1.60
59°	1.66
60°	1.73
61°	1.80
62°	1.88
63°	1.96
64°	2.05
65°	2.14
66°	2.25
67°	2.36
68°	2.48
69°	2.61
70°	2.75
71°	2.90
72°	3.08
73°	3.27
74°	3.49
75°	3.73
76°	4.01
77°	4.33
78°	4.70
79°	5.14
80°	5.67
81°	6.31
82°	7.12
83°	8.14
84°	9.51
85°	11.43
86°	14.30
87°	19.08
88°	28.64
89°	57.29

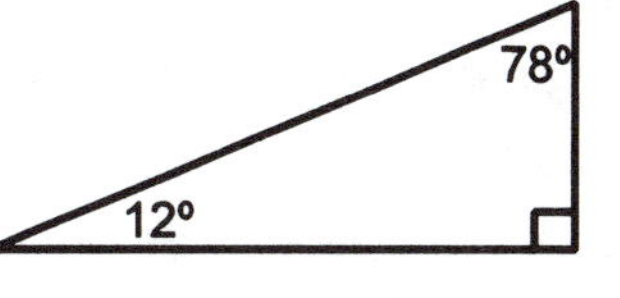

The entries in the table on the previous page are rounded off to the hundredths place. I actually didn't have to draw 89 different triangles. I kind of cheated. When I drew the triangle for 12°, for example, I could reuse that same triangle for finding tan 78°. (The acute angles of a right triangle are complementary (i.e., they add to 90°).

The only thing I had to do was adjust which side was the opposite and which side was the adjacent side.

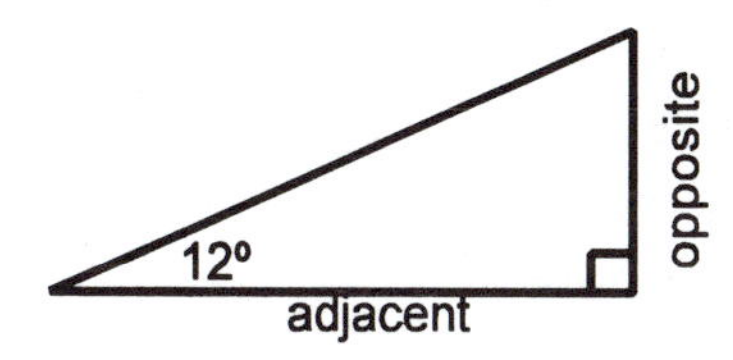

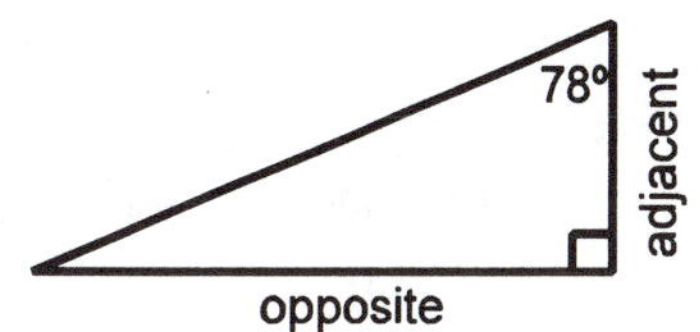

Now to answer your question . . .

What question?

About five pages ago you asked me how I got

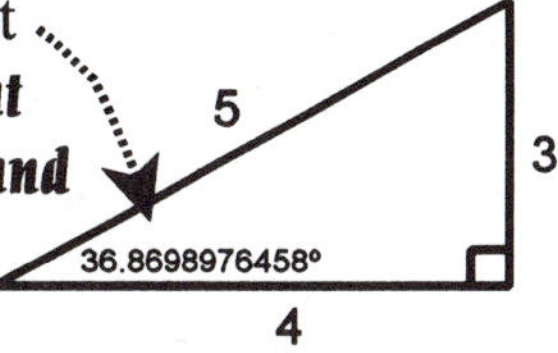

Oh yes. That question. I understand now that you took the tangent of A and got 3/4 which is 0.75 and that you probably looked it up on your tangent table on the previous page. But when I try that, all I get is that tan 37° is equal to 0.75.

You got tan 36.8698976458° = 0.75. How did you do that trick?

I told you that the table on the previous page is rounded off to the nearest hundredth. It's just a table for beginning trig students. What if I told you I just used a bigger, more detailed table?

A	tan A	
36.8698976452°	0.749999999982	
36.8698976453°	0.749999999985	
36.8698976454°	0.749999999988	
36.8698976455°	0.749999999991	
36.8698976456°	0.749999999993	
36.8698976457°	0.749999999996	
36.8698976458°	0.749999999999	⟸ That's the closest for tan A = 0.75
36.8698976459°	0.750000000002	
36.8698976460°	0.750000000004	
36.8698976461°	0.750000000007	
36.8698976462°	0.750000000010	

I think you're pulling my leg. That "bigger, more detailed table" would have something like 10^{12} entries. Suppose your super table had really tiny type and had 1000 entries per page. $10^{12} \div 10^3$ is 10^9. So your more detailed table would have a billion pages. That's some book. Tell me another fairy tale.

Okay. Suppose I told you that I used my scientific calculator.* To find tangents on a scientific calculator, all you do is type in the angle and then hit the tan button and out pops the answer.

To find the tangent of 45° using a calculator, you type in 45 and then hit the tan button and out will come 1.0000000. Of course, we knew that it would be equal to one since the 45° right triangle is one of the three famous triangles and the side opposite the 45° and the adjacent side are both equal to 1 and so tan 45° = 1/1 = 1.

Ha! Nice try. You almost had me believing you for a minute. I'm not your average gullible reader. Those hand-held calculators can only display eight digits. The stuff in that "bigger, more detailed table" that you wrote on the previous page has 12 digits.

If you want more accuracy, I can give you 15 digits.

A	tan A
36.869897645795	0.749999999998663
36.869897645796	0.749999999998691
36.869897645797	0.749999999998718
36.869897645798	0.749999999998745
36.869897645799	0.749999999998773
36.869897645800	0.749999999998800
36.869897645801	0.749999999998827
36.869897645802	0.749999999998855
36.869897645803	0.749999999998882
36.869897645804	0.749999999998909
36.869897645805	0.749999999998937
36.869897645806	0.749999999998964
36.869897645807	0.749999999998991
36.869897645808	0.749999999999018
36.869897645809	0.749999999999046
36.869897645810	0.749999999999073
36.869897645811	0.749999999999100
36.869897645812	0.749999999999128
36.869897645813	0.749999999999155
36.869897645814	0.749999999999182 ⇐

Confession Time: I have to admit it. I like playing. Too many people in our world go around with ☹ faces and that's a shame. Especially with something as fun as mathematics can be, there's no reason to be glum. I'll also confess that I got my tangent values from my word processing program on my desktop computer.

---

* Hand-held calculators come in three basic varieties. There's the bottom-of-the-line one that you can get for under $5 which has buttons for +, –, ×, ÷, √‾. That one will be enough for all of algebra and geometry. The next step up is the scientific calculator (under $20) which has the additional buttons of sin, cos, tan, log, and ln. That's the last one you will have to purchase to do all of trig and all of calculus. The extra-fancy graphing calculators, which cost about four times as much as a scientific calculator can be fun to play with, but you really don't need one. I've never owned one.

Thank you for answering my question. Now you can get back to the story if you want to.

Fred sat up in his boat and looked around. There was no sign of rescue. What he did see, however, was an old-fashioned wooden ship. It must be some kind of pleasure-cruise ship. I can see people running around on the deck. And look! There's even a diving board set up so that the passengers can take a dip in the lake whenever they want to. And if my eyes don't deceive me, there's Tanya on the diving board right now. She took beginning algebra from me a couple of years ago and now is in my second semester calculus class. Oh how Tanya used to tease me about my height. I've been 36 inches tall for a long time and she's grown about four inches since I've known her. It wasn't that long ago that she remarked to me that she was now twice as tall as I am. She's very proud of her six feet.

She's standing on the very edge of the diving board. Some people are encouraging her to jump. I guess they want her to hurry up so that they can take their turn.

Fred had an idea. He pulled out his protractor. He was going to "measure her" just as she hit the water. I wonder why she's waiting? Those rude people in back of her just gave her a poke. There she goes.

Fred quickly used his eagle eyes and 99¢ protractor and took the measurement. It was 2°.

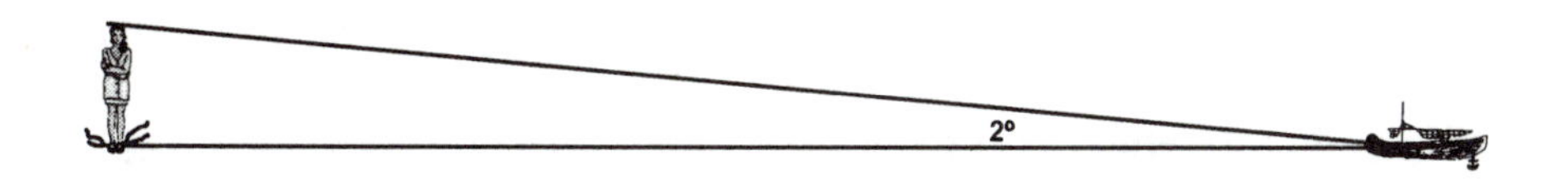

Your Turn to Play

1. This might be a fairly difficult question, but give it a try: How far is Fred from Tanya? You'll need the tangent table on p. 336.

2. You are standing a mile (5280 feet) away from a building that looks like the Empire State building. With your protractor you estimate that the angle from your feet to the top of the building is 14°. How tall is the building? Again, use the tangent table on p. 336.

14° ? 5280'

COMPLETE SOLUTIONS

1. The triangle we are trying to "solve" looks like:

Using the definition of tangent, $\tan 2° = \frac{6}{?}$

By algebra, $? = \frac{6}{\tan 2°}$

and if we approximate tan 2° using the tangent table, $? = \frac{6}{0.03} = 200$ ft.

2.

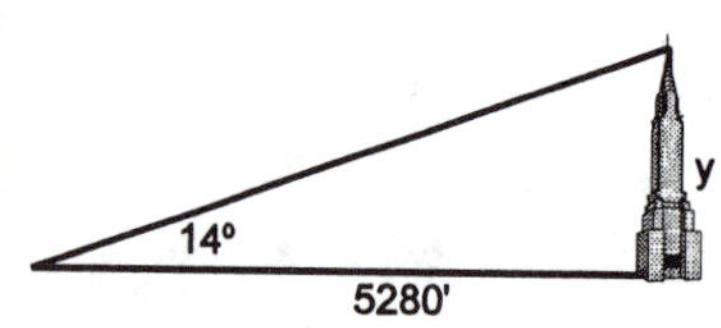

We'll let the height of the building = y. That looks a lot more like algebra than using the "?" as we did in the previous problem.

By def of tangent, $\tan 14° = \frac{y}{5280}$

By algebra, $y = 5280(\tan 14°)$

Using the tangent table, $y = 5280(0.25)$

$y = 1320$ feet

The Question & Answer Time in which you, the reader, may ask any questions that come to mind (before I start asking questions on the next page).

Q: How come you're doing this? Is this new?

A: Yes, this is new.

Q: And why are you doing this?

A: You'd better start asking your real questions or you'll run out of space on this page.

Q: Okay. I understand that I can now "solve" any right triangle in which I know the acute angle and one of the two legs ("opposite" and "adjacent"). What do I do if I'm confronted with knowing the acute angle and the opposite side and want to find the hypotenuse? The good ol' tangent function only works with opposite divided by adjacent. How do you get the answer?

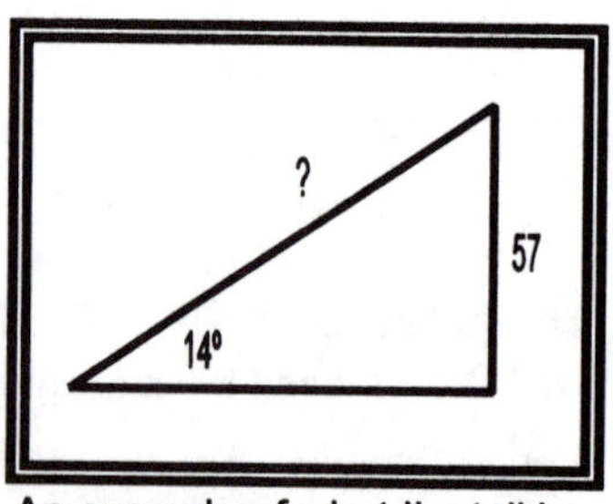

An example of what I'm talking about

A: You ask such wordy questions. Let's continue this on the next page.

A: (continued) That's what the rest of trig deals with. We're in geometry right now and if I tell you that the sine function is defined as $\frac{\text{opposite}}{\text{hypotenuse}}$ (which it is), then what will there be left to talk about in the trig book?

Q: And, let me guess, the cosine function is $\frac{\textbf{adjacent}}{\textbf{hypotenuse}}$

A: Good guess!

Q: And so on the scientific calculator, the sin ***button doesn't mean stealing or lying or murder or adultery. It means sine.***

A: Right again.

Q: I've got a question that students always ask: Will this stuff—sine and cosine—be on the test? I mean are you going to stick it in the Cities?

A: I wasn't planning on doing that, but now that you mention . . .

Q: I didn't, I mean, please, mercy. Save it for the trig book.

A: Okay.

Q: Can I ask any questions I want? I've never done this question stuff before. It reminds me of a press conference. I'd better ask my question before I run out of room on this page. I don't think you'll give me a third page.

A: Pardon me for interrupting, but if you're really concerned about running out of room, it might not be a bad idea to not spend so much time talking about running out of room and ask your questions.

Q: Okay. To the point. Why did it look like Tanya's arms were wrapped around herself? And why wasn't she wearing a swimsuit?

A: Her arms were tied. And she wasn't wearing a swimsuit because she wasn't going swimming.

Q: It was two pages ago that Tanya jumped into the water. Why hasn't she come to the surface? I'm starting to get worried.

A: It's too late for that. Actually, it's always the wrong time to commit the sin of worrying.

Q: You mean the sine of worrying? Sorry. Really bad joke. I think I'm starting to get the picture. Was Fred right when he thought this was a pleasure-cruise ship?

A: Nope.

Q: Did he (and you and me) just witness an execution?

A: The first one in the history of geometry books.

Q: And the flag on that ship. Is it what I think it is?

A: Yup.

Emden

1. Find the mean proportional to 5 and 30. The mean proportional is sometimes called the **geometric mean**.
2. Find the exact values of b and c.

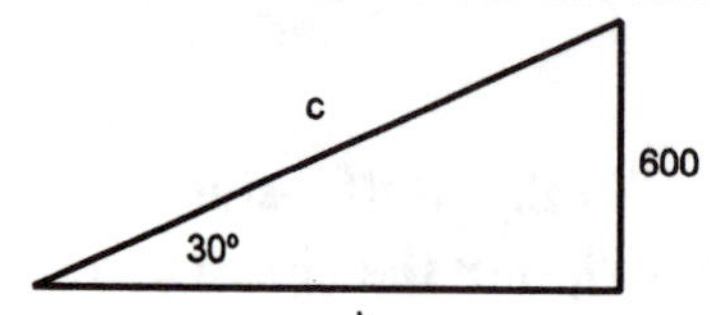

3. Find the approximate value of m∠B rounding your answer to the nearest thousandth of a degree.

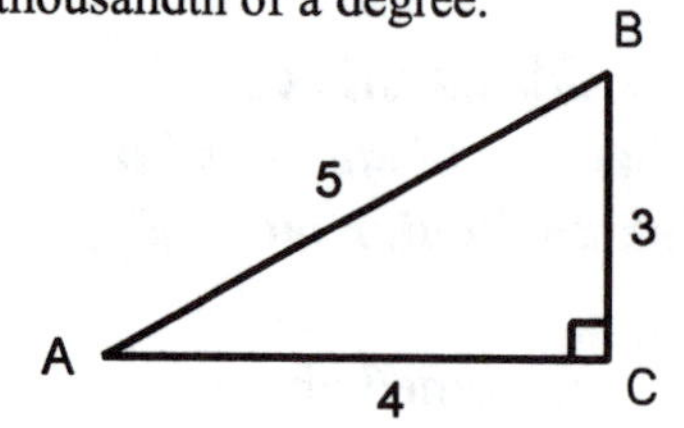

4. Find the value of m∠D to the nearest degree.

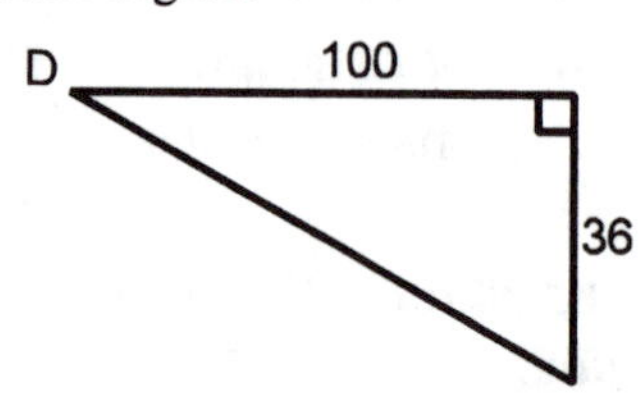

5. Find the approximate value of m∠G rounding your answer to the nearest millionth of a degree.

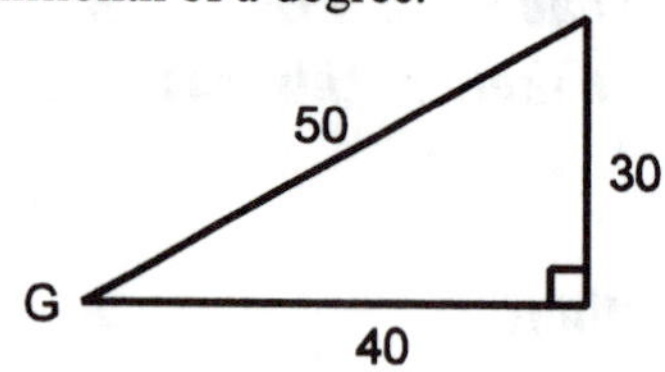

answers

1. $\pm\sqrt{150}$ or $\pm 5\sqrt{6}$

2. c = 1200; b = $600\sqrt{3}$ (Using the tangent tables won't give the exact answer for b.)

3. On p. 333, Fred found that m∠A was approximately 36.8698976458° and since the acute angles of a right triangle are complementary, m∠B = 90° – 36.8698976458° = 53.130103° which rounds to 53.130°

4. From the tangent table on p. 336, we note that the tangent of 20° = 0.36, which happens to be the case in the triangle we're looking at. m∠D = 20°.

5. By the SSS Similarity Theorem, this triangle is similar to the famous 3–4–5 right triangle. Hence corresponding angles will be congruent by definition of similar triangles. Looking at the 3–4–5 right triangle on p. 333, we note that the corresponding angle (to ∠G) has a measure of 36.8698976458°. Rounding this to the nearest millionth of a degree, m∠G ≐ 36.869898° ("≐" means rounded off to)

Galena

1.

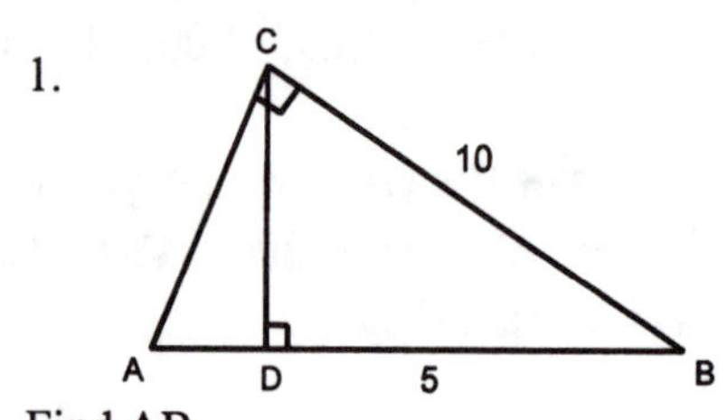

Find AB.
Find AD.
Find CD.

2. Find the exact area of ΔDEF.

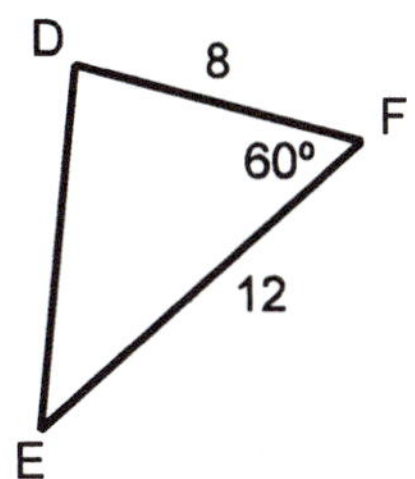

(Hint: Drop the altitude from E. You don't need trig in this problem.)

3. Find the exact value of c.

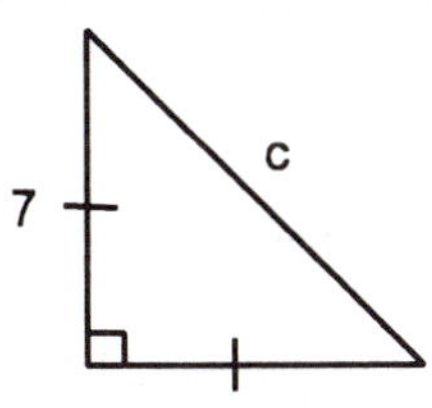

4. Find the length of the legs of this triangle.

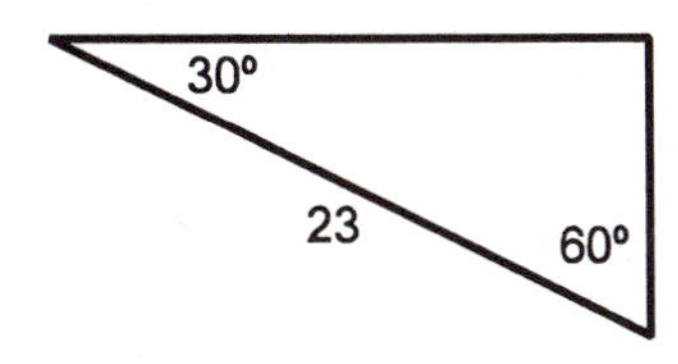

5. You're on the football field. At the other end of the field, 100 yards away, is the tallest football player you've ever seen. You measure with your protractor and find that the "angle of elevation" (that's what they call it in trig) to the top of his head is 2°. How many FEET tall is he? (There are both feet and yards in this problem. Convert everything to feet before you head to the tangent table.)

answers

1. AB = 20; AD = 15; CD = $\sqrt{75}$ or $5\sqrt{3}$. (Note: CD can't be negative since it is a length.)

2. The length of the altitude (using the properties of the 30°–60°–90° triangle) is $6\sqrt{3}$. Using A = ½ bh formula for the area of a triangle, aΔDEF = (½)(8)($6\sqrt{3}$) which is $24\sqrt{3}$.

3. That is a 45° right triangle. The hypotenuse is $\sqrt{2}$ as large as either leg. The answer is $7\sqrt{2}$.

4. $\frac{23}{2}$ and $\frac{23\sqrt{3}}{2}$

5. I wonder if he plays basketball when he's not playing football. Nine feet tall.

Urbana

1. In the proof of the Pythagorean Theorem (on p. 326), we took the equations of steps 3 and 4 which are

$$x + y = c$$

$$\frac{x}{b} = \frac{b}{c} \quad \text{and} \quad \frac{y}{a} = \frac{a}{c}$$

and obtained by "Algebra" in step 5

$$a^2 + b^2 = c^2.$$

That was a big leap of algebra. Fill in some of the missing lines of algebra.

2. Find the geometric mean between 1 and 29.

3. Find the exact area of ΔGHI. Do not use the tangent table since that often yields only approximate answers.)

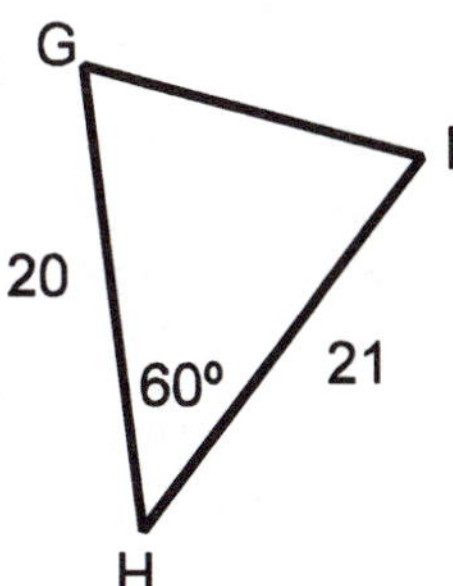

4. You are looking at a tall statue of Ozymandias (286 feet tall) before it fell into ruin. With your protractor

you note that the angle from your feet to the top of his head is 55°. How far are you standing from this vainglorious blowhard?

Ozymandias

I met a traveler from an antique land
Who said: Two vast and trunkless legs of stone
Stand in the desert. Near them, on the sand
Half sunk, a shattered visage lies, whose frown,
And wrinkled lip, and sneer of cold command
Tell that its sculptor well those passions read,
Which yet survive, stamped on these lifeless things,
The hand that mocked them, and the heart that fed,
And on the pedestal these words appear:
"My name is Ozymandias, King of Kings:
Look upon my works, ye Mighty, and despair!"
Nothing beside remains. Round the decay
Of that colossal wreck, boundless and bare
The lone and level sands stretch far away.

— Percy Bysshe Shelley, 1792–1822

(and in case you're wondering, it's pronounced Bish. In elementary school, they probably called him Bysshe, the Fish.)

5. (continuing the previous problem) If you were to tie a rope from the top of Ozymandias' head to your feet (in order to keep the statue from tipping over) I would normally use the sine function ($\frac{\text{opposite}}{\text{hypotenuse}}$) and sine tables to figure out the length of the rope. However, you, the reader, made me promise not to have any sine problems in this book.

Describe how you could find the length of the rope without resorting to using the sine function.

odd answers

1. First, cross multiply the two proportions: $cx = b^2$ and $cy = a^2$
Then divide each of those equations by c: $x = b^2/c$ and $y = a^2/c$

Add the two equations together: $x + y = \frac{b^2 + a^2}{c}$

Using the first given equation we may substitute c for x + y: $c = \frac{b^2 + a^2}{c}$

Multiply both sides by c: $c^2 = b^2 + a^2$.

3. Dropping an altitude from G, we find its length by the properties of the 30°–60°–90° triangle to be $10\sqrt{3}$. By $A = \frac{1}{2}bh$, we have aΔGHI = $(\frac{1}{2})(21)(10\sqrt{3}) = 105\sqrt{3}$.

5. From the previous problem you learned that the distance from your feet to the statue was 200 feet. (Oops! I guess I shouldn't have given that away since it's an even problem.) In any event, since you know the lengths of the legs of the right triangle, you may use the Pythagorean Theorem to compute the length of the hypotenuse.

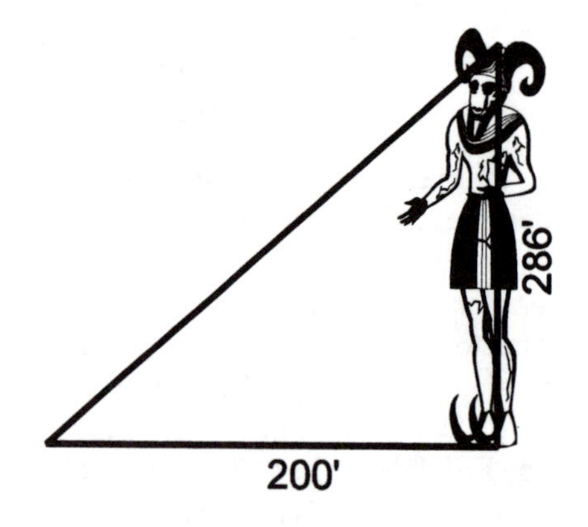

Walnut Grove

1. Using the tangent tables find AC.

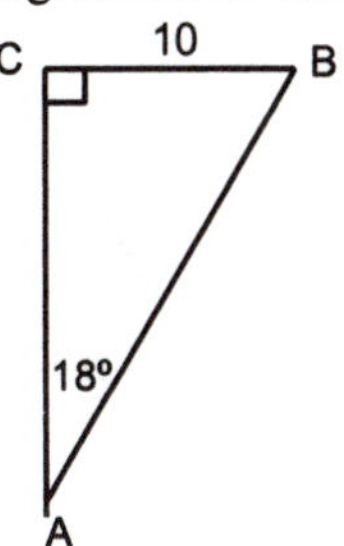

2. Find the exact value of aΔPQR. Do not use the tangent tables since that often gives only approximate answers.

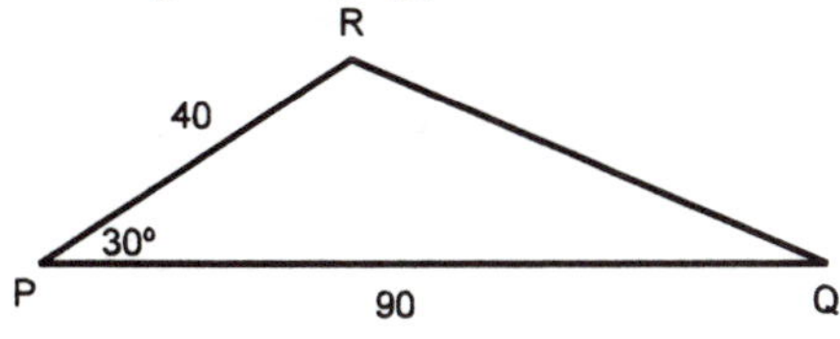

3. Find the lengths of all the segments in the diagram.

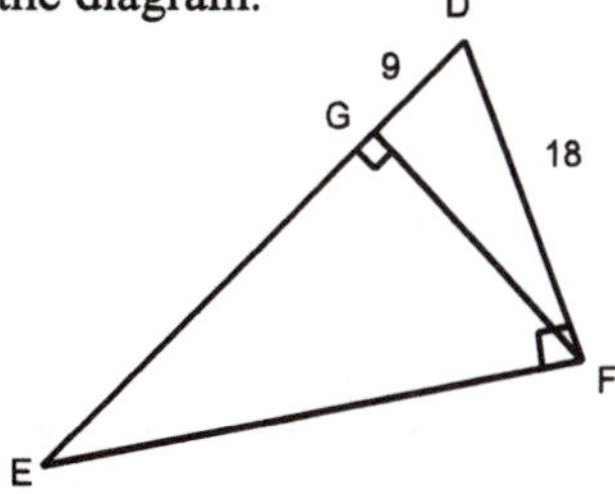

4. What is the exact perimeter of an equilateral triangle whose altitude is equal to 12?

(Hint: If you were to approximate your answer it would be 41.569.)

5. There are three altitudes in any triangle. Find the lengths of the two longer altitudes in the 3–4–5 triangle.

odd answers

1. $\frac{10}{0.32} \doteq 0.31$ ("$\doteq$" means rounded off to)

3. DE = 36; EG = 27; FG = $\sqrt{243}$ or $9\sqrt{3}$

5.

AB = 5

Just take a peek at the diagram.

The three altitudes of ΔABC are $\overline{AC}$, $\overline{BC}$, and $\overline{AB}$.

$\overline{CD}$ is shorter than either $\overline{AC}$ or $\overline{BC}$. (For example, CD < AC since $\overline{CD}$ is a leg and $\overline{AC}$ is a hypotenuse in the triangle on the left.)

So the lengths of the two longer altitudes are just 3 and 4.

Karnack

1. Find the mean proportional to 11 and 17.

2. Find the exact values for b and c.

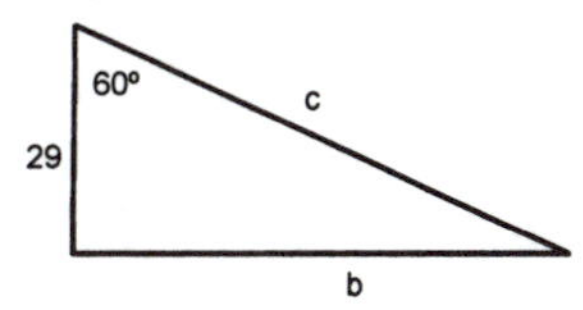

(Do not use the tangent table since that often only gives approximate answers.)

3. Find the exact value of aΔABC. Do not use the tangent tables since that often gives only approximate answers.

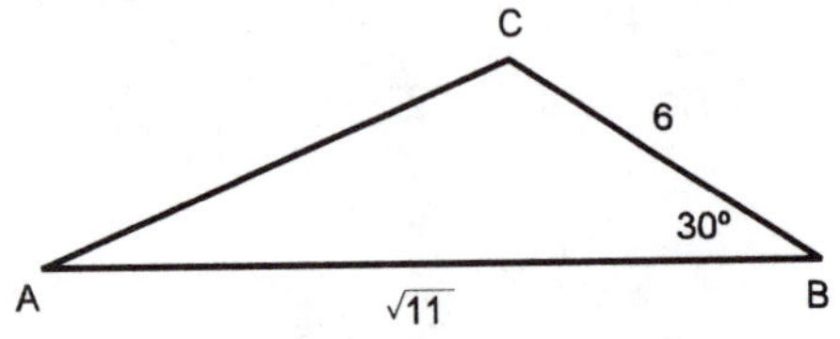

4. Find the exact value of DF. Do not use the tangent tables.

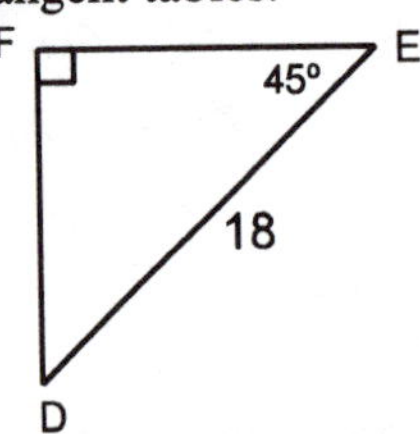

5. You are a surveyor working on a large flat piece of land. Suppose you have lots of rope, a hammer and stakes and scissors.

In the discussion of the Three Famous Right Triangles we showed how to lay out on the ground a 3–4–5 right triangle.

How would you go about laying out a 45° right triangle?

Lancaster

1.

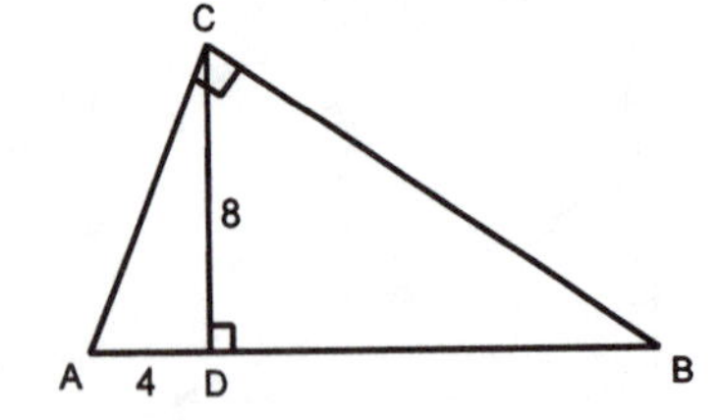

Find BD.

Find AB.

Find AC.

2. Find the exact value of aΔJKL. Do not use the tangent table since that often gives only approximate answers.

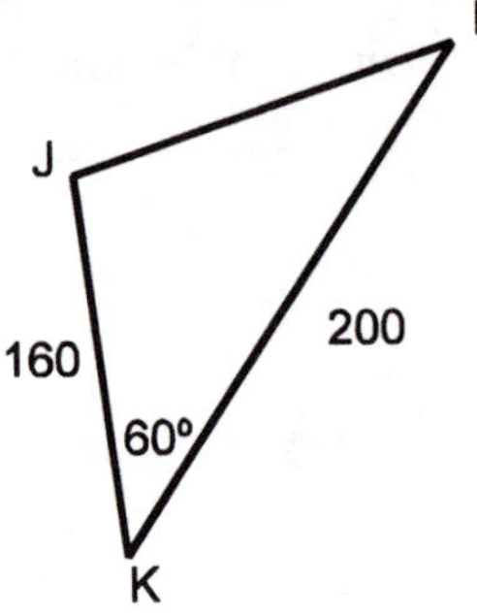

3. Find the approximate value of m$\angle$E rounding your answer to the nearest millionth of a degree.

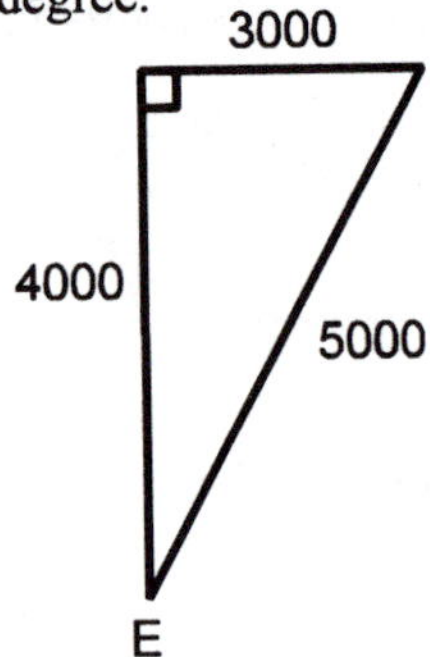

4. Find the lengths of the sides that are not given.

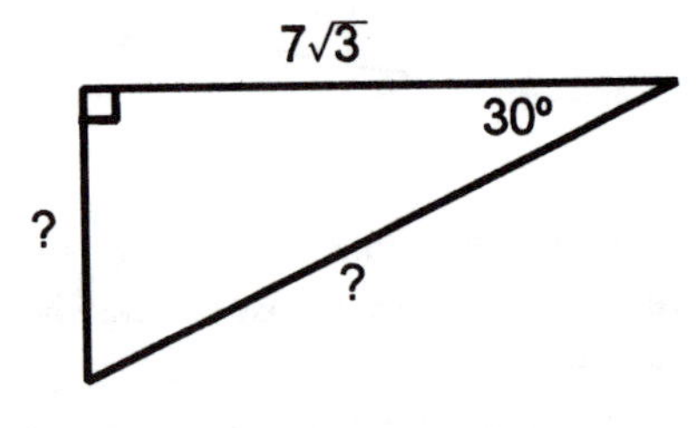

5. What is the exact value of tan 60°?

Chapter Ten

Circles

Fred was so envious of all those people playing on that big wooden boat. He just knew that in addition to that wonderful diving board on the side of the ship, there must be all kinds of games like shuffleboard and tag that you could play on the deck. To climb up to the crow's nest near the top of the mast would be fun. In fact, I can see some sailor up there right now, and the sailor is looking right at me. He's pointing at me!

They're turning the ship toward me. I'm going to be rescued.

The minutes passed as the ship headed straight toward him. Fred stood up in his boat, waved, and shouted, "Hi guys! Hi guys! Can I come and play on your boat?"

Straight toward him.

"Hey guys! Slow down a bit. You're coming too fast."

Straight toward him.

If geometry ever had a boat-bisection theorem . . .

. . . this would have been the moment.

The surface of the lake looked like toothpick soup with one little piece of meat (Fred) in it.

"Hook the little runt and let's see what we've got," ordered Captain H. Several sailors threw harpoons, but they all missed their target. It's easier to hit a whale than a 37-pound minnow. Fred grabbed onto the rope attached to one of the harpoons and climbed up on the deck.

"Thanks for saving me. That was really nice of you to come to my rescue. Please don't worry about my boat. I bought collision insurance. I don't know what the deductible is, but I'll take care of it. What funny costumes you all are wearing! They make you guys look like pirates. What a neat idea to have theme days on your pleasure-cruise ship. I like to dress up in costumes sometimes too. In the classroom I once dressed up as the Greek mathematician Euclid. He's the guy that wrote *The Elements* back in 300 B.C. I see by your name tag that your name is Hook. That's a clever name. I get it. Captain Hook. Just like in Peter Pan."

"Would you please shut %$&*%$ %#&*%$ up!" To emphasize his point, the captain clamped his hand over Fred's mouth.

There's something wrong here. I was just trying to be friendly and he swears at me. And his hand is stinky.

"Empty out the little bugger's pockets. Let's see what we've got."

The sailors reached in and pulled out:

- ☞ Fred's 99¢ protractor,
- ☞ A copy of *Prof. Eldwood's Tangent Tables for Angles Between 36.8698976452° and 36.8698976462°*,
- ☞ P's letters to Fred,
- ☞ A lock of llama hair tied with a ribbon and the letter λ on the ribbon,
- ☞ A ball of cookie dough that his mother gave him when he was little.

Hook carefully examined each item and then threw it overboard. "Junk!" he exclaimed. "Don't you carry any cash, or credit cards? No cell phone? No jewelry? And this stupid bow tie of yours? Who dresses you kid?" He pulled off Fred's bow tie and threw it into the water.

"Okay, squirt. Here's your last chance to live." The Captain picked up Fred and held him close to his face.

Ugh! His breath. It smells like Joe's apartment the month he forgot to clean his cat's litter box. I wonder if Mr. Hook has gum disease or something. I should recommend my dentist to him. And remind him to floss each night.

"Listen up, brat! If you give me the wrong answer, you'll be making a one-way trip to Davy Jones's locker." To emphasize his point, he grabbed Fred by an ankle and held him over the edge of the ship. "Here's a real simple question. How much money do your parents got? A little ransom money could buy your life."

At this point many individuals would have been tempted to lie and say that they had fabulously rich parents with tons of cash, stocks, gold, baseball trading cards. And if they were really adept at prevarication, they would add that their parents were off on vacation for many months somewhere in the Himalayas where they couldn't be reached.

Fred was a little bit more straightforward. "They're both dead."

The Captain laughed. He pulled Fred back on board and dropped him on the deck. "That's the first time I've ever heard that kind of answer. Everybody usually invents some stupid story about their rich parents who can't be reached for weeks. Aren't you afraid to die?" He held his cutlass to Fred's nose.

Before Fred could answer, a woman's voice intervened. "Hey Hookie! Stop scaring the boy."

Fred looked up. Her name tag read, "Mrs. Harry Hook."

"I could use the little half-pint," she continued. "I always wanted a little cabin boy. You know, like they have in the pirate movies." She was going to grab him by his ear (not very evident), by his hair (no handhold there), by his nose (too sharp—she might cut herself), but decided to seize* him bodily. She carried him as as you or I might carry a package of spaghetti.

"Oh #*#% !" she muttered. "Hookie forgot to take the pooch inside."

What we tend to think of as a pooch

A pirate's pooch

✶ SEIZE—Your English teacher told you, "I before E, except after C, or except when pronounced as A in neighbor or sleigh."

After Snarls (Hook's affectionate name for his dog) had eaten several of the crew, he decided to keep the dog on a chain. Which he had nailed to the center of the deck.

Snarls was a very fast dog. He could appear to be sleeping, but if you stepped within the radius of his chain, you were dog food. If point O was the center of the circle and you were at point P, and if r was the radius of the circle, then in symbols:

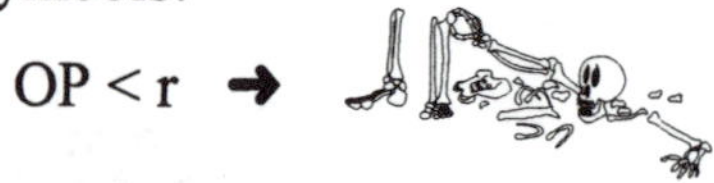

Officially, since some people think that this is a geometry textbook, it would probably be advisable to define circle, radius, etc.

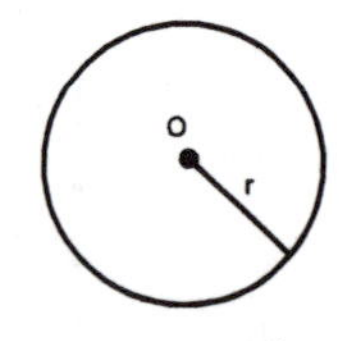

Definitions: Given a positive number r and a point O, a **circle** is the set of all points a distance r from O. The distance r is called the **radius**. The point O is called the **center of the circle.**

Definitions: A **chord** of a circle is a segment whose endpoints are on the circle. A **diameter** is a chord which passes through the center of the circle. A **secant** is a line which contains a chord. A **tangent** is a line which intersects a circle at only one point.

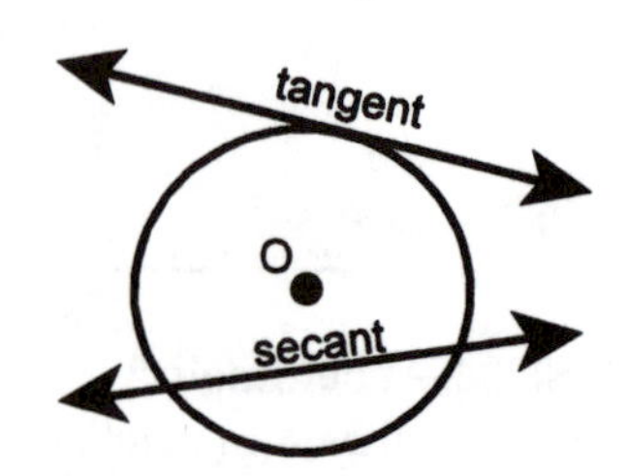

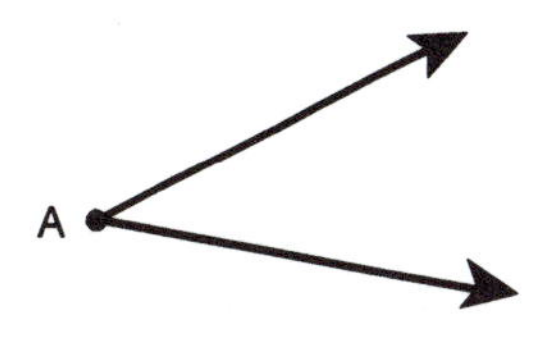

Many chapters ago when we looked at angles, we defined an angle as a pair of rays that shared a common endpoint. ∠A was a geometrical object. Associated with each angle (by the Angle Measurement Postulate) was a number, m∠A.

You could talk about the sides of ∠A or about any points in the **interior** of ∠A.* But you can't speak about the interior of a number. That would be silly. There's no way to get in the interior of m∠A.

You can talk about adding m∠A to m∠B, or about taking the square root of m∠A, but it would be difficult to take the square root of a pair of rays. One student once suggested that the square root of A

would be 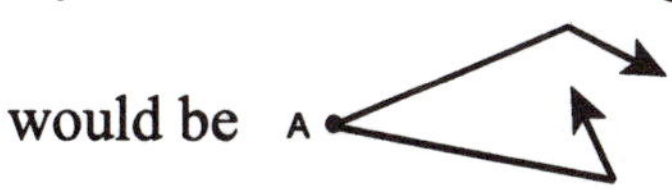

However, with the language of circles, things are a lot more sloppy. A radius, for example, can be a segment from the center of a circle to the circle—or it can be the length of that segment.

A diameter can be a chord containing the center of the circle—or it can be the length of that chord. If we write $d = 2r$, then it wouldn't be too hard to guess that we're talking about diameters and radii (plural of radius) as lengths. If we say that a radius is perpendicular to a particular chord, it's obvious that we are in Geometryland rather than Algebraland.

As a result of this looseness of language, it is all right to say as a reason in a proof: "All radii of a circle are =" which is true by definition of a circle, or you may say "All radii of a circle are ≅."

Here is a definition that we won't use in this book, but is used in everyday life, such as when you draw targets.
Definition: Two circles are **concentric** if they have the same center.

✶ A point P is in the interior of ∠A if it lies on a segment whose endpoints are on the sides of ∠A.

A P

Your Turn to Play

1. Prove: <u>Theorem</u>: If a radius is perpendicular to a chord, it bisects it.

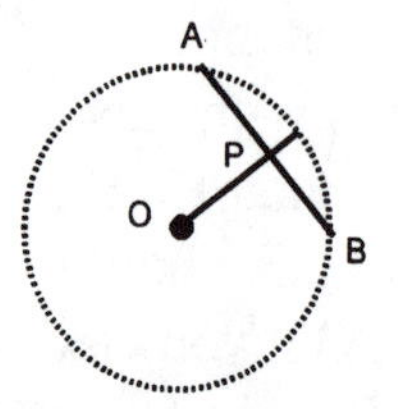

2.

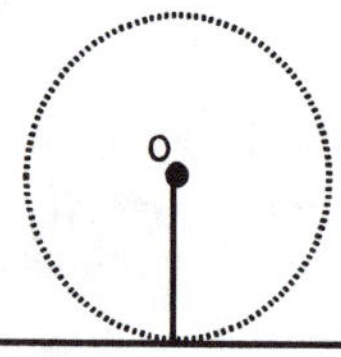

Fill in the missing parts of the proof of <u>Theorem</u>: A tangent is perpendicular to a radius drawn to the point of tangency.

Statement	Reason
1. Circle O with ℓ tangent to O at point P.	1. ?
2. Assume that . . . ?	2. Beginning of an indirect proof
3. Let $\overline{OR}$ be the ⊥ from O to ℓ.	3. Theorem: You may drop a ⊥ from a point to a line.
4. Locate point S on $\overleftrightarrow{PR}$ such that the distance from S to R is equal to PR.	4. Ruler Postulate

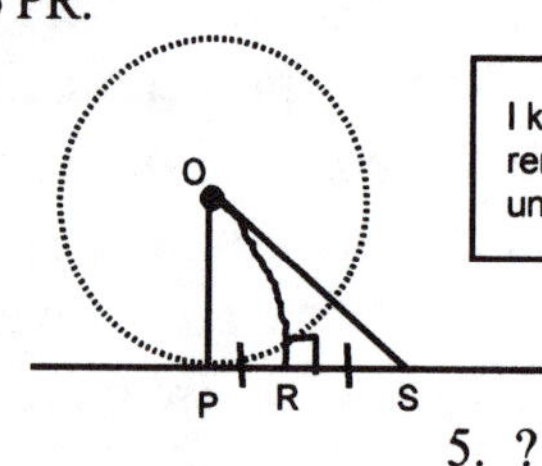

I know this diagram looks weird, but remember, we assumed something untrue in step 2.

Statement	Reason
5. OR = OR	5. ?
6. $\overline{OR} \cong \overline{OR}$ and $\overline{SR} \cong \overline{PR}$	6. ?
7. ∠PRO ≅ ∠SRO	7. ⊥ lines form ≅ right ∠s
8. ΔPRO ≅ ΔSRO	8. ?
9. $\overline{OP} \cong \overline{OS}$	9. ?
10. OP = OS	10. Def of ≅ segments
11. S is on circle O	11. Def of circle
12. ℓ is not tangent to circle O	12. Def of tangent
13. ?	13. Contradiction in steps 1 and 12 and hence the assumption in step 2 is false. ⊠

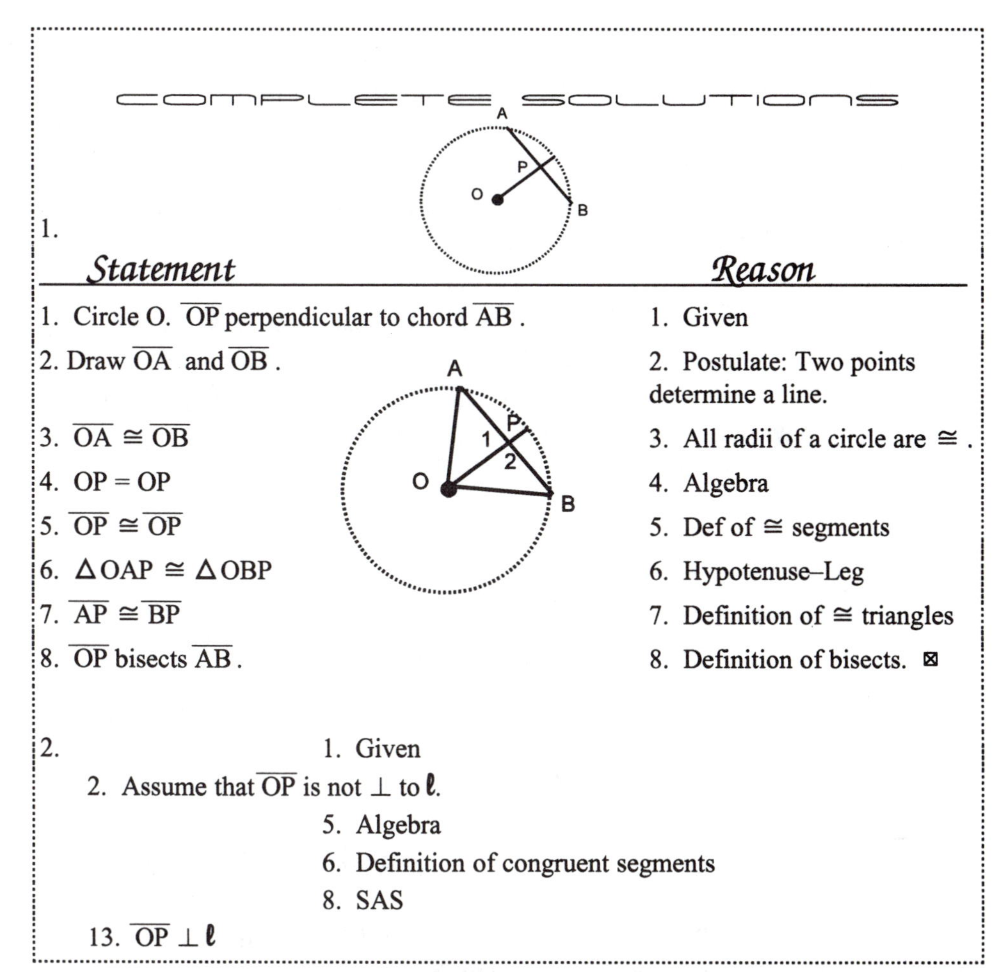

COMPLETE SOLUTIONS

1.

Statement	Reason
1. Circle O. $\overline{OP}$ perpendicular to chord $\overline{AB}$.	1. Given
2. Draw $\overline{OA}$ and $\overline{OB}$.	2. Postulate: Two points determine a line.
3. $\overline{OA} \cong \overline{OB}$	3. All radii of a circle are $\cong$.
4. OP = OP	4. Algebra
5. $\overline{OP} \cong \overline{OP}$	5. Def of $\cong$ segments
6. $\Delta OAP \cong \Delta OBP$	6. Hypotenuse–Leg
7. $\overline{AP} \cong \overline{BP}$	7. Definition of $\cong$ triangles
8. $\overline{OP}$ bisects $\overline{AB}$.	8. Definition of bisects. ⊠

2.
1. Given
2. Assume that $\overline{OP}$ is not $\perp$ to ℓ.
5. Algebra
6. Definition of congruent segments
8. SAS
13. $\overline{OP} \perp \ell$

In *LOF: Beginning Algebra*, Fred's rib had been broken. Mrs. Hook's tight grip was making it hurt again. It felt a little better when she dropped him to the deck.

"How in blazes are we gonna get around that dog!" she exclaimed.

Fred, ever willing to be helpful, suggested, "Why don't we just walk around the circle?"

"Hey pipsqueak, don't you realize that the diameter of that Circle of Death is the width of the ship? Look! You can see where Snarls has gnawed into the sides of the ship."

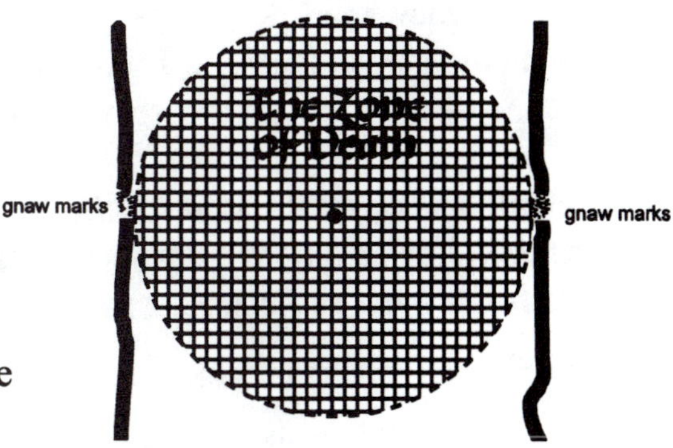

She handed Fred a pail of soapy water and a mop. "It's time you earn your keep, swabby.* Mind you don't get too close to Snarls."

The mop was twice as tall as Fred. She pulled it out of his hands and replaced it with a toilet bowl brush. The other sailors who were watching started to make fun of him. One of them took away his toilet brush and handed him a 4" cotton swab and jeered, "Now that's more your size, little swabby!"

Fred dipped it into the pail and began to clean the deck. As he worked he realized why some of the sailors had inadvertently entered Snarls' Circle: You can't see the circle. All that is visible is a sleeping dog and the two gnaw marks on the sides of the boat.

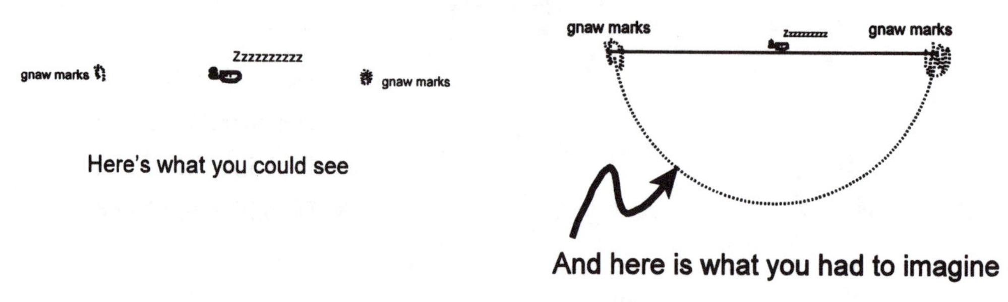

I guess those sailors that were eaten never studied geometry. If you're standing on the deck and all you can see is the endpoints of a diameter, there's an easy way to see whether you are outside the semicircle (half circle).

Fred set down his cotton swab, excused himself, and headed off to the ship's library. He needed a book. Any book. He selected Prof. Eldwood's *Introduction to the Poetry of Armenia*, 1849, and took it back to his workplace.

He placed the book on the deck and aligned the bottom edge of the cover with the left gnaw mark.

Then using the Inscribed Angle Theorem, which is the most important theorem of this chapter, he . . .

Wait a minute, Mr. Eager Author. We haven't had that yet! You've let your story get ahead of the geometry. I, for one, can't see how a book on Armenian poetry is gonna keep our little hero out

---

* In the Coast Guard, they often call new sailors swabbies.

of the jaws of Snarls. The only thing I can think of is that the book has square corners. So somehow, a right angle will be related to a round semicircle. Do a little geometry for a moment while I stand here next to Fred and make sure that he doesn't get munched by Hook's doggie.

Okay.

Before I get to the Inscribed Angle Theorem, I have to first talk about arcs.

I get it. Like a semicircle is an arc.

That's true. Arcs come in all shapes and sizes:

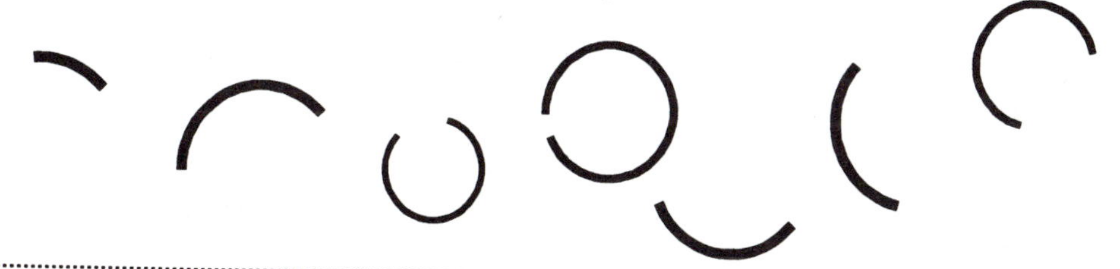

Your Turn to Play

1. What is wrong with, "Definition: An arc is part of a circle"?

2. What is wrong with, "Definition: An arc is a part of a circle that is all connected together"?

3. What is wrong with, "Definition: An arc is two points, A and B, of a circle and all the points C such that A–C–B"?

4. <u>Definition</u>: An angle is a **central angle** of circle O, if its vertex is at O. (Place that definition on your page of definitions that you are keeping. It's a "good" definition.)

What is wrong with, "Definition: An arc is the set of all points of a circle that are on the interior of a central angle"?

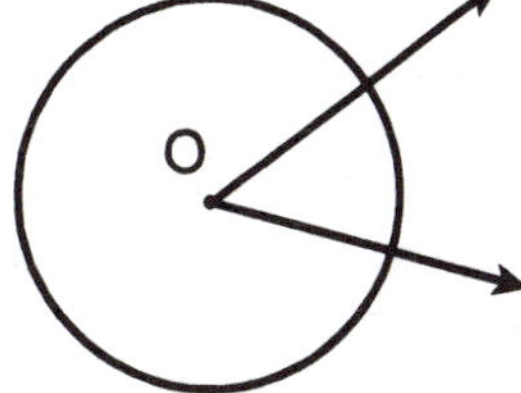

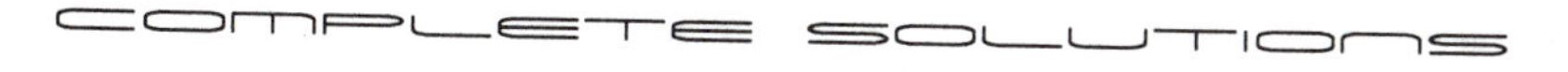

1. What does it mean to say "part of a circle"? This is part of a circle, but we don't want to call it *an* arc.

2. We have never defined what "all connected together" means, and so we can't use that in a definition.

3. When we write A–C–B, we have defined that to mean that A, C and B are collinear and that AC + CB = AB. Actually, this is a good definition . . . of a chord.

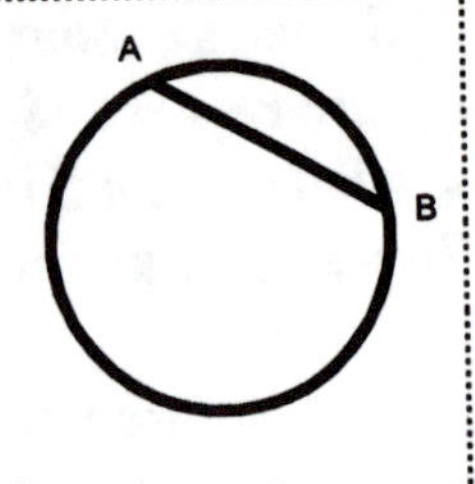

4. Recall that in the footnote on p. 351, we defined what it means for a point to be in the interior of an angle. Our difficulty is that our angles in geometry are between 0° and 180°. With the definition given in the question, we couldn't have arcs like:

Here are the definitions we'll use:

Definition: A **minor arc** is all the points of a circle that are on the interior of a central angle. A **major arc** is all the points of a circle that are not on a minor arc. Both minor arcs and major arcs are called **arcs**.

We will also call *all* the points of a circle an **arc**.

And one last thing to make our definition of arc complete. We will also call a semicircle* an arc.

Notation: The minor arc consisting of all the points that are on the interior of ∠AOB is written $\overset{\frown}{AB}$.

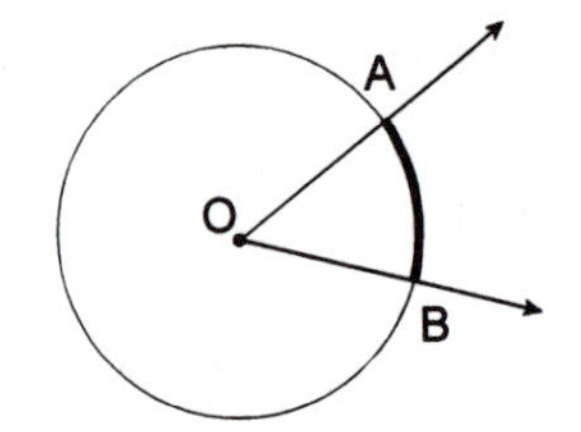

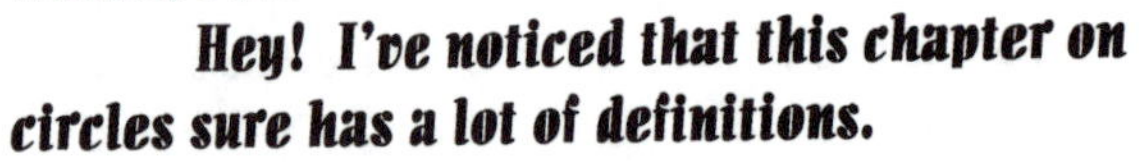

Hey! I've noticed that this chapter on circles sure has a lot of definitions.

It sure does. Here are some more.

★ Just for completeness we will officially define a semicircle. Definition: A **semicircle** is the set of all points of a circle that are on one side of a line that contains a diameter.

And for more completeness: Definition: **On one side of a line ℓ** is defined to be a half plane whose edge is ℓ.

And finally: Definition: A **half plane whose edge is ℓ** is the set consisting of some point P, not on ℓ, together with all other points Q such that $\overline{PQ}$ does not intersect ℓ.

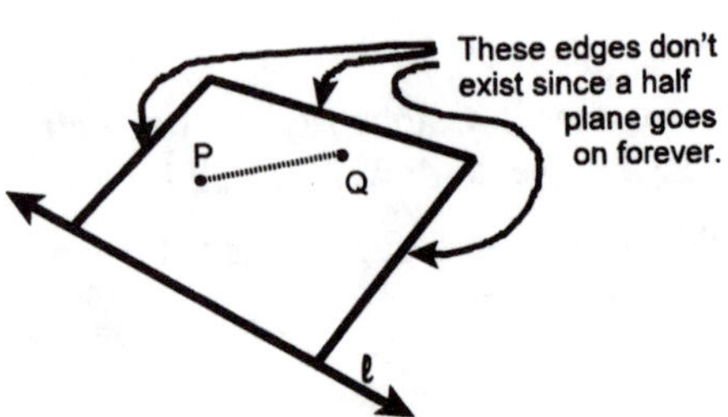

Now you are complete.

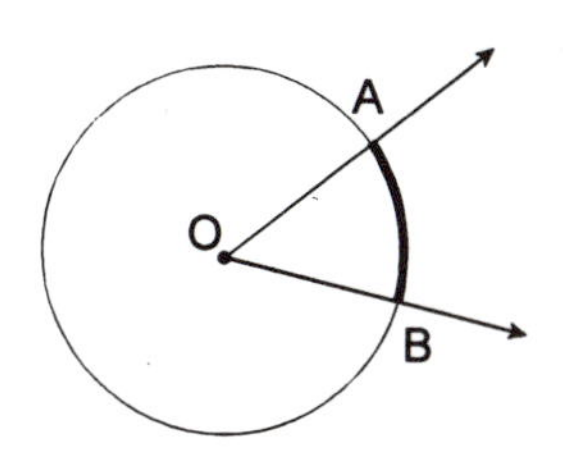

Definition: The **measure of minor arc** $\overset{\frown}{AB}$ (written $m\overset{\frown}{AB}$) is equal to m∠AOB.

That makes everything easy. An arc that is a quarter of a circle will have a measure of 90°. We'll define the measure of a semicircle as 180° and the **measure of a major arc** as 360° minus the measure of its corresponding minor arc.

Then, for example, the measure of an arc that is three-fourths of a circle will be 360° – 90° = 270°. And, of course, the measure of a whole circle will be defined as 360°.

What's nice about this is that we don't have to reinvent a lot of things. The measures of two arcs are equal iff the measures of their corresponding central angles are equal.

And now we can state the Inscribed Angle Theorem.

Don't you think that it would be sensible if you first told me what an "inscribed angle" is?

That means I have to give you yet another definition. I thought you were getting overloaded with them in this chapter.

Those definitions you just gave of the measure of an arc were pretty rinky-dink* stuff. If an inscribed angle isn't much more difficult than a central angle, then this whole chapter on circles makes me think I'll be ready for calculus before you know it. Okay. Fire away with your definition.

Definition: An ∠BAC is an **inscribed angle** of a circle iff A, B and C are on the circle.

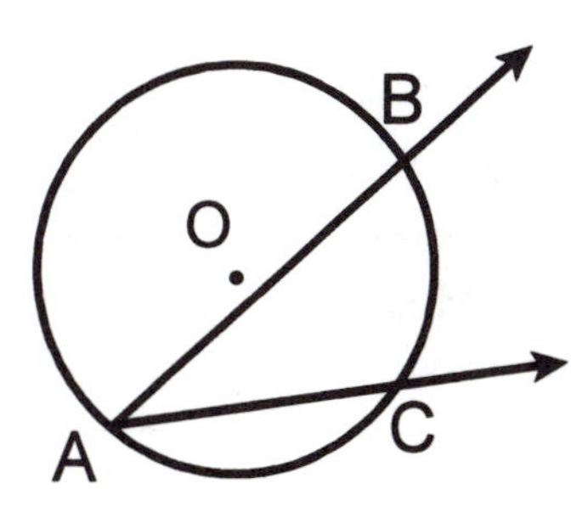

✶ The adjective *rinky-dink* entered the English language about 1910. It's an Americanism. Oh what we've done to the King's English!

Please draw some circles and measure ∠BAC and its corresponding central angle, ∠BOC. These two angles **intercept** (cut off) the same arc. There are three ways you might draw your sketches.

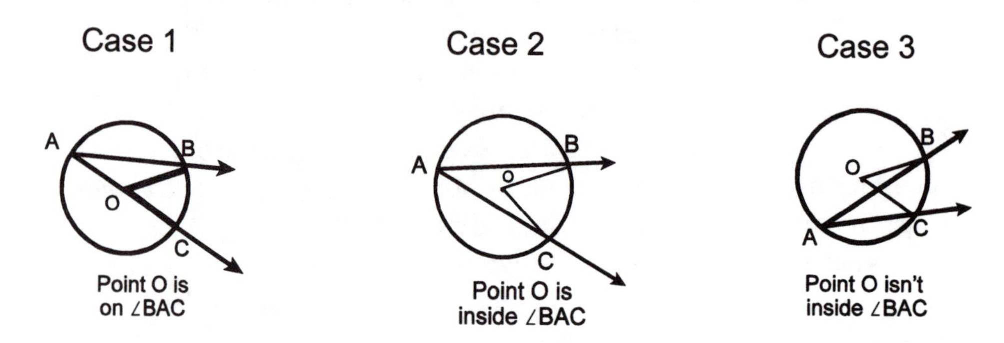

If you own one of those plastic protractors, measure the inscribed angle (∠BAC) and its corresponding central angle (∠BOC) in each of the drawings you made. (You *did* make those drawings, didn't you? Most people do learn more easily if they are actively involved in their education rather than just passive recipients. There's an old observation that in many classrooms the professor reads out of his old lecture notes and the students copy down what they hear and the material never passes through the mind of either professor or student.)

What you will probably have noticed in the sketches you made was that the inscribed angle is half of its associated central angle.* We arrive at the most important theorem of this chapter:

Theorem: (Inscribed Angle Theorem) The measure of an inscribed angle is half of the measure of the intercepted arc.

(Since the measure of an arc is, by definition, the same as the measure of its central angle, the statement of the theorem is exactly what we've been talking about in those sketches you made.)

Most of the proof of the Inscribed Angle Theorem is old stuff. We use the Exterior Angle Theorem and the Isosceles Triangle Theorem ("The base angles of an isosceles triangle are congruent"). What's new is that we have to prove this theorem *by cases*.

---

* Properly speaking, the *measure* of the inscribed angle is half of the *measure* of the associated central angle.

I object! We've never done this "proving by cases" thing before. Why are you suddenly introducing something I haven't had before?

Suddenly? I've been trying to lead up to this "proving by cases" as gently as possible by having you draw those sketches and mentioning that there were three possibilities when you draw an inscribed angle and its corresponding central angle.

Yeah but this "proving by cases" is still something I haven't had before. I have to crank up my brain once again. Rearranging my thinking—that's what you're trying to do to me! Admit it!

Yup.

And that involves effort on my part.

Yup.

Is there any end to this "effort" stuff? I mean, like is there any end to it before I die?

Sure. Learning does take a lot of effort. If Life & Learning are like a highway, many people, after they finish their classroom years, take their foot off the gas and let their car roll to a stop. It's much easier to watch television than it is to open a book. The difference between those who have parked their car and those who continue to read and learn is not very apparent when they are 25 years old. But as the years roll by, the gap begins to widen. By the time the people in the two groups are in their 40s, they still *look* the same, but when they open their mouths, the difference becomes embarrassingly apparent. All the parked-car group can talk about is what is right in front of their faces (the movies, the newspaper, what's happening with their friends or at work)—a very small world.

Wait! Stop. I was just kidding about objecting to learning something new. Any idiot can see that newness is what is to be expected when you read something above the level of a romance novel. I was just trying to get you a little excited and I guess it worked. Let's get back on track and show me the proof of the Inscribed Angle Theorem using cases.

I'm sorry. I had just recently read an article which stated that only six percent of American adults have read at least one book in the last year. The pain of that tragedy was on my mind. You kinda got me at a tender spot. I should have known better. After all, I know that *you* read.

Okay. Back to thinking/effort/work/learning.

On the next page is the proof that the measure of an inscribed angle is half the measure of the intercepted arc.

Proof of the Inscribed Angle Theorem

Case 1: Point O is on ∠BAC.

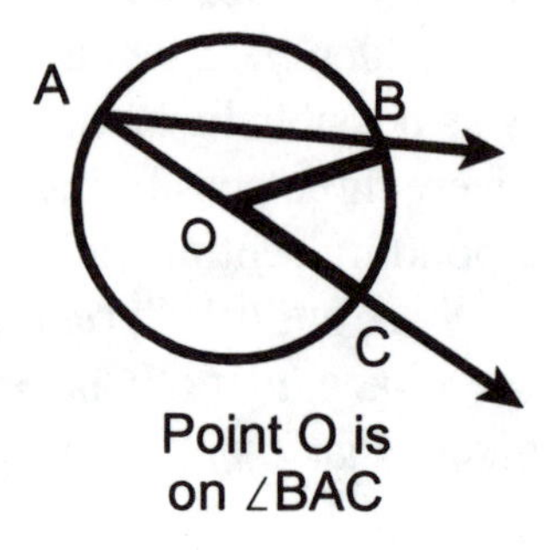

Point O is on ∠BAC

AO = BO since all radii are equal. By the Isosceles Triangle Theorem, m∠A = m∠B. By the Exteior Angle Theorem, m∠A + m∠B = m∠BOC. By algebra, m∠A = ½ m∠BOC. Since by definition, m∠BOC = $m\overparen{BC}$, we're done.

Case 2: Point O is inside ∠BAC.

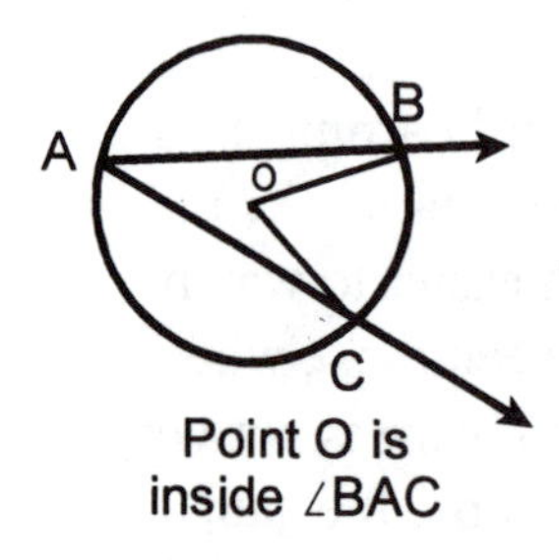

Point O is inside ∠BAC

Draw the diameter through A and O.

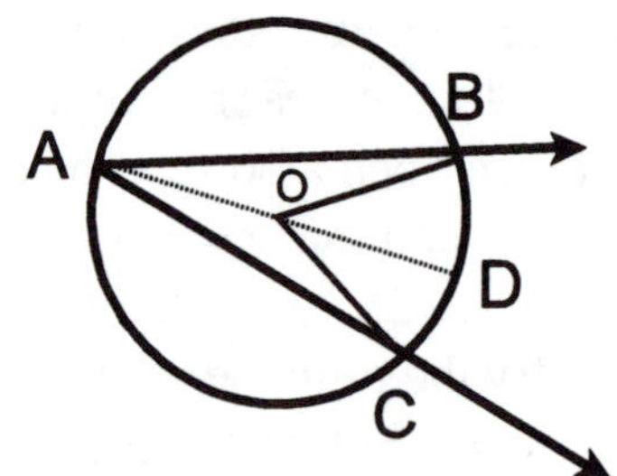

Then using case 1 (twice) we have that m∠BAD = ½ m∠BOD and m∠CAD = ½ m∠COD. Then using the Angle Addition Postulate, we nail down the proof of case 2.

Case 3: Point O isn't on the interior of ∠BAC

This is almost the same proof as case 2. Again draw the diameter $\overline{AD}$. Then using case 1 (twice)* and subtraction instead of addition, we have our result.

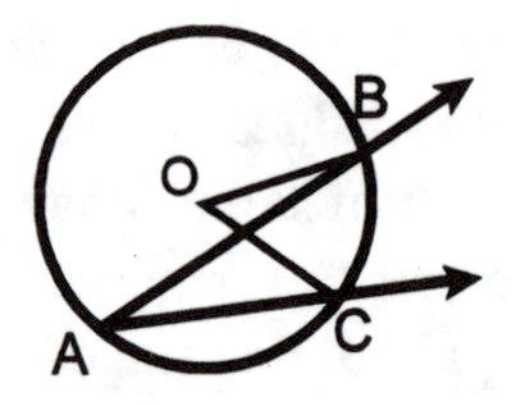

Point O isn't inside ∠BAC

✶ m∠BAD = ½ m∠BOD and m∠CAD = ½ m∠COD

Your Turn to Play

1. Here is an overhead view of a movie theater. The patron at point P can see the whole screen with a viewing angle of x°. How would you find all the other seats in the theater that offer the same viewing angle?

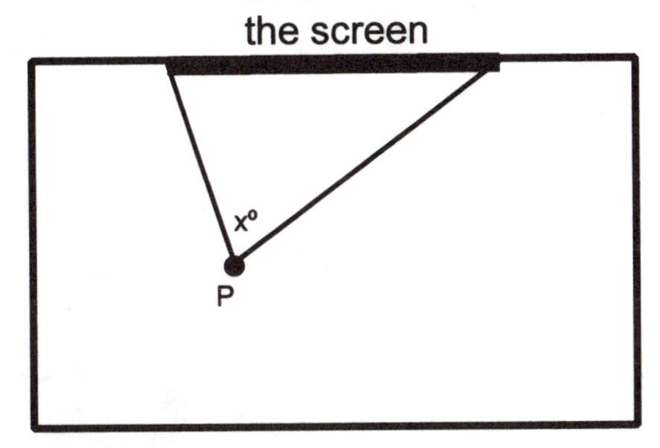

2. Draw, if possible, an inscribed angle that intercepts a major arc.

3. Using a metal carpenter's square, how could you locate the endpoints of a diameter of a circular piece of wood?

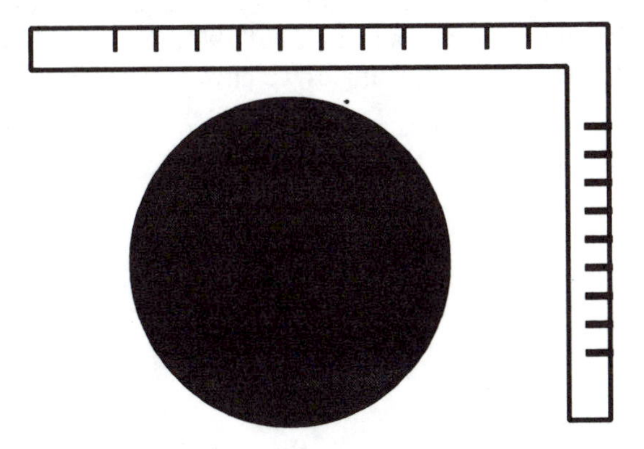

4. Draw, if possible, a minor arc that has a measure of 190°.

5. You have decided to hang your favorite owl picture on your living room wall. Instead of the usual method, you drive two large spikes into the wall and rest the picture on the spikes. This gives your picture a little wiggle room. What is the shape of the curve that the lower lefthand corner of the picture makes?

COMPLETE SOLUTIONS

1. First draw a circle through P and the endpoints of the screen. (In the next chapter, you'll learn how to construct a circle through any three noncollinear points.) Then any point on the circle (on the "P side" of the screen) will work.

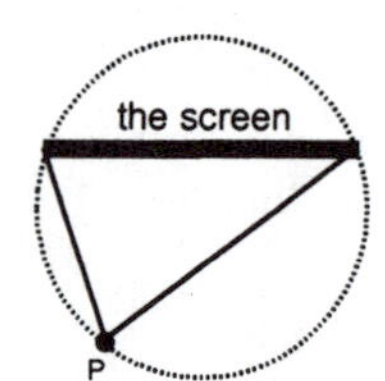

2. An inscribed angle can intercept a major arc.

3.

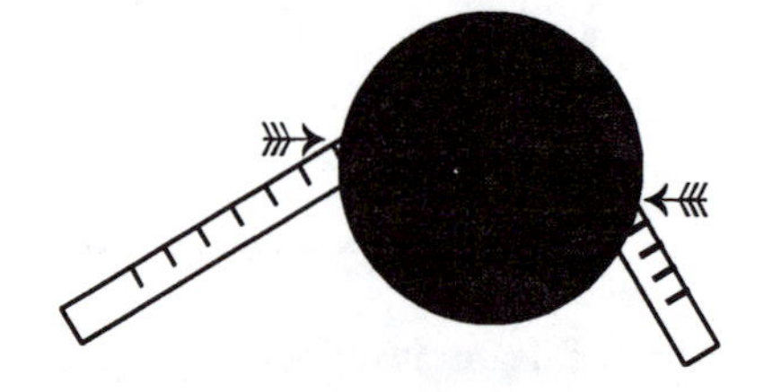

4. Since the measure of a minor arc is defined to be the measure of its associated central angle, if we had a minor arc with a measure of 190°, we would have a central angle with a measure of 190°. The largest angles we have in geometry are 180°.

5. Since the corner of your owl picture is a 90° angle, it would intercept a semicircle whose endpoints were the two spikes. The corner itself would trace out the rest of the circle which would be another semicircle.

This is a safe place to be

So after Fred had aligned the bottom edge of *Introduction to the Poetry of Armenia* with the left gnaw marks, he looked along the spine of the book and saw where it pointed in relation to the right gnaw marks. As he worked, he slid the book forward and stopped when both gnaw marks lined up with the book's edges.

This was as far as he would go

Fred thought about his life as it used to be. He was a professor at KITTENS University teaching his favorite subject to students he cared for. He thought about his friends, especially Betty and Alexander. He began to miss his office in room 314 of the Mathematics Building. And Joe and Darlene. And the Wistrom brothers. And he would never see his true love, P, again.

Yes, P. The delightful anticipation of their getting to know each other over the years. Of being married when they were in their early 20s. Of their children. He pictured a bunch of kids playing in his office. If P didn't want to live in his office, maybe they would shop together and find their dream house near the campus. With a house to live in instead of an office, they could plant trees and have a garden. Lots of flowers.

All gone.

Fred stared down at the cotton swab he was using to scrub the deck and noticed it was getting a little worn. He turned it over and began to use the fresh end.

But this is my reality now. I guess I better make up my mind about it. I remember reading about victims of severe physical trauma, like losing a leg, that it takes months for the new reality to sink in.

But crying a bit is also a very natural response. He scrubbed the deck with the soapy water and with his tears.

I guess I had better start working on fitting in with my new surroundings. Fred took a piece of paper out of his pocket and made an ersatz name tag.

Prof. Fred Gauss, Cabin Boy

One of the sailors looked at what Fred was doing and yelled, "Hey, runt! What's that piece of paper hangin' from your shirt?" He walked around in front of Fred and ordered, "Stand up and lemme look at you. What are you doing? Making yourself into a paper tree or something?"

As Fred stood up he realized that the sailor was standing behind the *Introduction to the Poetry of Armenia.* Before Fred could exclaim anything about the Inscribed Angle Theorem, the sailor shared the same fate as Jezebel (II Kings 9:36, which is almost everyone's favorite dog food story).

Fred went back to work.

He had been swabbing the deck for four minutes, not counting the break he had taken to get the book out of the ship's library. It was time to break up the monotony.

A sea chanty! That's what all the sailors do. They sing to pass the time as they labor. I'll prove to my shipmates that I'm an "old salt." What song shall I start out with? "Three Men in a Tub" is kind of babyish. I've got to show them I'm tough.

His little voice broke into song, making up the words he couldn't remember.*

♪ ♬ Fifteen hairs on a dead man's chest
Yo ho ho and a bottle of pop.
Drink up one bottle and down with the rest
Yo ho ho and a bottle of pop. ♪ ♬

He couldn't remember any more lines, so he sang those four lines over and over. The other sailors put on earphones and turned up the volume on their portable radios to drown out the sound of Fred's "caterwauling" as they called it.

One of the sailors who had been assigned to clean the seagull's deposits off the railings found a better use for one of his rags.

Fred continued with what he called "sea chanty humming." He knew that no one would object to that since with his honker, he had what he considered to be outstanding nasal reverberations.

---

* From the 1901 Broadway musical, here's the first of the six verses:

Fifteen men on a dead man's chest
Yo ho ho and a bottle of rum
Drink and the devil had done for the rest
Yo ho ho and a bottle of rum.
The mate was fixed by the bosun's pike
The bosun brained with a marlinspike
And cookey's throat was marked belike
It had been gripped by fingers ten;
And there they lay, all good dead men
Like break o'day in a boozing ken
Yo ho ho and a bottle of rum.

Fred scrubbed diligently for another three minutes. He didn't want anyone to think that he was a slacker. He had cleaned a circle that had a four-foot diameter. And, surprisingly, not a single theorem, definition or postulate arose in his work.

Hey! Don't you, Mr. Author, be slacking off now! This would be the perfect time to define the circumference of the circle. You know, tell everyone that C = πd.

Let me help you. I've always wanted to be a writer. Why don't we throw in a definition like, "Circumference is the distance around a circle." I could even make an illustration. I could draw a real pirate.

But where's the circumference? I don't get it.

Just take his eye patch. That's round. We could define circumference as the distance around his eye patch.

I could even make up a little *Your Turn to Play* ***with nifty questions like, "1. If d = 7 feet, then what is C? Answer: 1. 7π."***

And a really tough question: "2. If the radius is equal to 20 meters, then what is the circumference? Answer: 2. 40 π meters." That's because the radius is half of the diameter, so I had to double the 20 meters to get the diameter, and then I could multiply by pi to get the distance around.

You're frowning. I can tell. Don't you like my eye patch idea?

The eye patch is swell. I wish I had thought of that myself. But I'm having difficulty with getting a definition of *circumference*. All the lines we have in geometry are straight. We know the distance between points A and B on a line.

But curvy distance?

How can we get a handle on that?

Don't you have any ideas?

Sure I do, but I can't mention them in a geometry book. Getting from C = 2πr (I prefer that to C = πd) is a trip into deep waters.

Fred's already out to sea—at least on the pond on the KITTENS campus. Feel free to take me out into "deep waters" as you call it.

Okay. But you have to call out "Uncle!" when the sailing gets too rough. Okay?

Okay. When you lose me, I'll call out "Uncle."

We start out with regular polygons.

Unc... What's a *regular* polygon? I know what polygons are. You defined that back in Chapter 7 on p. 246. They look like:

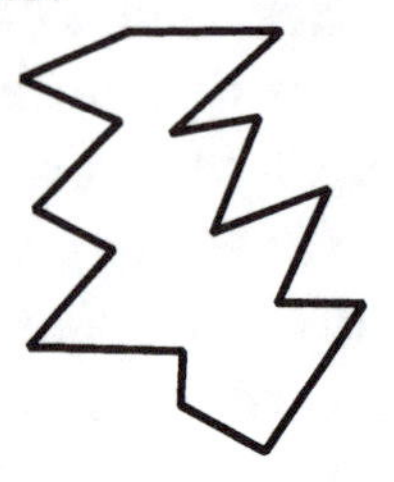

Regular polygons are polygons where all the sides are congruent and all of the angles are congruent. They look like:

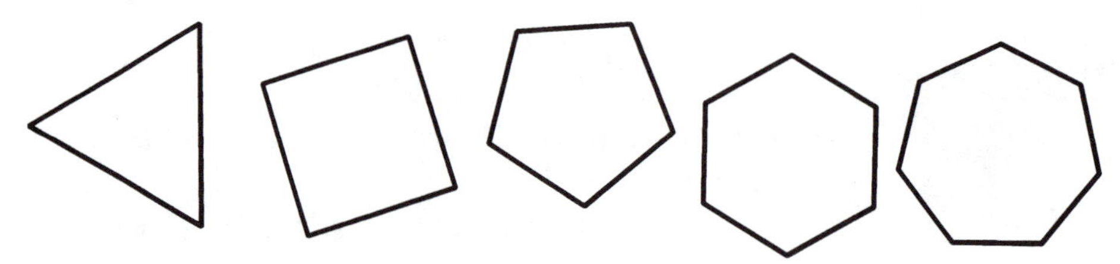

So far you haven't lost me.

Start with any circle O with a radius of r.

I forget. What are you trying to do?

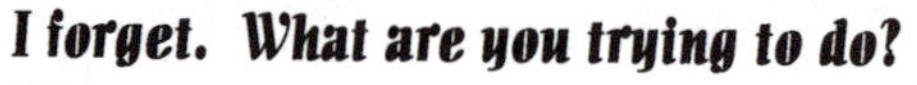

I'm trying to define *circumference* which can't be done in a geometry book.

Okay. Keep going.

In that circle inscribe a regular polygon with three sides.

That's an equilateral (and equiangular) triangle.

Right.

Then do a bunch of math and figure out the perimeter (distance around) the triangle.

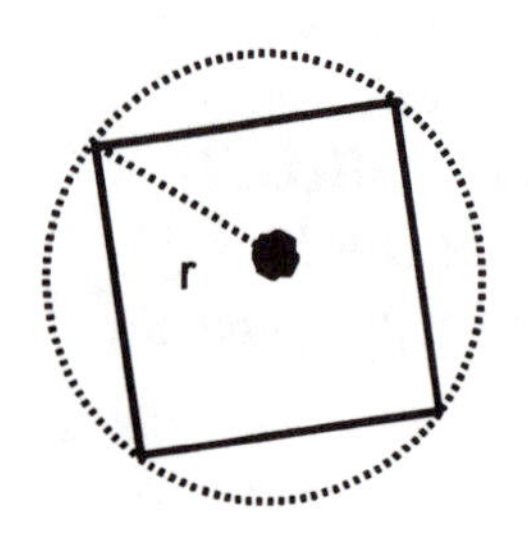

Then switch to a four-sided regular polygon.

And again, find the perimeter.

And a five-sided regular polygon.

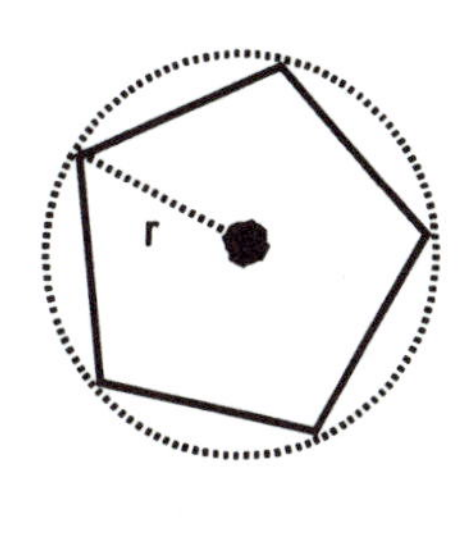

And a six-sided.

And a n-sided regular polygon.

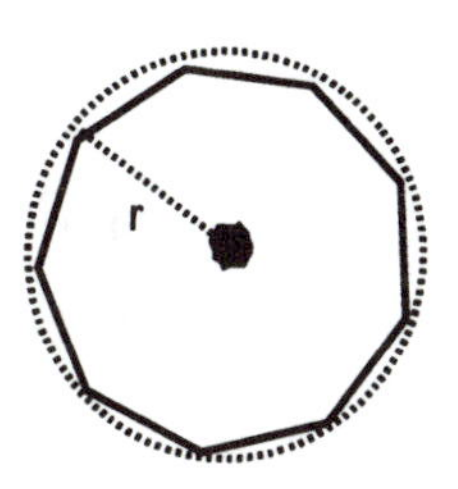

And in each case find the perimeter.

Hey! I get it. As n gets larger and larger, then the perimeter of the regular polygon gets closer and closer to the circumference of the circle.

It's called "taking the limit." We define the circumference of a circle to the limit of the perimeter of an n-sided regular polygon as n gets larger and larger.

I'm still with you . . . barely. But what's this "limit" stuff?

It's calculus. In fact, the concept of limit is the only really new concept in all of calculus. Once we define what *limit* means, we spend all the rest of calculus playing with that idea and its consequences. We take limits of functions, f(x), as x goes to infinity and we take limits of functions, f(x), as x approaches some finite number *a*. We talk about functions in beginning algebra, advanced algebra, and in trig, so that by the time you hit calculus, you have some feeling for what f(x) means. Then all we have to do is let x approach *a* (which we symbolize as $x \rightarrow a$) and then we're really cooking.

So what's all the big fuss over some definition of limit? You haven't heard me cry "Uncle!" yet. Lay the definition of limit on me.

But, but, but. . . .

Lay it!

We say that the **limit of function f(x) as x approaches *a*** is equal to some number M, which we symbolize as $\lim_{x \to a} f(x) = M$, is true iff

for every number $\varepsilon > 0$, there exists a number $\delta > 0$, such that $|f(x) - M| < \varepsilon$ whenever $0 < |x - a| < \delta$. (ε and δ are the Greek letters epsilon and delta.)

Uncle! Uncle!

I can squeeze the definition down to less than two lines:
$\lim_{x \to a} f(x) = M$, iff for every number $\varepsilon > 0$, there exists a number $\delta > 0$, such that $|f(x) - M| < \varepsilon$ whenever $0 < |x - a| < \delta$,

but we spend roughly THREE SEMESTERS of calculus working on making those two lines intelligible and comfortable. We give zillions of examples, such as in chapter 16 when, in order to get their cotton-candy breakfast, the tin-wood duck and Fred walked on stepping-stone paths on the side of the road that led to the Emerald City in the kindergarten play that. . . . Well, you get the idea. It's all there in *LOF: Calculus*.

Now, can you see why I didn't want to define *circumference* when Fred was scrubbing his four-foot diameter circle? Or define π (pi) as the ratio of the circumference of a circle to its diameter ($\pi = C/d$)? π is roughly equal to 3.1416.

And the area of a circle is just as bad. How do you define the area of a circle? You take a circle and inscribe a square in it. (We know the area of a square by one of our postulates.)

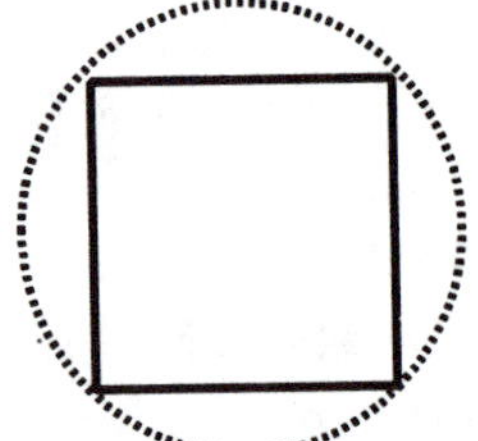

Then you subdivide the square into smaller squares.

And subdivide again.

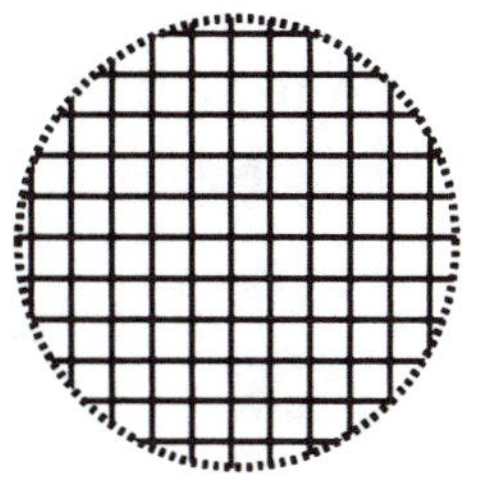

As the number of squares, n, gets larger and larger, the sum of the areas of the squares starts to approach the area of the circle.

Yeah. The limit of the sum of the areas of the squares approaches the area of the circle. But wouldn't it be easier just to say $A = \pi r^2$ like we learned in elementary school?

$A = \pi r^2$ does have a nice ring to it. And the problems are real easy. If r = 5 inches, then the area of the circle is 25π square inches. And, if we approximate π by 3.14 *, then the area of the circle is approximately 78.5 square inches.

But your elementary school teacher never told you *why* the area of a circle is equal to πr^2.

And I can't either, until you get to calculus and get that limit definition $\lim_{x \to a} f(x) = M$, iff for every number $\varepsilon > 0$, there exists a number $\delta > 0$, such that $|f(x) - M| < \varepsilon$ whenever $0 < |x - a| < \delta$

I think you talked me into it. I'll wait.

That's good, because something very important was happening on board.

"What ho!" came the cry from the crow's nest. "Captain, there's a big bad ship following us!"

Captain Hook scowled. "Can't you get the lingo right!? You're supposed to yell, 'Enemy ship astern,' and then say how many leagues away it is."

From the crow's nest, "Enemy ship astern, but I forget what a league is."

---

* The San Francisco Exploratorium has a Pi Day which they celebrate on March 14th.

Hook called for his spyglass. In his agitation the first mate dropped it twice. By the time Hook got a chance to look through it, instead of seeing he saw instead:

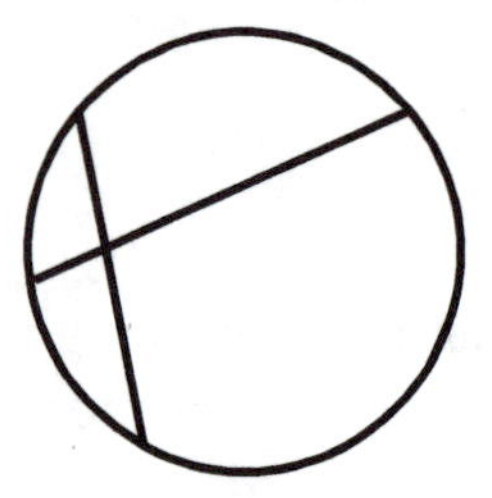

The cross hairs were all messed up

"Confound it, man! You broke my spyglass, the one my wife gave me for Christmas." He threw it down on the ground and it rolled to Fred's feet.

Fred, of course, picked it up and looked through it. In contrast to the captain, he was delighted in what he saw. He saw two intersecting chords. They weren't necessarily even the same length. To the geometrically uneducated, this meant nothing except a broken spyglass. But to Fred it was beautiful. It was the hypothesis (the given) for *one of the most unexpected theorems of geometry.*

If you, the reader, would take a sheet of paper and your compass and draw a nice big circle, one that almost fills the paper, and then draw any two intersecting chords in the circle, you have this opportunity to discover this theorem before you see it on these pages. Each chord breaks apart into two smaller segments.

With your ruler, measure each of those four smaller segments. Then play with those four numbers.

While you're doing that, we'll take a look at what was fast approaching Hook's ship.

"It's . . . it's . . . " Hook gasped. "It's a Double Pirate Ship."

A Double Pirate Ship is the nightmare of every good and honest pirate. It is the subcategory of pirates who only prey on other pirates, a little like lawyers who sue other lawyers. Or as the birds that eat dead buzzards often explain, "Somebody's gotta eat 'em."

And since pirates carry all kinds of valuables—gold, jewels, slaves—somebody has got to steal from them.*

As Captain T. Tock and his band of double-pirates boarded Hook's ship, Hook cried out in a little, squeaky pitiable voice, "It's not fair! We never did any harm to you. Be merciful. Why are you picking on us?" Of course, these were phrases that he had heard a hundred times when he had boarded other ships.

The fighting was fierce but brief. Soon Hook and his men surrendered. Tock lined his captives on the fo'c's'le** (pronounced FOLK-sel) and asked his first mate, "What do they look like?" Tock himself couldn't see that well. Since he was a double-pirate, he wore two eye patches.

" 'Tis a motley crew, sir," the first mate reported. "They're a scurvy lot. Shall we teach them to swim?"

Tock nodded. Hook and his crew soon joined Tanya.

Fred was sitting cross-legged on the deck and was still looking through the spyglass

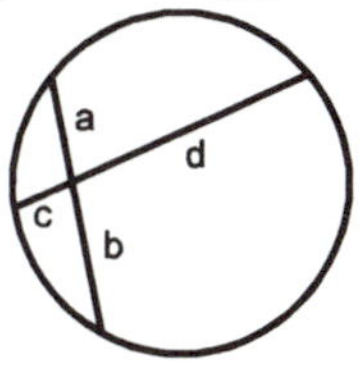

Earlier he had done a little editing of his name tag.

Prof. Fred Gauss, ~~Cabin Boy~~

✶ If this has never occurred to you, maybe you are not one of the big breakers of the Eighth Commandment.

✶✶ Yes, this is the correct spelling. It's a word with three apostrophes in it. The fo'c's'le is a kind of raised deck at the front ("aft of the bow") of the vessel. It is also spelled *forecastle*.

"What's you looking at, boy?" Tock asked as he grabbed the spyglass out of Fred's hands.

"Isn't it amazing?" said Fred.

After removing one of his eye patches, Tock examined the little telescope. It was made of black plastic and had an inscription on the side of it: HE WHO HAS EYES, LET HIM SEE. "What's so amazing, kid?"

Fred invited him to look through it.

Tock looked. Then banged it against the mast and looked again. He did this several times and each time the cross hairs were bounced around.

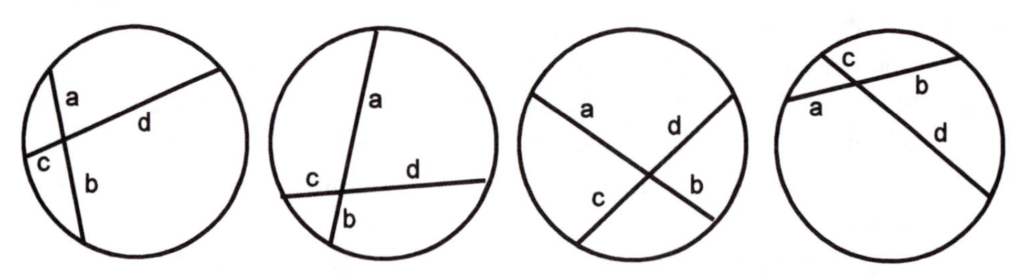

"Say Ticky, see any pretty girls through that thing?" asked Mrs. Tock. Only his wife ever called Capt. T. Tock by the name Ticky. His parents had named him Timothy Tock, but in elementary school, T. Tock was changed by all the other schoolboys into Tick Tock. His parents didn't like the new name, because it reminded them of those bloodsucking bugs. That was all the encouragement that Timothy needed. He changed his name to Tick.

He handed the toy back to Fred and said to his wife, "I see *you* Honey."

She smiled and kissed him. She pointed to Fred and said, "Ticky, what a cute little boy. Can I have him? He'd make a darling little cabin boy. Look, he even comes with a name tag. His first name is Proof. Poor kid can't spell worth a darn though."

She knelt down next to him and smiled. "Hi Proofie! Do you know how old you are? Show me some fingers."

Fred had always had problems with people underestimating his age because of his small stature. He held up three fingers on each hand.

"Oooo, look Ticky! The little peanut knows his numbers. Not too bad for a three-year-old."

"Can I keep my telescope?" Fred asked. If, by some miracle he ever resumed his position as Senior Professor of Mathematics at KITTENS University occupying the Sir Isaac Newton Chair, this little spyglass would be a good teaching prop for introducing the Intersecting Chords Theorem.

<u>Theorem</u>: (The Intersecting Chords Theorem) If any two chords of a circle intersect, then ab = cd.

I hope you, the reader, were able to discover this theorem three pages ago when I asked you to draw a big circle with your compass and play with the lengths of the four segments. If you did, then you deserve some recognition for your efforts. If this is your book, please feel free to fill in your name on this certificate of honor.

In Recognition of Your Untiring Efforts
in Pursuit of the Acquisition
of the Knowledge of
Geometry
fill in your name on the line
is hereby awarded the title of
D.I.C.T.A.T.O.R.
Discoverer—Intersecting Chords Theorem—
A True Original Ratiocinator

The proof of the Intersecting Chords Theorem is pretty short and straightforward.

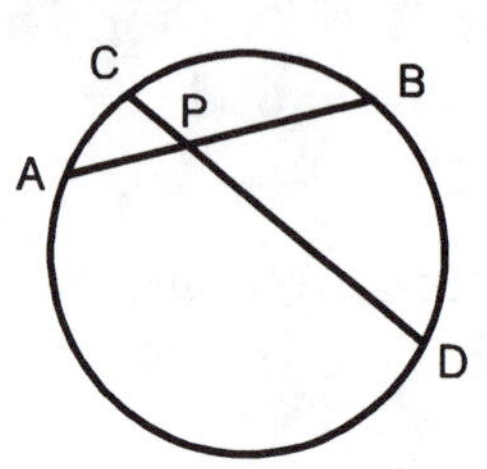

Draw $\overline{AC}$ and $\overline{BD}$.

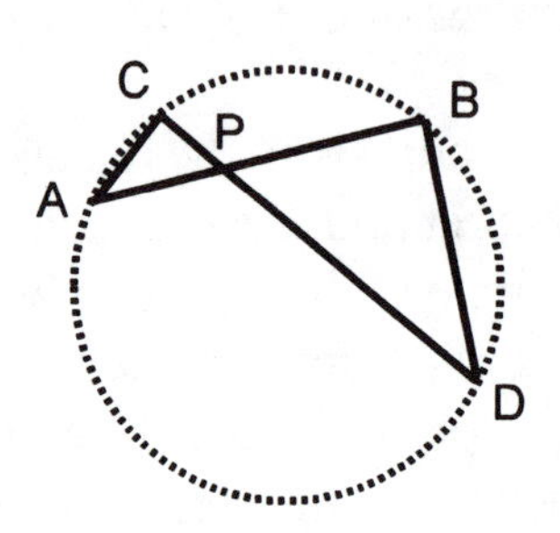

Note that $\angle C$ and $\angle B$ both intercept $\overset{\frown}{AD}$ and therefore are both half of $m\overset{\frown}{AD}$. Thus, $\angle C \cong \angle B$.

The same is true for $\angle A$ and $\angle D$. (They are both half of $m\overset{\frown}{BC}$.)

By the AA Similarity Theorem, $\triangle ACP \sim \triangle DBP$.

By definition of similar triangles, $\dfrac{AP}{DP} = \dfrac{CP}{BP}$

Cross multiplying, (AP)(BP) = (CP)(DP)

which, in terms of our original lettering, is ab = cd. Q.E.D.

If you wanted to get really fancy, you could have the chords* intersect outside of the circle at a point P and you could still prove that (AP)(BP) = (CP)(DP).

Both of these cases would be covered by the terribly wordy: <u>*Generalized Intersecting Chords Theorem*</u>: *If two chords, or their extensions, intersect, then the product of the distances from the endpoints of one of the chords to the point of intersection is equal to*

✶ Technically speaking, it would be the lines containing those chords that intersect outside the circle.

the product of the distances from the other chord to the point of intersection. Yuck! Gag! That kind of prolixity is enough to turn almost any English major into a math major.

Did you ever notice what a stranglehold the English department has on the math department? In math classes we have to use English, but when is the last time you ever saw $x^2 - 25y^2$ ever factored into $(x + 5y)(x - 5y)$ in an English class?

And, what is worse, since they control the spelling, we have to capitalize *English*, but *mathematics* isn't capitalized.✶ (In German, things are more fair. They capitalize all nouns.)

Your Turn to Play

1. If a pizza has a diameter of 18", what is its circumference? Give both the exact answer and a decimal answer rounded to the nearest tenth of an inch.

2. One not-so-famous bridal shower game is to hand the bride-to-be a length of string and ask her to make a circle with it on the carpet that is the size of the groom's tummy. Brides, thinking of their big, strong, handsome fiancés, often wildly overestimate. When the length of the string is measured, usually a titter ripples through the assembled party guests. What diameter should the bride make the circle so that things "come out right?"

3. Stanthony's pizzas are a popular favorite for wedding receptions. His 20" (diameter) pizza costs $13. His 28" cost $24. Which is a better buy? (Hint: compute the cost per square inch.)

4. The brother of the bride at some wedding receptions is called upon to cut the wedding pizza. If he gets distracted by a pretty bridesmaid or because of the ingestion of too much C_2H_5OH (a depressant affecting the central nervous system known as ethanol) he might miss the center of the pizza.

Find the values of a, b and c.

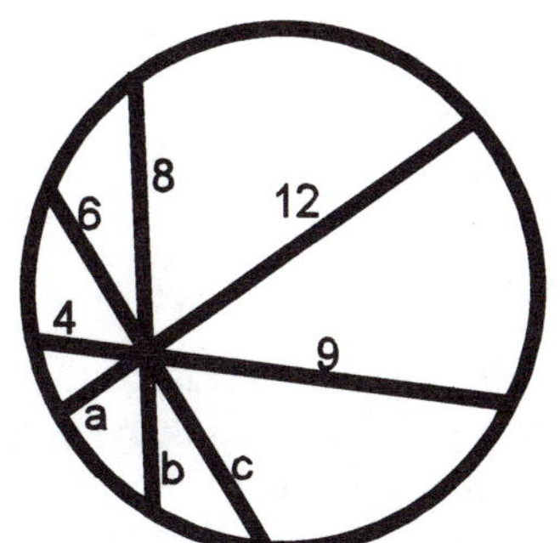

✶ That's why it's important to begin sentences with the word *mathematics*. For example: Math is great.
Math is neat.
Math's the subject that can't be beat.

4. Find PD.

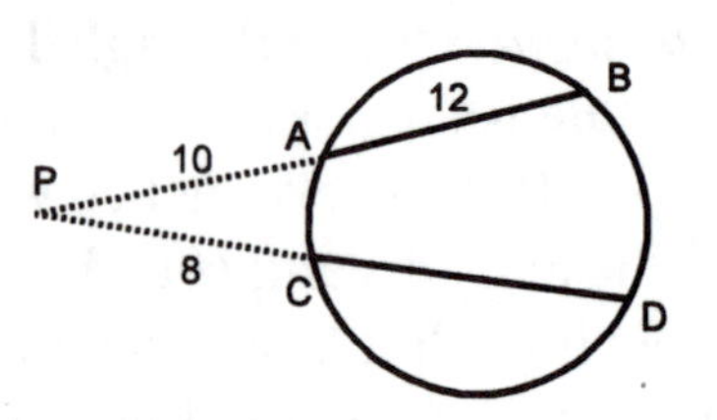

1. Since C = πd, and since d = 18", the exact value for C is 18π. To find the approximate decimal value of C, we'll use 3.1416 as an approximation for π. Then C ≈ (3.1416)(18) = 56.5488 ≐ 56.5 square inches.
(≈ means "approximately equal to," and ≐ means "rounded off to.")

2. Suppose the groom's waist size was about 30". Since π ≈ 3, the approximate diameter she should make the circle is about 10".

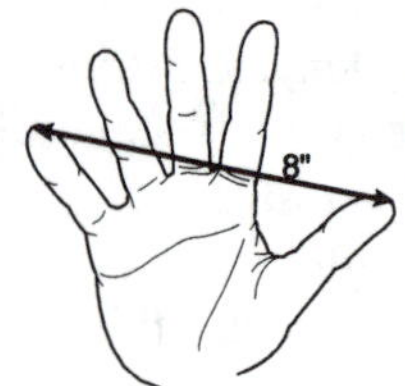

Since a bride's hand span is about 8"—it varies from bride to bride—she should make a circle with a diameter one or two inches larger than her hand span.

3.

Diameter of Pizza	Corresponding Area	Price	Price/sq in
20"	400π	\$13	$\frac{13}{400\pi}$ ≈ \$0.010345
28"	784π	\$24	$\frac{24}{784\pi}$ ≈ \$0.009744

So the smaller pizza costs a little more than a penny a square inch and the larger pizza costs a little less than a penny a square inch. As they say in the pizza world, when in doubt, always buy the larger pizza.

4. The 4—9 chord tells us that the product of any two segments that make up a chord passing through that point must equal 36. So the a—12 chord must have a = 3. The c—6 chord would have c = 6. The b—8 chord would have b = 36/8. This is true since 36/8 × 8 = 36.

5. (PA)(PB) = (PC)(PD). From the diagram, (10)(22) = (8)(PD). PD = $\frac{220}{8}$ = 27.5

"Say Proofie," Mrs. Tock asked, "do you know any proofs? I bet you hear that all the time with your funny first name." She picked him up and carried him back to Tock's ship, the *Crocodile Express*.

Fred was bewildered. Asking him if he knew any proofs was like asking Stanthony if he knew how to make pizza or asking Martin Luther if he had any interest in religion.

"I'll tell you what a proof is, little boy. Then if anyone ever asks you, you can tell them. Now pay attention and be sure to ask me any questions if I get too complicated for you." She smiled, set him down in a chair in her cabin, and proceeded with her explanation.

Drawing a picture on her blackboard, she said, "Now look. This is a picture of the sun. It comes up every morning. We call that dawn."

"Or sunrise," Fred chimed in.

"Oh, very good. You are a smart little fellow. Now let me continue. You know that we have seen the sun come up every morning for the last gazillion years. See, that *proves* the sun will come up again tomorrow."

Fred's eyes crossed. What kind of tommyrot is she telling me? That's like the tooth fairy or bunnies that lay eggs. She has no idea of the difference between inductive and deductive reasoning. The deductive reasoning, which is the kind we use in geometry, is where we use proofs. I would love to have her let me draw a picture of what a mathematical theory looks like with its beginning assumptions (postulates), its undefined terms, and definitions and then show how you "grow" theorems like a palm tree by doing proofs.

"Now," she continued, "do you have any questions? Have I gone too fast for you?"

"If I may? I'd like to seek a little clarification regarding the difference between **inductive** and **deductive reasoning**. You say that because the sun has risen in a quotidian* manner over recorded history, that proves it will rise tomorrow. That's an example of inductive reasoning. Scientists use that kind of reasoning all the time. But that kind of reasoning in which you start from experiments, trials, and observations

---

* kwo-TID-e-en Something that happens daily.

and then arrive at a conclusion, can only result in tentative conclusions. Scientists never *prove* anything. All they come up with is the statement that something is probably true. That the sun will come up tomorrow is 99.9999999999999999999999999999999999999999999999999999999999999999999999999999999999999999999999999999999999999999999999999999999999999999999999% certain, given its long history of arriving just at dawn each day. But some unknown star might come during the night and wack it to smithereens while we sleep and then there'd be no sunrise tomorrow. Or, as my Sunday school teacher once said, maybe the Morning Star might appear tonight and all the old things would pass away.

"Let me take a more immediate example," Fred continued. "How many people do you know that have eye patches?"

Mrs. Tock thought for a moment and responded, "There's Alfie, Bertrand, Cassius. . . . In fact, all the pirates on this ship have them. I even have one here in my pocket. In my experience, every pirate has one."

"That's my point, Mrs. Tock. You have come to a conclusion based upon your observations, just as every scientist does."

She smiled at being placed in such distinguished company.

"So you feel that it's probable that every pirate owns an eye patch?"

"Sure."

"But that doesn't make it certain. In fact, Mrs. Hook didn't own one. When scientists look at the world and try to draw conclusions (using inductive reasoning) they first get what they call a hunch. The ornithologist (one who studies birds) upon first seeing that there aren't any dodo birds (*Raphus cucullatus*) running around any more might have the **hunch** that they are extinct. When several other ornithologists report not seeing any dodo birds, then the hunch might become a **hypothesis**. When no one reports seeing any of them for several years, then their extinction might become a **theory**. If no one has seen any for a century, then the conclusion might be upgraded to a **law**.

hunch ➡ hypothesis ➡ theory ➡ law is the way it's expressed in increasing degree of probability.

"But sometimes even some laws that have been verified for centuries turn out to be wrong. Newton's Laws (1642–1727) were believed to be absolutely true until Einstein showed otherwise in the early 1900s.

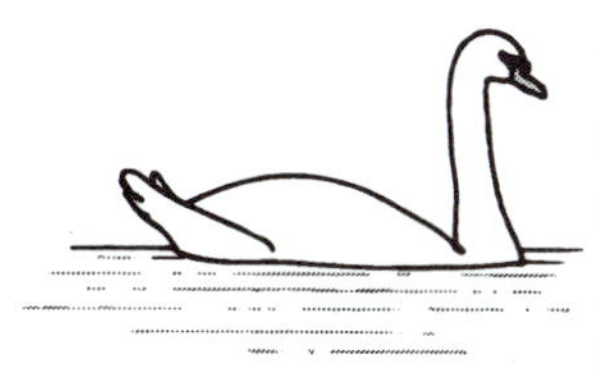

"Or, for example," Fred continued, "Europeans for centuries *knew* that all swans are white. That's all they had ever seen . . . until they discovered Australia. There are black swans in that Land Down Under."

Mrs. Tock stood at the blackboard in utter amazement. She wondered where this kid got all his information.

Fred asked her to draw a circle on the board and put one "dot" on the circle. She complied.

Then he asked her to draw another circle and put two dots on the circle. After she did that, he asked her to connect the dots. Last, he asked her to count the number of regions in her picture.

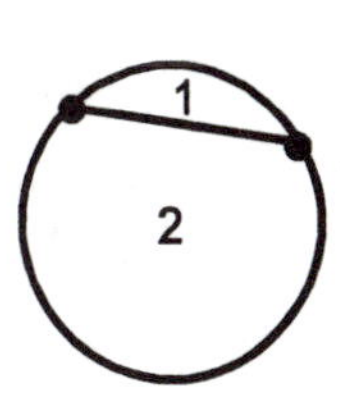

Fred then started a little chart:

Number of Dots	Number of Regions
1	1
2	2

Mrs. Tock liked this. It was easy.

He asked her to draw a third circle with three dots on it and connect the dots. Then she was to count the regions again.

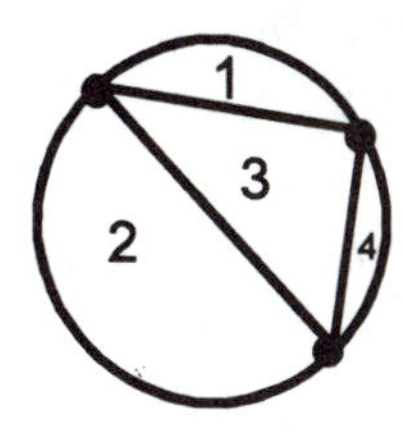

And then with four dots:

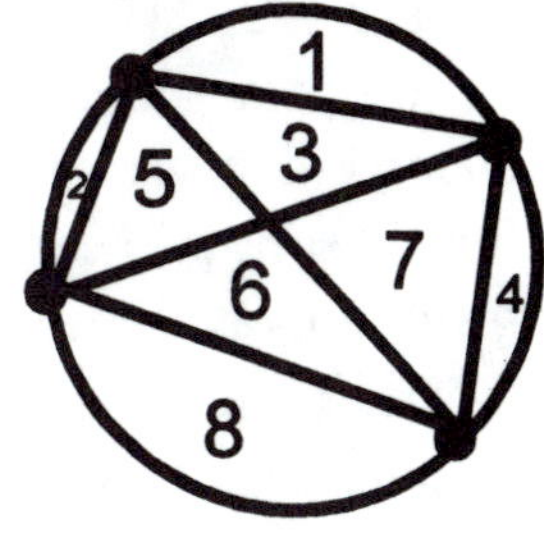

And five dots:

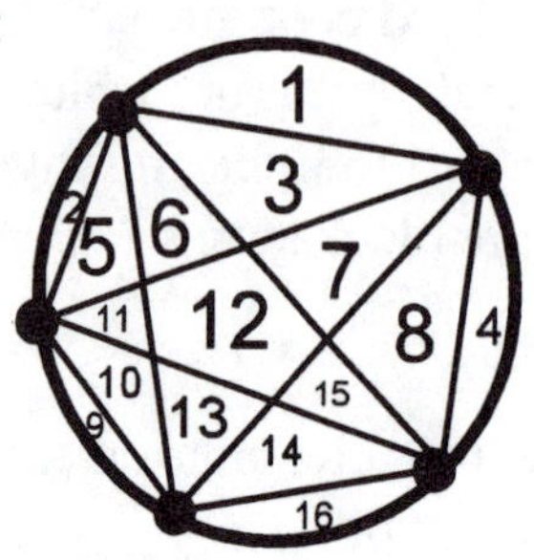

Fred enlarged his chart:

Number of Dots	Number of Regions
1	1
2	2
3	4
4	8
5	16

Mrs. Tock said, "I get the idea. There's no need to have me draw more of these circles. Any idiot can see that adding one more dot just doubles the number of regions you get."

"And this," said Fred, "is an example of inductive reasoning. You start with experiments, trials, or observations. When you drew the first three pictures and saw that the number of regions was 1, 2 and 4, you may have had the hunch that the next number would be 8.

"After you drew the fourth circle and saw the pattern of 1, 2, 4 and 8, then you may have formed the hypothesis that adding one more dot just doubles the number of regions you get.

"After you drew the fifth circle and saw that the number of regions formed the pattern: 1, 2, 4, 8, 16, . . ."

"Stop!" shouted Mrs. Tock. "Hey kid. I'm not some dumbo. You don't have to beat it into the ground. I know what you're going to say. You're going to have me draw another circle with six dots on it and have me count the regions. Then when I get 32 regions, you'll say I've got a theory that $R = 2^{n-1}$ (where R = number of regions and n = number of dots). I don't want to draw any more circles. I believe. I believe. It's a law. Okay, kid?"

"I'm sorry. I don't mean to be disrespectful. But if you'll pardon a neologism: no-kay."

Mrs. Tock wasn't exactly sure what Fred meant. She shook her head to indicate her lack of understanding.

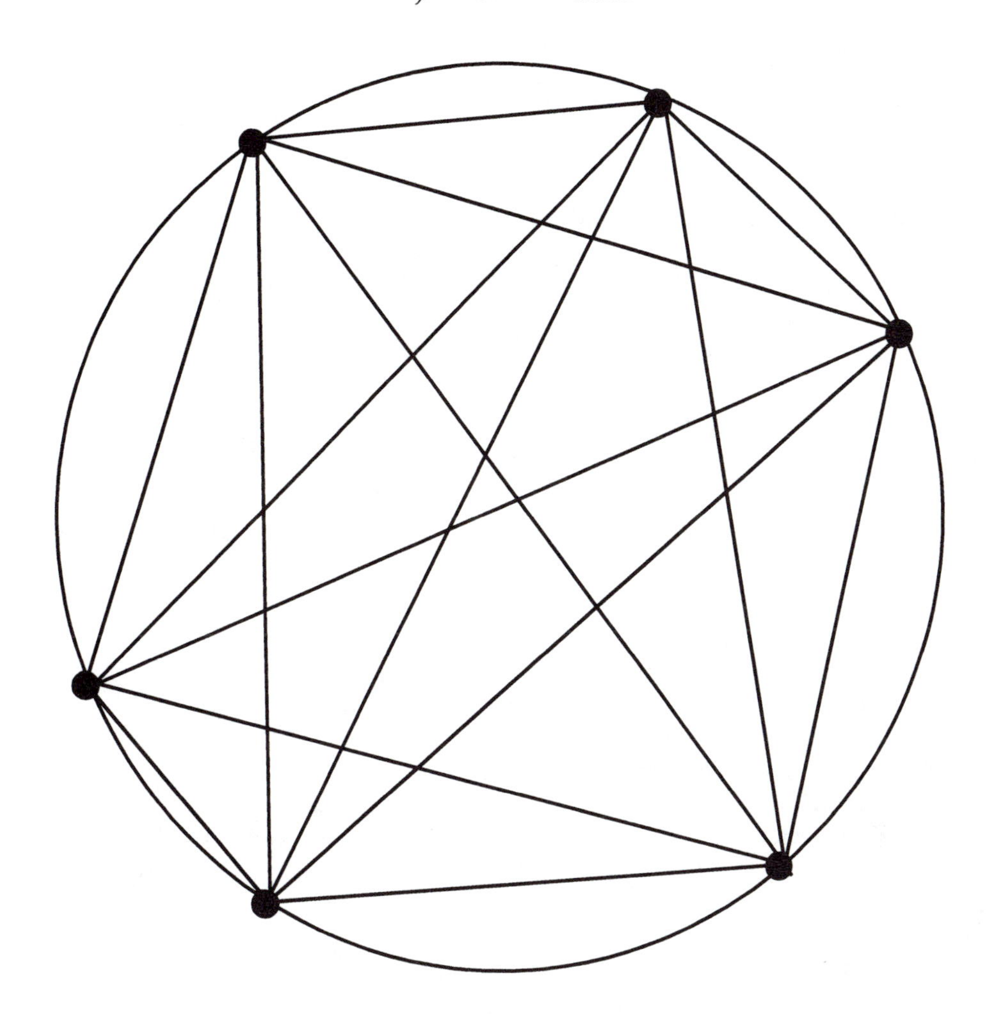

"I mean that I know that I'm just a lowly cabin boy who would never think of being disputatious or contentious in any way. But I aver that you are incapable of showing to me that six dots will give me 32 regions."

"You sure as shooting don't sound like a three-year-old."

"I bet you my freedom you can't do it."

Mrs. Tock erased the board and drew a very large circle. It would take a lot of work and careful drawing to sketch n = 6 implies R = 32.

Fred was free.

Fairhaven

1. Prove that if a radius bisects a chord, it is perpendicular to that chord.

2. This is a **sector**. It is in the shape of a piece of boysenberry pie.

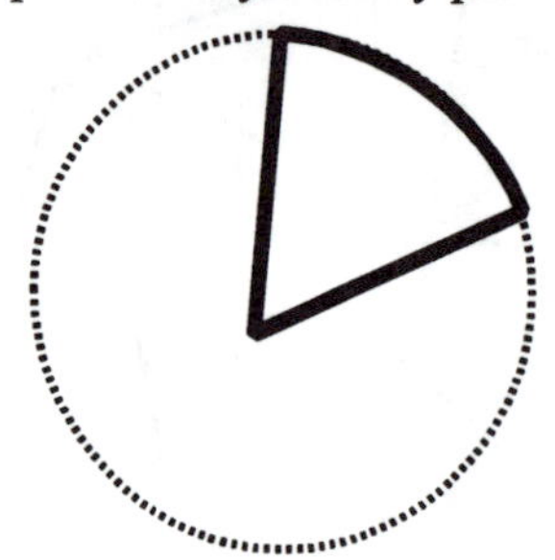

If the radius of the circle is 5" and the central angle of the sector is 40°, what is the area of the sector. (Hint: First compute the area of the whole circle and then figure out what fraction of the whole circle's area is occupied by the area of the sector.)

3. What is the area of a sector with a radius of r and a central angle of θ?

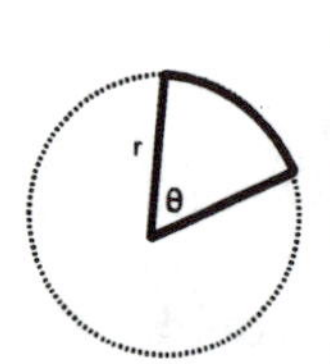

(θ, theta, is the Greek letter commonly used in trig and calculus to measure angles. It's pronounced THAY-ta.)

4. If the radius of a circle is 5 inches, what is the length of the arc that a central angle of 40° intercepts? (Hint: First compute the length of the whole circle—its circumference. Then figure out what fraction of the circumference is intercepted by the 40° central angle.)

5. What is the length of the arc of a circle of radius r that is intercepted by a central angle of θ?

6. Find the value of x.

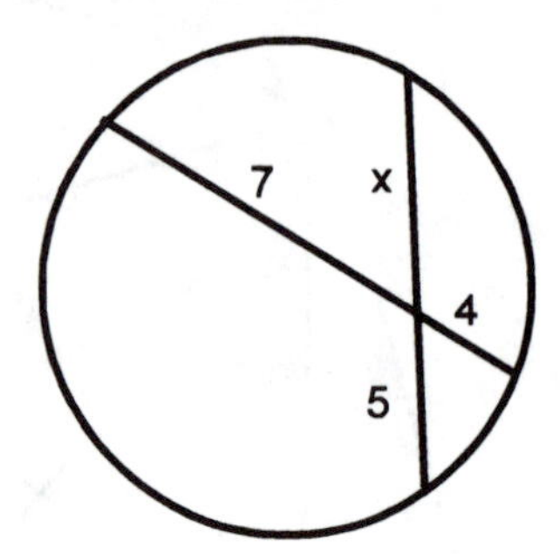

answers

1.

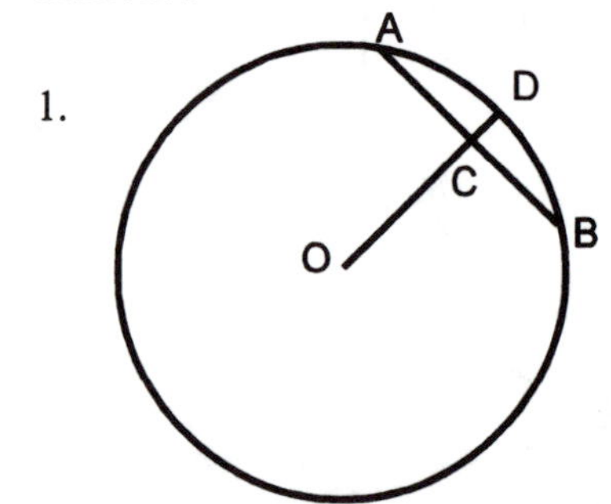

S	*R*
1. $\overline{OD}$ bisects chord $\overline{AB}$	*1.* Given
2. $\overline{AC} \cong \overline{BC}$	*2.* Def of bisects
3. $\overline{OA} \cong \overline{OB}$	*3.* All radii are $\cong$
4. $\overline{OC} \cong \overline{OC}$	*4.* Every segment is congruent to itself.
5. $\triangle OAC \cong \triangle OBC$	*5.* SSS
6. $\angle OCA \cong \angle OCB$	*6.* Def of $\cong$ ▲s
7. $\angle OCA$ and $\angle OCB$ form a linear pair.	*7.* Def of linear pair
8. $\overline{OC} \perp \overline{AB}$	*8.* If two lines intersect and form a linear pair of $\cong$ ∠s, then the lines are perpendicular. (p. 164)

2. The area of the whole circle is 25π. The area of the sector is 40/360 of the area of the whole circle = $(40/360)25\pi$

3. The area of the whole circle is πr^2. The area of the sector is $(\theta/360)\pi r^2$.

4. The circumference is 10π. The length of the arc intercepted by a 40° central angle is (40/360) of the whole circumference = (40/360)10π.

5. The circumference is 2πr. The length of the arc intercepted by a central angle of θ° is (θ/360)2πr. (What's fun in calculus is that we'll find the length of all kinds of wiggly curves—not just circles.)

6. x = 28/5 or 5 3/5 or 5.6

Valdez

1. Prove that the perpendicular bisector of a chord passes through the center of the circle.

2. Prove that no circle contains three collinear (all-on-the-same-line) points.

3. Prove that two distinct circles can intersect in at most two different points.

4. Prove that three noncollinear points can lie on at most one circle.

5. If you were given three noncollinear points, how might you go about finding the center of the circle which passes through those three points?

answers

1.

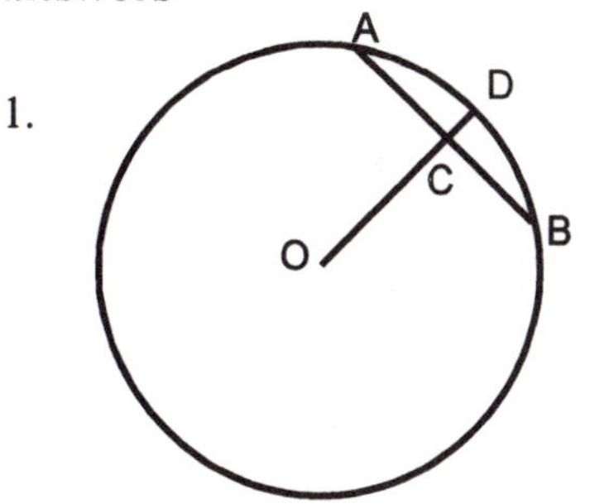

S	*R*
1. $\overline{CD}$ is the perpendicular bisector of $\overline{AB}$.	*1.* Given
2. AO = BO	*2.* All radii are =
3. O is on the perpendicular bisector of $\overline{AB}$	*3.* Theorem: Any point that is equidistant from the endpoints of a segment must lie on the perpendicular bisector. (The Converse of Cupid's Bow Theorem)

2.

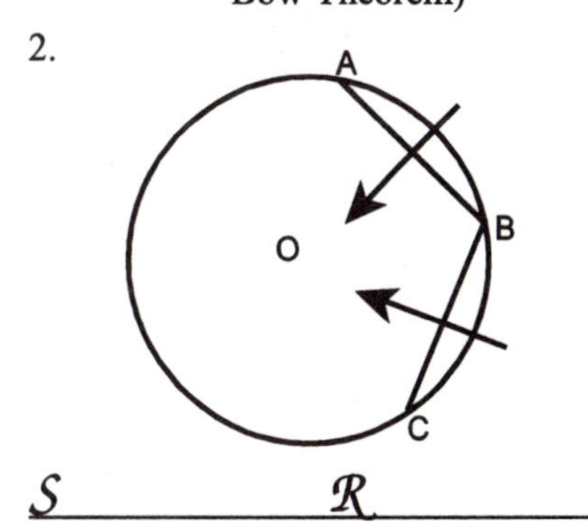

S	*R*
1. Three distinct points A, B and C on circle O.	*1.* Given
2. Assume that A, B and C are collinear.	*2.* Beginning of an indirect proof.
3. The ⊥ bisectors of $\overline{AB}$ and $\overline{BC}$ are parallel.	*3.* Theorem: Two lines perpendicular to the same line are parallel.
4. The ⊥ bisectors of $\overline{AB}$ and $\overline{BC}$ both contain point O.	*4.* Theorem proved in the previous problem.
5. A, B and C are not collinear.	*5.* Contradiction in steps *3* and *4* and hence the assumption in step *2* is false.

3.

S	*R*
1. Assume two distinct circles intersect in three (or more) points, A, B and C.	*1.* Beginning of an indirect proof.
2. The ⊥ bisectors of $\overline{AB}$ and $\overline{BC}$ pass through the centers of both circles.	*2.* Theorem proved in problem 1
3. The centers of the two circles must be the same.	*3.* Postulate: If you have two points, there is at most one line through them.

(Actually, we're using the logically equivalent contrapositive: If more than one line passes through a set of points, the points must all be the same.)

4. The radii of the circles must be the same.	*4.* Def of radii (= the distance from the center to any point on the circumference)
5. Two circles with different centers can intersect in at most two points.	*4.* Contradiction in steps *1, 3* and *4* (We have two circles in step *1* and one circle in steps *3* and *4*) and therefore the assumption in step *1* is false.

4. We just proved this in problem 3.

5. The perpendicular bisectors of $\overline{AB}$ and $\overline{BC}$ will intersect at the center of the circle which contains A, B and C.

Walpole

1. Invent a definition to describe when point P is in the interior (inside) of circle O. There are several possible answers.

2. Fill in the missing steps in the proof of *If a line ℓ is perpendicular to a radius $\overline{OA}$ at A, then ℓ is tangent to the circle.*

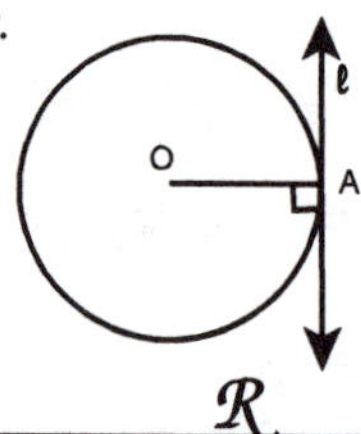

S	*R*
1. Circle O. Radius $\overline{OA}$. ℓ ⊥ $\overline{OA}$ at point A.	*1.* Given
2. Assume that ℓ intersects circle O in two points, A and B.	*2.* ?
3. Draw $\overline{OB}$	*3.* ?

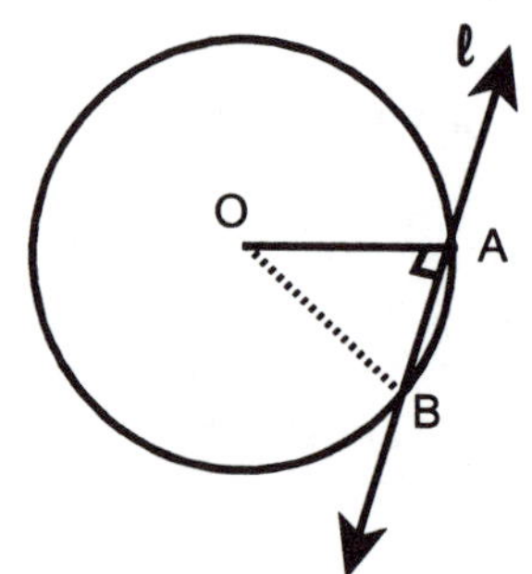

4. ?	*4.* All radii are congruent
5. ∠A ≅ ∠B	*5.* ?
6. ∠A is a right angle.	*6.* Def of ⊥
7. m∠A = 90°	*7.* ?
8. m∠A = m∠B	*8.* Def of ≅ ∠s
9. m∠B = 90°	*9.* ?
10. ∠B is a right angle.	*10.* Def of right angle

11. No triangle can contain two right angles. *11.* Theorem (in problem 1 of Elmo on p. 152)

12. ℓ does not intersect circle O in two points.

12. Contradiction in steps 6, 10 and 11 and therefore the assumption in step 2 is false.

13. ℓ is tangent to circle O. *13.* ?

3. Prove that two chords (in the same circle) are congruent if the measures of the arcs associated with those chords are equal.

4. Prove that if two chords (in the same circle) are congruent, then the measures of the arcs associated with those chords are equal.

5. If a chord of a circle is 6 feet long, what is the smallest area that that circle might be? What is the largest area that circle might be?

odd answers

1. One definition might be: P is in the interior of circle O iff OP is less than the radius of the circle.

A second definition might be: P is in the interior of circle O iff $\overline{OP}$ does not intersect the circle.

A third definition might be: P is in the interior of circle O iff P is on a chord of the circle but is not one of the chord's endpoints.

A fourth definition might be: P is in the interior of a circle iff P is on a diameter of the circle but is not one of the diameter's endpoints.

A fifth definition might be: P is in the interior of a circle iff P is never on any tangent to the circle.

All these definitions are correct, but some of them would be much harder to use than others.

3.

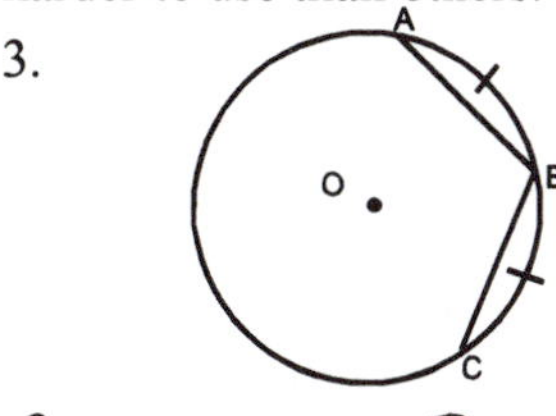

S	*R*
1. Circle O with $m\overset{\frown}{AB} = m\overset{\frown}{BC}$	*1.* Given
2. Draw $\overline{OA}$, $\overline{OB}$ and $\overline{OC}$.	*2.* Two points determine a line.
3. $m\angle AOB = m\overset{\frown}{AB}$ $m\angle BOC = m\overset{\frown}{BC}$	*3.* Definition of the measure of an arc
4. $m\angle AOB = m\angle BOC$	*4.* Algebra
5. $\overline{OA} \cong \overline{OB} \cong \overline{OC}$	*5.* All radii are congruent
6. $\Delta AOB \cong \Delta BOC$	*6.* SAS
7. $\overline{AB} \cong \overline{BC}$	*7.* Def of ≅ ⚠

5. The smallest circle that contains a 6-foot chord is the one in which the 6-foot chord is the diameter. Then the radius is 3 feet and the area is 9π square feet.

There is no limit to the size of a circle containing a 6-foot chord:

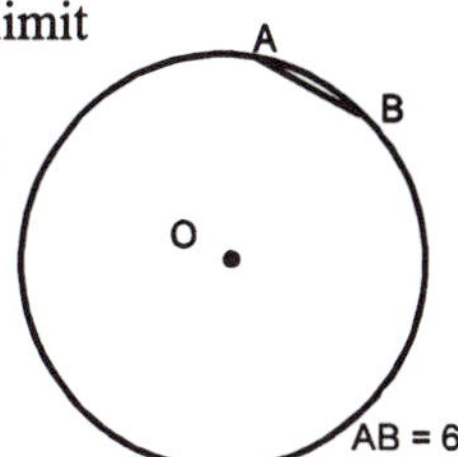

Zamora

1. Invent a definition to describe when two circles are tangent. There are several ways this might be defined.

2. If circles O and O' are tangent at point P, must it always be true that O, O' and P be collinear?

3. Can two different concentric circles ever be tangent? Argue why your answer is correct. (If you say they can be tangent, then showing an example would be a great argument. If you say they can never be tangent, then explain why that is the case.)

4. From a point that is outside a circle, there are two tangents that may be drawn to the circle. Prove that they are congruent.

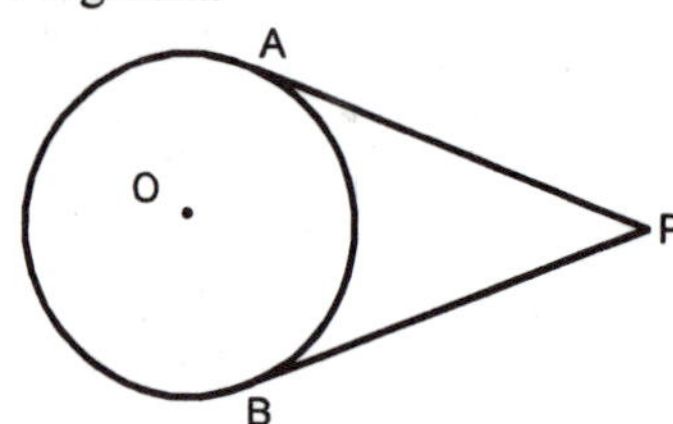

5. Given $m\widehat{BC} = 50°$.

Find, if possible:

a) $m\angle A$

b) $m\widehat{AB}$

c) $m\widehat{AC}$

d) $m\widehat{AB} + m\widehat{AC}$

odd answers

1. First possible definition: Two circles are tangent iff they share exactly one point in common.

Second possible definition: Two circles are tangent iff there exists a line which is tangent to both circles at the same point.

3. If two concentric circles (with center at point O) were tangent, then they would share a common point by either definition given in problem 1. Call this point P. Then OP would be the radius of both circles. But knowing the center of a circle and its radius uniquely defines a particular circle which contradicts the fact that we were given two distinct circles.

5. a) 25° There's no way to determine b) and c). For d), we note that since $m\widehat{BC}$ is 50° and the measure of the whole circle is 360°, then $m\widehat{AB} + m\widehat{AC}$ must be 360° – 50° = 310°

Lamkin

1. Prove that if two chords are congruent, they are equidistant from the center of the circle. (You may use the lemma that we've never gotten around to proving: *Midpoints of congruent segments divide them into congruent segments*.)

2. Can a central angle intercept a major arc?

3.

Remember clocks with hands? What is the measure of the arc intercepted by the hands when it's 1 o'clock?

4. What time (or times) is it when the hands of the clock intercept an arc of 150º?

5. Please be careful on this question: What time (or times) is it when the hands of the clock intercept an arc of 270º?

McCamey

1. Prove that if two chords are equidistant from the center of the circle, they are congruent.

2. Explain in everyday English whether a secant to a circle can ever be a tangent to that circle.

3. Can a radius of a circle ever be a chord of that circle?

4. Can a chord of a circle ever be a diameter of that circle?

5. Find AB. (Hint: The answer is *not* equal to 19.2)

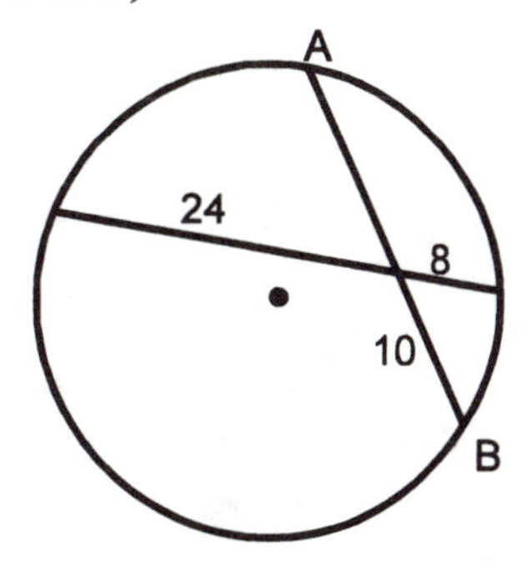

Fred was free because there are only 31 regions possible when you draw all the lines between six dots on a circle. The apparent doubling pattern doesn't hold. Inductive reasoning, in which we take our observations and try to arrive at some conclusion, can only give us *probable* results.

In contrast, deductive reasoning is the kind of reasoning we have been doing here in geometry. We start with postulates, such as "Two points determine a line," and from them obtain results (theorems) which *must be true*. If you believe the postulates, then you must accept the theorems as true.

For the record:

Number of dots	1	2	3	4	5	6	7	8	9	10	11	12	13	14
Number of regions	1	2	4	8	16	31	57	99	163	256	386	562	794	1093

Chapter Eleven

Constructions

Mrs. Tock had lost her bet with Fred. He no longer faced the prospect of serving as a cabin boy for the rest of his life. He would no longer have to sing sea chanties unless he wanted to. Instead of swabbing the deck, he would be faced with the more pleasant task of working at his desk in his office.

Fred wanted to dance. He felt so light on his feet—which isn't hard when you're 37 pounds.

"As a parting gift, and a celebration of my manumission," Fred said to Mrs. Tock, "I want to give you a board game that I've created."

She didn't understand part of what he was saying but held out her hand anyway and received:

"That's short for 'The Traditional Geometrical Construction Game,' but we didn't have room for all of that on the box," Fred explained. "I think you'll really enjoy it."

"Thanks, Proofie. There are many long hours out here on the open sea with little to do. Now I guess it's time to keep my part of the bet and let you go." She picked him up and held him over the edge of the ship and let him drop.

Telegram for Dr. Schmidt! Telegram for Dr. Schmidt!

Telegram

from: Elaine Marie, Polka Dot Publishing

to: Author

Message: We are concerned. What have you done? You didn't do what we think you just did, did you? When you had Tanya walk the plank in chapter nine we didn't object, although there was much editorial discussion over the topic.

Even in chapter ten when Hook and his crew were "taught to swim"—as you expressed it—we offered no objection. But this is different. This is FRED. And this is the LIFE of Fred series.

And besides, we really like that cute little guy.

Best wishes,
Elaine marie

Reply Telegram

from: Author

to: Elaine Marie, Polka Dot Publishing

Message: Thank you for your concern. The crash helmet and life preserver for Fred that you included with your telegram will not be necessary. By the way, where did you ever find a square helmet?

All the best to you!
Stan

As Fred fell from Mrs. Tock's hand toward the inky, icy waters in the pond near the Great Lawn on the KITTENS campus, he quickly recalled the one fact that he needed to know about gravity on planet earth:

Gravitational acceleration = –32 ft/sec/sec. It is a constant. Physicists have found that number by doing experiments. Namely they drop things and measure the velocity at different points in time.

The velocity of a dropped object increases by 32 ft/sec for each second that passes. (It's negative since things fall downward toward the earth—their height decreases over time.) If, for example, you had a velocity of –1000 ft/sec at some moment in time, then a second later, your velocity would be –1032 ft/sec. And a second later, it would be –1064 ft/sec.

Acceleration = –32 ft/sec/sec. Fred took the antiderivative* of that and found that his velocity was: $V = -32t$ (where t is the time in seconds since she dropped him). Note that at $t = 0$, which is the moment she dropped him, his velocity was equal to zero. At $t = 1$ second, his velocity was –32 ft/sec. At $t = 2$ seconds, his velocity was –64 ft/sec.

He took another antiderivative and found that his height above the inky, icy waters (y), was: $y = -16t^2 + h$ (where h is the height of the ship). Note that at $t = 0$, his height, y, is equal to h. This means that he was h feet above the surface of the water when he was dropped.

✶ That's calculus. After geometry, a couple of algebras (beginning and advanced), and trig, you will be ready to start calculus. In the first third of calculus, you will learn to take derivatives. In the second third you will learn to take antiderivatives.

Taking derivatives is *really easy*. For example, the derivative of $5t^8$ is $40t^7$. You multiply the coefficient by the exponent (that gives you the 40) and you decrease the exponent by one (that gives you the 7). The derivative of $300t^5$ is $1500t^4$. The derivative of $8t^{20}$ is $160t^{19}$.

What you can *do* with derivatives is what makes it exciting. There are a zillion things you can use them for. And antiderivatives are even more useful. There are two zillion uses for them.

While Fred falls, let us shift our attention for a moment back to Mrs. Tock onboard the *Crocodile Express*. She was opening The Con Game box.

"Oh, look Ticky, at the game that our former cabin boy left us to play with," Mrs. Tock shouted to her husband.

After this brief interlude, we now shift our attention back to Fred who has just hit the water. He had fallen for one whole second. From the equation $V = -32t$ (where V is his velocity and t is the number of seconds after he was dropped), Fred calculated that his speed as he hit the water was –32 ft/sec. That's not terribly fast.

From the equation $y = -16t^2 + h$ (where y is his distance above the water, t is the number of seconds after he was dropped, and h is the height from which he was originally dropped), he put $t = 1$ and $y = 0$ and found that $h = 16$ feet. Fred had dropped 16 feet. That isn't an especially long dive.

And the pond into which he had been dropped wasn't especially deep either. In fact, he could stand on the bottom.

As he walked toward shore, he saw a great crowd of people engaged in all sorts of activities. The Rescue Workers of America were there. They had put up a big sign: *This is an Official Rescue Site.* The Caboodle reporters were there taking flash photos and sticking microphones in people's faces. The KITTENS police force (viz., Sergeant Friday) was there to keep order.

As Fred came ashore, three Rescue Workers rushed to him. One threw a blanket around him. The second obtained his name and then stamped his forehead.

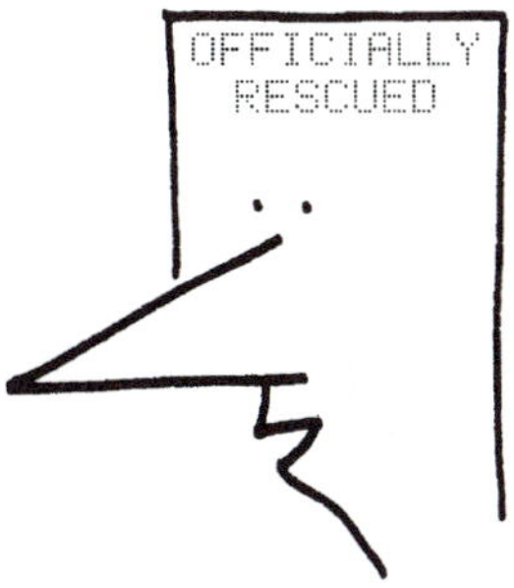

OFFICIALLY RESCUED

The third handed him: ① a cup of hot chocolate; ② some stationery that was preaddressed to congressmen and senators so that the victims—whom they called rescuees—could demand more federal money for the RWA; ③ a book and a board game—"to help pass the time."

The book was entitled, *What Every Rescued Person Should Know About the Wonderful Work of the Rescue Workers of America*. It contained a history of the organization, photos of the president and his family, and a convenient envelope for making donations.

The board game was . . . The Traditional Geometrical Construction Game. They had handed Fred his own game, the one he had created. Then a Rescue Worker escorted him to the Rescued Persons Detention Area. Fred was told that the guards were there to prevent the Rescued Persons "from wandering off before they were fully rescued."

Inside the barbed wire compound, Fred saw people sitting in groups on the ground playing the game that he had created. That made him feel good.

"Want to join me?"

Fred looked. It was Tanya! She wasn't dead. Not at all. He sat down next to her and asked, "Where's your board game?"

"They took it," she said, pointing to a group of wet pirates (Hook and his crew) seated on the far side of the detention area. They had amassed a huge stack of the board games that they had stolen from all the other victims. "I heard that the RWA will let us go later this afternoon after they've had a chance to take all the pictures that they need for their publicity and fund-raising efforts. Meanwhile, since they haven't taken your copy of the game, why don't we play it to pass the time?"

Fred passed the box to her. He noticed that the pirates were asking the Rescue Workers for second cups of chocolate and extra blankets. Tanya opened the box and poured out the contents: a compass, a pencil and a straightedge. In addition, there were a sheet of instructions, a ream of blank paper and a deck of Construction Cards. She put the deck between them and turned over the first card.

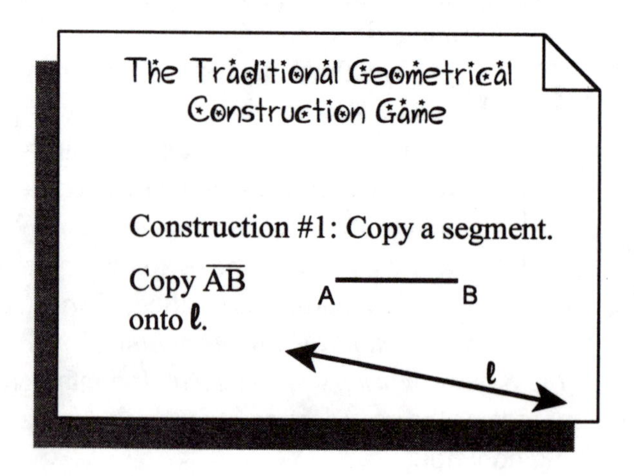

The Traditional Geometrical Construction Game
Construction #1: Copy a segment.
Copy $\overline{AB}$ onto ℓ.
A
B
ℓ

"Oh, that's so simple-dimple," Tanya exclaimed. "I'll just use some scissors and cut out the segment and paste it onto the line."

Fred winced. "Maybe we should take a peek at the instructions first, since there are no scissors supplied with the game."

"Who needs to read the rules? Since we don't have any scissors, I'll just use this ruler that is in the game and make a couple of pencil marks on it to match up with $\overline{AB}$ and then transfer that to ℓ."

"The whole point of the Con Game is that it is a *game*," Fred explained. "It's most fun when you play by the rules. Let's take a look at the instructions. Then we'll know how to play."

"I got it! If I crease the paper just right, I can make $\overline{AB}$ go kerflop right on top of ℓ."

Fred silently handed her the instruction page:

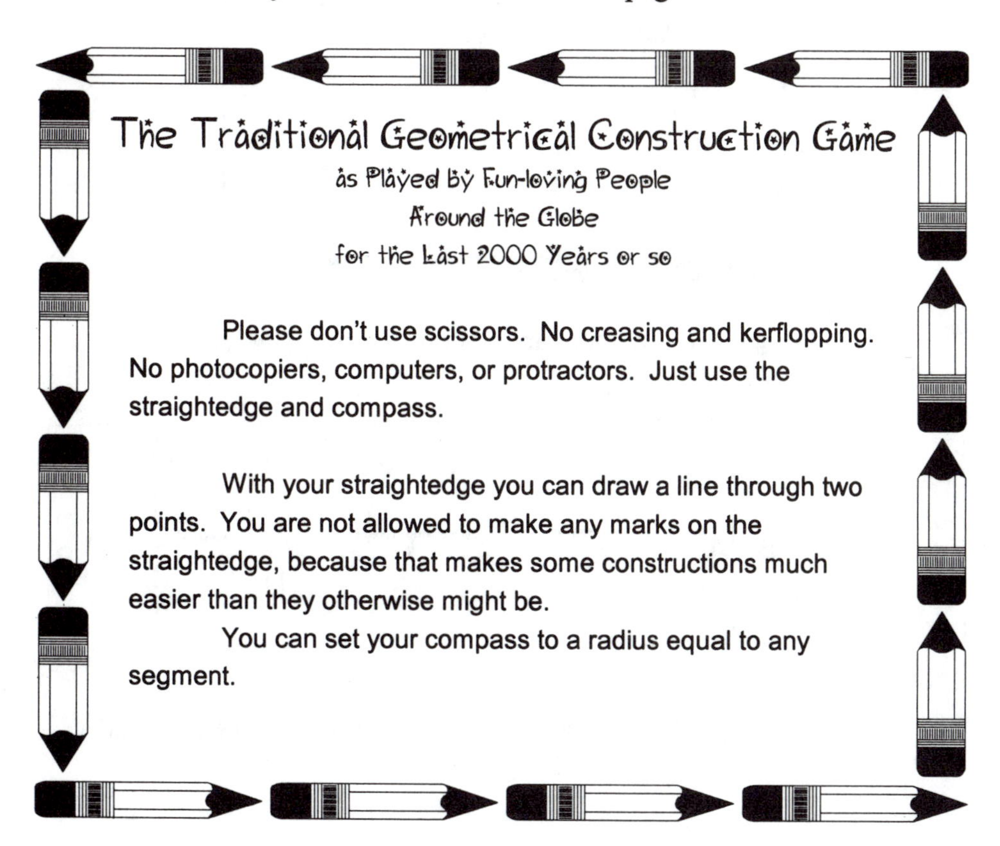

The Traditional Geometrical Construction Game

as Played by Fun-loving People

Around the Globe

for the Last 2000 Years or so

Please don't use scissors. No creasing and kerflopping. No photocopiers, computers, or protractors. Just use the straightedge and compass.

With your straightedge you can draw a line through two points. You are not allowed to make any marks on the straightedge, because that makes some constructions much easier than they otherwise might be.

You can set your compass to a radius equal to any segment.

Tanya read the rules. "Why didn't you tell me about these instructions?" she asked.

Fred didn't know how to respond.

Tanya took one of the blank sheets of paper and said, "This is easy-breezy. First you pick a point C on ℓ, set your compass to $\overline{AB}$ and use C as the center.

"There's no strain on the brain," she declared. "Are they all this easy?" She flipped over another card.

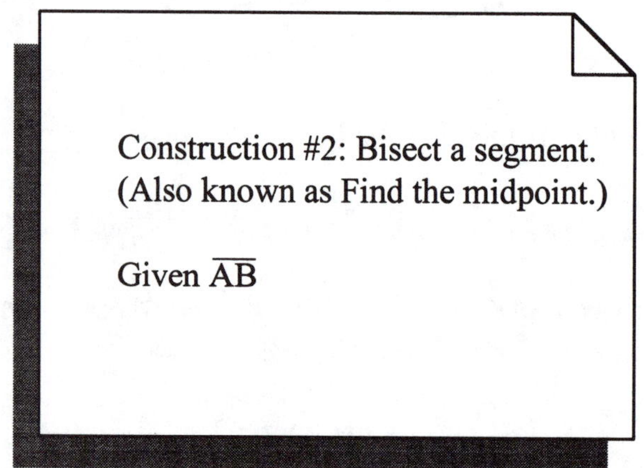

She thought for a moment and announced that this second construction was a "lazy daisy."

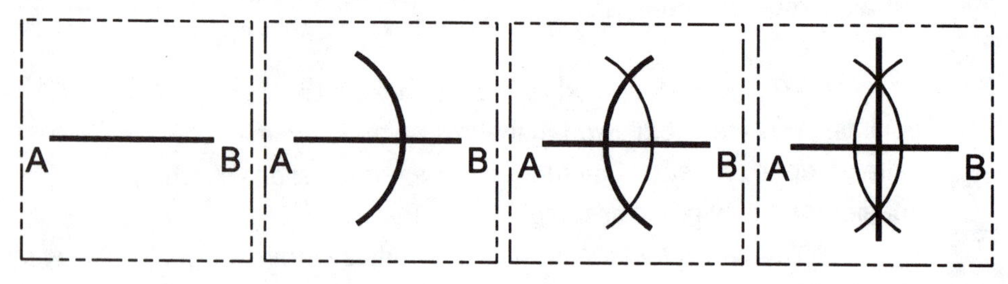

She looked at her construction for a moment, took the card and wrote on it.

Construction #2: Bisect a segment.
(Also known as Find the midpoint.)

Given $\overline{AB}$

Also good for constructing the perpendicular bisector of a segment.

Your Turn to Play

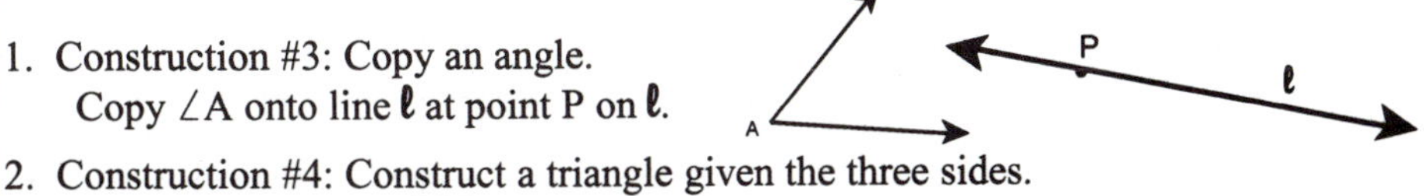

1. Construction #3: Copy an angle.
   Copy ∠A onto line ℓ at point P on ℓ.
2. Construction #4: Construct a triangle given the three sides.
   Given: $\overline{AB}$, $\overline{CD}$ and $\overline{EF}$.
3. Construction #5: Construct an equilateral triangle. (Also known as construct a 60° angle.)
4. Construction #6: Construct a parallel line.
   A line through P that is parallel to ℓ.

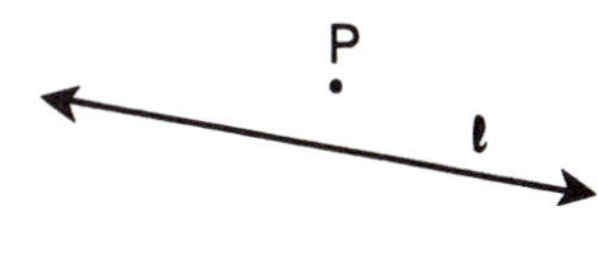

COMPLETE SOLUTIONS

1. Copy an angle (Const. #3)

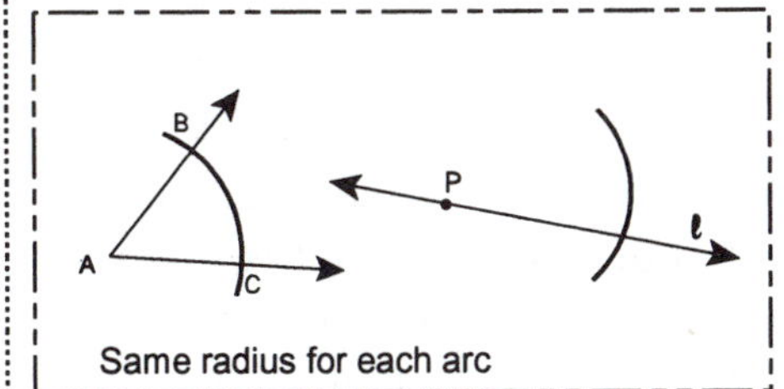

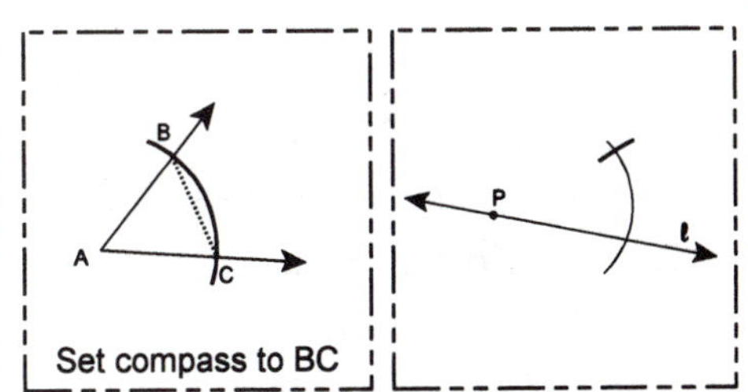

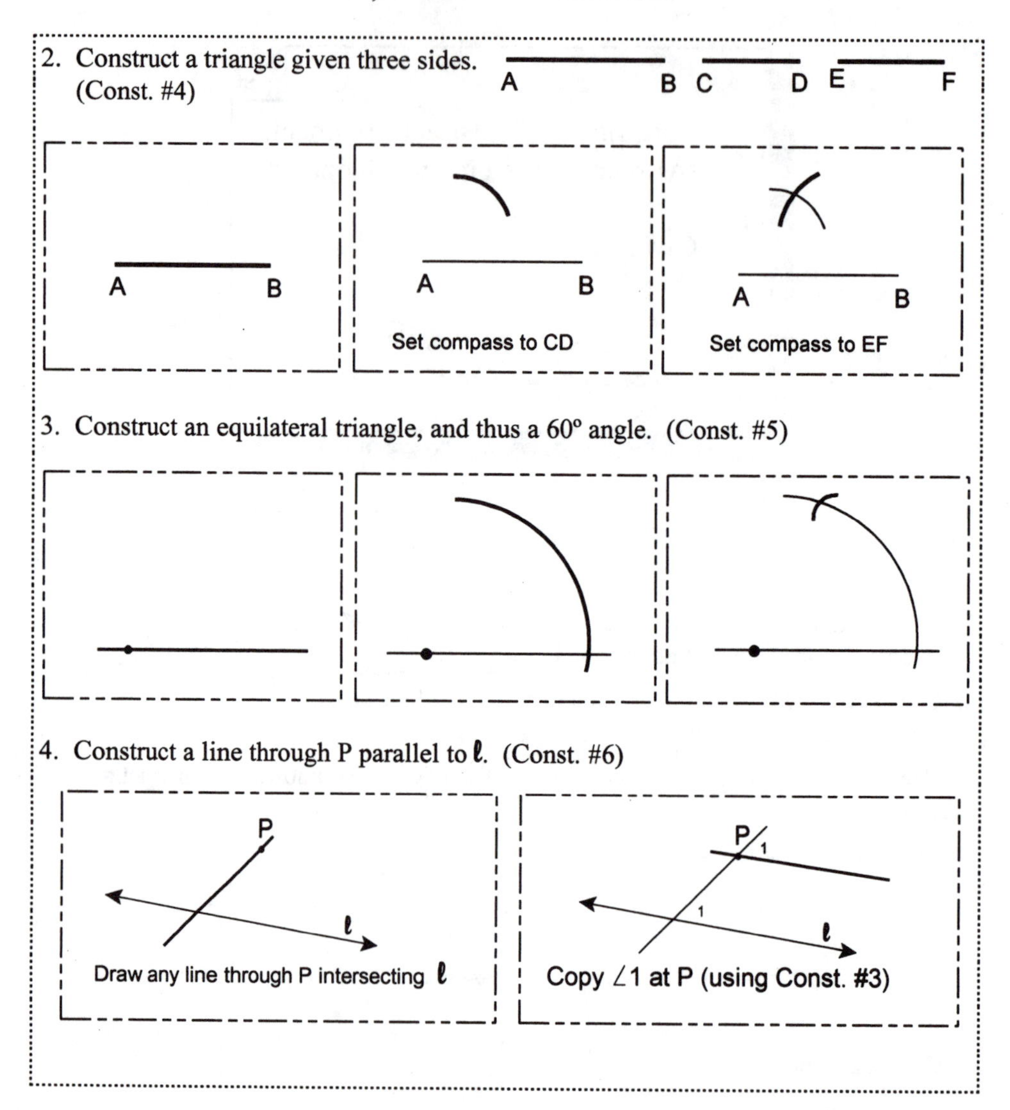
2. Construct a triangle given three sides.
(Const. #4)
A B C D E F
A B
A B
Set compass to CD
A B
Set compass to EF
3. Construct an equilateral triangle, and thus a 60° angle. (Const. #5)
4. Construct a line through P parallel to ℓ. (Const. #6)
P
ℓ
Draw any line through P intersecting ℓ
P
1
1
ℓ
Copy ∠1 at P (using Const. #3)

"You notice," Fred said, "that Construction #5 (construct an equilateral triangle) was a **rusty compass** construction. That's a construction using a compass that has rusted into a fixed radius that can't be changed. We'll rename that construction of an equilateral triangle as Construction #5R to indicate that it was a rusty compass construction.

"Not all constructions can be done with a rusty compass. For example, I can't think of a way of copying an angle (Const. #3) without having to change the radius of the compass in the middle of the construction."

"I was hoping all these constructions wouldn't be so chinchy-winchy," said Tanya.

"Even without the rusty compass restriction, some of the constructions in this game can be quite difficult," Fred warned. "And you never know, just by looking at the problem, how hard it will be to find the answer. Sometimes it requires a bit of ingenuity."

"I don't mind if they're roughie-toughie. That can make them more fun and less boring."

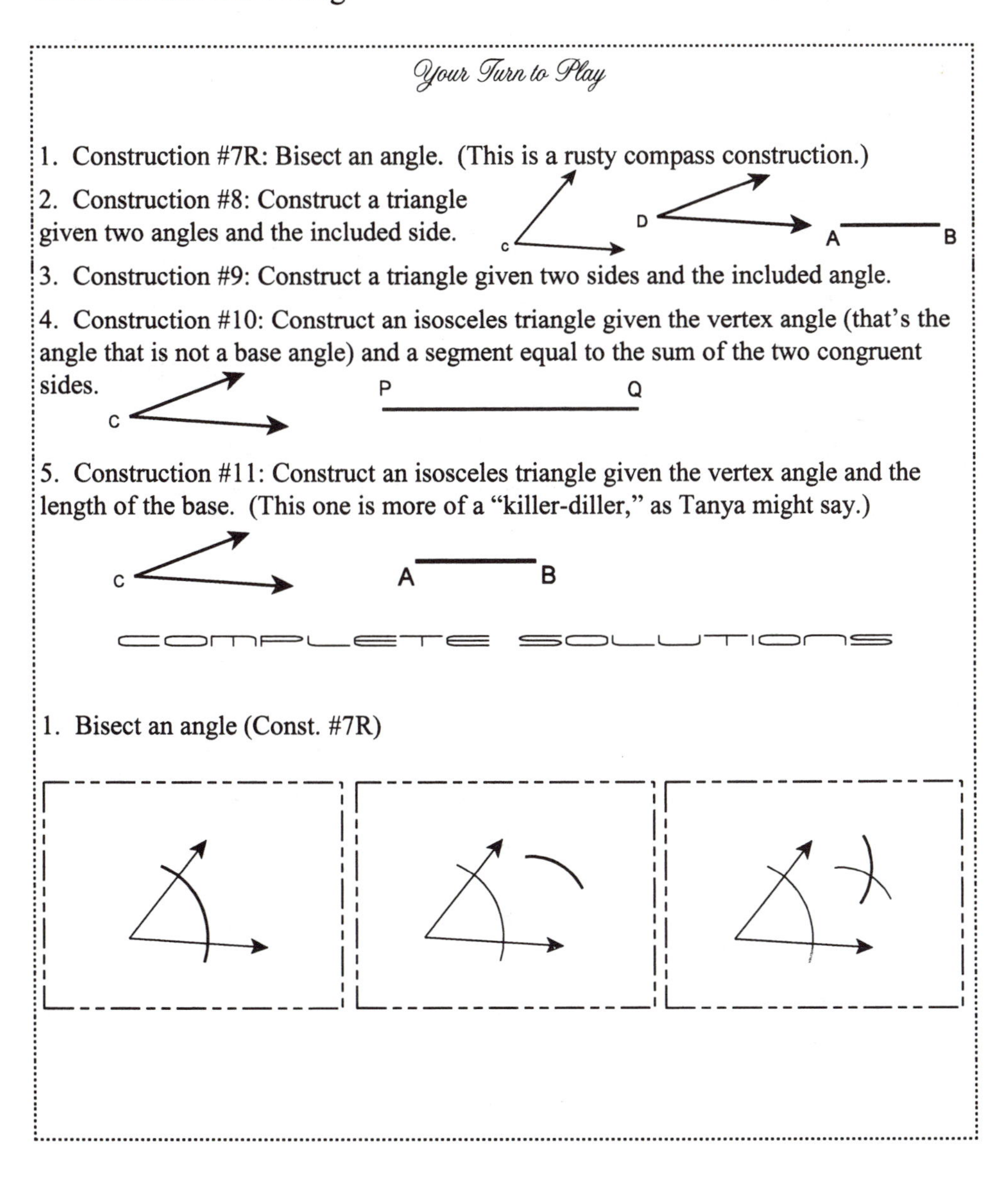

Your Turn to Play

1. Construction #7R: Bisect an angle. (This is a rusty compass construction.)

2. Construction #8: Construct a triangle given two angles and the included side.

3. Construction #9: Construct a triangle given two sides and the included angle.

4. Construction #10: Construct an isosceles triangle given the vertex angle (that's the angle that is not a base angle) and a segment equal to the sum of the two congruent sides.

5. Construction #11: Construct an isosceles triangle given the vertex angle and the length of the base. (This one is more of a "killer-diller," as Tanya might say.)

COMPLETE SOLUTIONS

1. Bisect an angle (Const. #7R)

2. Construct a triangle given two angles and the included side. (Const. #8)

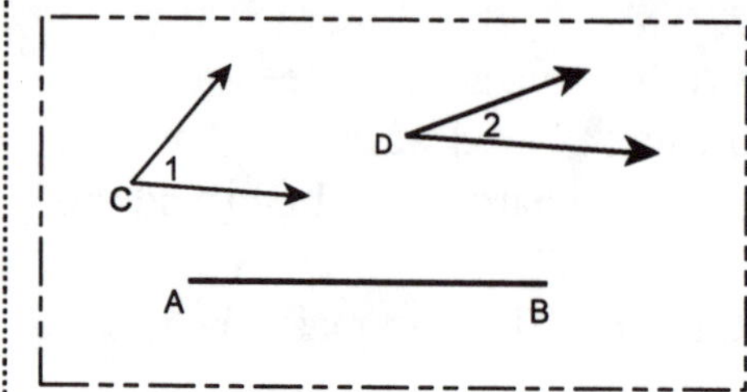

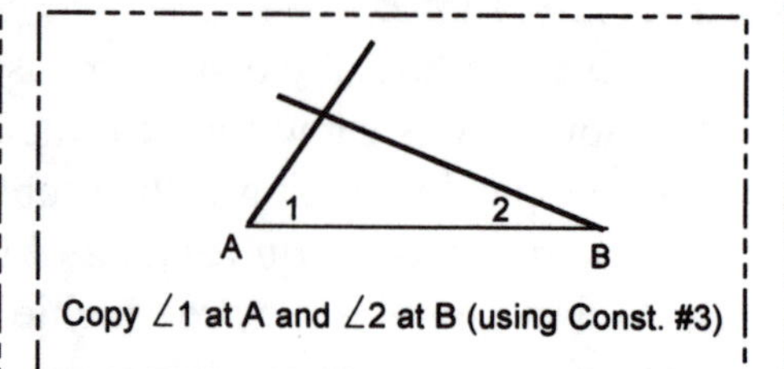

Copy ∠1 at A and ∠2 at B (using Const. #3)

3. Construct a triangle given two sides and the included angle. (Const. #9)

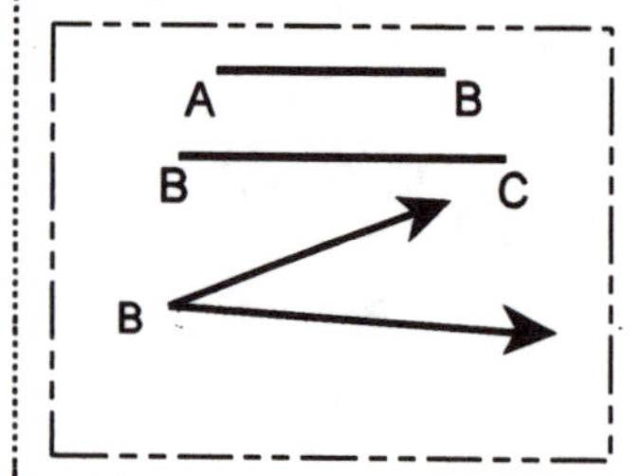

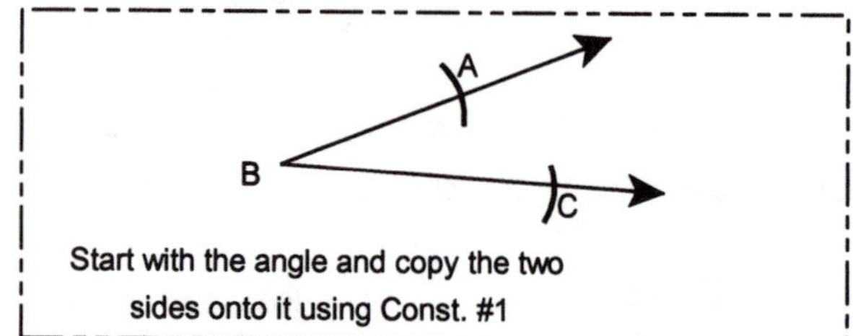

A
B
C
Start with the angle and copy the two
sides onto it using Const. #1

4. Construct an isosceles triangle given the vertex angle and a segment equal to the sum of the two congruent sides. (Const. #10)

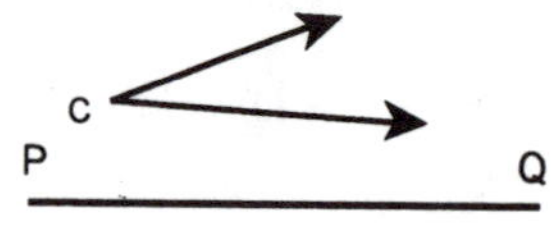

Bisect $\overline{PQ}$ (Const. #2) and then construct the triangle using the two sides and the included angle (Const. #9).

5. Construct an isosceles triangle given the vertex angle and the length of the base. (Const. #11)

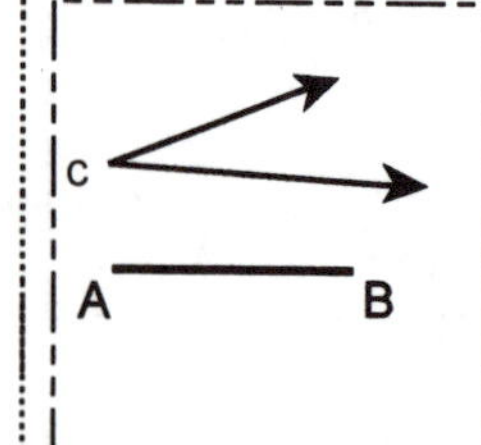

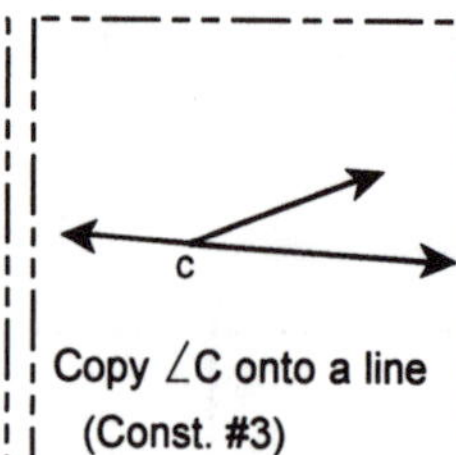

Copy ∠C onto a line

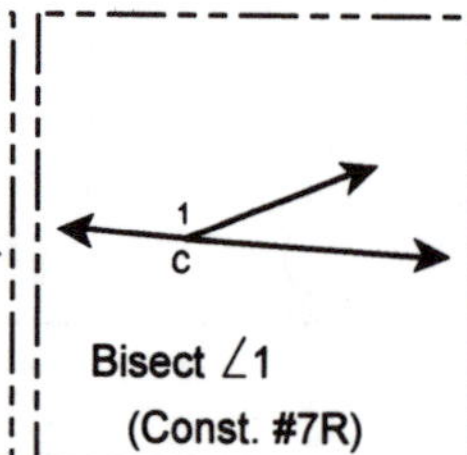

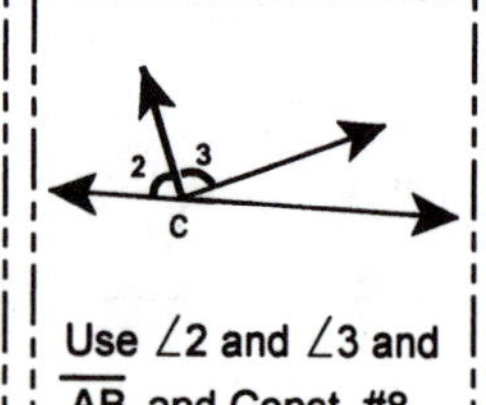

Use ∠2 and ∠3 and
AB and Const. #8

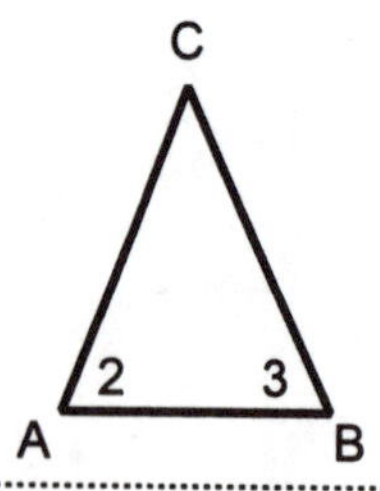

We used the fact that the sum of the angles of any triangle is equal to 180°.

"Have you noticed how beautifully the RWA has laid out the barbed-wire fencing in our detention camp?" Fred asked Tanya.

"Yeah, it's a rectangle, but what's the big dealie-wheelie?"

"But look at that rectangle. It's gorgeous. Not too fat and not too skinny. Mathematicians call it the **golden rectangle**.

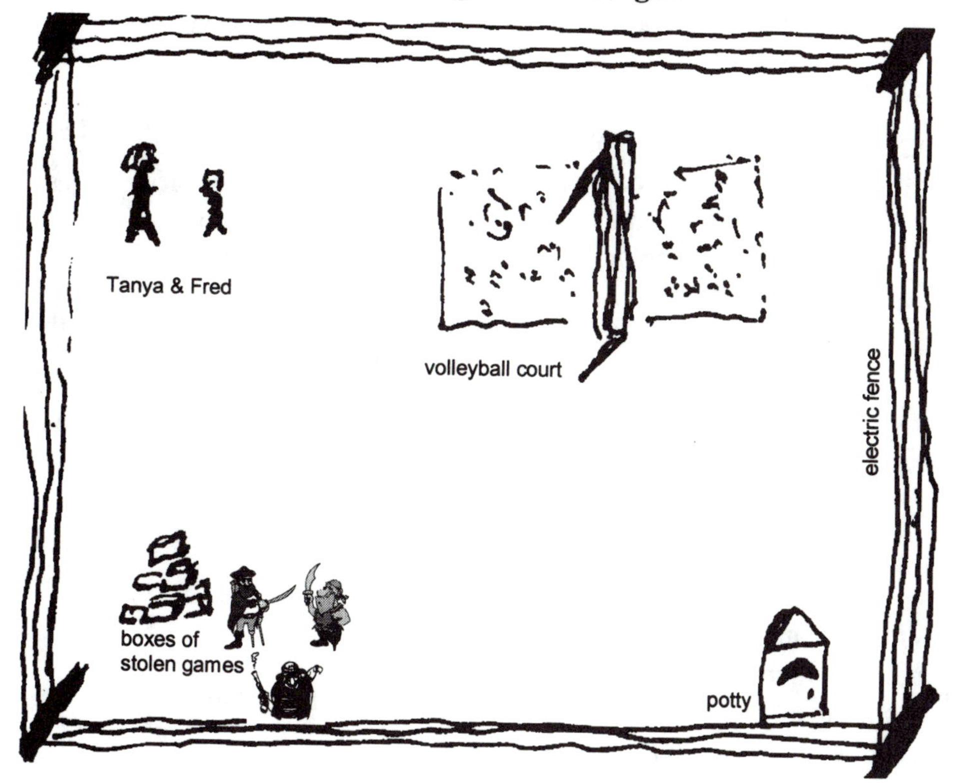

"That rectangle is just like the shape of the country home picture that I have on the wall in my office," he continued. "When P and I get married someday, that will be the kind of second home in the country that we'll have. The frame of that picture is a golden rectangle. Back in1917 psychologist Edward Lee Thorndike did some experiments to find the most pleasing rectangular shape. He found out that is was the shape of my country home picture. Thorndike checked people's eyeballs when they looked at various rectangles. They could see golden rectangles at a single glance, but when they looked at rectangles that were long and

skinny like ➘ they couldn't take it all in without shifting their eyes back and forth. Also rectangles that were more nearly square weren't as attractive to Thorndike's subjects.

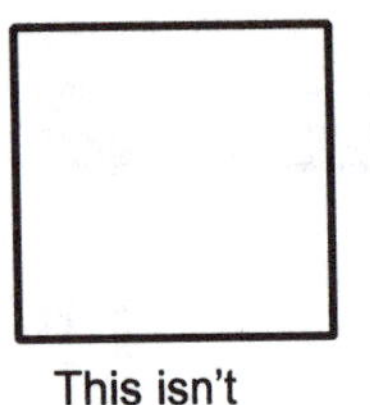
This isn't the most attractive

Thorndike wasn't the first to recognize that humans really like the golden rectangle. The Greeks in the fifth century B.C. knew this. They built their Parthenon in Athens using the golden rectangle."

the Parthenon

Fred was ebullient when it came to discussing the golden rectangle.

"Just take my country home picture and measure the lengths of the sides. The ratio $\frac{b}{a}$ is called the **golden ratio** and is equal to $\frac{\sqrt{5}+1}{2}$ and is usually represented by the Greek letter ϕ ("phi").

a

b

"And if you chop off a square from a golden rectangle, the leftover piece is a new smaller golden rectangle. And you could then chop a square off the smaller golden rectangle and get an even smaller golden rectangle."

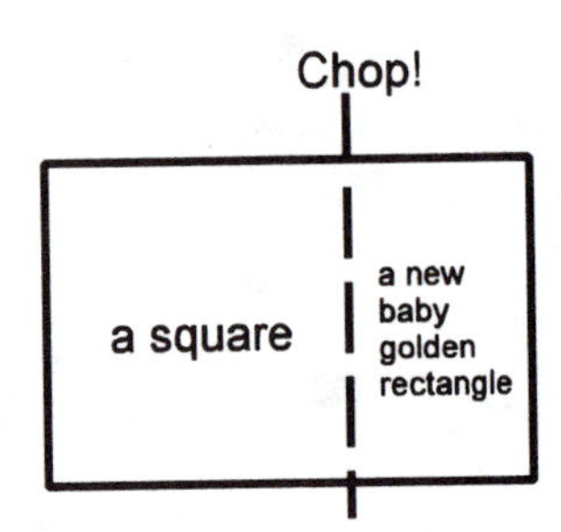

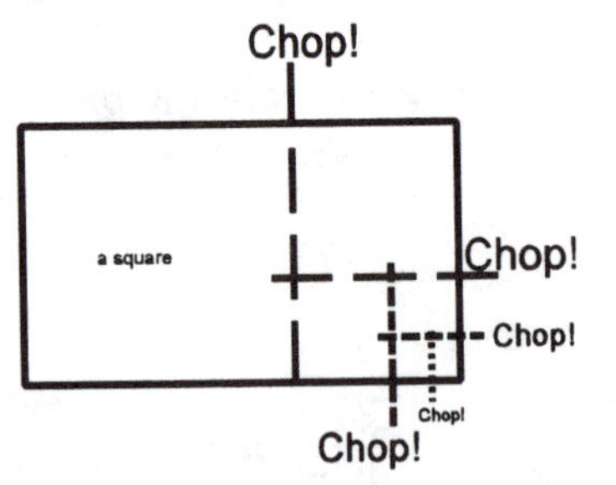

Fred was making all kinds of chopping motions with his hand.

"Hey, I read the rules in your construction game. It said that there is no 'creasing and kerflopping.' Doesn't that also include no 'kerchopping'?" Tanya wanted to get back to doing constructions with ruler and straightedge. "And, besides, how are you ever going to construct a ratio of $\phi = \frac{\sqrt{5}+1}{2}$? Isn't that an impossible construction?"

Fred smiled. "No it isn't impossible. Constructing ϕ and the golden rectangle is pretty easy. First, you need a square."

Construction #12: Construct a square.

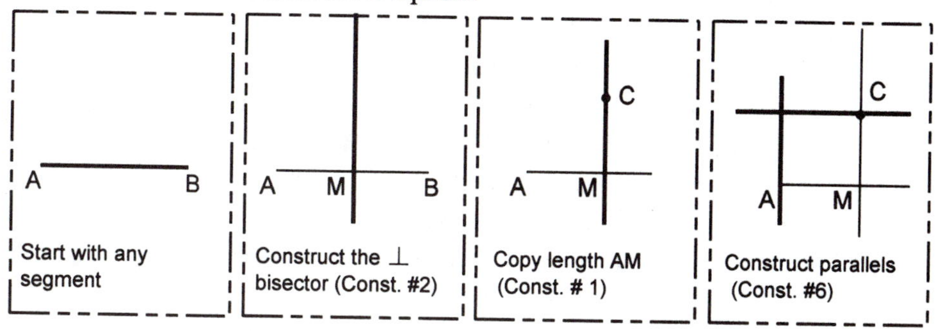
A
B
Start with any segment
A
M
B
Construct the ⊥ bisector (Const. #2)
A
M
C
Copy length AM (Const. # 1)
C
A
M
Construct parallels (Const. #6)

Construction #13: Construct a golden rectangle.

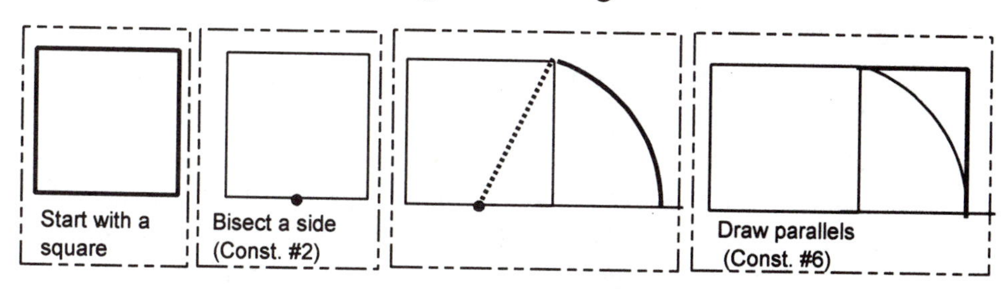
Start with a square
Bisect a side (Const. #2)
Draw parallels (Const. #6)

"So constructing the ratio $\frac{\sqrt{5}+1}{2}$ with straightedge and compass can be done," Fred said.

[Here's the only proof in this chapter.]
Suppose we start with a square of length *s*.

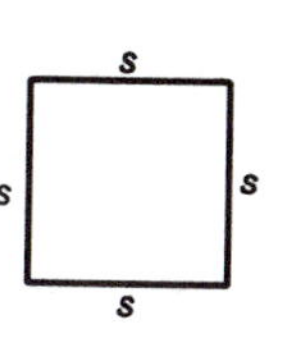
s
s
s
s

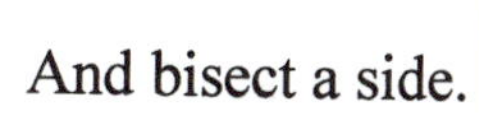

And bisect a side.

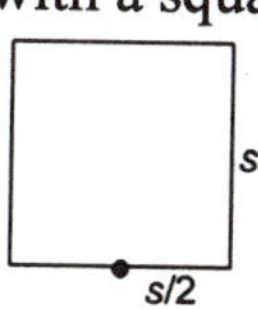
s
s/2

Pythagorean Theorem

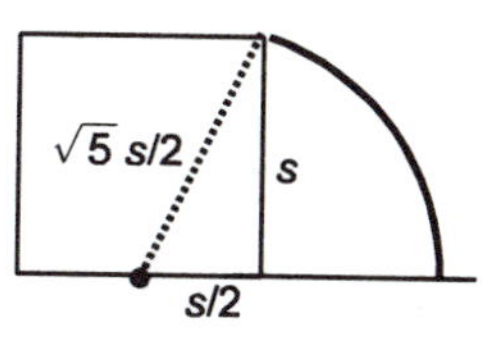
√5 s/2
s
s/2

So the length of the bottom of the constructed rectangle is $s/2 + \sqrt{5}\,s/2$ which is $\frac{\sqrt{5}+1}{2}\,s$

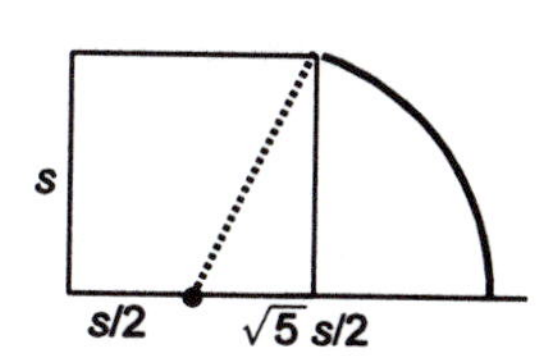
s
s/2
√5 s/2

"But there really are things you can't construct using just a straightedge and a ruler," Fred continued. "One of the most surprising is that you can't trisect an angle. That means cut it into three equal parts. "There are certain angles you can trisect, such as the 90° angle, but, in general, you can't trisect an arbitrary angle."

"Stand back buddy-wuddy," said Tanya. "I'd like to give it a try."

"Before you begin," Fred said, "let me give you a little history of this problem. Trisecting an angle is one of the 'great unsolved problems of antiquity.' One other of the 'great unsolved problems of antiquity' was to construct a square equal in area to a given circle."

"I wanna try that one also. Squaring the circle seems easy-squeezie."

Undeterred, Fred went on with his history. "For 2000 years people played with trying to trisect angles and square circles using only a straightedge and a compass. Then in the 1800s something happened."

"What?"

"Mathematicians proved that it was *impossible* to trisect an angle or square the circle."

"You mean never-ever?"

"I mean never-ever. Occasionally newspapers will print articles with titles like TRISECTION OF ANGLES FINALLY DONE, showing some fancy 18-step construction, but since it's been proven that this can't be done, these soi-disant* constructions are a bunch of bunk. There have been so many weird attempts at trisecting an angle that in 1994 one guy wrote a book** that catalogs a lot of them. And, of course, you can trisect an angle if you cheat. If you make marks on the straightedge, then you can trisect any angle in three steps,*** but that's prohibited by the rules of the game."

"I'm boredy-wardy. I want to do constructions. No extra stuff."

"Okay."

✶ swa-dee-ZAHN What they call themselves; pretended; supposed

✶✶ Underwood Dudley, *The Trisectors*, published by the Mathematical Association of America

✶✶✶ See the pictures at the bottom of pages 98–99 in the January 2000 issue of SCIENTIFIC AMERICAN.

Eloy

1. Construction #14: Erect a perpendicular to ℓ at a given point P on ℓ.
2. Construction #15: Trisect a segment. (This *is* possible to do.)
3. Construction #16: Divide a given segment into 273 equal parts.
4. Construction #17: Construct a triangle given sides a, b, and the altitude h to side a.
5. Construction #18: Given three noncollinear points, construct the circle which passes through them and locate the center of the circle.

answers

1. Erect a $\perp$. (Const. #14)

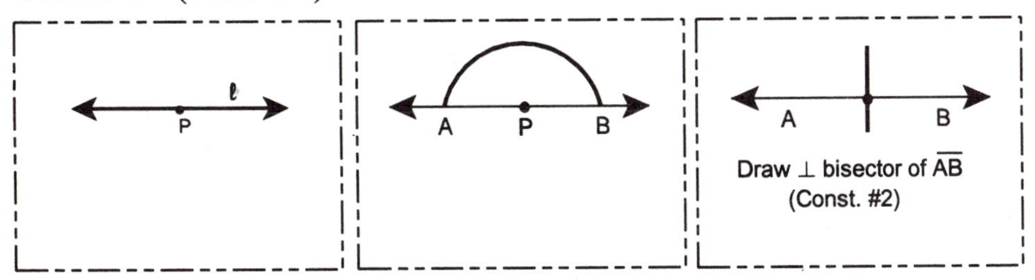

2. Trisect a segment. (Const. #15)

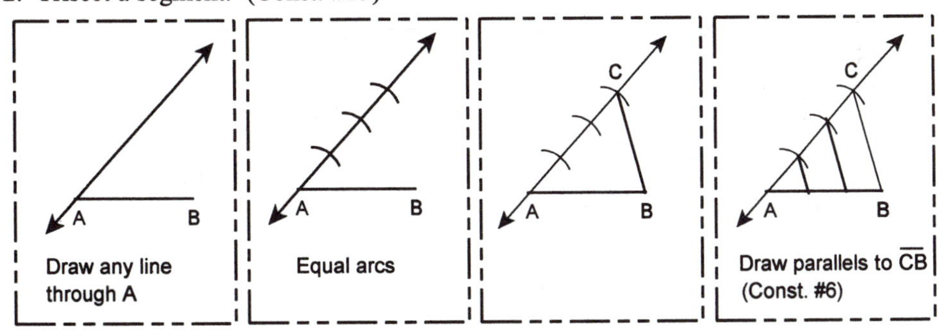

3. Divide a segment into 273 equal parts. (Const. #16)

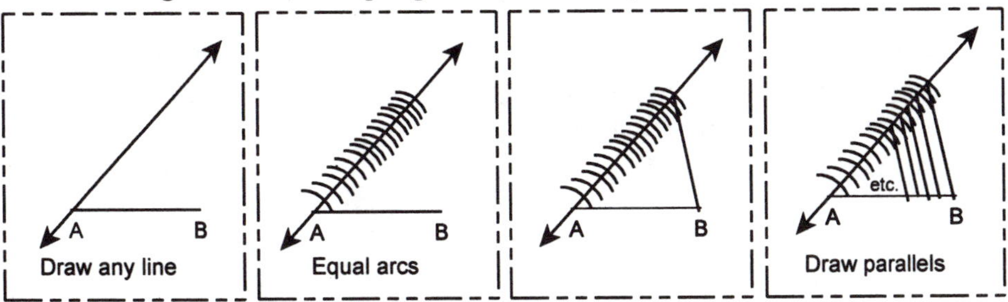

4. Construct a triangle given sides a, b, and the altitude h to side a. (Const. #17)

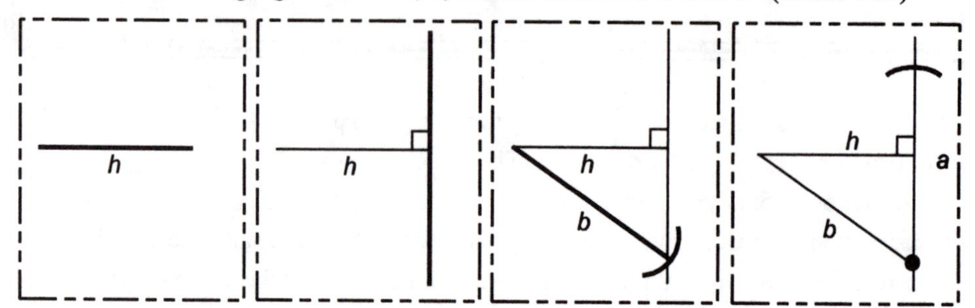

5. Given three noncollinear points, construct the circle which passes through them and locate the center of the circle. (Const. # 18)

Recall that the perpendicular bisector of a chord must pass through the center of the circle.

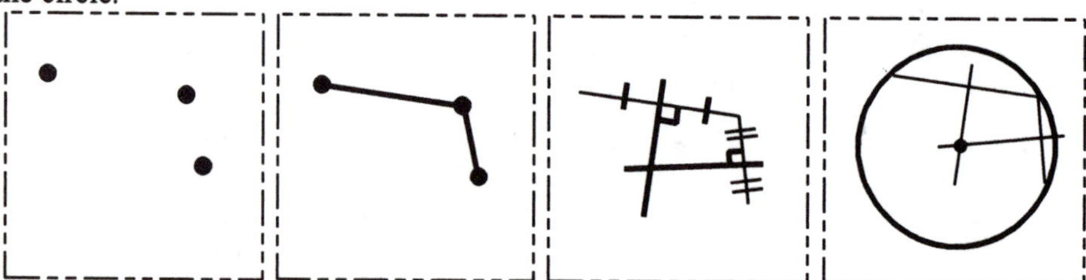

Park Ridge

All of the Park Ridge Constructions will be rusty compass constructions. Your compass is frozen at a fixed radius of r.

1. Construction #19R: Erect a perpendicular to ℓ at a given point P on ℓ.
2. Construction #20R: Bisect $\overline{AB}$ if $r > AB/2$.
3. Construction #21R: Bisect $\overline{AB}$ if $r < AB/2$.
4. Construction #22R: Drop a perpendicular from P to ℓ when $r >$ the distance from P to ℓ.
5. Construction #23R: Drop a perpendicular form P to ℓ when $r <$ the distance from P to ℓ.
6. Construction #24R: Construct a line parallel to ℓ through given point P.
7. Construction #25R: Construct the complement to a given angle. (Two angles are complementary if their measures add to 90°.)
8. Construction #26R: Locate the center of a circle.

answers

1. Erect a perpendicular to ℓ at a given point P on ℓ. (Const. #19R)

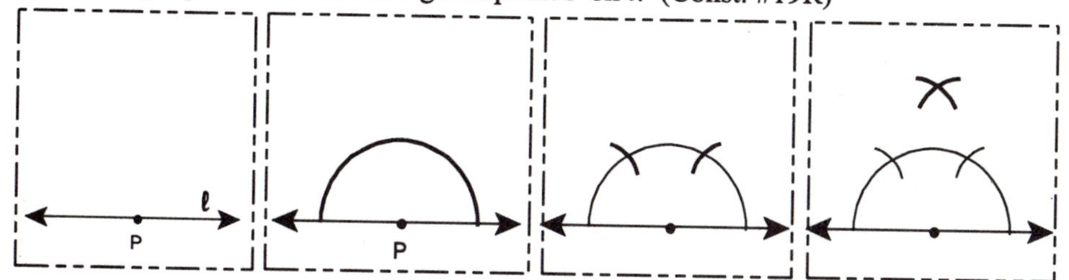

2. Bisect $\overline{AB}$ if $r > AB/2$. (Const. #20R) This is just Construction #2.
3. Bisect $\overline{AB}$ if $r < AB/2$. (Const. #21R)

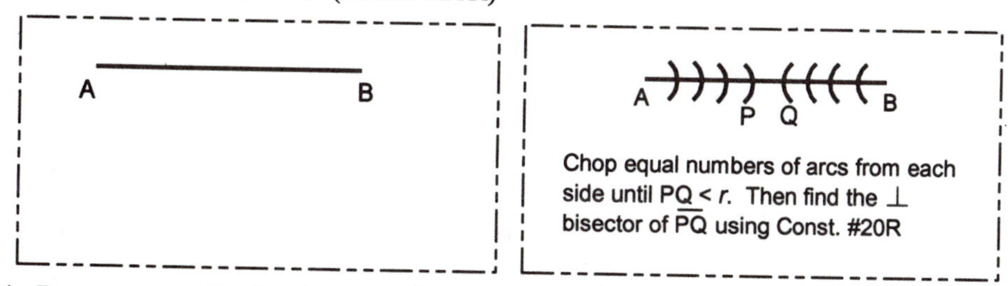

A
B
A
P Q
B
Chop equal numbers of arcs from each side until PQ < r. Then find the ⊥ bisector of PQ using Const. #20R

4. Drop a perpendicular from P to ℓ when $r >$ the distance from P to ℓ. (Const. #22R)

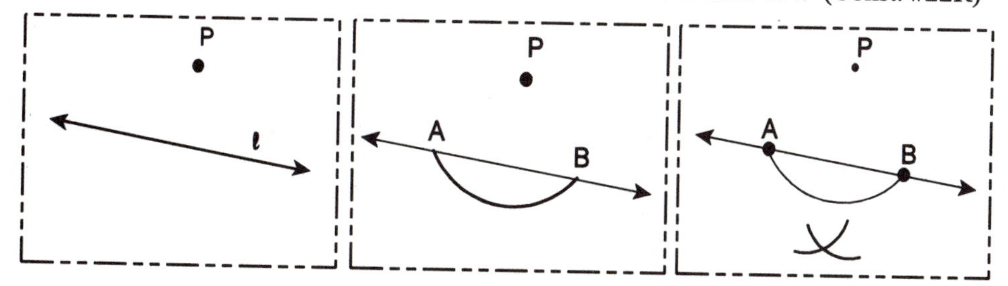

5. Drop a perpendicular form P to ℓ when $r <$ the distance from P to ℓ. (Const. #23R)

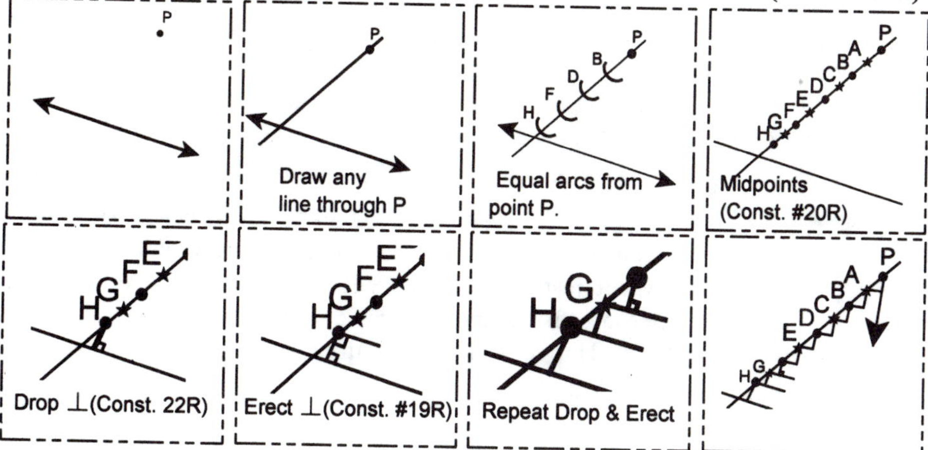

6. Construct a line parallel to ℓ through given point P. (Const. #24R)
Drop a ⊥ from P to ℓ. (Const. #22R or #23R).
Erect a ⊥ at P. (Const. #19R)

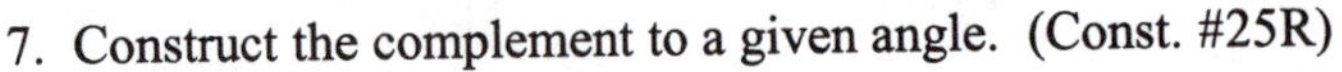

7. Construct the complement to a given angle. (Const. #25R)

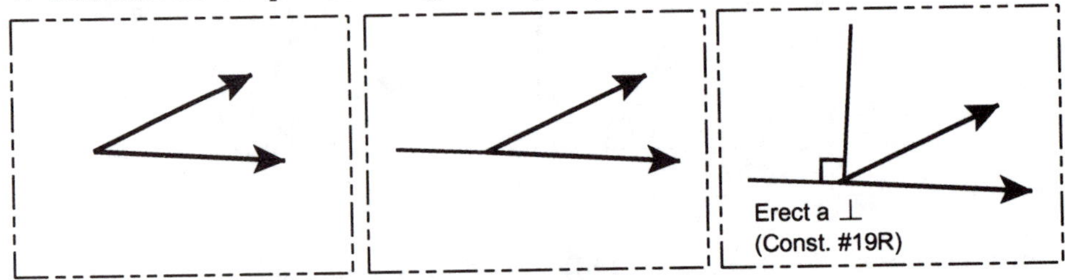

8. Locate the center of a circle. (Const. #26R)

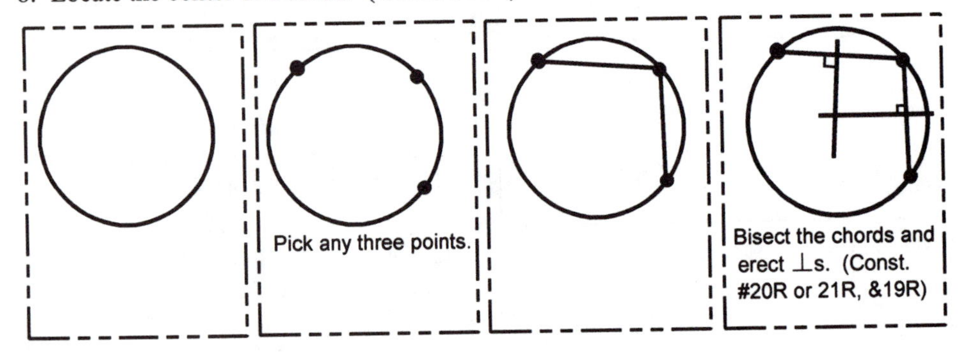

Pick any three points.
Bisect the chords and erect ⊥s. (Const. #20R or 21R, &19R)

Ramsey

1. Construction #27: Inscribe a square in a given circle.

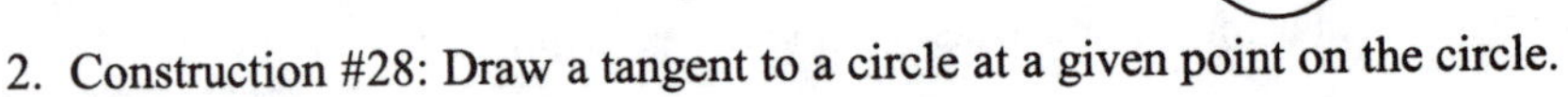

2. Construction #28: Draw a tangent to a circle at a given point on the circle.

3. Construction #29: Draw a tangent to a circle from a point P which lies outside the circle.

4. Construction #30: Given a circle and a point P outside the circle, draw a circle with center at P which is tangent to the given circle.

5. Construction #31: Inscribe a regular hexagon (6 congruent sides) in a given circle.

6. Construction #32: Inscribe a regular octagon (8 congruent sides) in a given circle.

7. Construction #33: Given two circles of different radii, draw a tangent to both circles. (You are not allowed to just put the straightedge against the edges of the two circles. By the rules of the game, the straightedge can only be used to draw a line through two particular points. This construction is fairly tough.)

odd answers

1. Inscribe a square in a given circle. (Const. #27)
Locate the center of the circle (Const. #26R). Draw any diameter. Find the ⊥ bisector of that diameter (Const. #2).

3. Draw a tangent to a circle from a point P which lies outside the circle. (Const. #29)
Locate the center of the circle, point O, (Const. #26R).

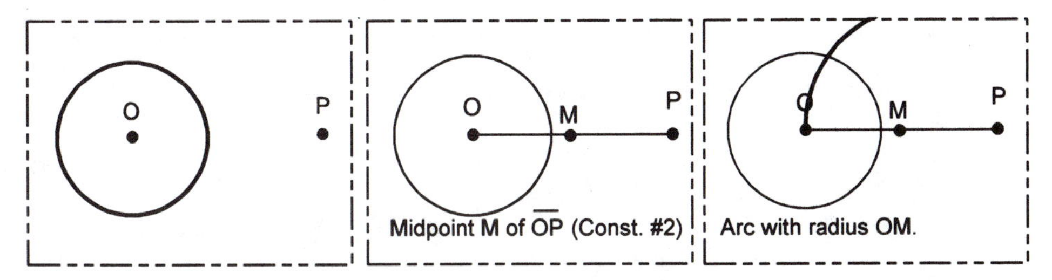

This construction works since the point T at which the arc intersects circle O will be a point on the semicircle through O and P. Hence ∠OTP will be a right angle forcing $\overline{PT}$ to be tangent to the circle.

5. Inscribe a regular hexagon (6 congruent sides) in a given circle. (Const. #31)
Just pick a point on the circle and start drawing with your compass set at the same radius as the circle. You can't miss.

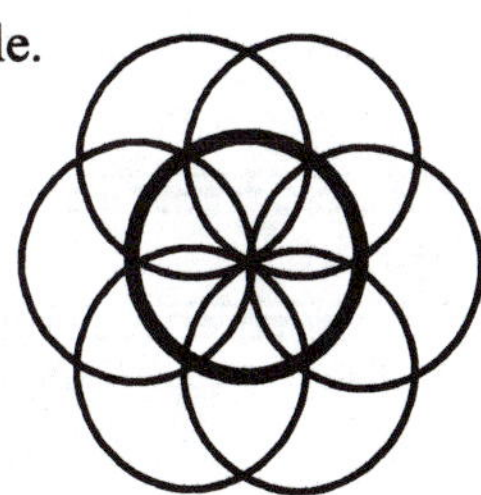

7. Given two circles of different radii, draw a tangent to both circles. (Const. #33)
Sometimes to figure out how to do a construction, we start with the finished product and try to work backward to see how to get there.

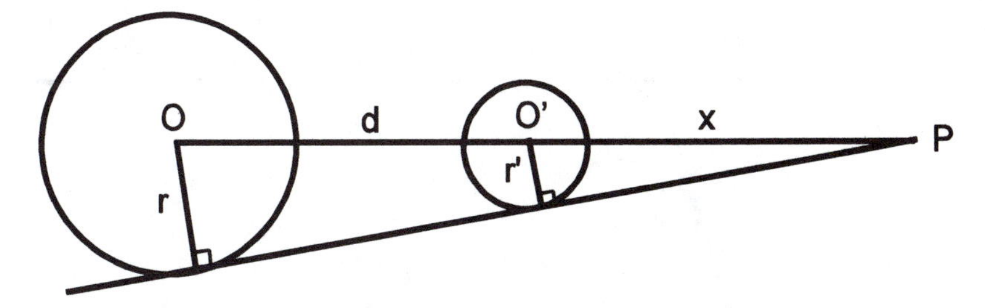

We know the distance d (between the centers of the circles). We know the two radii. What we'd love to find is the distance x (= O'P). Knowing x and given the two circles we could find the location of P by copying (Const. #1). Once we know the location of point P, we can draw the tangent to each circle using Const. #29.

How do we find x, given d, r and r' ?

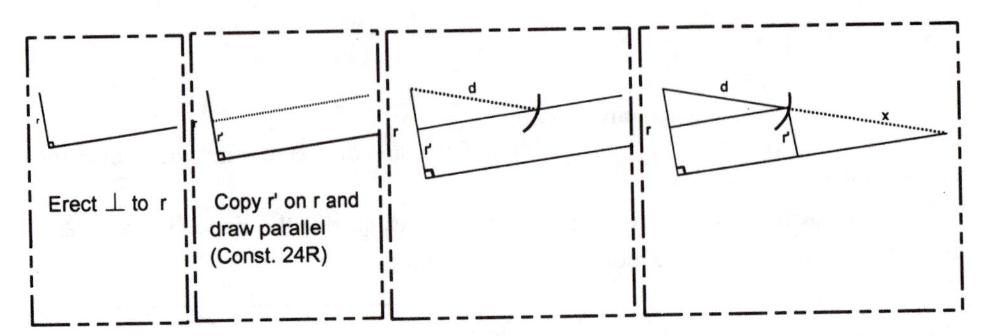
Erect ⊥ to r
Copy r' on r and draw parallel (Const. 24R)
d
x
r
r'

Now having found the length of x, the official procedure then is:

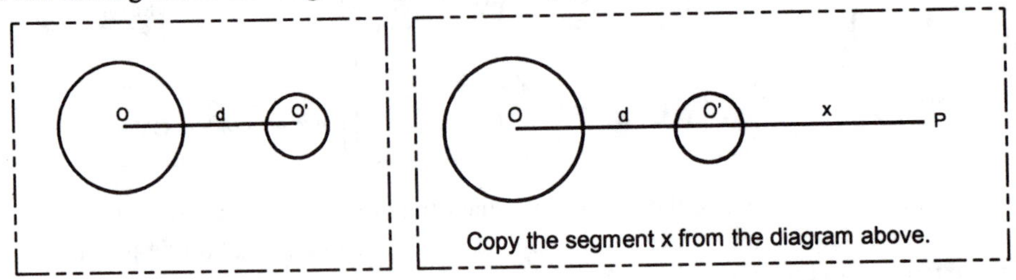
O
d
O'
x
P
Copy the segment x from the diagram above.

And then use Const. #29.

Ingersoll

All of the city of Ingersoll Constructions deal with concurrent lines in a triangle. (Three or more lines are concurrent if they all meet at the same point.)

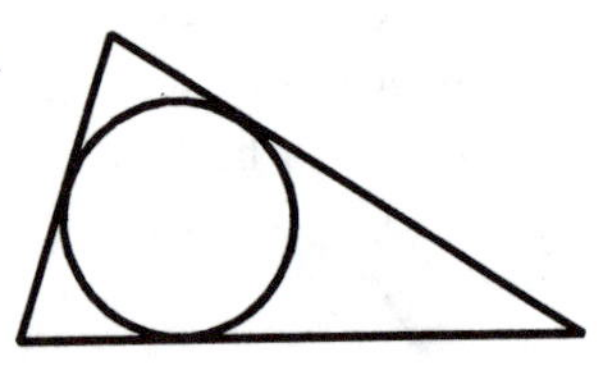

1. Construction #34: Given a triangle, inscribe a circle in it.

   The center of the inscribed circle is called the **incenter** of the triangle.

   Note that since all radii are equal and since the sides of the triangle are tangent to the circle, we have:

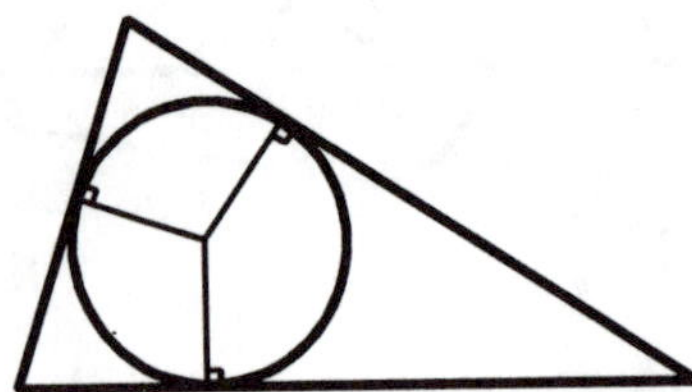

2. Construction #35: Construct an isosceles triangle given the base of the triangle and the radius of the inscribed circle.

3. Construction #36: Given a triangle, circumscribe a circle around it.

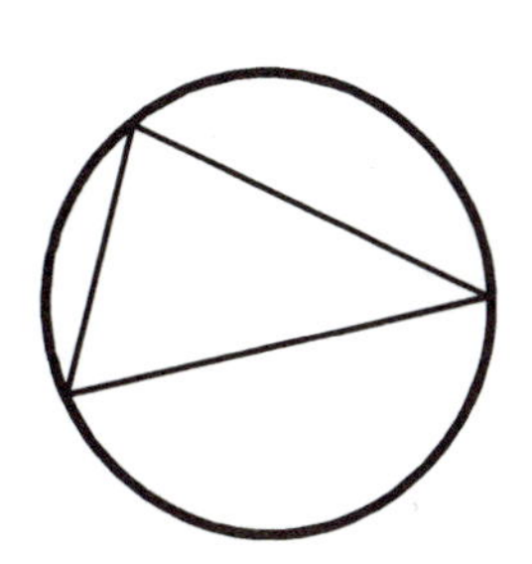

The center of the circumscribed circle is called the **circumcenter** of the triangle. Since all radii are equal, the circumcenter is equidistant from all three vertices of the triangle.

4. Construction #37: Construct an isosceles triangle given the base and the radius of the circumscribed circle.

5. In addition to the angle bisectors of a triangle being concurrent (Const. #34), and the perpendicular bisectors of the sides of a triangle being concurrent (Const. #36), it is also true that the altitudes of any triangle are concurrent.

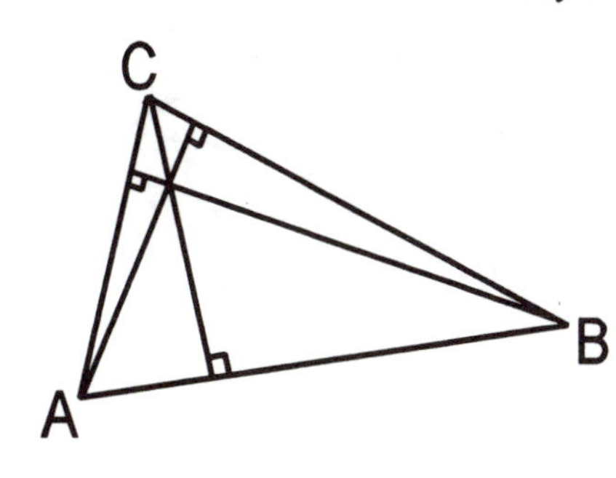

The proof of the concurrence of the altitudes in any triangle is one of those how-did-they-ever-think-of-that type of proofs. Many a geometry teacher has been stuck at the blackboard trying to prove this theorem when the "trick" is forgotten. You may consider yourself a *very* bright geometry student if you can show the altitudes are concurrent without looking at the answer supplied below.

6. In addition to
- ❋ the angle bisectors being concurrent
- ❋ the perpendicular bisectors being concurrent
- ❋ the altitudes being concurrent

. . . it is also true that the medians of any triangle are also concurrent. (Median = segment from vertex of a triangle to the midpoint of the opposite side.) None of these four *had* to happen. When the universe was first put together, things might have been arranged differently and maybe, for example, the angle bisectors of any triangle might have been concurrent only under very special circumstances. But since we live in the world we now inhabit, it is often considered a sign of mental health that we accept reality as it is. (Neat thought, isn't it?)

Reality theorem: The medians of any triangle are concurrent.

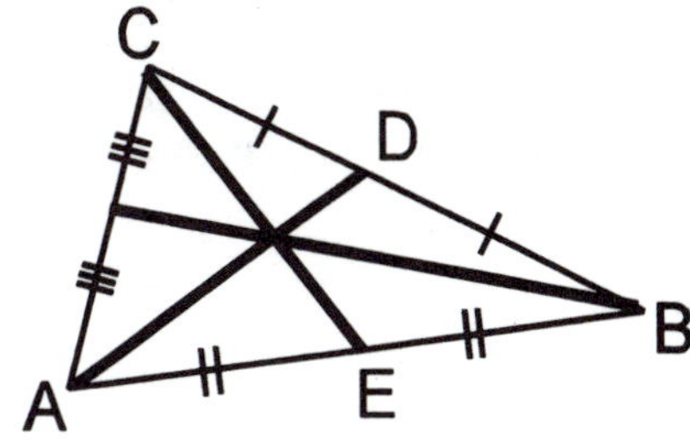

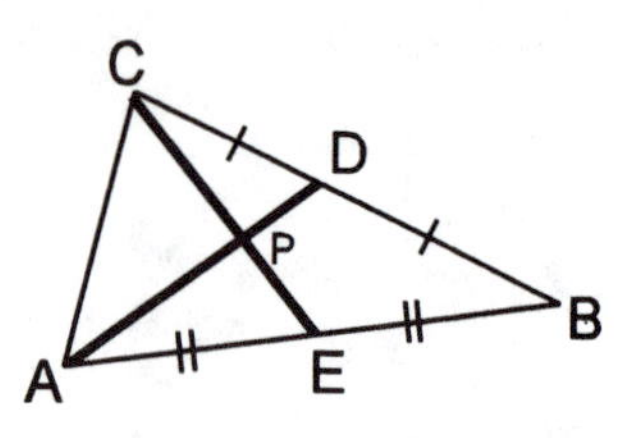

Fill in the missing steps:

Proof: We start with any ΔABC and any two medians. Find the midpoints of $\overline{CP}$ and $\overline{AP}$.

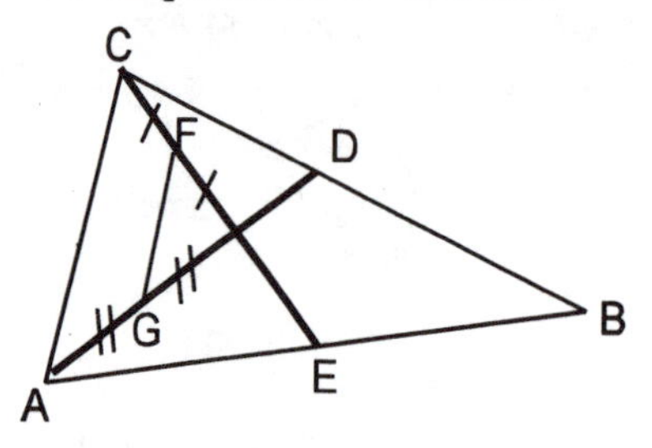

By the Midsegment Theorem, $\overline{FG} \parallel \overline{AC}$ and also FG = ½(AC).
Now mentally erase $\overline{FG}$ for a moment.

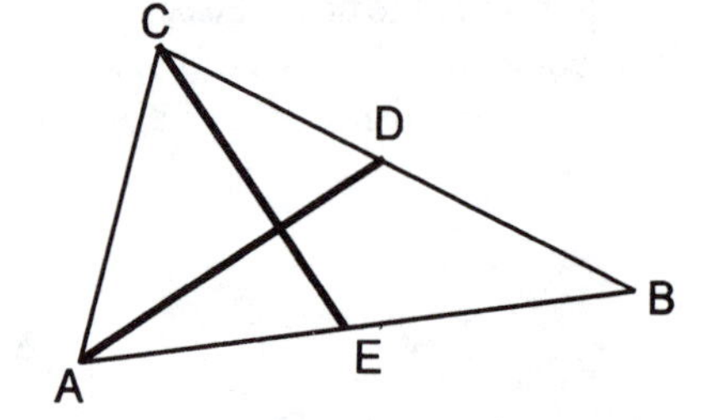

Now it's your turn. Make the argument that $\overline{DE} \parallel \overline{AC}$ and DE = ½(AC).

From the previous four lines, we have:

a) $\overline{FG} \parallel \overline{DE}$ and
b) FG = DE

Sticking $\overline{FG}$ back into the diagram, we look at quadrilateral FGED.

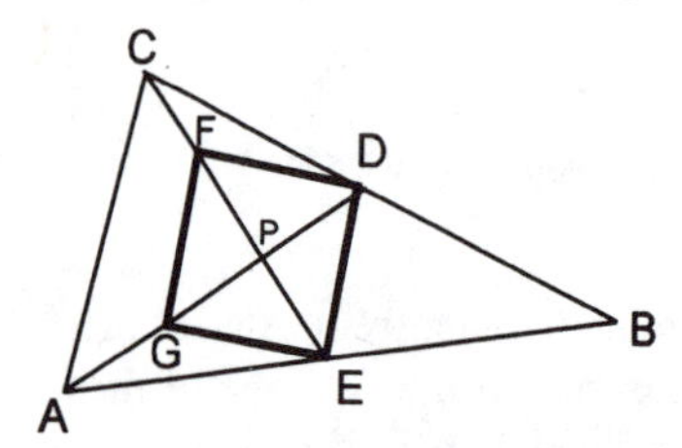

Why is it true that FGED is a parallelogram?

Since the diagonals of a parallelogram bisect each other, we have GP = PD, and since G is the midpoint of $\overline{AP}$, we have AG = GP = PD. The same is true for the other median: CF = FP = PE.

We have shown that any two medians intersect two-thirds of the way to the opposite side.

Then the median from B must intersect $\overline{AD}$ two-thirds of the way from A to D. That point is P. Thus all three medians must meet at P.

⊠

odd answers

1. Given a triangle, inscribe a circle in it. (Const. #34)

Recall from problem 2 on p. 176 that any point on an angle bisector is equidistant from the sides of the angle. That means that all three angle bisectors of the triangle must pass through the incenter. Bisecting an angle is Const. #7R. This establishes that the angle bisectors of the angles of a triangle are concurrent.

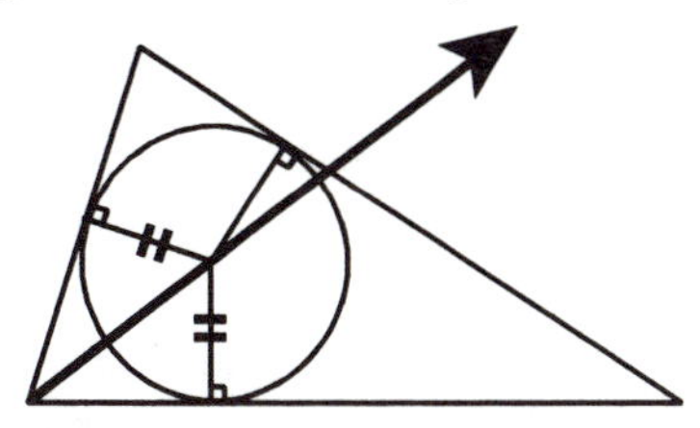

3. Given a triangle, circumscribe a circle around it. (Const. #36)

Recall that any point on the perpendicular bisector of a segment is equidistant from the endpoints of the segment. Hence each of the perpendicular bisectors of the sides of the triangle must pass through the circumcenter. This establishes that the perpendicular bisectors of the sides of a triangle are concurrent.

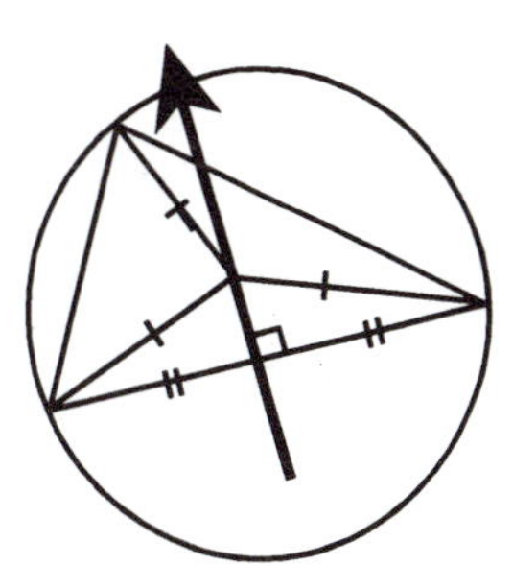

5. You take the original Δ ABC and make a huge triangle around it by drawing parallel lines. $\overline{DE}$ is drawn parallel to $\overline{AB}$, etc.

Since $\overline{DE} \parallel \overline{AB}$, the altitude to $\overline{AB}$ must be $\perp$ to $\overline{DE}$ (Segments $\perp$ to one of two parallel lines are $\perp$ to the other.)

If you look carefully, you can see a pair of parallelograms: ▱ ABCE and ▱ ABDC. They are parallelograms by definition: Their opposite sides are parallel. With the theorem (from Fred's famous monograph: "Why Parallelograms Are the Neatest") that the opposite sides of a parallelogram are congruent we have both $\overline{EC} \cong \overline{AB}$ and $\overline{AB} \cong \overline{CD}$. By the transitive property of congruence, $\overline{EC} \cong \overline{CD}$.

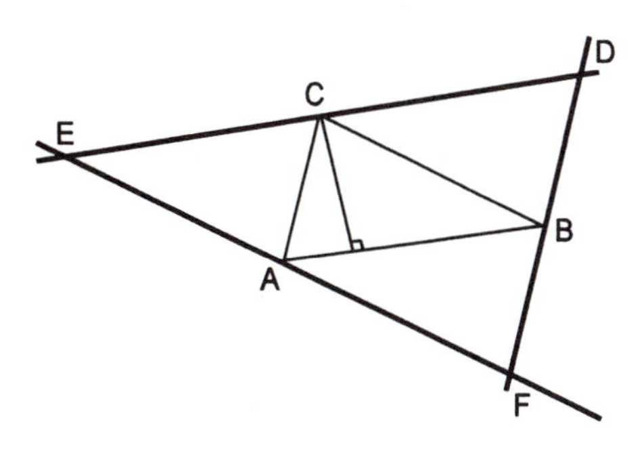

That makes the altitude of the smaller triangle (Δ ABC) into the perpendicular bisector of the larger triangle (Δ DEF).

We already know that the perpendicular bisectors of any triangle are concurrent (Const. #36), and we're done.

Jean

1. Construction #38: Construct a right triangle given the hypotenuse and one leg.

2. Construction #39: Construct a right triangle given the median and the altitude to the hypotenuse.

3. Construction #40: Construct the geometric mean to two given segments. In algebra the **geometric mean** of two numbers a and b is the number x such that $a/x = x/b$. If you are given $\overline{AB}$ and $\overline{CD}$, you are looking for a segment with a length x so that $\frac{AB}{x} = \frac{x}{CD}$ The Intersecting Chords Theorem will come in handy.

4. Construction #41: Construct an isosceles triangle given the altitude to the base and the perimeter. Let $\overline{AD}$ be the altitude from vertex A. Let EF be the perimeter.

5. Perhaps you thought that rusty compass constructions were a bit hard. Tanya did. She got out a can of spray lubricant and blasted the compass until every bit of rust was gone. By the time she was done, the compass was so greased that it could barely hold together.

She had invented the **collapsible compass** constructions. The rules for collapsible compass constructions are: ① You can take any two points and use one of them for the center and the other for the radius; and ② the compass collapses the second that you pick it up off the paper. That means that you can't take a distance AB and make a circle at point O with radius AB.

Collapsible Compass Construction #42CC: (Copying a segment) Given segment $\overline{BC}$, and given point A, construct $\overline{AF}$ such that AF = BC.

In fact, this construction using the collapsible compass rules is so tough that I'll give you the construction and give you most of the proof that the construction works. All you need do is supply the very end of the proof.

Given segment $\overline{BC}$, and given point A, construct $\overline{AF}$ such that AF = BC.

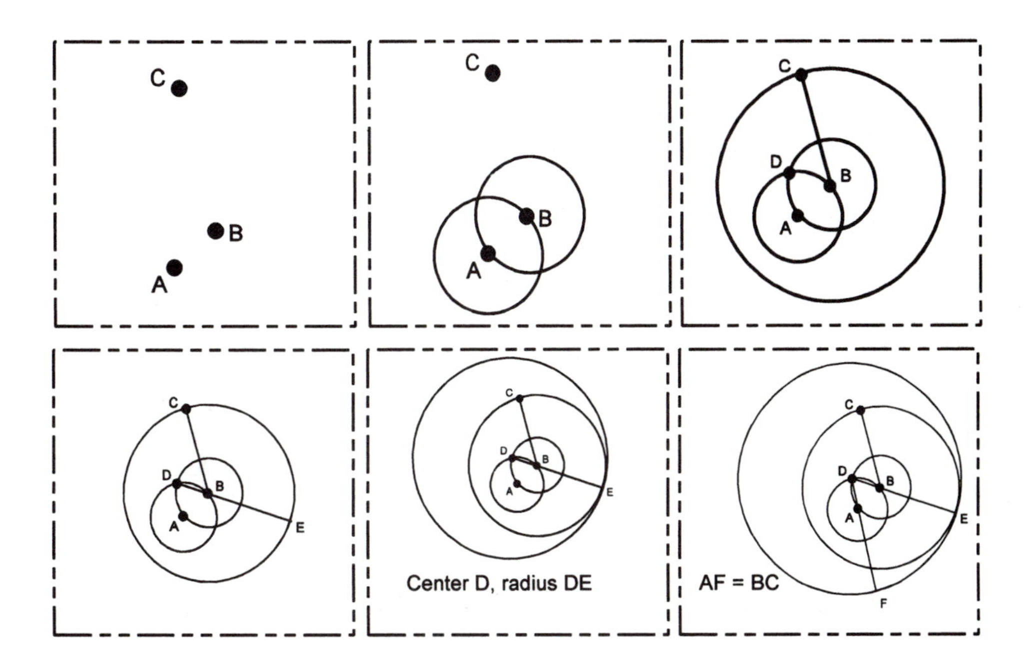

The little circles centered at A and B have the same radius.

DA = AB in circle A.

AB = BD in circle B.

∴ DA = BD ∴ is the symbol for *therefore*.

DF = DE since they are radii of circle D.

Subtracting the above two equations: DF = DE

minus DA = BD

AF = BE

And replace BE by BC and we get................. AF = BC

(which was what was wanted)

Your question: Why can we replace BE by BC?

Lamont

1. Given any angle, divide it into six smaller angles which add up to the original angle, if this is possible. If it is not possible, then give an argument why it is not possible.

2. Given any angle, divide it into four smaller angles which add up to the original angle, if this is possible. If it is not possible, then give an argument why it is not possible.

3. Construction #43: Construct a 15° angle.

4. Construction #44: Construct a 75° angle.

5. Construction #45: Construct a 24-gon (a regular polygon with 24 congruent sides). (Hint: If you had a regular 24-gon inscribed in a circle and drew a pair of radii from adjacent vertices of the 24-gon, what is the measure of the central angle that those radii would make?)

As you may have noticed, we never mentioned the ultimate construction. Just as it is impossible to trisect any arbitrary angle, it is also impossible to construct a regular nonagon (9-gon) using straightedge and compass. What is truly surprising is that it is possible to accomplish:

Construction #46: Construct a regular heptadecagon (17-gon).

In the hundreds of years before Christ, the Greeks couldn't do it. In fact, in the more than 2000 years since the ancient Greeks, no new constructions of any kind had been found. Then on March 30, 1796, some 19-year-old kid figured out how to do construction #46. He went even a little farther. Given any circle, he showed how to inscribe a regular (all sides congruent) 17-gon in that circle.

Many mathematical historians say that this kid became the best mathematician that our world has ever known. (Second and third places usually go to Isaac Newton and Archimedes.)

This son of a gardener, canal tender, and bricklayer has been called "The Prince of Mathematicians." Biographer Eric Temple Bell writes, "In all the history of mathematics there is nothing approaching the precocity of [this fellow] as a child." Before the age of three, for example, he corrected an error in his father's long computation of a weekly payroll. He eventually made major contributions to almost every area in mathematics. His name was Karl Friedrich Gauss.

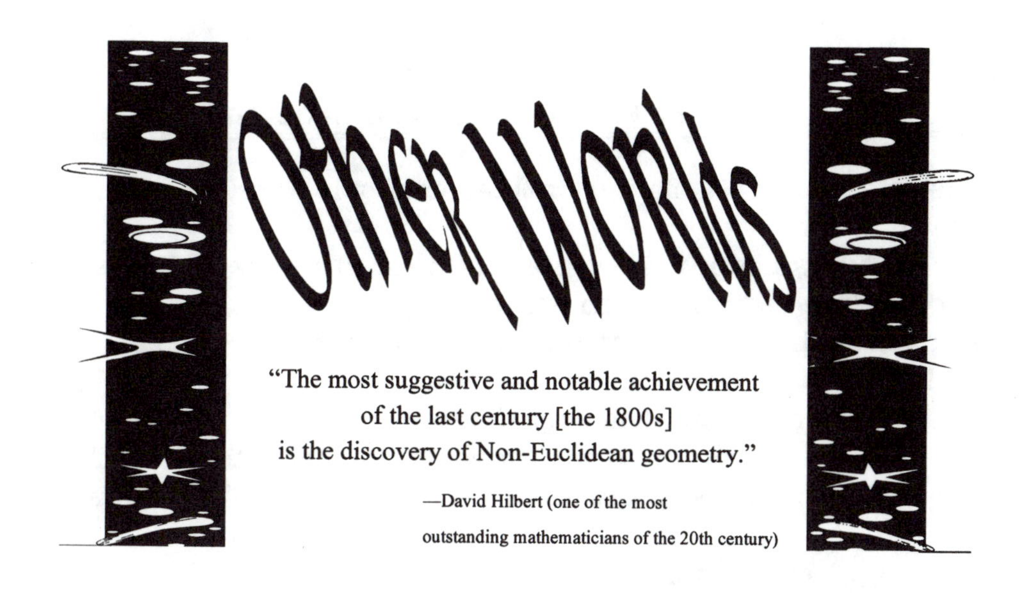

Other Worlds
"The most suggestive and notable achievement of the last century [the 1800s] is the discovery of Non-Euclidean geometry."
—David Hilbert (one of the most outstanding mathematicians of the 20th century)

Chapter 11½
Non-Euclidean Geometry

The parallel postulate says that if you have a line ℓ and a point P not on ℓ, then there is at most one line through P that is parallel to ℓ.

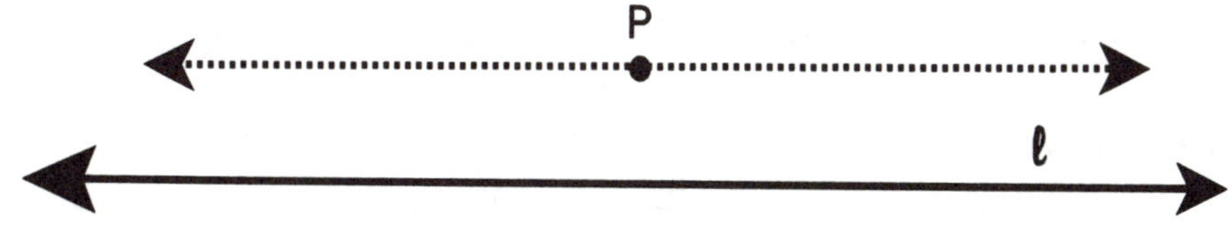

Imagine if you faced this multiple-choice question:

> Given a line ℓ and a point P not on the line, then how many lines pass through P that are parallel to ℓ?
>
> A) You can't draw any lines through P parallel to ℓ
>
> B) You can only draw exactly one line
>
> C) You can have many lines through P that are parallel to ℓ.

Don't you wish that all of life were this easy? When you have something so obviously true, one of the pleasant pastimes of people playing with geometry is to try to prove that it is true.

Ever since Euclid wrote *The Elements* back in 300 B.C. people have been trying to prove that there is at most one line through P that is parallel to ℓ. That's what Joe was trying to do back in chapter four.

But Joe wasn't the first to try.

The first one (that we know of) that tried to prove the parallel postulate was Claudius Ptolemy. He tried in about 150 A.D.

AND HE FAILED.

In the 1200s, the Persian mathematician Nasir-ed-din made a pretty good try.

AND HE FAILED.

John Wallis in the 1600s offered a neat "proof" of the parallel postulate on the basis of the other postulates of Euclid. The only problem with his proof was that he, in effect, assumed all of Euclid's postulates—including the one he was trying to prove.

AND HE FAILED.

In the early 1800s, a kid in his late twenties had the thought, "If for more than 2000 years people haven't been able to prove the parallel postulate from the other postulates, then, maybe, it can't be proven from the other postulates." The technical word is that, perhaps, the parallel postulate is **independent** of the other postulates. That's what K. F. Gauss* thought. He wrote to his friends and told them about his ideas for a new geometry. A geometry in which the correct answer to the multiple-choice question on the previous page is: C) You can have many lines through P that are parallel to ℓ. Gauss was probably the first person on the planet to begin working on a geometry that was non-Euclidean.

---

* No. K. F. doesn't stand for Kentucky Fried.

A long-time buddy of Gauss named Janós Bolyai also starting thinking about it and in 1823 wrote to his father, "Out of nothing I have created a strange new universe." (Actually, he didn't write *exactly* that since he wrote in Hungarian, but that's a fair translation.)

And then at about the same time as Bolyai, the Russian, Nicolai Ivanovitch Lobachevsky, also started drifting off into the weird new geometry.

Throw out the old parallel postulate! Gauss, Bolyai and Lobachevsky replaced it with: If you have a line ℓ and a point P not on ℓ, then there are tons of lines through P that are parallel to ℓ.

This is bizarre.

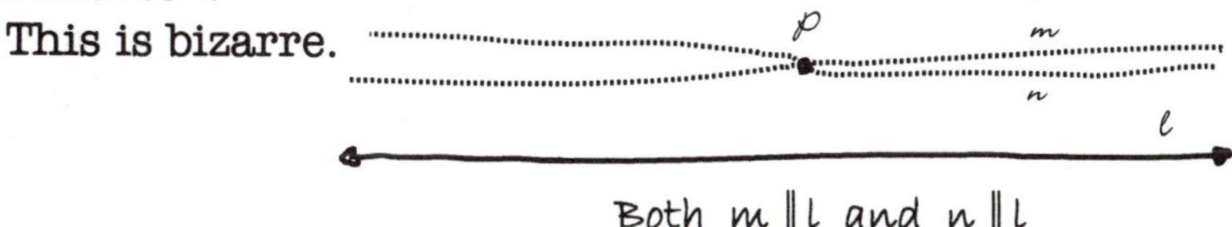

I can just imagine his mother saying to him, "Nicolai Ivanovitch Lobachevsky, what are you doing? Can't you even draw straight? You'll never amount to anything!"

To which he may have replied, "Mom. Could you lay off the 'Nicolai Ivanovitch Lobachevsky'? My friends all just call me 'Nick'. "

"I don't care what your friends call you. You'll always be my sweet little Nicolai Ivanovitch Lobachevsky. Now why don't you stop fooling around with those crazy drawings and go outside and do something useful. The tundra needs mowing."

In 1829–1830, Lobachevsky was the first to publish a paper that systematically laid out the framework for non-Euclidean geometry. It wasn't a best-seller in Russia. And in other countries it was even more of a flop (since it was written in Russian). In 1840 he wrote a book in German, Geometrische Untersuchungen zur Theorie der Parallellinien (Geometrical Researches on the Theory of Parallels). Then in 1855 (and after he became blind) he wrote in French, *Pangéométrie.*

And so, without the aid of psychotropic drugs, these three men—Gauss, Bolyai and Nick—entered a new world.

By replacing Euclid's parallel postulate with a new one ("If you have a line ℓ and a point P not on ℓ, then there are many lines through P that are parallel to ℓ") they built a foundation for a new mathematical theory. They didn't change the undefined terms. The definitions stayed pretty much the same.*

It was the theorems that changed. For example, with the new postulate you could prove that the sum of the angles of any triangle was always *less than* 180°. You could prove that if two triangles were similar, then they also had to be congruent. Triangles with two right angles became possible.

The strange thing is that with all of this goofiness, contradictions never appeared. The new theory of non-Euclidean geometry was **consistent**. Mathematicians from around the world** worked to show that this new geometry was just as consistent as the old Euclidean geometry.

When we look at the postulates that make up a mathematical theory, we think that it is *nice* if the postulates are all independent of each other—you can't prove any one of them from the others. But when it comes to the postulates being consistent, then it is not only just nice, but *necessary*. In the symbolism of chapter 8½, you

---

✶ With this new postulate there were a few new shapes that popped into being such as the biperpendicular quadrilateral and the Saccheri quadrilateral, but for the most part most of the definitions remained undisturbed. An acute angle, for example, was still an angle whose measure is less than 90°.

✶✶ If you are good at acting, try reading the list of contributing mathematicians with the appropriate accents: Eugenio Beltrami, Arthur Cayley, Felix Klein, Henri Poincaré.

don't want statement P to be true and also statement ¬P to be true. If they were both true, then (P & ¬P) would be true. And going a little deeper into symbolic logic for just a moment, if you build a truth table for (P & ¬P) ➜ Q, you will find that it is a tautology (always true).

Translating back into plain old English, (P & ¬P) ➜ Q means that if you start with a contradiction, (P & ¬P), you can prove anything!

Any mathematical theory that has a contradiction in its postulates will be a theory in which every statement in the whole world is a theorem. This kind of theory is what you would call a totally useless theory.

Virtually all of the advances in our civilization have been incremental—step-by-step, each one building a little bit on the previous one. The creative individual sees what has been done so far and then gives it a little twist to create something just a little different. Gauss, Bolyai, and Lobachevsky changed "exactly one line" to "more than one line" in the parallel postulate, and a hundred and seventy years later we're still talking about their great contribution to geometry. Some books call this non-Euclidean geometry Lobachevskian geometry in his honor. (There are so many things in math and physics named after Gauss, that it's only fair that one of the other guys gets something named after him.)

I know what you're thinking: If all those guys can get fame and glory by changing "exactly one line" to "more than one line," then perhaps I can grab a little honor by going in the opposite direction. What about selecting the other alternative in the multiple-choice question on the first page of this chapter? A) You can't draw any lines through P parallel to ℓ. That's a clever idea. It's too bad that Georg Friedrich Bernhard Riemann in the mid-1800s beat you to it. He created Riemannian non-Euclidean geometry.

In this *second* non-Euclidean geometry we find that in every triangle the sum of the angles is greater than 180°. We can prove that every pair of lines intersect—parallel lines don't exist. Riemann also needed to change another postulate a little bit: Through every two points there is *at least* one line. It's just a minor change, but it keeps the postulates consistent.

To put things in perspective, you already knew that there was another geometry. In chapter 7½ we introduced a finite geometry (that only contained three points). Now with Lobachevskian geometry and Riemannian geometry you now know about four different geometries. Of course, there are obviously other finite geometries such as the one that contains five points instead of three. And, actually, there is a whole class of different Riemannian geometries. Riemann's first geometry, which he thought of in 1854, in which there are no parallel lines was just the beginning.

And then there's affine geometry.

And projective geometry.

And non-Archimedean geometry.

And non-Desarguesian geometry.

You may be thinking: So what? This may be all fun-and-games, but let's get back to reality. If I'm going to design a kitchen or a baseball field, and figure out the areas and angles, I'm going to use the stuff that I learned in the first eleven chapters of this book: good ol' Euclidean geometry. That's the way God made the world with 180° triangles and the parallel postulate and that's good enough for me.

You are in good company. Archimedes and Pythagoras would agree with you. Isaac Newton and your uncle who is a carpenter also know that there is exactly one line through a point P that is parallel to a given line. George Washington and Thomas Jefferson both knew that Euclidean geometry describes the world as it really is.

Any of these people would challenge Gauss, Bolyai, Lobachevsky, or Riemann to pull out a sheet of paper and show how there could be anything but exactly one line through P that is parallel to ℓ. Everybody knows that is how reality is built.

Everybody knows . . . Everybody knows . . . Everybody knows . . . Everybody knows . . . Everybody knows . . . Everybody knows . . . Everybody knows . . . Everybody knows . . . Everybody knows . . . Everybody knows . . . Everybody knows . . . Everybody knows . . . Everybody knows . . . Everybody knows . . . Everybody knows . . . Everybody knows . . . Everybody knows . . . Everybody knows . . . and it isn't true.

"We" used to think that the earth was flat, and that if you sailed west from Europe, you would eventually fall off the edge. And if you ask your aunt, she probably believes that the parallel postulate really describes the universe we live in.

Things changed about 1915. A guy in the Swiss patent office made some minor adjustments to Isaac Newton's view of the universe. He created what he called the General Theory of Relativity. And when he, Einstein, looked at the whole big expanse of the universe, he found that he had to cross out alternative B in the multiple-choice test. ~~You can draw exactly one line through P parallel to ℓ.~~ On sheets of paper, in laying out kitchens and baseball fields, you can't tell the difference between the Euclidean and the non-Euclidean approaches. However, when you start looking at the big picture in which a zillion miles is like an inch, then it turns out that reality is a little different from what you had expected.

A good education is one that molds your brain to reality, not vice versa.

Chapter Twelve
Solid Geometry

Just as Tanya was finishing up the last construction, the pirates came by and took her straightedge and compass and the box they came in. As they headed back to toss their booty on the pile of stolen games, one of the pirates suggested that they make up a new pirate song. He had gotten tired of singing all the old chanties.

"What are you going to call this new song?" Captain Hook asked him.

"In the light of our zero tolerance for other people owning anything," the composer-sailor answered, "I think that my new song will be called **No Thing Un-thieved**."

"Forget it!" said the captain. "I can hardly thay that—I mean *say* that."

The pirates put up a sign on the volleyball net. It read:

Fenced Goods, Inc.

Slightly Used Board Games	Hot Chocolate
Your Choice	Comes with cup
$5	$2.95

Fred looked at Tanya and wondered if she was troubled by the theft of the construction game that had kept her pleasantly occupied for hours.

"Hey, it's okay Fred. I was done with the sun."

Done with the sun! Fred thought to himself. *Tanya's right. It's almost sunset.* He knew from *The Kitten Caboodle*'s sunrise/sunset charts that it was nearly 5:30. *I've got to see P at 5:45!*

He rushed to the RWA guard near the gate and explained the situation to her. The guard told Fred that she would have to talk with her senior officer to see if they could make an exception for this emergency and let Fred out of the Official Rescue Site before they took the group photos. Those photos were important, she explained to Fred, since they would be used in their fund-raising brochure to show the wonderful work that RWA did.

As Fred waited by the electrified fence for the guard to come back, he thought of all kinds of reasons why they should let him go:

♥ They were keeping him from his true love. Surely they couldn't do that if they were really a caring organization.

♥ He was really short and he wouldn't show up in any group photo that they took.

♥ If they made him stay and have his picture taken, he would promise that he would stick out his tongue every time they pointed the camera at him.

The guard came back. "Sorry buster. No dice. If we let one of you guys go, then we'd have to let you all go. And then, just tell me, who would we have to take care of? Would you care for another cup of hot chocolate?"

Ever polite, Fred said, "No thank you," and headed back to Tanya.

No matter which direction Fred headed, there was an electric fence keeping him from seeing P. With 14 minutes left before his date with P, he turned to Tanya and said, "I feel like a caged rat."

She smiled. "You're not a caged rat. You're a fenced rat. You have been doing geometry in two dimensions for so long, that you've forgotten that we don't just live on a flat plane."

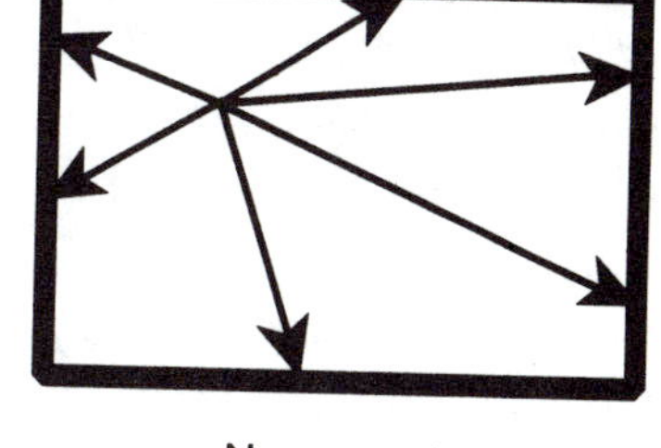

No way out

Fred looked up at the clear blue sky, and said "You mean. . . ."

"Of course. There's no top to stop."

She picked him up and tossed him over the fence. When he hit the ground, his feet were a blur. All his years of jogging were about to pay off.

5:32 p.m. To C.C. Coalback rentals to report that his boat was "lost at sea." Pays the $18.27 for the Naval Chow box that he had opened.

5:35 p.m. Crosses the KITTENS campus to the Math Building.

5:36 p.m. Up to the third story and down the hall to his office.

5:37 p.m. Realizes that the pirates had taken his keys along with everything else.

5:38 p.m. Down the hallway to the department secretary's office. Sign on the door reads, "Out to lunch."

5:39 p.m. Down to the basement and knocks on the door of *Samuel P. Wistrom, Chief Educational Facility Math Department Building KITTENS University, Inspector / Planer / Remediator for offices 225–324.*

5:40 p.m. His butler answers the door and asks, "Who may I announce is calling? Mr. Wistrom is dining right now and requests that visitors wait in his parlor for a postprandial audience."

5:41 p.m. Fred no longer wondered how Sam had done on his recent gambling trip to Reno. Fred: "I'm sorry. I will be unable to accede to Mr. Wistrom's kind offer of hospitality." He was trying to talk like the butler.

5:42 p.m. Goes outside and scales the face of the brick building and climbs into his office through the window.

5:43 p.m. Off with his old clothes which smelled of lake water. Into the tub. Turns on the water. Brushes his teeth while the tub fills. Flosses. Combs his hair. The tub fills. Soaps up. Shampoos his hair. Out of the tub and dries off. Combs his hair again. Wipes out the tub. Puts new laces in his black leather shoes. Dresses to the nines.* Dashes out the door.

5:44 p.m. Stops by the KITTENS Campus Florist and purchases a boutonniere for himself, a corsage for her, a flower for her hair, a bouquet for the Commonsizerinski house, and a miniature rose bush for her room.

5:45 p.m. Takes a deep breath and rings the doorbell at the square house.

Mr. T Commonsizerinski opened the door and looked at Fred. He turned and yelled up the stairs, "Hey P. Your little friend is here to play with you." He turned to Fred and said, "Looks like you're selling flowers for some fund-raiser. How much for the bouquet?"

Fred handed him the bouquet and said, "It's for you and Mrs. L. Take it. It's a gift."

"Aw. Naw, kid. This big a bunch of flowers must have cost your organization a bundle. Here's ten bucks for the cause."

Fred pocketed the money without protest. It was too much to explain. He was just glad that T liked the flowers. It was the KITTENS Campus Florist's top-of-the-line Big Bomb Bouquet® (the BBB, as they call it), which contained every flower in the shop: Ageratum, Alyssum, Asters, Bachelor Button, Begonia, Browallia, Buttercup, Calendula, Celosia, Chamomile, Chrysanthemum, Cleome, Coleus, Cosmos, Dahlia, Daisy, Dianthus, Dusty Miller, Gazania, Geranium, Gladiolus, Gypsophila, Impatiens, Lavender, Lily, Lobelia, Marigolds, Nasturtium, Nicotiana, Nierembergia, Orchid, Ornamental Pepper, Pansies, Petunias, Phlox, Portulaca, Rhododendron, Rose, Rudbeckia, Salvia, Snapdragon, Statice, Strawflower, Sunflower, Verbena, Vinca, Viola, Violet, and Zinnia, all for \$324. Fred thought that was a reasonable price for 40 pounds of flowers. He also liked the fact that 324 was 18 squared.

P came bounding down the stairs. "Something smells. Do you have a new aftershave Pop?"

✶ *Dressed to the nines* is an idiom that dates back to before 900 A.D. It means to put on your good clothes. That used to include, if you were a man, a top hat and tux. Nowadays things are a bit more relaxed and you are dressed to the nines if you're wearing shoes that can be shined.

Fred instinctively put his hand to his cheek. Someday I'll have some whiskers, Fred thought to himself, and then I can wack them off and splash on aftershave.

"Hey, Dad. I thought you said that Freddie was here. And who sent this forest of flowers?"

A timid "me" sounded from behind the foliage.

"Before you and my daughter take off," T said, "I want to talk with you a moment, son. I've got some things I want to say." He took Fred by the arm and lead him off to a special room at the back of the house.

It was not the usual kind of room that you might find in a typical three-bedroom, two-bath Kansas bungalow. It had a metal door which shut behind them with a sound like that of a truck door being slammed shut. The room was empty except for a large wooden table and two heavy wooden chairs. T sat in one and motioned for Fred to sit in the other.

Fred's eyes were level with the table top, and T asked him whether he would prefer a highchair.

"No sir. But I would prefer a higher chair."

T left to fetch three phone books to bring Fred up to a reasonable height. While he was gone, Fred gazed out over the tabletop. So this is what it'd be like if I lived on a plane. There was a sheet of paper on the tabletop, but when Fred looked at it from the side, it looked like a line segment. You know, there's no way to tell if that sheet of paper is rectangular or square or even circular. If I lived on a plane, things would be so constricted for me. They really could imprison me in a rectangle. And the only flowers I would ever know would be pressed flowers. And if I met P in this flat land, she would look like everything else, because everything would look like a line segment if you viewed it edgewise.*

✶ Actually, there are ways of telling different shapes apart from each other. A school teacher in London named Edwin Abbott Abbott wrote a book entitled *Flatland* in 1884 that describes how you might see the difference between various shapes if there were a fog over the land. Then if a very pointy isosceles triangle were coming toward you, it would still look like a line segment except that the middle of the segment would be much more distinct than the ends of the segment. If you looked at a circle, its shading would be slightly different; the ends would fade off more slowly.

In case you're wondering, Edwin Abbott Abbott's father's name was Edwin Abbott. On the title page of *Flatland* is, "With Illustrations by the Author, A SQUARE." I wonder if in elementary school he was nicknamed Edwin Abbott Squared, which then was shortened to A SQUARE.

Fred's ruminations were cut short when T returned, and the metal door slammed shut again.

With his chin now above the tabletop, Fred looked around the windowless room. There was nothing to see. The beige walls were featureless. The whole room demanded, "Concentrate on what's happening right here at the table!" It was T's interrogation room.

T sat down across from Fred and handed him a pencil.

A test?

"Okay son. I'm a father that cares about his daughter. I need to know some things before you head out the door with her. First, write out an autobiography."

Fred began, **Once upon a time, a long time ago, on the western slopes of the Siberian mountains there lived....***

"Your handwriting looks just like my daughter's. And she tells me that you were both born six years ago last Friday. And that you are both really smart in math. You can skip the rest of the autobiography. I'm sure it's just like every other six-year-old's with the usual stuff of loving parents, potty training, favorite toys, and thoughts of heading off to kindergarten. Say. Have you been to kindergarten yet?"

"Well, it's kind of a long story," Fred answered with his hands becoming a little damp. "I tried it for a while, but it didn't seem to work out."

"That's okay, kid. Sometimes they stick you in school before you're ready. Let's change the subject. Turn the paper over and write out exactly when and what you'll be doing with my daughter tonight."

We're heading to PieOne Pizza. We'll meet Alexander and Betty there at 6 p.m. We will eat pizza and talk about fun stuff. We will drink Sluice® and will have salad and the pizza of your daughter's choice. Then we will come back to your house...."

"That's fine son. You don't have to write anymore. I met Alexander and Betty and they seem like responsible adults who will properly chaperon you kids."

What Fred didn't get a chance to write was ...**come back to your house and then you and I could discuss my plans to marry your daughter.**

---

* These are the first words of *Life of Fred: Calculus* which tells of the earliest years of our hero.

T indicated that Fred could leave. Fred hopped off the chair and headed back into the hallway where L and P were waiting. Fred wanted to tell them about what he thought it would be like to live on a plane instead of in three dimensions. Before he could get his words out, L had something to say to him.

"Now you both be real careful crossing the street. And here's some money, P, so you can help pay for the dinner. And be home by nine."

Fred's head spun with thoughts as L offered her motherly injunctions. He had learned to be really careful in crossing streets since he had been going jogging at dawn for years and had seen how little regard some drivers have for small pedestrians. Having P help pay for the dinner made him smile on the inside. Just this morning he had received $70,000 in reward money. And this was in addition to the years of saving most of his KITTENS salary.

Being home by nine reminded him of the Cinderella story, except that he and P would have three hours less. There were so many things that Fred wanted to talk about with her such as how many children they would like to have after they got married and why he was so happy that they were not confined to living on a two-dimensional surface.

For eleven chapters in this geometry book, we have remained in the plane. We are now liberated into three dimensions. In the old days in plane geometry, two lines were either parallel or they met. Now there is a third alternative. In solid geometry, we have three possibilities. If the two lines are coplanar (in the same plane), then they either intersect or are parallel. If the two lines are not coplanar, then they must be skew. In chapter three we drew the picture of two rocket ships in space to illustrate skew lines.

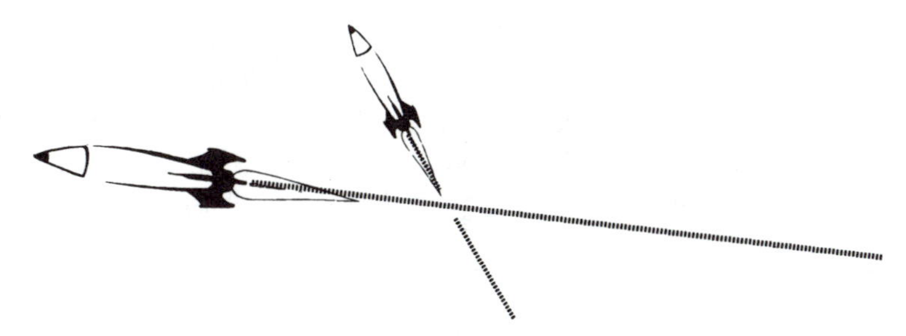

Instead of just working with points and lines, we will now use all three undefined terms: points, lines, and planes. As Fred and P walked toward PieOne, Fred shared his excitement about being in three dimensions. They talked about all the things that can happen in this new world with points, lines, and planes. Here are some of the things they challenged each other with. As usual, please don't look at the answers until you're tried to answer the question on your own. It's a lot more fun that way—and you'll learn a lot more besides.

Your Turn to Play

1. We need a postulate that corresponds to, "Two points determine a line," or more formally, "If you have two distinct points, then there is exactly one line that contains them."

Would this be a good postulate: "If you have three distinct points, then there is exactly one plane that contains them"?

2. Invent a definition of skew lines.

3. Can two skew lines lie in the same plane?

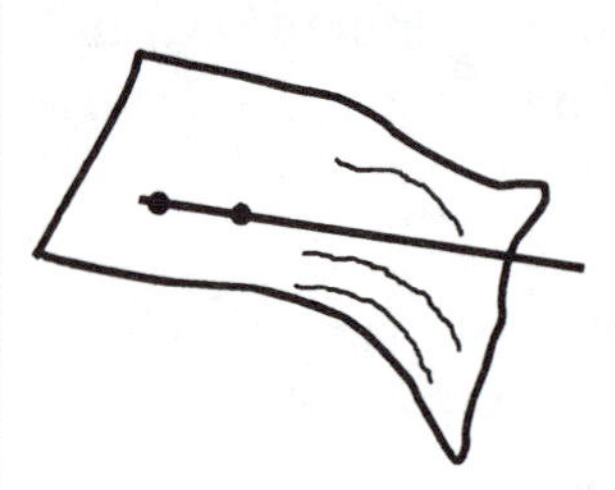

4. "We need to do some ironing," P announced to Fred. "Now that we are no longer limited to a single plane, we have to make sure that our lines stay in the plane they start out in." She took a piece of chalk out of her pocket and drew on the sidewalk. "We wouldn't want a line to start out in a plane and then have the line and plane part company. We need to have the planes be ironed flat like a nicely ironed handkerchief."

Fred had trouble with the image of a plane being like an ironed handkerchief. When he ironed his handkerchiefs, he always ironed them after folding them three times so that they could fit in his pocket. Apparently, P ironed them flat and put them in her dresser like sheets of paper. Fred had been on his own for a little more than five years and had learned the iron, fold-and-iron-again, fold-and-iron-again technique necessary for proper handkerchief ironing. P was just learning to iron.

Can you guess how we will word the Flat Plane Postulate? Of course, we can't just say, "All planes are flat," since we haven't defined what the word *flat* means.

5. Prove: <u>Theorem</u>: If two lines intersect, then they lie in the same plane. This is the only theorem in this chapter that will be proven. The emphasis instead will be on learning to visualize things in three dimensions and learning the basic volume formulas. To begin your proof, let $\overleftrightarrow{PQ}$ and $\overleftrightarrow{PR}$ be given. Since both lines share the point P, they intersect.

COMPLETE SOLUTIONS

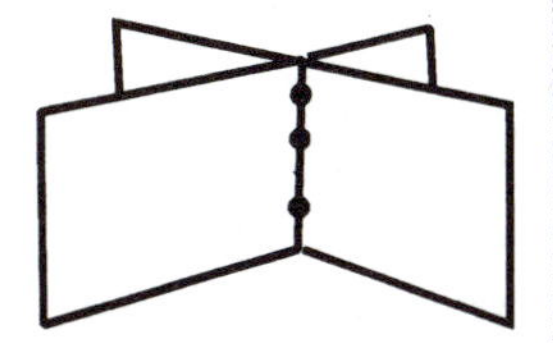

1. There would only be one little drawback to having this as a postulate: It isn't true. Sometimes three points do not determine a unique plane. If the three points are collinear (they lie on a single line), then there are many planes which can contain them.

Here is our official new postulate:

Postulate 15: Three noncollinear points determine a plane.
Or: Given any three noncollinear points, there is exactly one plane which contains them.

2. Your definition may differ from mine. Definition: Two lines are **skew** iff they are neither parallel nor intersecting.

3. The minute that they were in the same plane, they would either have to intersect or be parallel. In which case, they couldn't be skew.

4. Postulate 16: (The Flat Plane Postulate) If two points lie in a plane, then the line which contains them also lies in the plane.

5.

Statement	*Reason*
1. Two lines: $\overleftrightarrow{PQ}$ and $\overleftrightarrow{PR}$	1. Given
2. There is exactly one plane that contains P, Q and R. Call that plane, plane $\mathscr{A}$.	2. Postulate 15
3. All the points of $\overleftrightarrow{PQ}$ are contained in $\mathscr{A}$ All the points of $\overleftrightarrow{PR}$ are contained in $\mathscr{A}$	3. Postulate 16 (The Flat Plane Postulate)

(This proof gave us a chance to use both of our new postulates.)

P remarked to Fred that three-dimensional things were a wee bit different than the plane geometry she had studied. A line perpendicular to one of two parallel lines need not even hit the other parallel line, much less be perpendicular to it. Fred and P chatted about such things as they approached PieOne.

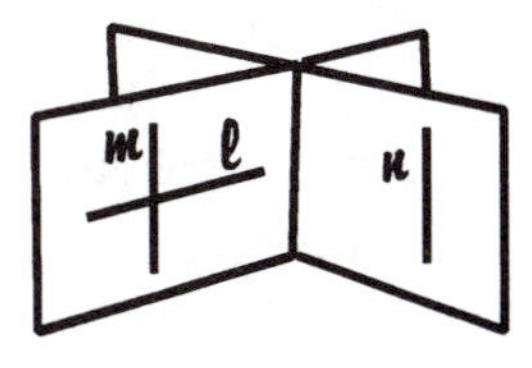

$m \parallel n$ and $\ell \perp n$ but ℓ doesn't intersect n

Now that Fred had gotten to know P a little better (it's always good to know what your potential wife thinks about parallel and perpendicular lines), Fred opened up a new topic.

"You and I both had a birthday last Friday. What did you do on your birthday?" Fred asked.

"Nothing much 'cause my mom had a cold last Friday. My folks were going to take me out to PieOne yesterday to celebrate my birthday, but they couldn't get in since some bigwig at the University was having his birthday party there. His party took up the whole restaurant. He must be some really popular guy."

Fred blushed. "And so, what did you do instead?"

"My folks took me to the pet store and said that I could pick out any pet I wanted."

Fred looked at her. "It must be really wonderful to have such loving parents." He thought of his own childhood. "And so, what did you pick out?"

"I told them I needed to think about my choice. I wondered whether I wanted a little animal like a hamster or some really huge one like a horse or something. Then today, you know what I saw in the dumpster behind the math building?"

Fred shook his head.

"There was this big ol' llama. Like dead. That someone had tossed in the garbage. That was so sad. I told them that I decided that I didn't want a pet right now. My Dad bought me a dolly instead, which is much better. I've named her Queenie."

They arrived at the door of PieOne and were greeted by the Stanthony, the owner. "Hi, Fred. Betty and Alexander are already here. I'm glad you like my pizza. How did you enjoy your birthday party here last night?"

P cocked her head and pointed an index finger at Fred.

Fred nodded. They were both embarrassed—she, for having inadvertently called him a bigwig, and he, for having been the one that kept her from celebrating her birthday at PieOne.

As their eyes adjusted to the dark inside the pizzeria, Fred took P's hand. The gold-plated grape leaves that Stanthony had hung from the ceiling in order to give PieOne an authentic Italian look brushed against their faces. The white glue that he had used to attach the leaves to the ceiling wasn't holding very well and some of the grape leaves were on the floor.

They spotted Betty and Alexander at a corner table. The couple were holding hands across the table and looking into each other's eyes and didn't see Fred and P coming. Fred hoped that some day he could do that with P (when his arms were long enough to reach halfway across a table).

Your Turn to Play

1. On the table where Alexander and Betty were sitting was a large sheet cake upon which were the words, "**Happy Belated Sixth Birthday to P and Fred**" along with a single candle. (If you think back to your kindergarten days, virtually all of your classmates were born within the same 365-day period and if there were at least 23 kids in your class, the probability that two were born on the same day is greater than 50%.* So it isn't so surprising that Fred and P are exactly the same age.) Looking at that candle on the sheet cake, how might you complete the following definition?

Definition: A line is perpendicular to a plane iff. . . .

2. Complete: Definition: The distance from a point to a plane is. . . .

3. Complete: Proposition: Two lines perpendicular to the same plane are. . . .

✶ The computation is often done in probability classes since the result is so surprising to many people: *In a group of twenty-three people chosen at random, the probability is greater than 50% that two of them share the same birthday.* The trick is to compute the probability that no two of the birthdays are alike. If we can show that that is less than 50%, we're done.

Here's how that's done. First, line up the 23 people in a straight line. The probability that the second person in line has a different birthday than the first person is 364/365.

The probability that the third person in line has a different birthday than either the first or second persons is 363/365. (He has 363 possible choices to avoid having the same birthday as either the first or second person.)

The chances that both the second person in line doesn't share a birthday with the first person, and that the third person doesn't share a birthday with the first two people is (364/365) × (363/365). (It's a theorem in probability that the chances of two different and independent events happening is the product of the chances of each event happening. If there's a fifth of a chance my car will break down today and there's a half a chance that the stock market will go down today, then there's a tenth of a chance that both things will happen.)

The probability that the fourth person in line doesn't share a birthday with any of the previous three people is 362/365. The probability that the second person doesn't share a birthday with the first AND the third doesn't share with the first and second AND the fourth doesn't share with the first three is (364/365) × (363/365) × (362/365).

Finally, that none of the 23 share a birthday is (364/365) × (363/365) × (362/365) × . . . × (343/365) which is approximately 0.4927 which is less than 50%.

4. Complete: <u>Definition</u>: Two planes are parallel iff. . . .

5. Is this true? <u>Definition</u>: Two planes are perpendicular iff one plane contains a line that is perpendicular to the other plane.

6. In one of the special banquet rooms at PieOne, Stanthony wanted to run a string of lights from one upper corner of the room down to the opposite diagonal corner of the room. People told him that this was silly, because it would make the room very difficult to use. He said that he had never thought of that. The room was 14 feet long, 10 feet wide and 8 feet tall. How long would that string of lights be? (This is not a one-step problem.)

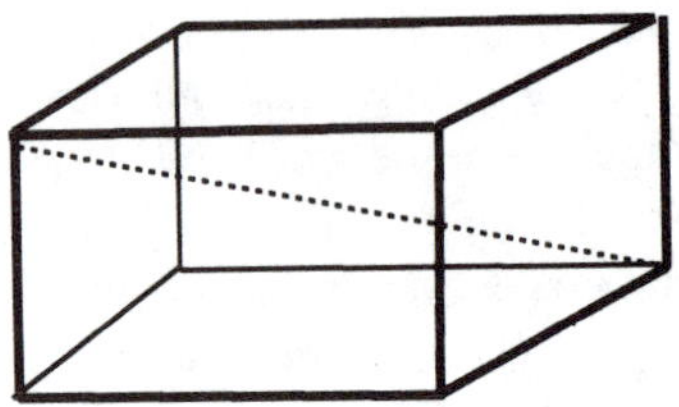

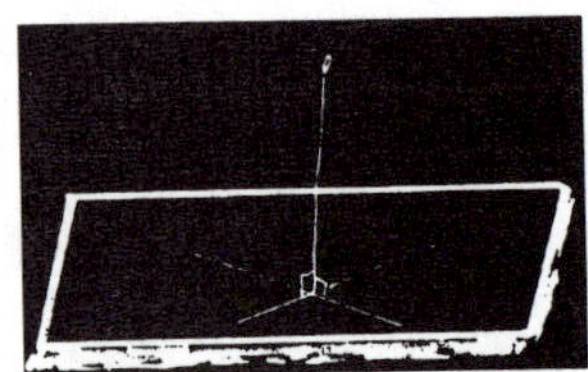
Birthday cake with candle

1. <u>Definition</u>: A **line is perpendicular to a plane** iff it is perpendicular to every line in the plane that intersects it.

2. <u>Definition</u>: The **distance from a point to a plane** is the length of the perpendicular from the point to the plane.

3. <u>Proposition</u>: Two lines perpendicular to the same plane are parallel. Think of two candles stuck in the same cake.

4. <u>Definition</u>: **Two planes are parallel** iff they don't intersect.

5. Yes. <u>Definition</u>: **Two planes are perpendicular** iff one plane contains a line that is perpendicular to the other plane.

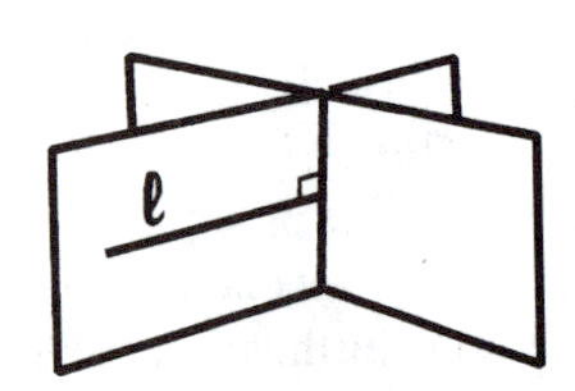

6.

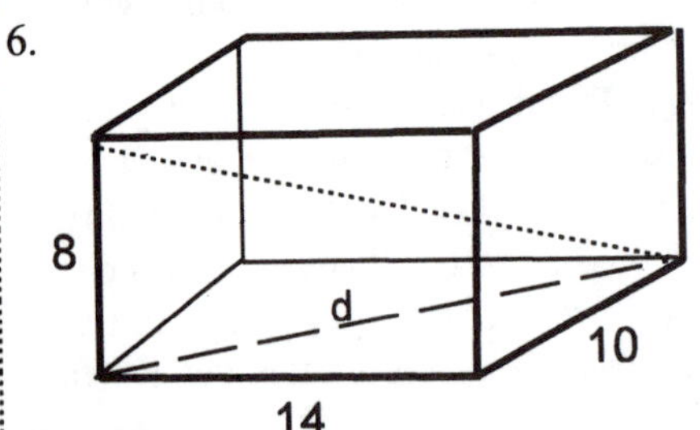

Step one: The diagonal across the floor is the hypotenuse of a right triangle whose legs are 10 and 14. By the Pythagorean Theorem, $d^2 = 14^2 + 10^2$. By algebra, $d = \sqrt{296}$.

Step two: If we let s be the length of the string of lights, then s is the hypotenuse of a right triangle whose legs are 8 and d. By the Pythagorean Theorem, $s^2 = 8^2 + d^2$. By algebra, $s = \sqrt{360}$ feet which is approximately 19 feet.

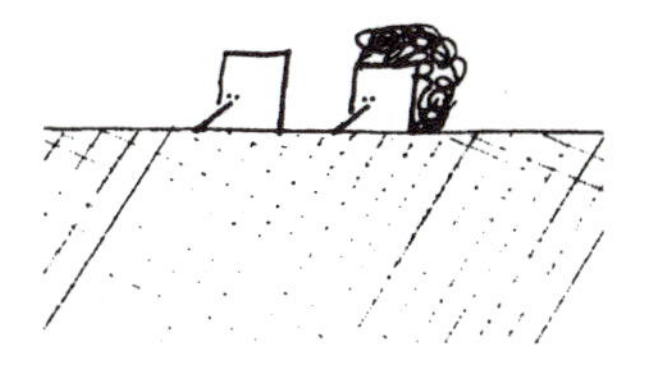

"Oh hi!" Betty exclaimed as the two adventurers came up to the table. "We were just discussing our wedding plans for this summer. Alexander thinks that after the wedding at the KITTENS chapel, we could hold the reception on a houseboat on the big lake on campus."

Fred hoped that P was listening to the part about *wedding*, but made a mental note to talk to Alexander later about the dangers of sailing on the big lake.

Three telephone books were already in place on each of the empty chairs, Fred and P hopped on their chairs, and the evening fun was about to begin.

Stanthony's pizza place was a little different than many others. There were no televisions mounted to the wall flashing football, baseball, ice hockey, or boxing. The music in the background was quiet Italian love songs that made conversation easy. The waiters cheerfully took the patrons' orders but didn't intrude. Stanthony had posted a large sign in the waiters' changing room. It was a statement by David Hume in *Letter to Michael Ramsay:*

THE FREE CONVERSATION OF A FRIEND IS WHAT I PREFER TO ANY ENTERTAINMENT.

After the salads, pizza, and Sluice were ordered, their conversation began in earnest.

Alexander began. "I've been inspired by the board game that you invented, Fred. Your construction game —I've seen it for sale almost everywhere. I heard that someone had set up a booth on the Great Lawn and was selling hundreds of them.

"I'm going to call my game Months. Players will shake dice at the beginning of the game to determine which month they will start in."

"But there are twelve months," Fred said. "How do you make shaking dice determine which month you start in? If you used two dice, you could get any number from 2 to 12, so you never could get January. And the chances of rolling a 7 are a lot more than rolling an 11 or 12. How are you going to fix that?"

"That's easy," Alexander said. He took out a stiff piece of paper and drew: "Now cut this out and fold it up and what do you get?"

P smiled. "It's a hexahedron, one of the regular polyhedrons."

Your Turn to Play

1. It's time to translate what P just said. What's the ordinary, everyday word for hexahedron?

2. Polyhedrons are solids with flat faces. A die (singular of *dice*) has six faces. How many edges does a cube have?

3. How many vertices (corners) does a cube have?

4. A hexahedron has the prefix *hexa.* Look back to the poetry lesson on p. 57 where we discuss, "How Many Feet in a Line." Does the *hexa* in hexahedron refer to faces, edges or vertices?

5. When P spoke about *regular* polyhedrons, she meant polyhedrons which have congruent faces, which have all the angles congruent, and which have the same number of edges meeting at each vertex. A regular **tetrahedron** (look again at p. 57) has how many faces? The faces of a tetrahedron are triangles.

6. Alexander drew on stiff paper the diagram that could fold up into a cube. Draw the diagram that would fold up into a tetrahedron.

7. How many vertices and how many edges does a tetrahedron have?

8. *Cube* was another name for a hexahedron. What is a more common name for a tetrahedron? (Think Egypt.)

9. Now let's do all of the above for the **octahedron**. Namely, (a) how many faces; (b) how many edges; (c) how many vertices; (d) draw the diagram that would fold up into an octahedron. There is no common name for an octahedron.

10. We have named three regular polyhedrons: the cube, the pyramid, and the octahedron. Now, for fun, make a guess as to how many regular polyhedrons there are. We know that there are at least three. Is there another? Are there five of them? Are there a dozen? Is there an infinite number of them?

COMPLETE SOLUTIONS

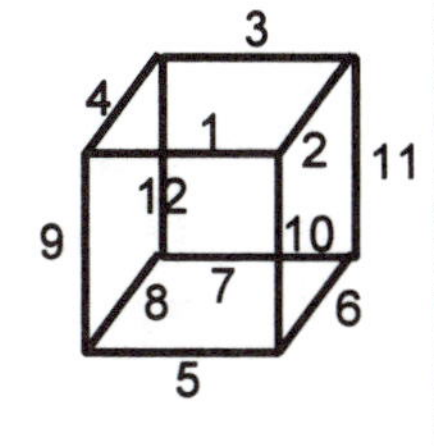

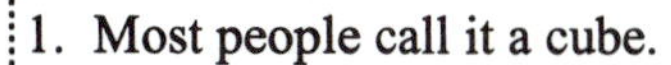

1. Most people call it a cube.

2. Counting the edges of a cube: I count four around the top; four around the bottom; and four vertical lines. A total of 12.

3. Eight vertices on a cube. Four around the top and four around the bottom.

4. *hexa* is the prefix for "six." Since a cube has 6 faces, 12 edges and 8 vertices, the *hexa* in hexahedron must stand for faces.

5. *tetra* is the prefix for "four." A tetrahedron has four faces.

6.

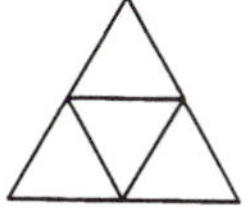

7. A tetrahedron has 4 vertices and 6 edges.

8. A more common name for a tetrahedron is a pyramid. The ones in Egypt often had square bases. Our regular tetrahedron has a triangular base.

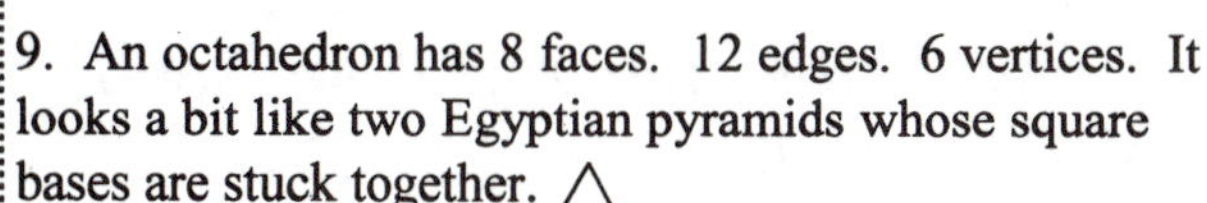

9. An octahedron has 8 faces. 12 edges. 6 vertices. It looks a bit like two Egyptian pyramids whose square bases are stuck together.

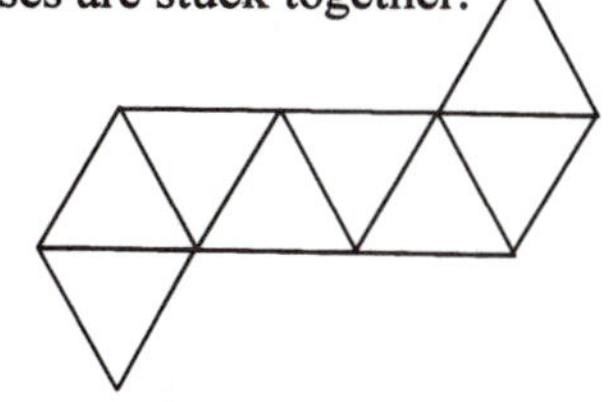

Your triangles might be in different positions than mine.

10. Pythagoras (sixth century B.C.) knew of four regular polyhedrons. In addition to the cube, the tetrahedron and the octahedron, he knew about the **icosahedron** which has 20 triangular faces. Its fold-up diagram looks like:

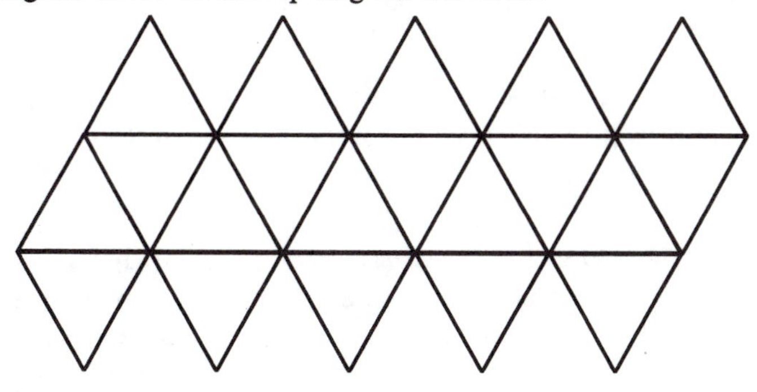

Hippasus (fifth century B.C.) found the fifth and last of the regular polyhedrons which Alexander is just about to describe.

"So what I'm going to use for my *Months* game is a die in the shape of the last regular polyhedron, the dodecahedron. It has 12 faces which is perfect for rolling out the 12 months.

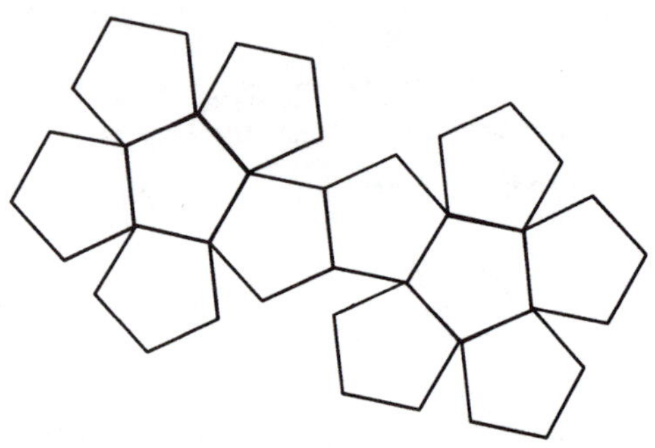

P was busy taking notes.
She created her famous . . .

Chart of the Regular Polyhedra*			
	How many FACES	How many VERTICES	How many EDGES
Tetrahedron	4	4	6
Hexahedron (Cube)	6	8	12
Octahedron	8	6	12
Dodecahedron	12	20	30
Icosahedron	20	12	30

"Would you look at this!" P exclaimed. "There's a pattern here." She placed her chart in the middle of the table so everyone could see it.

✶ *polyhedra* is another way to spell *polyhedrons*. You can also spell the plural of *formula* as *formulae*. These alternate spellings are less common than just adding an *s*, but they may make you appear a little smarter (and less intelligible) than you really are. One "safe" weird pluralization is that of *alumnus* (a graduate of some school). The plural of *alumnus* is *alumni*, not *alumnuses*.

Alexander, Betty, and Fred had seen this before and let P continue talking.

"Look. If you add up the number of faces and vertices for any of the regular polyhedra, they always almost equal the number of edges. In fact, the faces plus vertices are always two more than the edges."

FACES	VERTICES		EDGES
4	4		6
6	8		12
8	6		12
12	20		30
20	12		30

"What you've discovered is Euler's Theorem,* " Betty told P. "It can be written as F + V – 2 = E. The nice thing that Euler discovered was that it is not only true for the five regular polyhedrons, but for any polyhedron. If you take a lump a clay and throw it down on the table or hit it with a paddle until all the faces are flat, then F + V – 2 = E will be true."

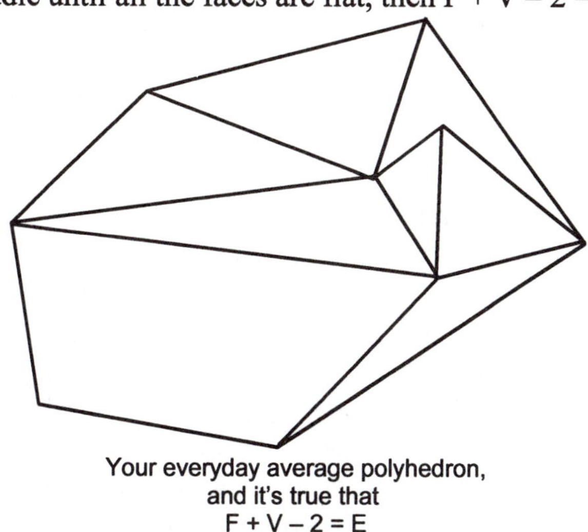

Your everyday average polyhedron, and it's true that F + V – 2 = E

* pronounced Oiler. Leonhard Euler was one of the greatest mathematicians of the 1700s.

Maybe P and I could head to jewelry stores, thought Fred, and we could count the faces, vertices, and edges on some gem stones. Fred already had in mind what P's engagement ring would look like.

The classic four-pound engagement ring

"You know," said Alexander, "there's only so much you can do with the five regular polyhedrons. Once you've identified the tetrahedron, the cube, the octahedron, the dodecahedron, and the icosahedron, and counted their faces (4, 6, 8, 12, 20), you pretty much run out of steam. There are so many other solids that we meet in everyday life that are much more common than these 'five regulars.' The solid that I saw for years—before I met you Betty—was the tin can. Every morning it was a can of spaghetti. Every lunch, a can of tuna. Every evening, a can of soup. Never a veggie. Never a salad. It was cheap and quick, but it was sending me to an early grave. I never paid much attention to what I was eating until you came along and suggested, in your sweetest way, that eating was more than just 'filling up'."

Betty smiled.

"As I look back," Alexander continued, "the only real memory I have of those insane eating days, was that of sitting there at the breakfast table one day and staring at the tin can and wondering what volume of spaghetti I had just inserted into myself. The volume of a cylinder. I just sat there and looked at the can. I couldn't remember the formula."

"That's easy," said P. "It's just base times height."

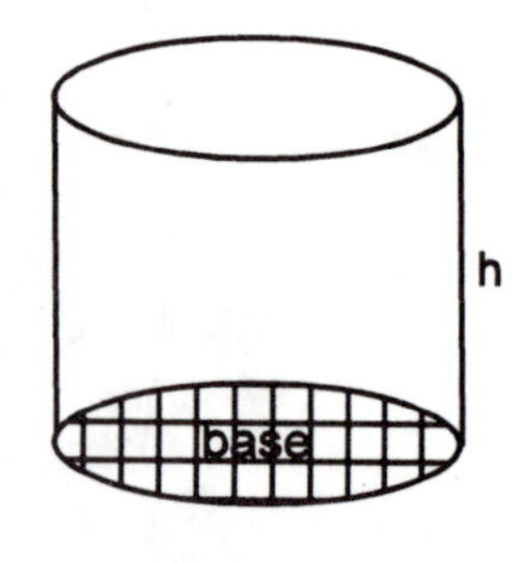

"I know," said Alexander. "But on my fiberless diet, my thinking was totally impacted. I just felt stupid. I know that sort of sounds silly, but. . . ."

"Not at all," Betty interrupted. "They've done all kinds of scientific studies. Eating decently and getting a reasonable amount of huff-and-puffing exercise helps you to think more clearly."

When Stanthony brought the salads that they had ordered, they thanked him for his splendid salads—not just iceberg lettuce with bottled dressing—and then they bowed their heads and thanked God for Stanthony.

Your Turn to Play

1. If the width of a tin can (a cylinder) is six inches, what is the area of the base?

2. What is the volume of a cylinder if the width of the base is 6" and its height is 7"?

3. If a soup can is five inches tall and has a volume of approximately 141 cubic inches, what is the radius of the cylinder to the nearest tenth of an inch?

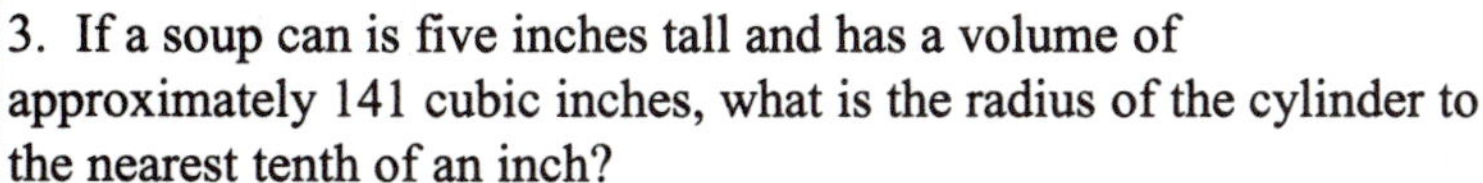

4. Alexander sometimes could get a special deal on tin cans that had been damaged in shipment. Once he bought a whole case of spaghetti in which each can was smashed over to the side. He called it the "leaning tower of pasta." The volume is still the same: the area of the base times the height. (The height is defined as the distance between the planes that contain the upper and lower circles.)

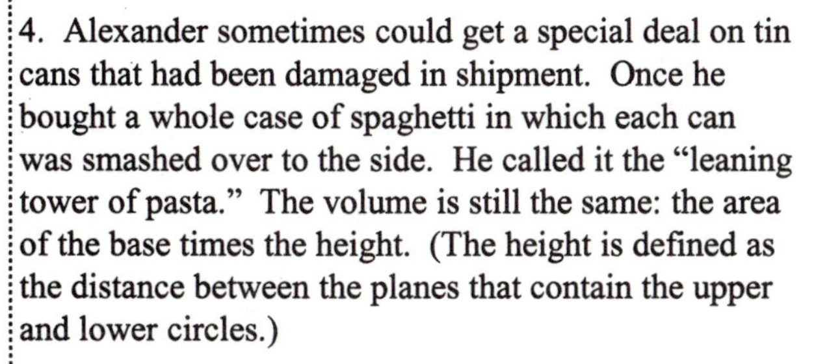

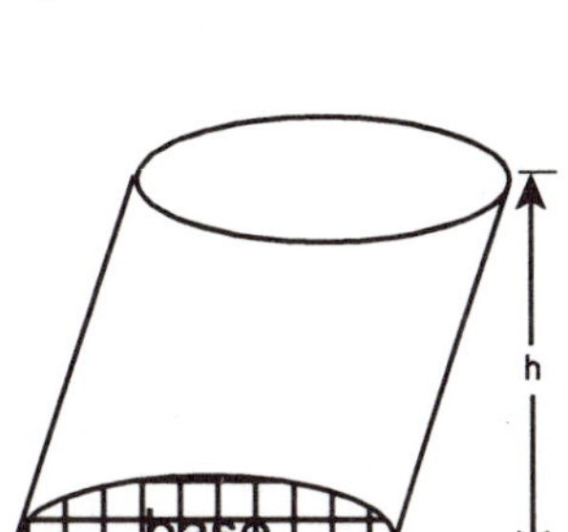

This is easy to see if you think of a can as a stack of infinitely thin poker chips. The volume of the can is the volume of the sum of the volumes of all the poker chips. Whether you stack your chips up straight or at an angle, the volume remains the same.

same volume

The fancy name for the poker chip principle is **Cavalieri's Principle**.

Now let's test your three-dimensional imagination. Suppose you had two tin cans of the same height, one that is straight up and down (this is called a **right cylinder**) and one that is slid over to the side (this is called an **oblique cylinder**). Look at the two pictures directly above. Suppose that you had to put labels on these two tin cans. Would the oblique cylinder require a label that had more, less, or the same area as the label for the right cylinder.

Rephrasing the question: Is the **lateral surface area** of an oblique cylinder larger, smaller or the same as the lateral surface area of a right cylinder of the same height?

COMPLETE SOLUTIONS

1. The base of a cylinder is a circle. If the width of the base is 6", then we are looking at a circle with diameter 6". The radius would be 3". Since the area of a circle is πr^2, the area of the circle in this case would be $\pi 3^2$ or 9π square inches.

2. The volume of a cylinder is base times height. In the previous problem, we found that the area of the base is 9π. Since the height is 7", the volume of the cylinder is 63π cubic inches.

3. The volume of a cylinder of radius r and height h is $\pi r^2 h$.

Starting with $V = \pi r^2 h$

Putting in the values we know $141 = \pi r^2 5$

Algebra $\dfrac{141}{5\pi} = r^2$

More algebra $\sqrt{\dfrac{141}{5\pi}} = r$

Approximating using a calculator $2.99605 = r$

Rounding to the nearest tenth of an inch $3.0 \doteq r$

4. The thing to remember is that we are dealing with two cylinders that are the same height. *If* the can were smashed over (as if an elephant stepped on it) the height would change, but the areas of the two labels would be the same.

But keeping the same height is a different story. The lateral surface area of an oblique cylinder could be almost infinite!

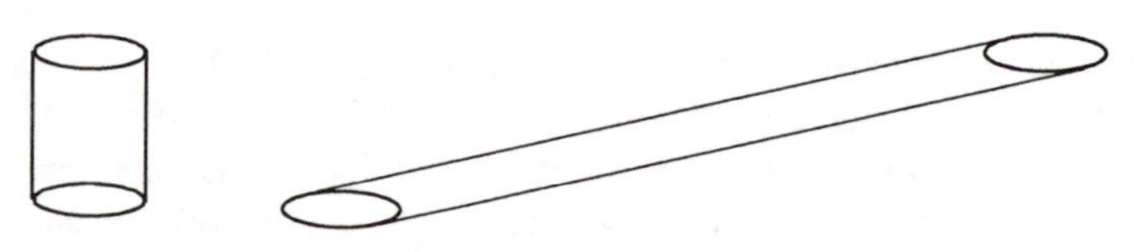

"When I was teaching about the volume of cylinders in geometry," Fred said, "Joe made a very interesting observation. He said that cans of soup on the grocery shelves waste a lot of space since circles (looking down at the cans) don't efficiently cover the plane.

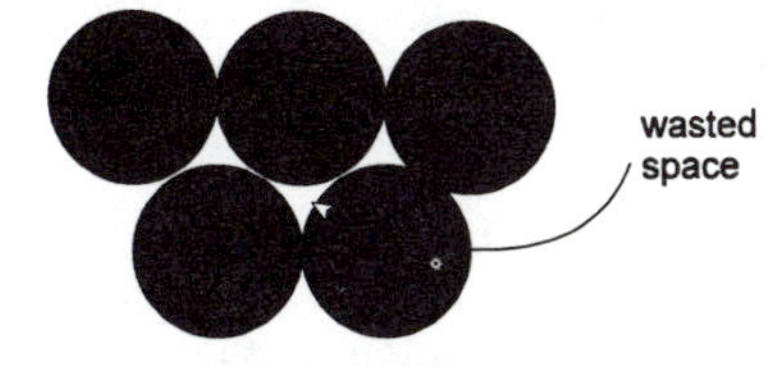

Top view of a bunch of tin cans

"Joe suggested that instead of cylinders with their rounded sides, the can manufacturers should make prisms. Prisms would probably fit together much more neatly than cylinders.

"Joe drew a whole bunch of prisms on the blackboard."

Your Turn to Play

1. (This is not especially easy.) Complete the definition: A prism is a polyhedron in which. . . .

2. The volume for any prism is the same as the volume for a cylinder: *base times height*. For a box (officially called a rectangular parallelepiped), this gives us the familiar formula:

Vol = length × width × height.

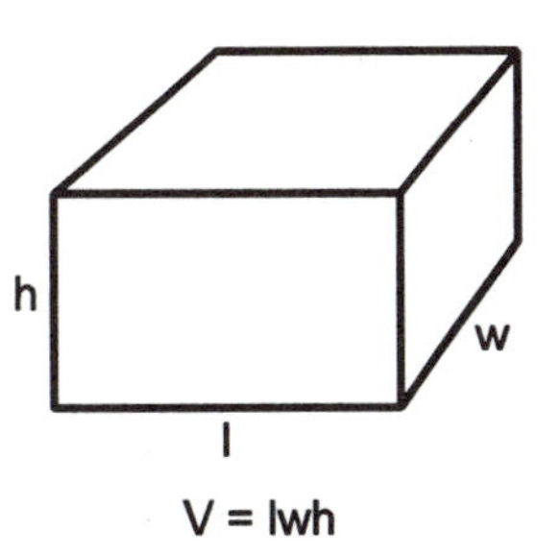

Suppose you had two cans of soup. One of them has the traditional cylindrical shape with a width of 3" and the other is Joe's new prism-shaped can. Let's say that it has a square base with a width of 3". Compare the volumes of these two cans.

3. Joe's suggestion for tin cans in the shape of prisms has one slight drawback. What is it?

1. <u>Definition</u>: A **prism** is a polyhedron in which two of the faces (called the **bases of the prism**) are congruent polygons lying in parallel planes and in which all of the vertices of the polyhedron are vertices of the two congruent polygons. The edges of the polyhedron join corresponding vertices of the bases and are all parallel. (Your definition may have differed a lot from mine. Some people mention that the faces of

the prism, except for the bases, all have to be parallelograms. Some people start with two parallel planes and a polygon in one of the planes. Then they define a prism as all the segments parallel to some given line which have one endpoint in or on the polygon and their other endpoint in the other plane.

2. If the width of the cylinder is 3", its radius is 1.5". The area of the base is the area of a circle ($A = \pi r^2$) which in this case is $\pi(1.5)^2 = 2.25\pi$. If the height of the cylinder is h, then the volume of the cylinder is $2.25\pi h$.

For the prism with the square base, if the width of the square is 3", its area is 9 square inches. If the height is h, then the volume if $9h$.

So how does $2.25\pi h$ compare with $9h$? If we divide one by the other, we obtain $\frac{2.25\pi h}{9h}$ which is approximately 0.7853982. So the traditional cylindrical tin cans have a volume of about 79% of that of Joe's new prism cans with square bases.

3. I'm not sure whether my can opener would work, especially on one like: 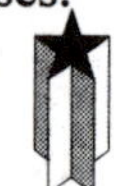

P giggled at the thought of tin cans in the shape of prisms. Then she said, "I know how to tell whether I'm inside or outside a can of spaghetti—just open my mouth—but how do you tell whether you are inside or outside of any polyhedron? Any general polyhedron can be devilishly complicated. They needn't look like some simple old prism."

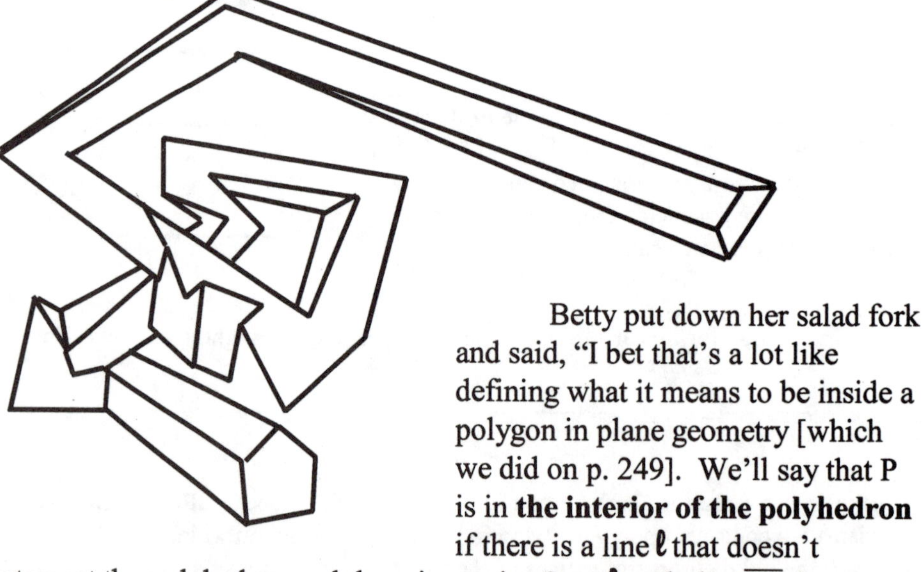

Betty put down her salad fork and said, "I bet that's a lot like defining what it means to be inside a polygon in plane geometry [which we did on p. 249]. We'll say that P is in **the interior of the polyhedron** if there is a line ℓ that doesn't intersect the polyhedron and there is a point Q on ℓ such that $\overline{PQ}$ does not intersect any of the vertices or edges of the polyhedron and $\overline{PQ}$ intersects the faces of the polyhedron an odd number of times. But that's an extremely tough definition. I bet that even most geometry teachers couldn't define what it means for a point to be on the interior of a polyhedron. That was a deep question you asked."

P smiled. She finished the last bite of her salad.

Fred had yet to take the first bite of his salad. He was still busy arranging the napkin in his lap and straightening his silverware. He had also been paying very close attention to every word that P had said. When she had asked about the way to define the interior of any polyhedron, he was impressed. He had never met any other six-year-old who understood geometry so well.

Alexander continued his story of how he ate before he met Betty. "I was starting to think that every kind of food came in tin cans. I once even had a dream that ice cream cones were sold in tin cans."

"That's silly," said P. "You buy them in boxes"

"I know, but this was a dream," Alexander replied. He drew a picture on the paper tablecloth:

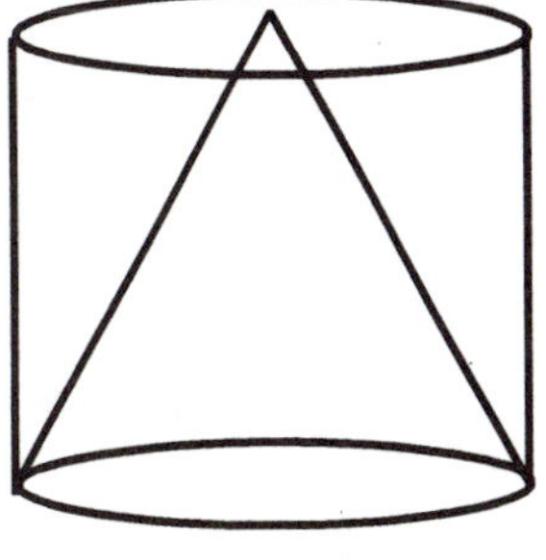

Alexander's
Cone-in-a-Can

Fred didn't know that he could fall more deeply in love with P, but it happened when P looked at Alexander's drawing and remarked that the volume of a cone is exactly one-third of the volume of the cylinder which contains it.

There are times in life when everything starts to fit together. This was one of those moments for the four at the table. They all saw the pattern that was taking shape:

Cylinder h Volume = base × height	Cone in a Cylinder h Volume = 1/3 × base × height
Prism Volume = base × height	

Your Turn to Play

1. What goes in the fourth box on the previous page?

2. Complete the definition: A pyramid is a polyhedron in which. . . .

3. I read once that some of the innkeepers in merry olde England used to serve beverages to their patrons in cone-shaped containers. That meant that the customers couldn't set their drinks down until they had finished them. Very clever. Nowadays, they couldn't get away with such an obvious ploy.* How much Sluice could a cone contain that was 4" across and 6" tall?

4. Of the Seven Wonders of the Ancient World,** only one of them is still standing, the Pyramids of Egypt. The biggest of them has a square base (755 feet long) and was originally 482 feet tall. In the 4800 years since it was built, it has weathered a bit and is now only 450 feet tall. What was its original and its present volume?

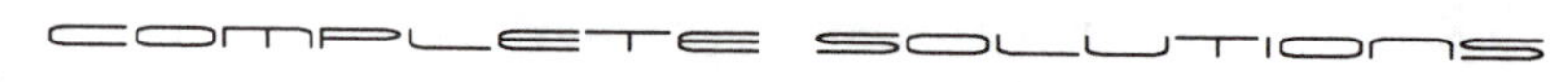

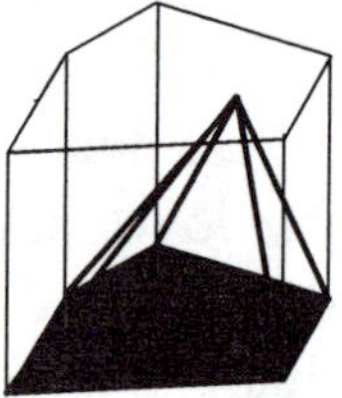

1. A pyramid in a prism. Its volume is 1/3 × base × height.

2. Definition: A **pyramid** is a polyhedron in which all but one of the vertices are coplanar.

3.

The radius of the base of the cone is equal to 2". The area of the base is πr^2 which is $\pi(2)^2 = 4\pi$ square inches.
The volume of the cone is $(1/3)(4\pi)6 = 8\pi$ cubic inches.

---

* Instead, they use modern ploys. In fast-food joints the owners have noticed that, in general, their customers don't come back for seconds and they don't leave tips, so once they've received their food, it's in the owner's best interest to have them eat quickly and leave. Have you noticed how many of these places play energetic music? It isn't played to entertain you, but to encourage you to eat faster.

** (1) The Pyramids of Egypt; (2) The Hanging Gardens of Babylon built by King Nebuchadnezzar to please his queen; (3) The Statue of Zeus at Olympia, 40 feet high and made of gold and ivory; (4) The Temple of Diana; (5) The Mausoleum at Halicarnassus, built for King Mausolus by his wife (where we get the word *mausoleum*); (6) The Colossus at Rhodes, a 105-foot-high, bronze statue; and (7) The Pharos of Alexandria, a lighthouse off the coast of Egypt.

4. The areas of the bases of both the original and the 4800-year old pyramid are the area of a square whose side is 755 feet. $755^2 = 570{,}025$ square feet.

The volume of a pyramid is one-third of the volume of the prism that encloses it = $(1/3) \times$ base $\times$ height.

For the original pyramid, Volume = $(1/3) \times 570{,}025 \times 482 = 91{,}584{,}015$ cubic feet.

For the slightly weathered version, Volume = $(1/3) \times 570{,}025 \times 450 = 85{,}503{,}748$ cubic feet. In 4800 years it has lost 6,080,267 cubic feet, or about 6.6% of its volume.

Stanthony brought the giant combination pizza to the table and cleared away their salad plates. Fred asked for a doggie bag for the 99% of the salad he hadn't finished.

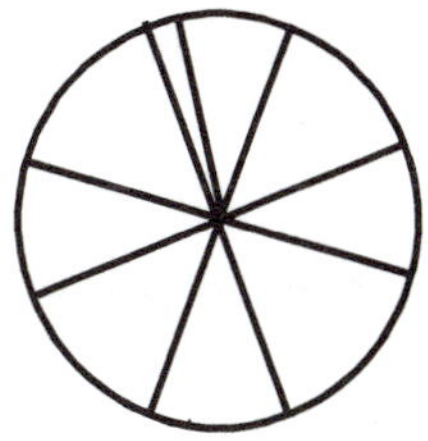

Stanthony knew Fred well enough that when he cut the pizza, he made one of the slices very thin for Fred. The central angle for Fred's piece was only 3°, so the area of his piece (which is in the shape of a sector) was (3/360) of the area of the whole pizza.

Everyone knew that that was Fred's piece. Everyone except P. They both reached for it, and their hands touched. Fred's heart skipped a beat.

"We could share it," P offered. Fred could only nod. He was in his speechless mode again.

"So, we'd like to get to know you a bit better," Betty said to P. She took the thoughts right out of Fred's brain. (There's no way she could have taken the words right out of his mouth.) Fred was delighted that he could hear something more about his beloved.

"Where were you born?" Betty continued.

"I'm not sure," was the unexpected response. "You see, I was adopted right after my birth by T and L Commonsizerinski. T and L have been just wonderful to me. I can't imagine any more loving parents than they have been. Betty, you should have seen the grilling that my dad put Fred through in his interrogation room just so Fred could take me out to pizza."

Fred found his voice. "You mean there have been other men, um, boys in your life?"

"Oh no. Not really. T told me he was just learning how to do it, so that when I'm in my late teens and the boys start flocking around my house, he'll be well practiced in doing his fatherly job.

"My mother tells me that six years ago," she continued, "the stork delivered me to a family that really didn't like children. And besides they already had a little boy. So I was forwarded, just like the mail, to T and L."

Fred smiled and said, "My folks gave me the old stork story, also. I guess they were also too embarrassed to discuss meiosis in which sperm or egg cells are created by chromosome conjugation followed by two cells divisions which then produces those gametes with half the normal number of chromosomes." He then blushed a little bit and said, "I hope I haven't been too sexually explicit."

"When I was three," P continued, "I told my Dad to forget all the stork stuff and tell me the true details of what happened. He let me read his diary of that Friday they found me."

Friday. Our eighth day in our tour of the South. Drove a solid bunch of hours south from Missouri. L not feeling very well. Probably dieting too much and her low blood sugar. She skipped lunch completely and about 4 in the afternoon she was getting kinda low.

I stopped at a King of French Fries and she said okay. Got out of the car to get her a bucket of fries that we could share. Something to sort of "grease her wheels." She's always tried to stay in shape (read *skinny*) in case we ever got pregnant. She was too tired to get out of the car.

I came back with this little new-born girl. She was in this used K.F.F. bucket. L asked me, "What kind of joke is this!?" I told her, "No joke. I found this little sweetie next to the dumpster behind the store."

L fell in love with the baby at first sight. The only thing that concerned her was the size and shape of our little girl's nose. "What a honker!" I remember her saying. We stopped off at the grocery store and bought some formula and a box of diapers.

P showed Fred the note that her father had found in the King of French Fries bucket. It was P's only link with her biological parents. It was written on a K.F.F. receipt.

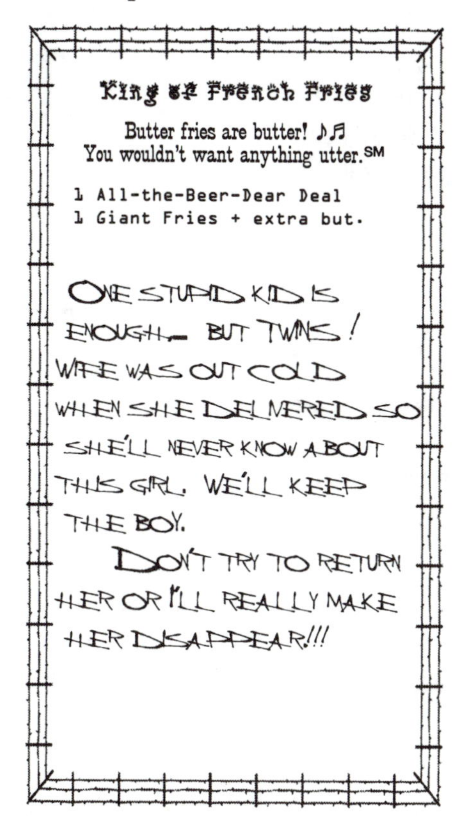

King of French Fries

Butter fries are butter! ♪♫
You wouldn't want anything utter.℠

1 All-the-Beer-Dear Deal
1 Giant Fries + extra but.

ONE STUPID KID IS
ENOUGH— BUT TWINS!
WIFE WAS OUT COLD
WHEN SHE DELIVERED SO
SHE'LL NEVER KNOW ABOUT
THIS GIRL. WE'LL KEEP
THE BOY.
DON'T TRY TO RETURN
HER OR I'LL REALLY MAKE
HER DISAPPEAR!!!

Fred handed the note back to P.

"Don't you think that it was all for the better that I was adopted?" P asked. "Imagine how bad it would have been if I had been reared in such a nasty home. That poor boy that they kept, I feel sorry for him."

P continued talking. Fred was a million miles away, passing through Hell and Heaven at the same time. That handwriting on the receipt. He recognized it. It was his father's.

Fairchild

1. Given a point on a line, how many lines can be drawn through that point which are perpendicular to that line?

2. A polyhedron has 16 faces and 10 vertices. How many edges does it have?

3. How many cubic *inches* are in a cube whose edges are each 5 *feet*?

4. Suppose we take some arbitrary region in a plane, and we take a second plane that is parallel to the given plane. Then we draw a segment one of whose endpoints is in the region and the second endpoint is on the second plane.

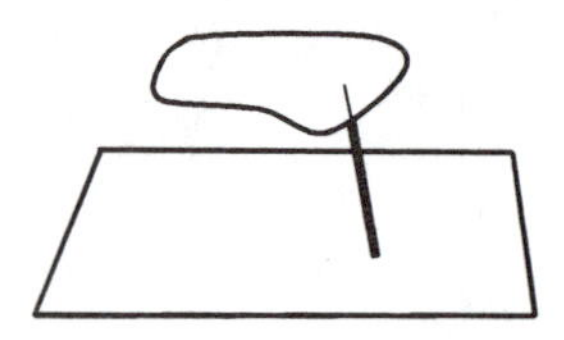

Then we draw all possible parallel segments with one endpoint in the region and the other endpoint on the second plane.

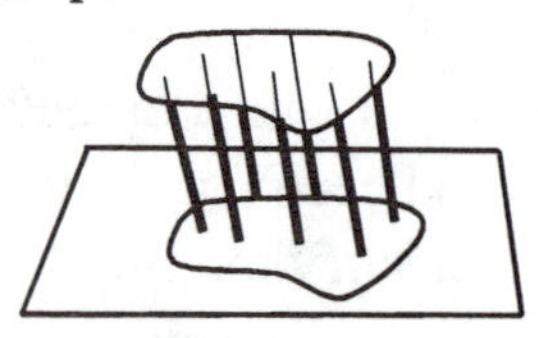

We arrive at a generalization of a cylinder.

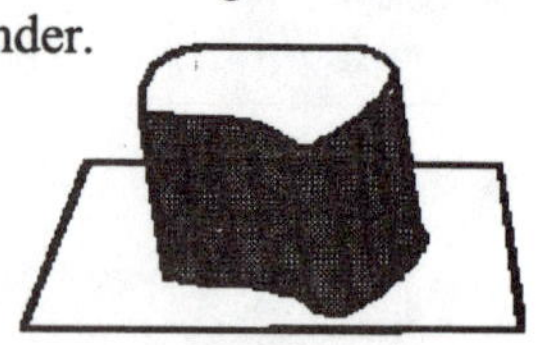

Guess what the volume would be.

5. Suppose we put a generalized cone inside the generalized cylinder of the previous problem.

Guess what the volume of this generalized cone would be.

6. What is the area of a sector of a 16" (diameter) pizza which has a central angle of 4°?

answers

1. An infinite number. Remember that we are in three dimensions in this chapter.

2. 24 edges

3. Since 5' = 60", the volume of the cube would be 60^3 or 216,000 cubic inches.

4. It would be the same as that of a cylinder or a prism: the area of the base times the height, the height being the distance between the planes containing the bases.

5. It would be the same as that of a cone or a pyramid: one-third of the area of the base times the height.

6. $(4/360)(\pi)(8^2) = 256\pi/360 = 32\pi/45$ square inches. (This is about 2.2 square inches, which is a little more than Fred normally consumes in a meal.)

Ursa

1. Given a point on a line, how many planes can be drawn through that point which are perpendicular to the line?

2. A polyhedron has 12 edges, and the number of faces is the same as the number of vertices. How many faces does it have?

3. One of the most famous solids which we haven't mentioned thus far in this chapter is the sphere. (Some people call it a ball.) Tennis balls are sometimes sold in tin cans, so that they stay pressurized and don't lose their bounce.

Imagine a sphere-in-a-can.

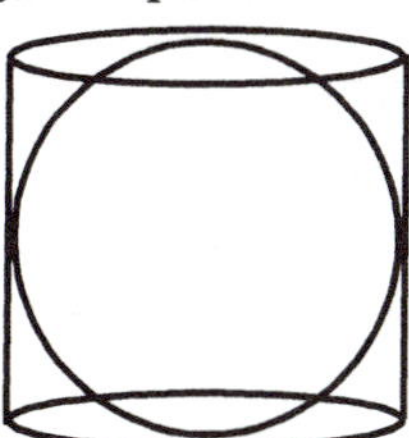

Now let's test your imagination for three-dimensional objects. What fraction (this is a pure guess on your part) of the volume of the cylinder is occupied by the sphere?

4. Suppose we have a sphere with a radius of r inscribed in a cylinder. What would be the volume of the cylinder?

5. From the answers to the previous two questions, what is the formula for the volume of a sphere with radius r?

answers

1. Only one.
2. seven
3. A sphere always occupies exactly two-thirds of the volume of the cylinder which contains it. That fact of nature was first discovered by Archimedes who lived in the 200s B.C. He is considered the greatest mathematician in the ancient world. Besides his work in geometry, he made significant contributions to physics, mechanics, and hydrostatics. But of all his work, he seemed to be most proud of having found the formula for the volume of a sphere. He asked his friends to mark his tomb with a sphere inscribed in a cylinder and marked with the ratio two-thirds.
4. The volume of the cylinder is given by the area of the base times the height. The area of the base is the area of a circle of radius r, which is πr^2. The height of the cylinder is $2r$. The volume of the cylinder is $2\pi r^3$.
5. Since Archimedes told us (and we will be able to prove it in calculus) that the volume of a sphere is two-thirds of the volume of the cylinder which contains it, $V_{sphere} = (4/3)\pi r^3$.

Walsh

1. If a line intersects a pair of parallel lines, must all three lines be coplanar?

2. If a line intersects a pair of skew lines, is it possible for all three lines to be coplanar?

3. Could a line which does not lie in plane $\mathcal{A}$ be perpendicular to two lines which lie in $\mathcal{A}$ and not be perpendicular to every line that lies in $\mathcal{A}$?

4. What is the volume of the earth if its radius is 4000 miles?

5. Let's imagine a sphere-in-a-can again.

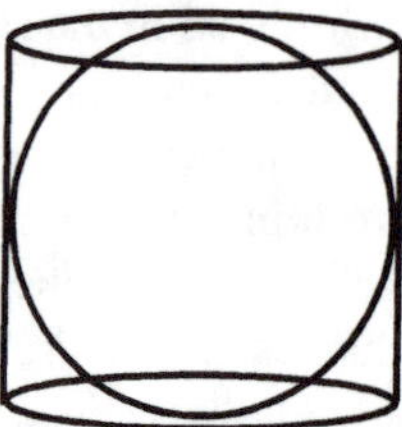

Now let's test your imagination for three-dimensional objects. What fraction (this is a pure guess on your part) of the surface area of the cylinder is occupied by the sphere? The surface area of the cylinder includes the top and the bottom and the curved side.

6. Suppose we have a sphere with a radius of r inscribed in a cylinder. What would be the total surface area of the cylinder? The areas of the top and bottom are easy to find. To find the surface area of the curved side, imagine cutting the can open with tin snips and flattening it out into a rectangle.

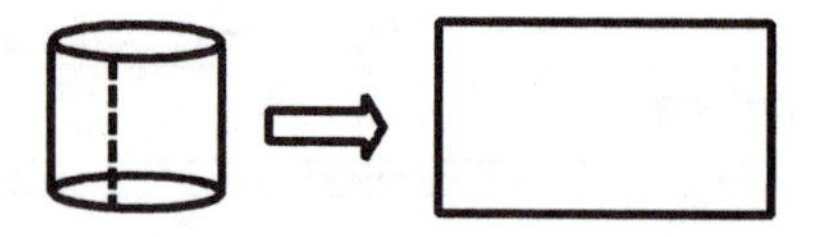

7. From the answers to the previous two questions, what is the formula for the surface area of a sphere with radius equal to r?

odd answers

1. Yes

3. Sure. If a line is perpendicular to two lines which lie in a plane, it will be perpendicular to every line that lies in the plane that intersects the given line, but it will be skew to every other line in the plane.

5. This, surprisingly, is also equal to two-thirds. The surface area of a sphere is exactly equal to the two-thirds of the cylinder which circumscribes it.

7. $(2/3) \times (6\pi r^2)$

Surface $\text{Area}_{\text{sphere}} = 4\pi r^2$. That means that the surface area of a sphere is exactly equal to four times the area of a circle which passes through the center of the sphere.

Daniel

1. Concurrent lines are lines that all share a common point. What might the definition of concurrent planes be?

2. How many concurrent lines could be parallel to a given line?

3. A polyhedron has 28 edges. The number of faces is twice the number of vertices. How many vertices does it have?

4. What is the volume of a can of spaghetti which has a radius of 2" and a height of 4"?

5. (Continuing the previous problem) Suppose that the can were filled with a single long piece of spaghetti. The strand of spaghetti was 0.1" in radius. How long would that strand be? (Hint: Consider the strand as a very tall skinny cylinder.)

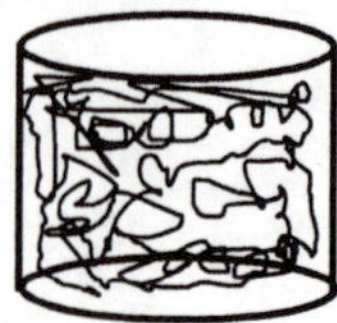

odd answers

1. Definition: **Concurrent planes** are planes that all share a common line.

3. 10

5. 1600"

Kathleen

1. Given a point in a plane, how many different lines can be drawn perpendicular to that plane through that point?

2. How might you define a line being parallel to a plane?

3. In three dimensions, if two lines are parallel to the same line, must they be parallel to each other?

4. Show that a polyhedron could never have the same number of edges as vertices.

5. Stanthony invented his famous Pile-of-Pizza® which consists of a very thin 16" (diameter) crust piled high with minced garlic, oregano, thyme, parsley, chives, basil, shredded Fontina cheese, tomatoes, shredded mozzarella, grated Parmesan, tender leaves of arugula, sautéed broccoli rabe, shredded radicchio, thin slices of prosciutto, cracked black pepper, mushrooms, artichokes, anchovy fillets, Gorgonzola cheese, red chile flakes, grilled lamb (This is a favorite in the Netherlands. They call it Shwarma.), goat blue, reblochon and emmental cheeses, Portuguese sausage (linguica) and Spanish sausage (chorizo), crumbled feta cheese, smoked gouda, pecorino romano, pepperoni, bacon, smoked sliced ham, Italian salami, smoked link sausage (thick sliced), ground beef, green bell peppers, sliced onions, chopped green onions, quartered zucchini rounds, black and green olives, chunky pineapple, sliced roma tomatoes, sun dried tomatoes, fire-roasted tomatoes, and jalapenos all piled into a cone-shaped mound that is 6" high. What is the volume of those toppings?

6. If those toppings (of the previous question) were leveled out on the pizza, what would be the height of the cylinder of toppings?

Langford

1. Given a point *not* on a plane, how many different lines can be drawn perpendicular to that plane through that point?

2. Given a point *not* on a plane, how many different planes can be drawn parallel to that plane through that point?

3. In three dimensions, if two lines are parallel to the same line, must those two lines be coplanar?

4. A polyhedron has 8 vertices and 22 edges. How many faces does it have?

5. Which, if any, of the regular polyhedrons are pyramids?

6. If lines ℓ and m are skew, and if m and n are skew, then:

    A) can ℓ and n be coplanar?

    B) can ℓ and n be skew?

    C) can ℓ and n intersect?

    D) can ℓ, m and n be concurrent?

    E) can ℓ be perpendicular to n?

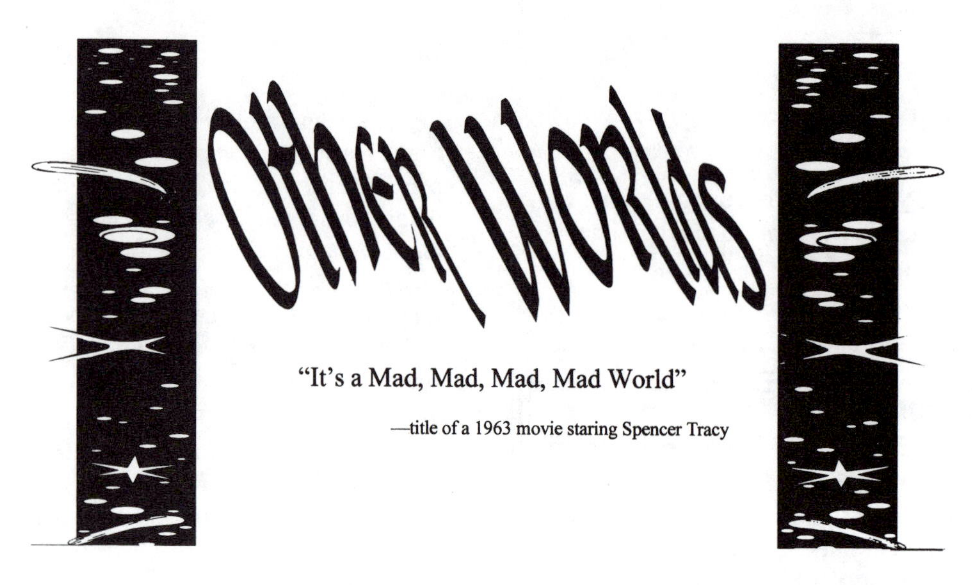

"It's a Mad, Mad, Mad, Mad World"

—title of a 1963 movie staring Spencer Tracy

Chapter 12½
Geometry in Four Dimensions

When Fred first sat in T's interrogation room, his eyes were level with the table top.
He had tried to imagine what it would be like to live on a plane instead of in three dimensions. It would be so terribly . . . flat. Everyone and everything would exist on some giant plane, like illustrations on a big piece of paper. To move from one place to another in that world, you could only slip-slide around. There would be no sun overhead, since there would be no "overhead."

In a two-dimensional world, there would probably be no railroads. Any train track running north-south would cut the world in half. The people on the west side of the tracks would have to go all the way around to get to the east side. Of course, a train wouldn't run on the tracks, but between them.

Worse yet, would be the idea of eating. Any food you ate, you would have to throw up. Food couldn't "pass through" you as it does in three dimensions, since that would cut you in half.

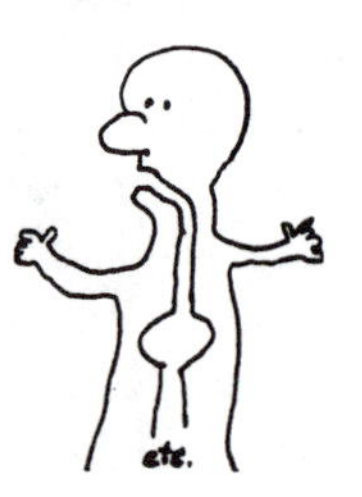

And, if you cooked pancakes, you could never flip them.

And kissing would also be out of the question unless you both had very small noses. In three dimensions have you noticed that when you kiss you turn your heads so that your noses don't get in the way?

Ah! Isn't it wonderful to live in a three-dimensional world? There are so many things that we can do that those poor slobs living on a plane can't do.

But it was just yesterday I heard a square talking with a pentagon. He said to her, "Ah! Isn't it wonderful to live in a two-dimensional world? There are so many things that we can do that those poor slobs living on a line can't do."

"Yeah," she replied. "Imagine living on a line. The only people you would ever meet would be the two people on either side of you. That would be your whole world. It would be like living in a world in which there was only north-south. They would have no concept of east-west. They would have no way of turning their heads to look at us as we stare at them. They don't even know that we exist."

"They would if I poked one of them," said the square.

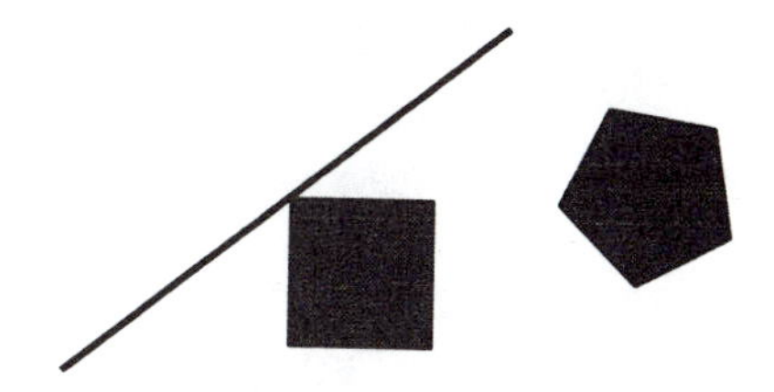

"Ouch!" exclaimed the point on the line who was hit by the corner of the square. The point looked to his neighbor on his left and to his neighbor on his right and wondered which one of them had poked him.

"That guy could never see me coming," the square said. "I hit him from his blind side. The only way he can see is up and down the line he lives on."

Those polygons try to feel better about their miserable two-dimensional condition by looking down on those less fortunate than they, those that live in only one dimension.

And similarly, you should hear how those that live in one dimension talk. They look down their noses at "that poor idiot who lives all alone in zero dimensions. He's got no one to talk to, whereas we've always got two neighbors for company: Mr. Left-of-me and Mr. Right-of-me."

Everyone, it seems, wants to feel smug by finding others who are worse off.

HOW TO TELL WHAT DIMENSION YOU LIVE IN

First you take a whole bunch of line segments. (If you can't do that, you are just a point living in zero dimensions.) Think of those segments as toothpicks.

Next, arrange as many of those toothpicks as you can so that they are all perpendicular to each other, so that every pair in your arrangement is perpendicular.

Hey, this is easy, you, the reader, say. ***I can put three toothpicks together, just like the corner of a room. We live in three dimensions.***

Have another toothpick.

Why?

Stick it in your diagram so that it is perpendicular to the three you have already assembled.

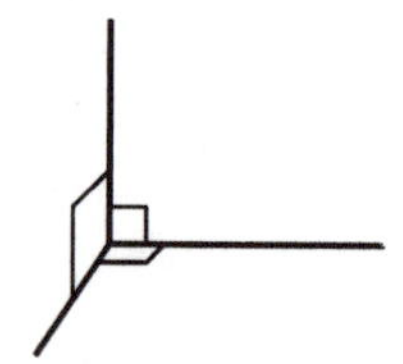

Three mutually perpendicular segments

No can do. Here, you take your toothpick and do it.

With pleasure. I am about to stick a fourth toothpick in, so that all four toothpicks are mutually perpendicular in pairs. Are you ready?

Sure. But you can't do it.

Yes I can.

That's impossible. Three dimensions are all there is. Everyone knows that. Here's your toothpick. Do it.

Okay. Here are the fourth toothpicks:

Wait a minute! I only see three toothpicks.

Of course. That's what you'd expect from those unfortunate people who live in only three dimensions.

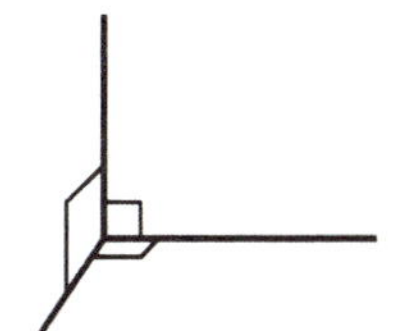

Four mutually perpendicular segments

Only three dimensions! Only? I suppose you're going to tell me that you are in four dimensions.

Can you see the fourth toothpick?

No.

I can.

Would you mind explaining yourself a little more slowly. You're not talking about that Einstein thing where "time" is the fourth dimension?

Oh no. This is a geometry book. We are looking at four *spacial* dimensions, four segments all mutually perpendicular in pairs. Please put down your phone. I can see you are thinking of dialing 9-1-1 and having me taken away in a straitjacket to the mental ward. Let's step down one dimension into the land of the square and the pentagon. I want to hear you, my reader, explain to them about three dimensions.

Okay, but you, Mr. Author, stay outta this discussion for a moment while I explain things to those two-dimensional people.

Okay. The show is all yours.

HOW THE READER EXPLAINS THREE DIMENSIONS
TO SQUARE AND PENTAGON

Okay, you guys. Can you hear me?

"Sure," responded square and pentagon.

Other Worlds

Chapter 12½ The Fourth Dimension

I'm going to explain to you about three dimensions. Pay close attention now.

"Wait a minute!" said the square. "There are only two dimensions. Here I've got two toothpicks and I've placed them so that they are mutually perpendicular. Here, take a third toothpick. I dare you to stick it in my diagram so that it's perpendicular to the two that I've already drawn."

Two mutually perpendicular segments

With pleasure. I am about to stick a third toothpick in, so that all three are mutually perpendicular in pairs.

"But that can't be done," said the pentagon.

I can do it.

"That's impossible. Two dimensions are all there is. Everyone here on the plane knows that. Here's your toothpick. Do it."

Okay. [At this point, the reader places a third toothpick perpendicular to the plane that the square and pentagon are in. Since the polygons can only see in the plane—north-south and east-west—they can't see the third toothpick that's in the up-down direction.]

Can you see the third toothpick?

"No."

I can see it.

"Where did you hide the third toothpick?" the square demanded, as he whispered to the pentagon that she should call 9-1-1 and have the reader taken away in a straitjacket.

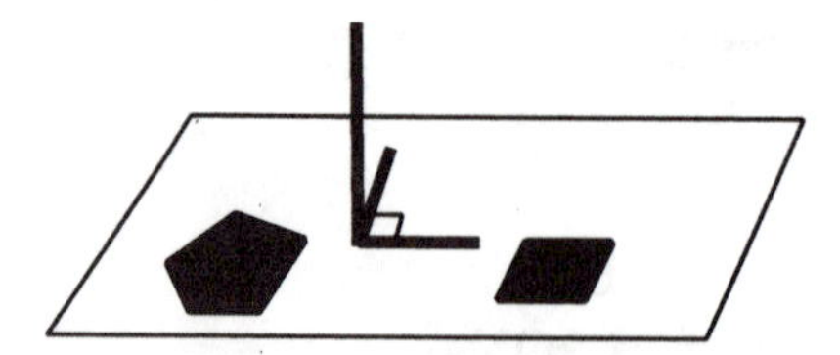

What the reader could see after adding the third toothpick

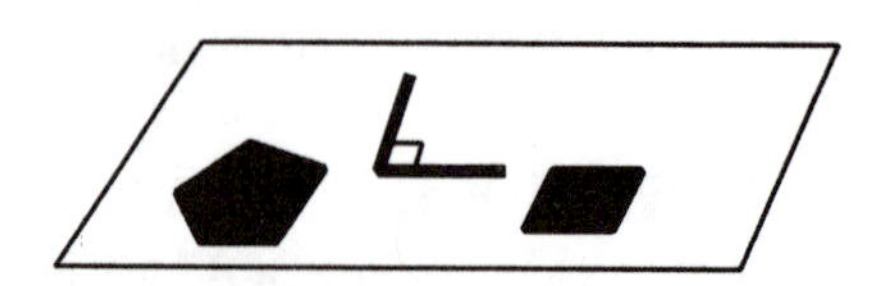

All that the polygons could see

Look. Let me draw you a diagram of what the three dimensions would look like if I had to draw them on a piece of paper in your cramped two-dimensional world. Take a look at this.

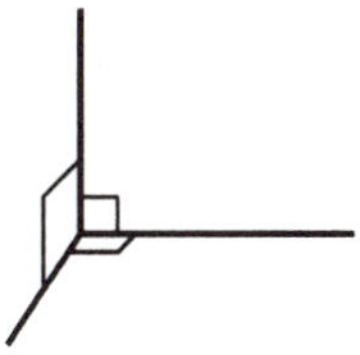

What the reader drew for square and pentagon to look at

"That's nutso," said the square as he looked at the drawing. Anybody can see that those three lines aren't all perpendicular. That line going off to the lower left is clearly not perpendicular to either of the other two. I happen to have a protractor and I will measure it for you and show you that it isn't 90°."

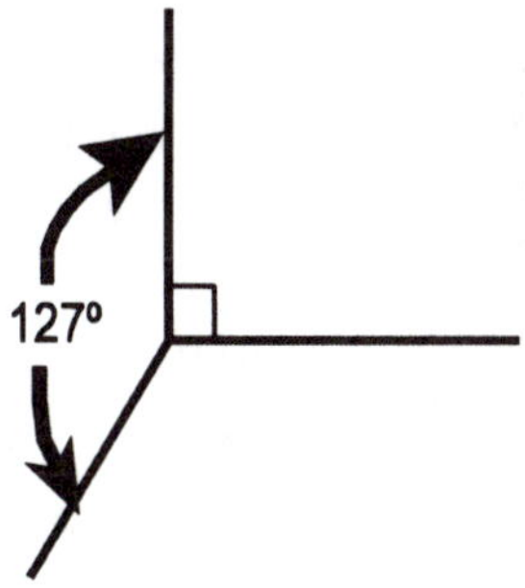

127°

But this is only a two-dimensional representation of a three-dimensional thing. You guys have got to imagine a little bit more than your measly flat little world that you inhabit. Take it from me: there is a third dimension.

"Sure there is. Sure there is," the square said in a soothing voice. "Now just stay calm while these men give you this little injection to quiet your nerves and help you into this white jacket."

THE READER RESUMES TALKING WITH THE AUTHOR

Those people in two dimensions are so closed-minded! You show them a diagram of what three dimensions would look like and they still don't believe. Anybody with half a piece of imagination could see how the third dimension would work.

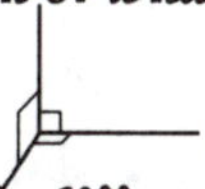

Now it's different with you, Mr. Fancy Pants Author. You never showed me any way that four dimensions would look crammed down into three dimensions, as I did for those unbelievers in two dimensions. That's because you can't.

I can't?

Of course not.

I didn't, because you never asked. Here's four mutually perpendicular segments drawn down onto the two-dimensional page of this book.

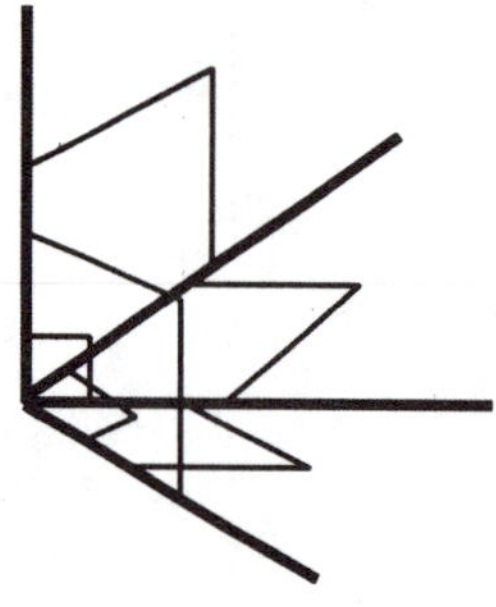
Four mutually perpendicular lines sketched on paper

But, but, but. . . .

I know. You want to tell me that anybody can see that these four lines aren't all perpendicular. You want to haul out your protractor and measure the angles and show me that those aren't 90° angles, but do you remember what you told the square when he made similar complaints about ?

Yeah. I told him that it was only a two-dimensional representation of a three-dimensional object. And now I bet that you're going to tell me that your diagram is only what four dimensions would look like if it were represented in two dimensions.

Yup. And anybody with half a piece of imagination. . .

Yes, and I remember saying those words. But still, the fourth dimension doesn't exist.

You sound just like the square when he wouldn't believe in the third dimension.

But I could prove to the square that there are three dimensions.

How?

Well, I could tell the square that I'm a living, breathing example of something in three dimensions. No, forget that. I know what you're going to say, Mr. Author. You're going to tell me that you are a living, breathing example of something in four dimensions.

But you wouldn't believe me if I did tell you that.

And I can't tell the square to look "up" to see me since he has muscles to move his eyes in only two directions, either north-south or east-west.

And you, my reader, only have muscles to move your eyes in three dimensions. The square didn't even have the word "up" in his vocabulary. He thinks you just invented that word. If I ask you to turn your head snortswise in order to look in the fourth dimension, you would claim that I just made up the word *snortswise*.

You did, didn't you? But still, having the word doesn't prove its existence. How could I prove to the square that there are more than just the two dimensions that he knows of?

Easy. Suppose that the square were in a closed room, like a prison, with no open doors or windows. For a square, such a jail is just a rectangle with wide sides.

Now, all you'd have to do to prove to him that there is a third dimension is just lift him up and place him outside the room. The walls aren't even 1/100" tall.

If his jailers are watching, the square will suddenly vanish from their sight, and when you place him down outside the prison, he will suddenly reappear.

As the jailers watched the square, they reported that the square did not move toward any of the four walls to attempt his escape. He was standing in the middle of his cell and then he disappeared from their sight.

Ha! Now I've got you, Mr. Pretending-That-There's-a-Fourth-Dimension. Have you ever heard of a case in which someone appears or disappears from a locked room without going through a window or something?

It doesn't happen that often, I'll agree. But the most popular book in the world does mention such an event. I'm not sure whether you can ascribe it to a snortswise move, but my mind is open to the possibility. (John 20:19)

If I were standing right in front of you and I moved just 1/100" snortswise, I would disappear from your sight.

One fun thing to imagine is the square and the pentagon "living" on one page of a book. On the next page would be an entirely different universe which they would not be aware of at all. That page might have sheep on it.

I just had a thought. I could prove to those two-dimensional folk that there is a third dimension. All I would have to do is take one of those sheep and flip it over. No amount of rotating in two dimensions would ever give a sheep that looked like:

Sheep-Rotation Attempt:

Famous saying: NO AMOUNT OF ROTATING WILL EVER FLIP A SHEEP.

Flipping flat sheep would certainly be a nice proof to those two-dimensional folk that a third dimension exists.

And what would it be like if you moved me snortswise (as you call it) into the fourth dimension and flipped me and set me down again? How would I be changed? I probably wouldn't notice any difference at all. I dare you. Do it. I don't believe in this four-dimensional stuff.

I really don't want to.

I double dare you.

Please. You don't know what you're asking. You might not like the results.

Result #1: The little mark that you have on the right side of your face would now be on the left side.

Result #2: Your friends would look at you and say that you looked different somehow. Everyone's face is slightly asymmetrical, some people's more than others'. People would now be seeing you as you see yourself in the mirror.

So far, I'm unimpressed. I triple dare you to take me into the fourth dimension and flip me over.

Result #3: When it came time for you to recite the Pledge of Allegiance, you might look a little funny. Your heart would be on the right side of your body. Your appendix would have also switched sides.

Result #4: Your handedness would have changed. If you used to be right handed, you would now be left handed.

Big deal. I even read somewhere that occasionally people have their intestines in the mirror image of the usual winding. It said that intestinal **situs inversus** ***occurs in about 0.02% of humans.***

This is my ultimate dare: I quadruple dare you.

Why is it that other authors don't experience such assertive readers? Okay, dear reader, here goes, but don't say you didn't ask for it.

Result #5: One of the fun parts of organic chemistry was the study of stereochemistry and how homochirality . . .

Whoa! I'd appreciate a little running translation of all the big words.

Okay. I forgot that you might not have had organic chemistry yet. Organic chemistry is usually taken in college after at least a year of inorganic [regular] chemistry. Let me begin again with a running translation in brackets.

Result #5: One of the fun parts of organic chemistry [that's the chemistry of carbon compounds] was the study of stereochemistry [the study of carbon compounds which come in two different mirror-image forms of each other. You may not remember, but back in the 1960s doctors prescribed Thalidomide as a sedative for pregnant women. That caused a lot of birth defects. It turned out that Thalidomide was a mixture of both right-handed and left-handed forms of the same molecule. One of these two mirror-image forms really messed up the fetuses.] and how homochirality [single-handedness] is the norm in biological processes [our bodies can usually only deal with only one form of the mirror-image carbon molecules].

Every human being can only digest dextro sugars [right-handed sugars]. If I flip you in the fourth dimension, you will be the only person on the planet who will be able to digest only the laevo form of glucose [left-handed sugar].

All the jelly doughnuts and caramel sundaes in the world won't supply you with a single calorie.

Hey. I could stand to shed a couple of pounds. This sounds like a tasty way to do it. No stomach stapling, no giving up sweet and greasy foods, no having to attend support groups.

I'm sorry. I guess I should be a little more clear.

Result #5: You will starve to death.

That's clear. Could you do something a little less lethal.

Original knot

Sure. Here's a piece of string. Tie it in an overhand knot. Notice that if you turn it over like a pancake, you get the very same knot back again. This knot has a handedness that you can't change in three dimensions.

Now if you will hand it to me, I can turn in over in the fourth dimension.

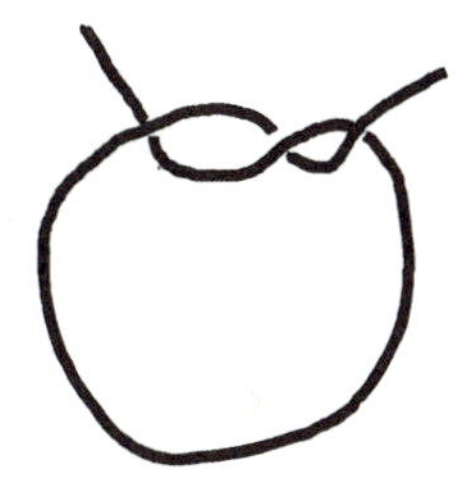

After a trip to the fourth dimension

CLIMBING UP THE DIMENSIONS: 0, 1, 2, 3, 4 . . .

How we measure something depends on what dimension we are in. In zero dimensions, we are just a point (●) and we can't go anywhere. In one dimension, we are on a line, and we could measure, say, one inch (—). In two dimensions, we are on a plane, and we could measure, say, one square inch (■). In three dimensions, we are in our everyday world, and we could measure, say, one cubic inch (◻).

The trick is to see the pattern in going from point to segment to square to cube. Once you see it, then it's an easy step to the fourth dimension.

If this were a classroom, then tonight's adventure would be to see if you could find the pattern in going from point ●

to segment —

to square ■

to cube ◻ .

It might take a half hour's thought to see the pattern in these stepping stones through the dimensions.

Some of the artist-geometry students, after seeing the pattern, could come back the next day with a drawing of a four-dimensional "cube." (It's called a tesseract.)

Before you turn to the next page and see how it's done, would you please stick a bookmark here and shut the book for a tenth of a second and try to see the pattern on your own. Please.

So, the question is how do you go from a point to the segment? Take *two* points and connect them.

How do you go from a segment to a square? Take *two* segments and connect them.

How do you go from a square to a cube? Take two squares and connect them. (Connect corresponding vertices.)

And how do you go from a cube to a tesseract? Take two cubes

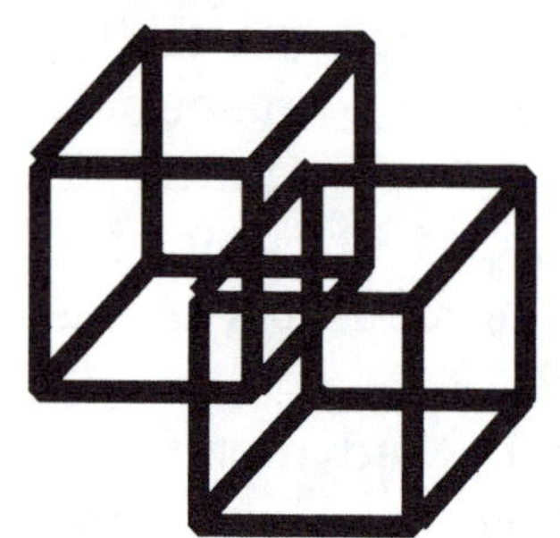

and connect corresponding vertices.

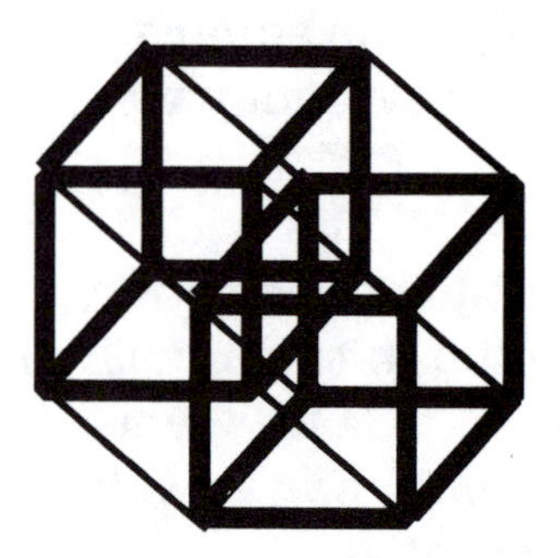

If you draw it with all the same thickness of lines and very symmetrically, it makes a pretty picture.

Start with this and then connect the vertices.

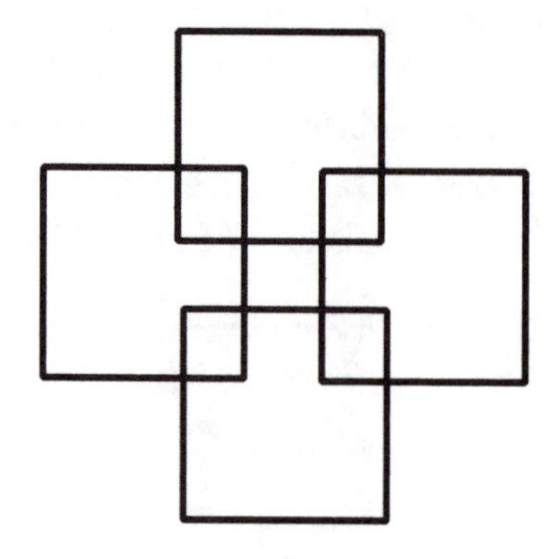

You get:

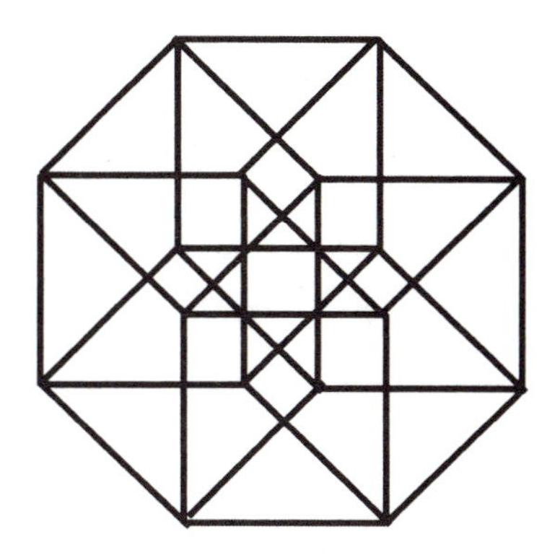

the tesseract

By gluing together toothpicks, students could make a three-dimensional model of a four-dimensional tesseract. You start by creating a cube. Then you make a second cube that is interlocking with the first cube. Then you connect corresponding vertices of the two cubes.

Your three-dimensional model will have more real right angles than the drawing I made above. But since you are only in three dimensions, you are somewhat limited. At every vertex, one of the four toothpicks won't be at right angles to the other ones.

If you look at just one vertex, it will look a lot like the drawing in the upper left-hand corner of p. 458.

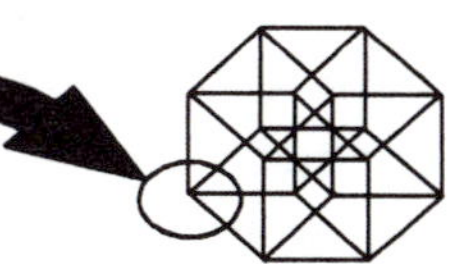

NOW LET'S DO A LITTLE COUNTING

Here's what you knew before we started this chapter:

		vertices	edges	faces	cubes
0 dim.	point	1			
1 dim.	segment	2	1		
2 dim.	square	4	4	1	
3 dim.	cube	8	12	6	1

Now we'll go a little further. It's not *that* hard to count the vertices on the tesseract at the top of the previous page. There are 16 vertices. And there are 32 edges (segments) in the diagram.

	vertices	edges	faces	cubes	tesseracts	hyper-tesseracts
point	1					
segment	2	1				
square	4	4	1			
cube	8	12	6	1		
tesseract	16	32			1	
hyper-tesseract						1

Wait a minute! Stop the bus! You were trying to lull me to sleep with all this counting stuff, and then you slip in this "hypertesseract" in your chart. Just when you have me thinking that just possibly, in the wildest stretch of my imagination, that there might be, just for purposes of discussion, a fourth dimension—now are you really talking about a fifth dimension?

Sure. Why limit yourself? All you do is take two tesseracts and join corresponding vertices. Child's play.

I once had a student who brought a toothpick model of a hypertesseract into class. It was beautiful.

Looking at the chart above, I'll bet that a hypertesseract has 32 vertices. I can see the pattern for the vertices column. It doubles. And that makes sense since to create the next

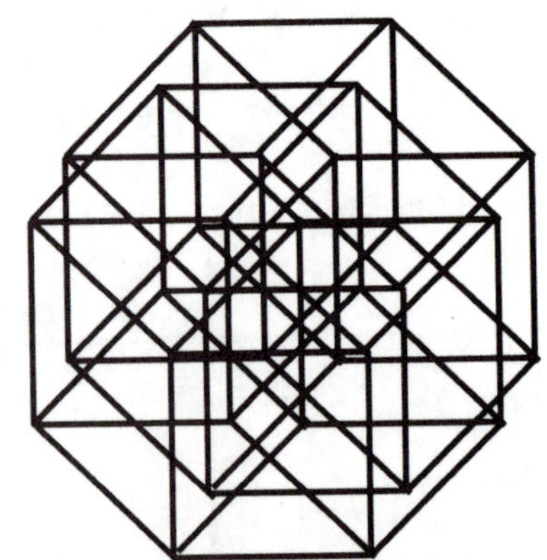

Here are two tesseracts.
If you join the corresponding vertices, you will have a cute little hypertesseract.

dimension, all you do is take two copies of the previous dimension and connect corresponding vertices. So the number of vertices would double.

Looking at that chart some more, I bet that there's 8 cubes in a tesseract. I see the pattern running down the diagonal: 2, 4, 6, 8. There's four cubes that are really easy to count in the tesseract. But the other four are a little more hidden. For example, if I may draw in your book:

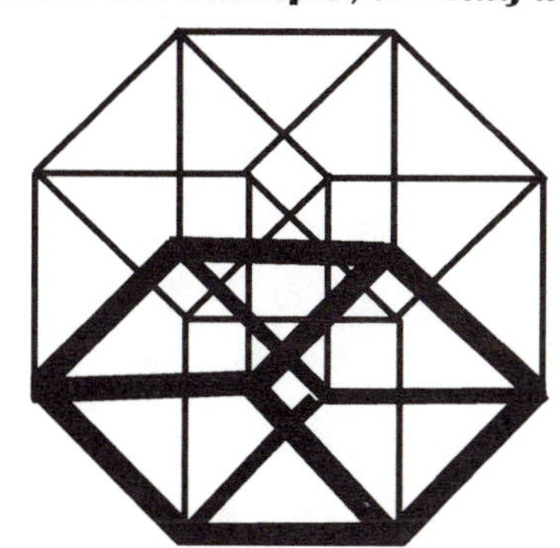

There are eight cubes. Our chart now looks like:

	vertices	edges	faces	cubes	tesseracts	hyper-tesseracts
point	1					
segment	2	1				
square	4	4	1			
cube	8	12	6	1		
tesseract	16	32		8	1	
hyper-tesseract	32				10	1

What is important to realize is that this chart is universal. You get the same numbers if you do it in Kansas or Kentucky or . . . Mars. It's like the multiplication table. $2 \times 3 = 6$ everywhere in the universe. If you look at a slightly larger portion of the chart, you can start to see all kinds of patterns.

The Chart of the Universe in Lots of Dimensions

	vertices	edges	faces	cubes	tesseracts	hyper-tesseracts	hyper-hyper-tesseracts	hyper-hyper-hyper-tesseracts
0 dim. point	1							
1 dim. segment	2	1						
2 dim. square	4	4	1					
3 dim. cube	8	12	6	1				
4 dim. tesseract	16	32	24	8	1			
5 dim. hypertes.	32	80	80	40	10	1		
6 dim. hyper^2 tes.	64	192	240	160	60	12	1	
7 dim. hyper^3 tes.	128	448	672	560	280	84	14	1

This chart can come in very handy. If someone walks up to you on the street and says, "I'm planning on building a casino in the 7th dimension. We are going to have to order dice for our craps table. Do you happen to know how many faces there are on a hyperhyperhypertesseract?"

You may smile and say, "There are 672 faces."

And for those really special occasions, here's a more complete chart:

	vertices	edges	faces	cubes	tesseracts	h-tes.	h^2-tes.	h^3-tes.	h^4-tes.	h^5-tes.	h^6-tes.	h^7-tes.	h^8-tes.	h^9-tes.	h^{10}-tes.
0	1														
1	2	1													
2	4	4	1												
3	8	12	6	1											
4	16	32	24	8	1										
5	32	80	80	40	10	1									
6	64	192	240	160	60	12	1								
7	128	448	672	560	280	84	14	1							
8	256	1024	1792	1792	1120	448	112	16	1						
9	512	2304	4608	5376	4032	2016	672	144	18	1					
10	1024	5120	11520	15360	13440	8064	3360	960	180	20	1				
11	2048	11264	28160	42240	42240	29568	14784	5280	1320	220	22	1			
12	4096	24576	67584	112640	126720	101376	59136	25344	7920	1760	264	24	1		
13	8192	53248	159744	292864	366080	329472	219648	109824	41184	11440	2288	312	26	1	
14	16384	114688	372736	745472	1025024	1025024	768768	439296	192192	64064	16016	2912	364	28	1

Since you know how hard it was to count the eight cubes that are in a tesseract by just looking at the diagram, you can imagine how hard it was for me to count the 13,440 tesseracts that there are in a 10th dimensional solid. Unless, of course, I had found the pattern which works for the whole chart. I will give that to you sometime later in this book.

Chapter Thirteen

Coordinate Geometry

Fred put down the piece of pizza he was holding and looked across the table at P. She was continuing on about the wonderful parents who had adopted her, about how bad her biological father must have been, and about how she pitied her poor twin brother having to grow up in such a family environment.

This is my beloved P, thought Fred. *This is the one I wanted to marry and have children with. I have never met any girl like her. This is the one I wanted to spend the rest of my life with.*

"And I wonder," said P, "where my brother is now."

Fred's brain shorted out at those words. There was so much to rethink. There was so much he needed to tell her. His body made its own choice. He passed out.

Betty and Alexander hopped out of the chairs and hurried over to Fred. They had both studied first aid and knew that in fainting cases you were not supposed to elevate the head (like they often do in the movies). Betty felt his forehead.

"He's burning up," she exclaimed. She wet a napkin and put it on his forehead. Alexander raised Fred's feet so that the blood would flow to his head.

Stanthony rushed over and asked, "Did I put too much hot peppers on the pizza?" Betty assured him that the pizza was just fine.

When Fred's eyes opened, his first words were, "I'm here."

Everyone thought that was a strange statement.

"Of course, you're here," Alexander assured Fred.

No one realized that Fred was responding to P's, "And I wonder where my brother is now."

"And where is *here*?" Alexander asked, repeating the kind of question that losing boxers are sometimes asked by the ringside physicians after a tough fight.

"Don't you know?" Fred answered. "We're at PieOne Pizza, which is at the corner of Pi Street and First Avenue. Such a silly question." He pulled a map out of his pocket and handed it to Alexander.

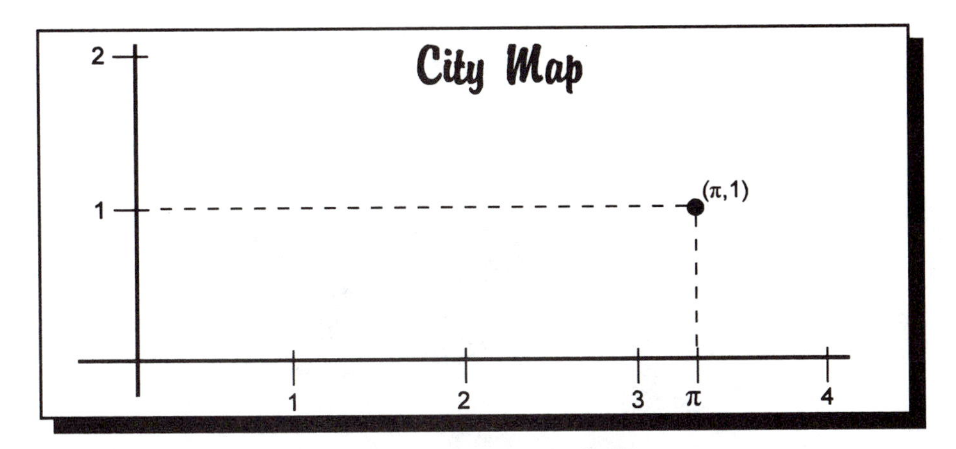

This city map was a mixture of geometry and algebra. The rectangle, the straight lines, the right angles are all from Euclidian geometry. The numbers are from algebra. You could call the mixture: geaolmgeetbrrya, but that would be silly. Instead, combining geometry with algebra is usually called **analytic geometry**.

The first person to really start looking at geometric figures from an algebraic point of view was René Descartes.* In his honor, the city map is an example of what is called the **Cartesian coordinate system**.** If you want to be less fancy, you can call it the **rectangular coordinate system**. If you want to be more fancy, you call it the **orthogonal coordinate system**. Some people just call it graphing.

In the old days before they did this mixing of the two subjects, some of the geometry theorems were stated a little differently. For example, the Pythagorean Theorem wasn't expressed: *In any right triangle, where the lengths of the sides are a, b and c (where c is the longest side),* $a^2 + b^2 = c^2$.

★ ren-NAY day-CART. The French write a lot of s's that they don't pronounce. In 1637, he wrote *La géométrie* which showed how to mix geometry with algebra. His book, along with Newton's *Principia*, are probably the two most influential pieces of scientific writing in the 17th century.

★★ The Latinized form of Descartes is Cartesius. And if you turn *Cartesius* into an adjective, you get *Cartesian*. Recent surveys indicate that less than 46% of all college graduates know this.

Instead, the Pythagorean Theorem was expressed without any reference to numbers at all: *The areas of the top two squares are equal to the area of the bottom square.*

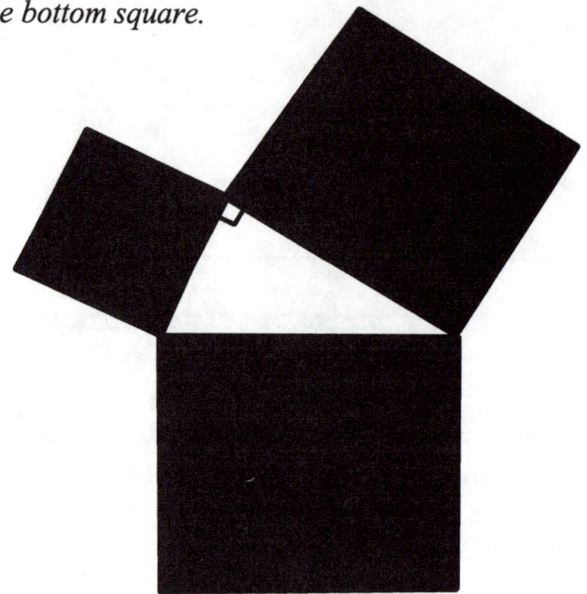

Alexander used the map as a fan to try and cool off Fred. Unfortunately, he waved it too close to Fred's face. The map caught the tip of his nose and cut the map.

"Oh dear," said P. "I have the same problem with my nose. When I wash my face, I have to be careful not to hurt my hands."

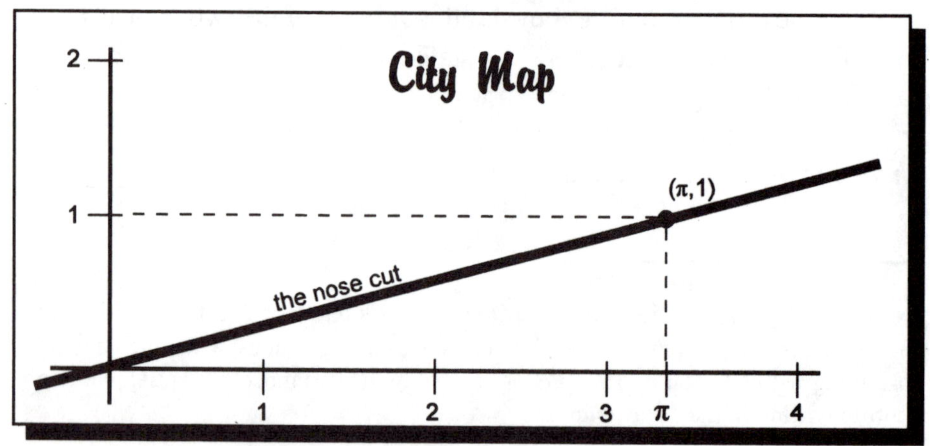

The cut on the map went through the origin. . . . Time out. Here's a quick review of all of the graphing vocabulary from algebra.

Quick Review Chart for Graphing

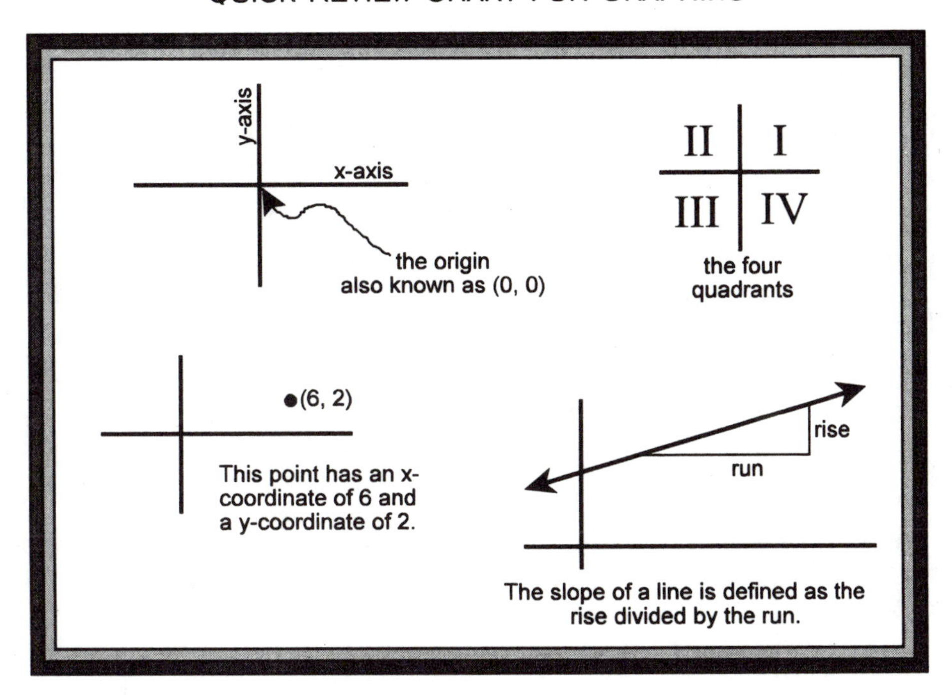
y-axis
x-axis
the origin
also known as (0, 0)
II
I
III
IV
the four
quadrants
●(6, 2)
This point has an x-coordinate of 6 and a y-coordinate of 2.
rise
run
The slope of a line is defined as the rise divided by the run.

Your Turn to Play

1. In what quadrant are the points: (3, –7), (–5, –1), and (4, 0)?
2. If a point is in QII, what can you say about its x-coordinate and its y-coordinate?
3. If line ℓ passes through (3, 4) and (10, 20), what is the rise (which is the change in the y-coordinate)?
4. What is the slope of ℓ in the previous problem?
5. [Fill in one word.] If the slope of line m is zero, then the ___________ must be zero.
6. [Fill in one word which begins with "h".] If the slope of line m is zero, then the line must be ____________________.
7. [Fill in one word.] If two lines have the same slope, then they must be ____________________.
8. [Fill in one word beginning with "r".] If a line is vertical, then its _______ is equal to zero. In this case, its slope is undefined, since we can't divide by zero.
9. Compute the slope of a line which passes through (5, 6) and (8, 2).

10. What is the slope of the "nose cut" line on the City Map (two pages ago)?

COMPLETE SOLUTIONS

1. (3, –7) is in QIV. (–5, –1) is in QIII. (4, 0) is on the x-axis and isn't in any quadrant.

2. The x-coordinate must be negative and the y-coordinate must be positive.

3. The y-coordinate changes from 4 to 20 and hence increases by 16.

4. The x-coordinate changes from 3 to 10 and hence increases by 7. This is the run. The slope (= rise/run) of ℓ is 16/7.

5. rise

6. horizontal

7. parallel

8. run

9. slope = rise/run = –4/3.
The slope is negative. Lines with negative slope run downhill: ↘

Those with positive slope run uphill: ↗

Those lines with zero slope are horizontal: →

Those with undefined slope are vertical: ↑

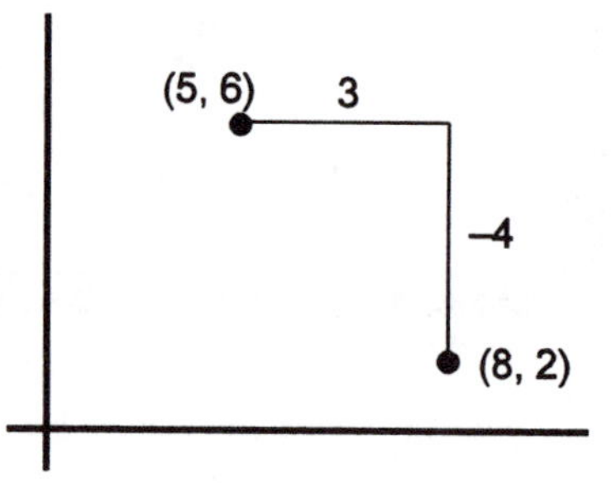

10. The line through (0, 0) and (π, 1) has a rise of 1 and a run of π. It's slope is $1/\pi$.

"There's a reason why our noses are so similar," Fred said to P as he got back into his chair.

"And our eyes are alike also," P said.

Fred now had the conversational opening he needed. Over the next five minutes, Fred carefully broke the news to her. He didn't know how she was going to receive it.

His last words were, ". . . and so, P, you are my sister."

She smiled and said, "Oh goody. That makes you my brother. My own dear brother. I never had one before." She hopped off her chair and ran over and gave Fred a sisterly hug.

Fred blushed.

All his thoughts of attraction to the opposite sex, of marriage and family, and of romantic love descended into the pit of oblivion called pre-adolescence.

The Hecks Kitchen grill would become the world's most expensive grill that a six-year-old ever gave to his sister.

Fred thought of a sign he might hang in his office:

REALITY IS TRULY WEIRD.

P took a bite of her pizza and asked Fred, "Say, brother, I guess I should get to know you better. Do you like sports?"

"Sure. I like everybody, including sports."*

Fred was evidently too far out of touch to participate in ordinary conversation at this point. While he experienced his post-traumatic symptoms, let's take a break and look at three formulas from analytic geometry that we'll use when we do geometry proofs by algebra later in this chapter.

Your Turn to Play

1. The first formula is **the distance formula**. Let's find the distance between the points (2, 3) and (7, 9).

What are the lengths a and b in the diagram?

By the Pythagorean Theorem, what is d (which is the length we're looking for)?

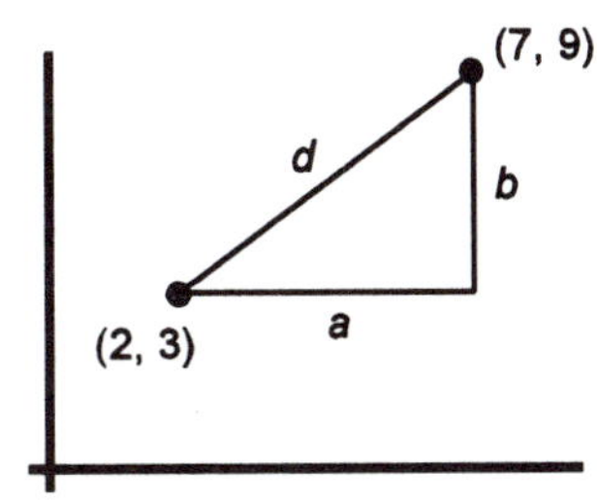

2. What is the distance between (4, 6) and (20, 30)?

3. What is the distance between (3, 7) and (8, 4)?

4. What is the distance between (5, 5) and (–3, –3)?

---

* A *sport* in biology is the child that is singularly different than the parents, a mutation.

5. Now doing the same thing with letters that you did with numbers, find the distance between the points (x_1, y_1) and (x_2, y_2)?

Those little 1's and 2's are called **subscripts**. Eventually in mathematics we sometimes start to run out of letters to use. If I had to name six points, I could call them (y, z), (w, x), (u, v), (s, t), (q, r), (o, p), but it's much neater and easier to call them (x_1, y_1), (x_2, y_2), (x_3, y_3), (x_4, y_4), (x_5, y_5), (x_6, y_6). I bet that without even looking, you can name the fourth point in that list.

6. The second formula is **the midpoint formula**. To find the midpoint between (5, 6) and (8, 2), you take the averages of the x-coordinate and of the y-coordinate.

The average of 5 and 8 is $\frac{5+8}{2}$ which is 6½.

The average of 6 and 2 is $\frac{6+2}{2}$ which is 4.

The midpoint is (6.5, 4).

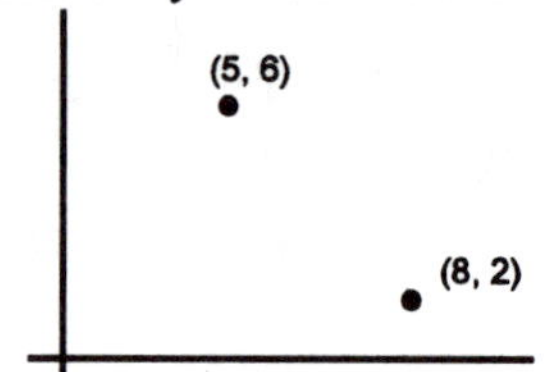

Find the midpoint between (3, 9) and (5, 19).

7. Find the midpoint between (4, −7) and (44, 77).

8. The slopes of parallel lines are always equal (or in the case of vertical lines, both slopes are undefined).

The third formula is that the slopes of perpendicular lines are negative reciprocals of each other. (Reciprocal means turn upside down. The reciprocal of 3/4 is 4/3.)

Examples: If line ℓ has a slope of 5/9, then the line ⊥ to ℓ has a slope of −9/5.
If line m has a slope of 4, then the line ⊥ to m has a slope of −1/4.
If line n has a slope of −1/7, then the line ⊥ to n has a slope of 7.

Another way of expressing this third formula is: The product of the slopes of two perpendicular lines always equals −1.

We prove this formula in *LOF: Advanced Algebra.* The proof uses similar triangles and involves combining five(!) equations. It's long but not hard.

There is one case in which the slopes of perpendicular lines are *not* negative reciprocals of each other. What is that case?

1. a is the horizontal change between (2, 3) and (7, 9). It is the change in the x-coordinates. It is the change going from 2 to 7. It is 7 − 2. $a = 5$.

b is the vertical change between (2, 3) and (7, 9). It is the change in the y-coordinates. It is the change going from 3 to 9. It is 9 − 3. $b = 6$.

By the Pythagorean Theorem, $d^2 = a^2 + b^2$.

$d^2 = 5^2 + 6^2$

$d^2 = 61$

$d = \sqrt{61}$

2. In going from (4, 6) to (20, 30), the change in the x-coordinate is 20 – 4 = 16. The change in the y-coordinate is 30 – 6 = 24.

$d^2 = 16^2 + 24^2$

$d = \sqrt{832}$

3. The distance between (3, 7) and (8, 4) is $d^2 = 5^2 + (-3)^2$. $d = \sqrt{34}$
4. The distance between (5, 5) and (–3, –3) is $d = \sqrt{(-8)^2 + (-8)^2}$ which is $\sqrt{128}$.
5. The distance between (x_1, y_1) and (x_2, y_2) is $d = \sqrt{(x_2 - x_1)^2 + (y_2 - y_1)^2}$.

This is called the distance formula.

6. The midpoint between (3, 9) and (5, 19) is (4, 14).
7. The midpoint between (4, –7) and (44, 77) is (24, 35).
8. If one of the two lines is horizontal and the other is vertical, then their slopes are zero and undefined. (Some people say their slopes are 0 and ∞, where ∞ stands for infinity. In any event, it's really tough to get a good clean simple answer when you multiply zero times infinity. You could argue that 0 × ∞ is equal to 0 since zero times any number is equal to zero. On the other hand, you could argue that 0 × ∞ is equal to infinity since infinity times any number should be infinite. Or, on the third hand, you could argue that 0 × ∞ is equal to –1 to make it fit with every other pair of perpendicular lines. Or, on the fourth hand, you could say that 0 × ∞ is undefined.)

P spoke about the first six years of her life. Of how she had learned to read at an early age. Of how T and L Commonsizerinski had taken her off to the public library so that she could get her first library card.

She thought Fred looked a little hungry and handed him a piece of pizza. Fred ate it as he listened to P.

She described how her mom had bought her a red wagon so that she could bring home lots of books from the library. Although math was her favorite, she read all kinds of subjects. There was a month in which she just read plays. Her favorite was "A Man for All Seasons."

She handed Fred another piece of pizza. He ate it.

Betty and Alexander were staring at Fred. They couldn't believe it. He had eaten two whole regular-sized pieces of pizza.

And then there was the month that P read all kinds of psychology books. She didn't especially like the "sicko stuff," as she called it, where the psychologists describe all the bad things that can happen to the mind. Instead she really enjoyed Abraham Maslow's *Toward of a Psychology of Being*, where he describes very healthy minds and how they function.

Doing her sisterly duty, she handed Fred a third piece of pizza.

P talked about all the history books that she had enjoyed. And the poetry she loved.

Fred was entranced. Here was someone who had read many of the same things that he had read. So many people that Fred met in his everyday life could only talk about what they read in the newspaper or saw on television.

Alexander looked at his watch and announced, "When does P have to get home?"

Betty and Alexander stood up. P and Fred hopped off their chairs. They all said goodnight. Fred grabbed his doggie bag containing his unfinished salad. Betty and Alexander headed out to their car. P and Fred went walking toward P's home.

As they walked the city streets they first talked about "A Man For All Seasons." They discussed how much the hero of the play really valued the presence of law in society even up to the moment they chopped off his head. They talked about how you decide which laws to obey when two laws are in conflict with each other.

They came to the edge of a construction area in the city. It covered many city blocks. Surrounded by a chain link fence, it was the new city park.

Since it was only eight o'clock, they decided to walk around the perimeter. One corner of the park was at 5th Street and 7th Avenue. A second corner was at 13th Street and 9th Avenue. A third at 15th Street and 15th Avenue, and the last one at 7th Street and 13th Avenue.

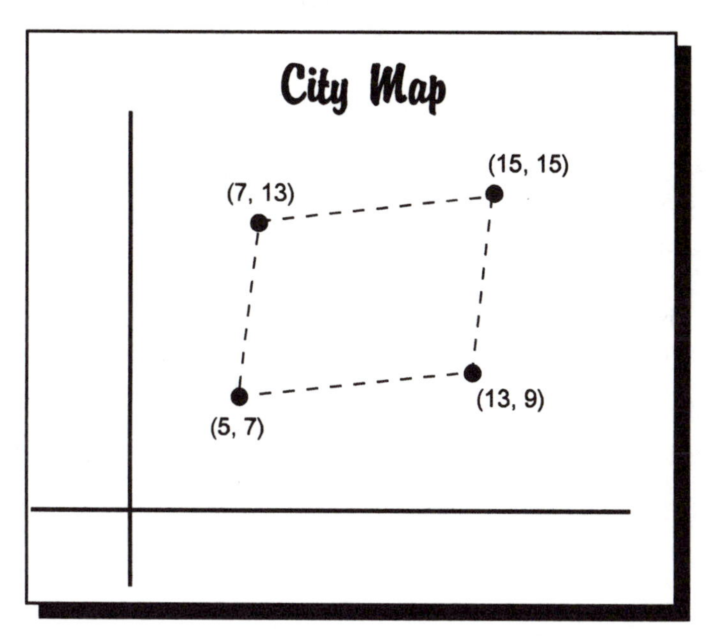

"They could call it Parallelogram Park," said P.

"I can think of three ways to prove that it's a parallelogram," announced Fred.

"Ha! I can think of four ways," said P.

Your Turn to Play

1. If the vertices of that quadrilateral were ABCD, then one way to prove that it is a parallelogram is to find the slopes of the opposite sides. If the slopes of the opposite sides are equal, then the opposite sides are parallel and you would have a parallelogram by definition.

Describe three other ways that you could prove that the park is a parallelogram. If you would like a hint (for two of the ways), look at P's essay, "Have You Got What It Takes?" on p. 198.

COMPLETE SOLUTIONS

1. In what P called the [frog] Theorem, if the opposite sides of a quadrilateral are congruent, then it is a parallelogram. By the distance formula, you can find the lengths AB and CD and show that they are equal. Then find the lengths BC and AD and show that they are equal.

Another way to establish that the park is a parallelogram is using what P called the [mouse] theorem: If a pair of sides are both congruent and parallel, then it is a parallelogram. So if you used the slope formula to find that $\overline{AB}$ and $\overline{CD}$ were parallel and the distance formula to show that AB = CD, then you would be finished.

The last way is to use the theorem in problem 3 of Camden at the end of chapter six: *If the diagonals bisect each other, then it's a parallelogram.* This might be the easiest way. (The distance formula involves square roots and it is usually the hardest to deal with.) All I have to do is show that the midpoint of (7, 13) and (13, 9) is the same as the midpoint (5, 7) and (15, 15).

(15, 15)
(7, 13)
(13, 9)
(5, 7)

Midpoint of (7, 13) and (13, 9) is $(\frac{7+13}{2}, \frac{13+9}{2})$ which is (10, 11).

Midpoint of (5, 7) and (15, 15) is $(\frac{5+15}{2}, \frac{7+15}{2})$ which also is (10, 11).

They talked about history ("How much did Roosevelt know before Pearl Harbor?"). They saw who could jump the farthest. They sang a couple of Gilbert and Sullivan songs.

Fred drew in the dirt a picture of a sphere of radius 5. He knew that P had seen graphs in two dimensions and he wanted to give her a peek at something graphed in three dimensions. The x-, y-, and z-axes, he explained to her, were like the corner of a room.

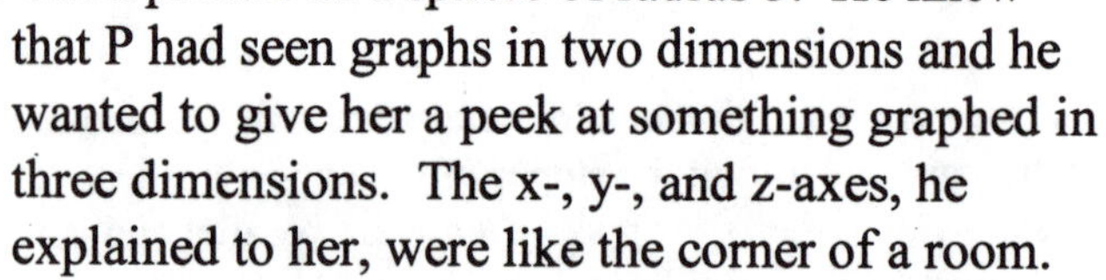

Z
Y
X
$x^2 + y^2 + z^2 = 25$

"That's like a circle in three dimensions," P said. "And since you have three dimensions, your points have three coordinates. The point (3, 0, 4) would be on that sphere since $3^2 + 0^2 + 4^2$ equals 25."

Your Turn to Play

1. Before we go drifting off into three dimensions with P and Fred, it might be appropriate to look at a circle in plane geometry.

A circle with radius equal to 5 is the set of all points in the plane that are 5 units away from the origin.

It is the set of all points (x, y), where the distance between (x, y) and (0, 0) is equal to 5.

Using the distance formula, find the equation for the circle with its center at the origin and whose radius is 5.

2. Find the equation of the circle of radius 5 whose center is at (13, 7).

1. The distance between (x_1, y_1) and (x_2, y_2) is $d = \sqrt{(x_2 - x_1)^2 + (y_2 - y_1)^2}$.
Substituting (x, y) for (x_1, y_1), and substituting (0, 0) for (x_2, y_2), we get

$$5 = \sqrt{(x - 0)^2 + (y - 0)^2}$$

$$5 = \sqrt{x^2 + y^2}$$

and squaring both sides of the equation $\quad 25 = x^2 + y^2$

In general, the equation of a circle of radius r centered at the origin is $x^2 + y^2 = r^2$.

2. $$5 = \sqrt{(x - 13)^2 + (y - 7)^2}$$

squaring both sides $\quad 25 = (x - 13)^2 + (y - 7)^2$

In general, the equation of a circle of radius r, centered at (a, b) is $\quad (x - a)^2 + (y - b)^2 = r^2$.

"I bet you never heard of four dimensions," P challenged Fred.

"Well. . . . " Fred began.

"A circle in four dimensions would look like $w^2 + x^2 + y^2 + z^2 = r^2$," P declared. "And instead of an ordered pair, like (5, 7), which is what you have in measly old plane geometry, or like an ordered triple, like (2, 3, 9), like you have in three-dimensional graphing which is done in calculus, you would have an ordered quadruple, like (4, 13, 2, 5). Isn't that neat?"

And since they were six-year-olds who were trying to outdo each other, Fred couldn't resist: "And an ordered 5-tuple would be used for graphing in five dimensions. And in five dimensions the distance between the points $(v_1, w_1, x_1, y_1, z_1)$ and $(v_2, w_2, x_2, y_2, z_2)$ would be

$$d = \sqrt{(v_2 - v_1)^2 + (w_2 - w_1)^2 + (x_2 - x_1)^2 + (y_2 - y_1)^2 + (z_2 - z_1)^2}$$

and the equation of a sphere in five dimensions would be. . . ."

"Aw, that's too easy," interjected P. "It would be. . . ."

Your Turn to Play

1. It's not fair if those kids have all the fun. What would be the equation of a sphere in five dimensions with a radius equal to *r*?

2. Make a guess. After calculus, mathematicians study spaces in higher dimensions. What is the largest dimension that they study?

COMPLETE SOLUTIONS

1. $v^2 + w^2 + x^2 + y^2 + z^2 = r^2$

2. If you guessed, "A thousand dimensions," you would have guessed too small. About a hundred years ago, David Hilbert went to the max when he created Hilbert Space. It has an infinite number of dimensions.

Fred dropped P off at her house.

After they said goodnight, Fred turned and walked toward his office in the math building. By the time he got to the campus, it had gotten fairly dark. He looked up and saw the stars. *We didn't even get to talk about astronomy,* he thought.

He climbed the two flights of stairs to his office. His tummy felt different. It didn't hurt. It was a new feeling—his tummy was full.

He flossed and brushed his teeth. He looked over to where Lambda had been. He missed her.

In bed, he pulled the blankets up to his chin and began his evening prayers. It had been a long day, this Thursday in his life. He smiled and was soon asleep.

Quinn

1. The **locus of points** (that's the set of points satisfying a given condition) that are a given distance from a point is a circle. What would be the locus of points whose x-coordinate and y-coordinate are equal?

2. What is the locus of points equidistant from two given points, P and Q?

3. Find the lengths of the sides of this quadrilateral.

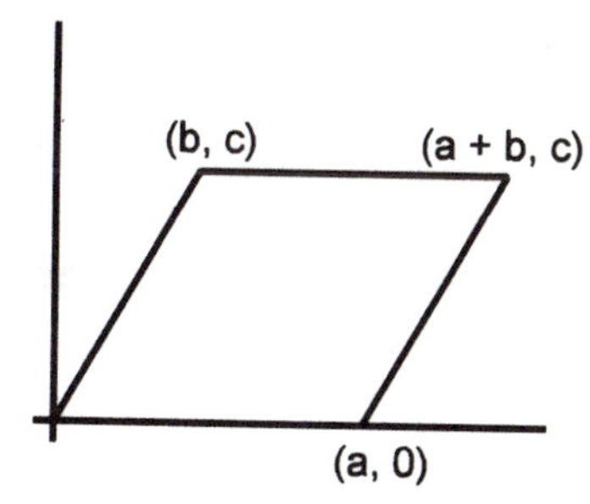

4. Find the midpoints of each of the diagonals of the quadrilateral in the previous problem.

5. Is the triangle with vertices (7, 0), (3, 4) and (2, –1) isosceles?

answers

1. This would be a line (in QI and QIII) at a 45° angle.

2. The ⊥ bisector of $\overline{PQ}$.

3. Two sides are equal to a and two sides are equal to $\sqrt{b^2 + c^2}$

4. Both are $(\frac{a+b}{2}, \frac{c}{2})$

5. Yes.

Zenda

1. A parallelogram has vertices (0, 0), (5, 0), (3, 18). Find the coordinates of the fourth vertex (in QI).

2. A parallelogram has vertices (0, 0), (4, 3), (2, 7). Find the coordinates of the fourth vertex (in QI).

3. A parallelogram has vertices (0, 0), (a, b), (c, d). Find the coordinates of the fourth vertex (in QI). (a, b, c and d are all greater than zero.)

4. Given three points (a, b), (c, d) and (e, f), describe two ways that you could show whether the points are collinear.

5. Is (3, –5) closer to (–2, 4) or to (11, 1)?

answers

1. (8, 18) This can be found by just drawing a graph. There is no need to use the distance or slope formulas.

2. (6, 10)

3. (a + c, b + d)

4. Consider the three segments determined by the given points.

One way would be to determine whether the slopes of any two of the segments were equal. (If two are equal, the third one would also have to be.)

A second way would be to find the lengths of the three segments. The sum of the two smaller numbers equals the largest iff the points are collinear.

5. (11, 1) [$\sqrt{106}$ vs. $\sqrt{100}$]

Ingot

1. What is the locus of points three inches from $\overleftrightarrow{PQ}$?
2. (–8376, 13) is in which quadrant?
3. Find the equation of the circle of radius 3, with its center at (6, 7).
4. Find the equation of the circle of radius 20, with its center at the origin.
5. Find the equation of the sphere of radius 7, with its center at (1, 2, 6).
6. In four dimensional space, what is the distance between the points (1, 2, –3, 5) and (3, –5, –5, 6)? (Hint: If you were to approximate your answer it would be 7.62.)

odd answers

1. Two lines parallel to $\overleftrightarrow{PQ}$.
3. $(x - 6)^2 + (y - 7)^2 = 9$
5. $(x - 1)^2 + (y - 2)^2 + (z - 6)^2 = 49$

Jeffersonville

For all the problems in this city, you are given the triangle with vertices (2, 5), (5, 1) and (9, 3).

1. Find the length of the longest side.
2. Find the midpoint of the shortest side.
3. Find the slope of the median to the shortest side.
4. Find the slope of the altitude to the longest side.
5. Explain two ways you could show whether this is a right triangle.

odd answers

1. $\sqrt{53}$
3. –3/5
5. One way would be to determine the slopes of the three sides and check to see if any pair of slopes were negative reciprocals of each other. (In which case, those two sides would be ⊥.)

A second way would be to find the lengths of the three sides and see if they satisfied $a^2 + b^2 = c^2$.

Karvel

1. What is the locus of points equidistant from the sides of the angle PQR?
2. If (c, d) is in QIII, what can you say about the product cd?

You are given the triangle with vertices (3, 7), (6, 1) and (7, 2).

3. Find the length of the longest side.
4. Find the midpoint of the shortest side.
5. Find the slope of the median to the shortest side.
6. Find the slope of the altitude to the longest side.

Lance Creek

1. (a, b) is on the y-axis. What can you say about either a or b?

2. Show that (7, 10), (10, 14), and (22, 30) are collinear.

3. If a line passes through the origin and through the point (a, b), what is its slope?

4. Find the missing coordinates.

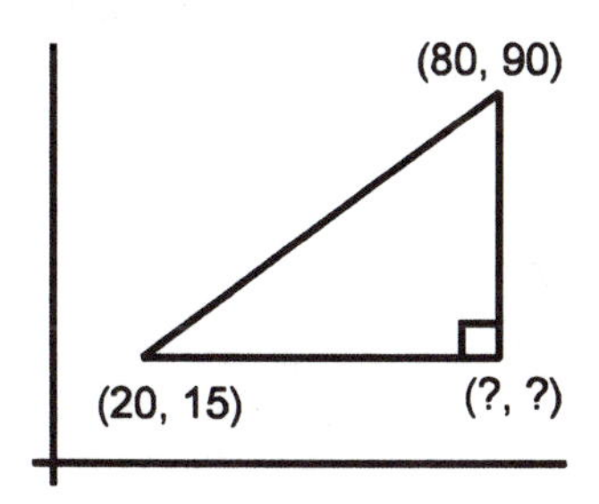

5. Fill in each of the blanks with either PQ, $\overline{PQ}$ or $\overleftrightarrow{PQ}$:

If (2, 5) and (–3, 7) are the endpoints of __?__, then –0.4 is the slope of __?__ and __?__ $= \sqrt{29}$.

FROM CHAPTER 12½

I promised I'd give you a formula which holds for the entire chart of dimensions.

There are lots of formulas which hold for parts of the table. For example, if you multiply the dimension (left column) times the vertices in the dimension above it, you get the edges. (I've marked an example of this in the chart below.)

	vertices	edges	faces	cubes
0	1			
1	2	1		
2	4	4	1	
3	8	12	6	1
4	16	32	24	8
5	32	80	80	40
6	64	192	240	160

$4 \times 8 = 32$

The formula that holds for anywhere on the chart: Take any entry plus twice the entry on its right and that gives you the entry below the one on the right.

a	b
	c

$$a + 2b = c$$

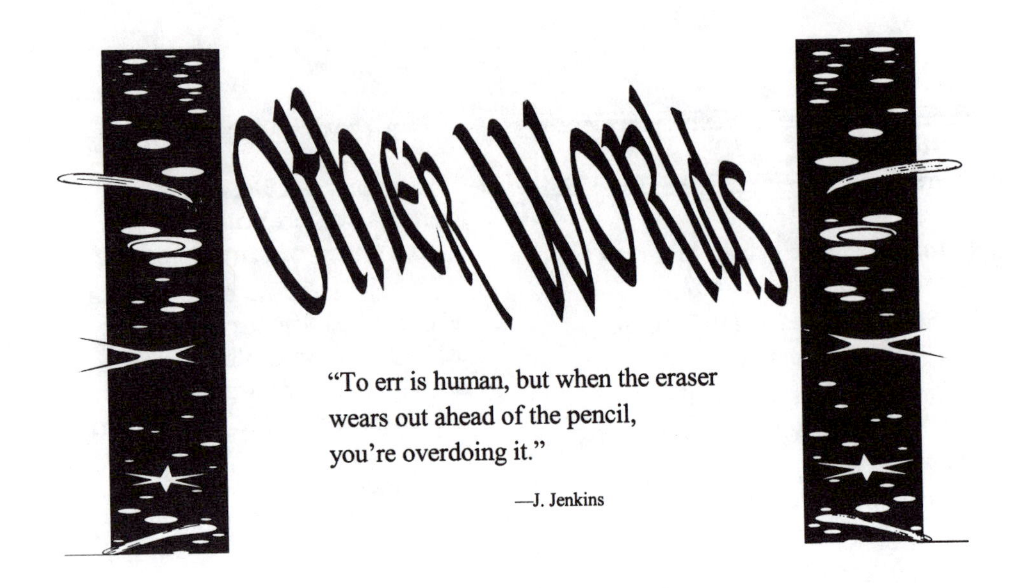
Other Worlds

"To err is human, but when the eraser wears out ahead of the pencil, you're overdoing it."

—J. Jenkins

Chapter 13½
Flawless Geometry

When Fred fell asleep at the end of the previous chapter, you officially finished geometry. You've done it. All the definitions. All the theorems. All the proofs.

If you don't believe me, just fill in your name on the line below and then read this certificate aloud.

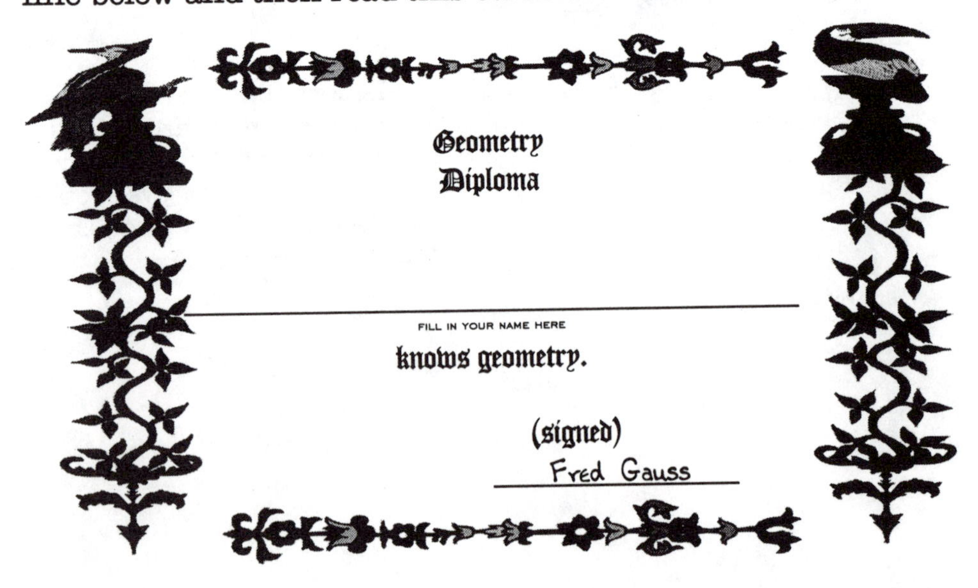
Geometry
Diploma
FILL IN YOUR NAME HERE
knows geometry.
(signed)
Fred Gauss

Chapter 13½ Flawless Geometry

This diploma is good anywhere in the world where it is accepted. With this (and a teaching contract) you could stand in front of a beginning geometry class tomorrow morning and begin telling the story of a Thursday in the life of Fred: *"Fred's two little eyes popped open. It was several minutes before dawn on an early spring morning."* Then you could tell about how he went jogging with Lambda. You could sing to your class the song that Fred wrote, "Another Day, Another Ray." And tell of the Great KITTENS Bank Robbery.

You are ready to teach geometry.

This is the geometry that Euclid wrote in *The Elements* in 300 B.C. and which has been the most popular textbook for 2100 years, all the way up to the end of the 1800s. You know all the magic phrases, such as "Two points determine a line," and "If two angles form a linear pair, then they are supplementary."

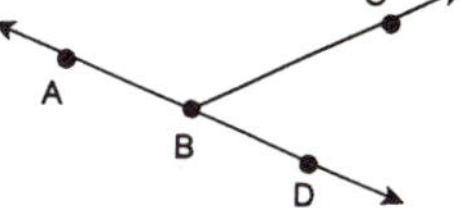

m∠ABC + m∠CBD = 180°

Geometry is one of the best fields in mathematics to teach reasoning. There are proofs in other parts of mathematics—in algebra, in logic, in set theory—but those proofs don't use diagrams like geometry does. In geometry you can *see* what is going on.

A Proof in Geometry

Theorem: The base angles of an isosceles triangle are congruent.

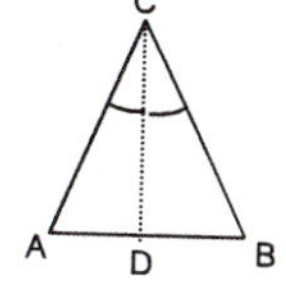

Condensed proof:

Draw the angle bisector. AC = BC by def of isosceles. CD = CD since every segment is congruent to itself. $\triangle ADC \cong \triangle BDC$ by SAS. $\angle A \cong \angle B$ by definition of congruent triangles.

Example of a Proof in Group Theory

[Note: In group theory we know that ai = ia = a, bi = ib = b. Think of "i" as the number 1. We know that $aa^{-1} = a^{-1}a = i$, $bb^{-1} = b^{-1}b = i$. However, we don't know that ab = ba.]

Group Theory Theorem: $a^2 = i$ and $a^{-1}b^2a = b^3$ imply that $b^5 = i$.
(Here is a condensed and almost impossible-to-read proof.)

1	$a^2 = i$ and $a^{-1}b^2a = b^3$	given
2	$a = a^{-1}$	mult line 1 by a^{-1}
3	$b^2a = ab^3$	mult line 1 on left by a
4	$a = b^{-2}ab^3$	mult line 3 on left by b^{-2}
5	$a^2 = (b^{-2}ab^3)(b^{-2}ab^3)$	mult line 4 by itself
6	$i = b^{-2}abab^3$	line 1 and 5 (also combining b^3b^{-2} into b)
7	$b^2 = abab^3$	mult line 6 on left by b^2
8	$b^{-1} = aba$	mult line 7 on right by b^{-3}
9	$b^{-2} = (aba)(aba)$	mult line 8 by itself
10	$b^{-2} = abba$	line 1 and 9
11	$b^{-2} = a^{-1}bba$	line 2 and 10
12	$b^{-2} = b^3$	line 1 and 11
13	$i = b^5$	mult line 12 by b^2 ⊠

Other Worlds

Chapter 13½ Flawless Geometry

Before you accept a geometry teaching position, there is just one little teensy-weensy thing that you should know. It's not a big deal and you can probably do your geometry teaching for years and never run into any real trouble.

> Classified Want Ads
>
> Geometry position available beginning September 1. Lovable students and excellent cafeteria. Send résumé to Mt. Ritzy Academy, Human Resources Division, Lake Quackaduck, Michigan. Please include the number of pages of *Life of Fred: Geometry* that you have read. Salary: $230,000 for the academic year.

How shall I express this?

You should know that Euclid and his *The Elements* are flawed. They are not quite perfect.

Some day Joe and Darlene may be sitting in your geometry class, and Joe may raise his hand with "a little question."

You call on him.

"I gotta question," he says.

Darlene at this point is trying to tell Joe to "cool it" and not make a fool of himself, but it's too late. You have already called on him.

"I gotta geometry proof that I'd like you to look at," he continues.

"Sure, Joe. Please come to the board and show us."

The sky is clear and blue outside. You had a good lunch in the cafeteria (saltimbocca and a chocolate milkshake). Your allergies aren't acting up. And yet . . . lightning is about to strike, as unexpected as a first-time heart attack. Joe is about to destroy 2100 years of Euclid.

Joe wrote on the blackboard:

My Proof of Something Weird

I'm going to prove that every triangle is isosceles.

And I'm only using the official reasons: ① Given; ② Postulate; ③ Definition; ④ Previously proven theorem; ⑤ Algebra; ⑥ Beginning of an indirect proof; and ⑦ Contradiction in steps ___ and ___ and therefore the assumption in step ___ is false.

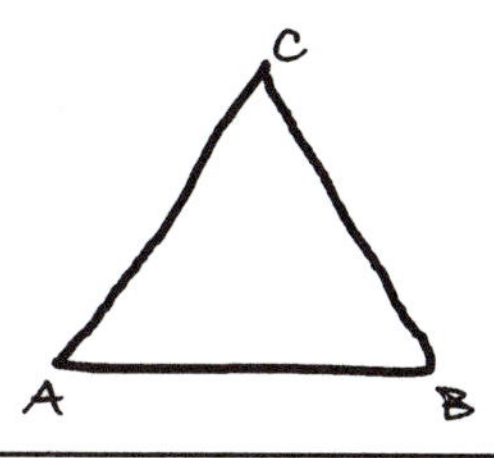

Statement	Reason
1. Any old triangle ABC.	1. Given
2. Draw the angle bisector of angle C. 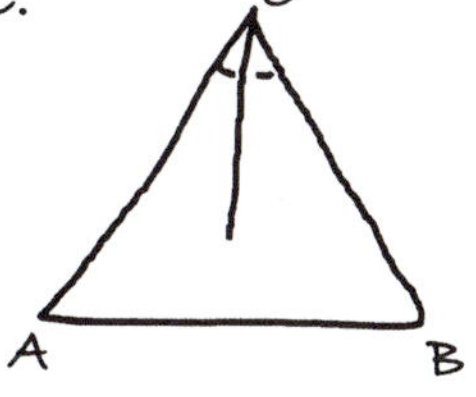	2. By the Angle Measurement Postulate, angle C has a measurement between 0 and 180. Again, by the Angle Measurement Postulate, you can find an angle with half of that measurement.
3. Erect the ⊥ bisector of $\overline{AB}$. 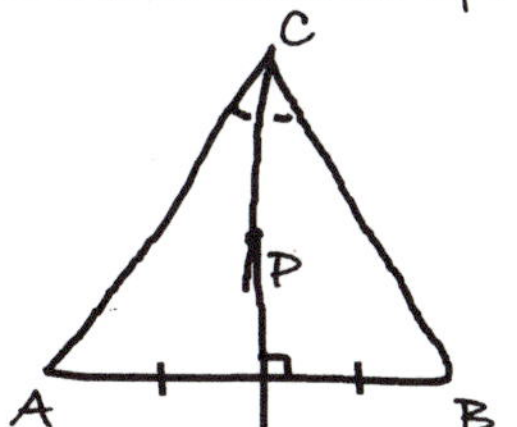	3. Theorem: Every segment has a midpoint; and Theorem: You can erect a ⊥ to a line at any point on the line.
4. From the point of intersection of the angle bisector and the ⊥ bisector . . .	

"Wait a moment," another student calls out. "How in the world do you know that they intersect? They might be parallel."

"Shucks. That would make my proof even easier," Joe replied. "If the angle bisector is parallel to the perpendicular bisector, then the angle bisector would have to be perpendicular to $\overline{AB}$. The theorem says that if a line is perpendicular to one of two parallel lines it is perpendicular to the other."

Joe drew a little diagram on the side blackboard showing an angle bisector that is ⊥ to $\overline{AB}$. "Then the triangles would be congruent by ASA and ∠A ≅ ∠B and △ABC would be isosceles."

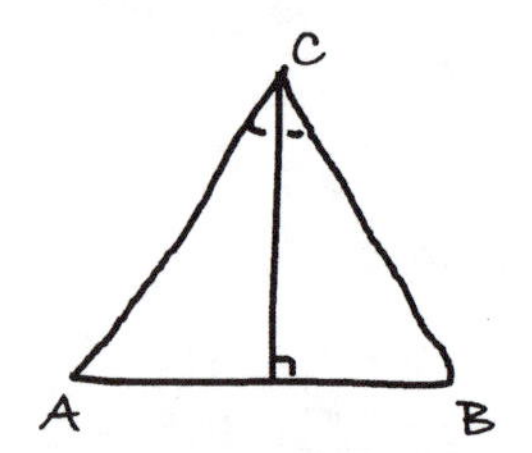

"Okay," the other student conceded. "You've handled that case. Go ahead with your proof."

4. From the point of intersection of the angle bisector and the ⊥ bisector, drop perpendiculars to $\overline{AC}$ and $\overline{BC}$.

4. Theorem: From a point P, you can drop a ⊥.

5. PD = PE

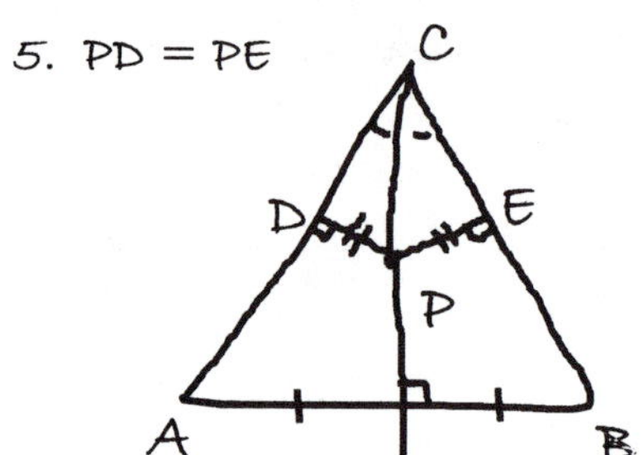

5. Theorem: Any point on an angle bisector is equidistant from the sides of the angle. (Proved in problem 2 in Cambridge at the end of chapter five.)

6. Draw $\overline{AP}$ and $\overline{BP}$.

6. Postulate: Two points determine a line.

7. AP = BP

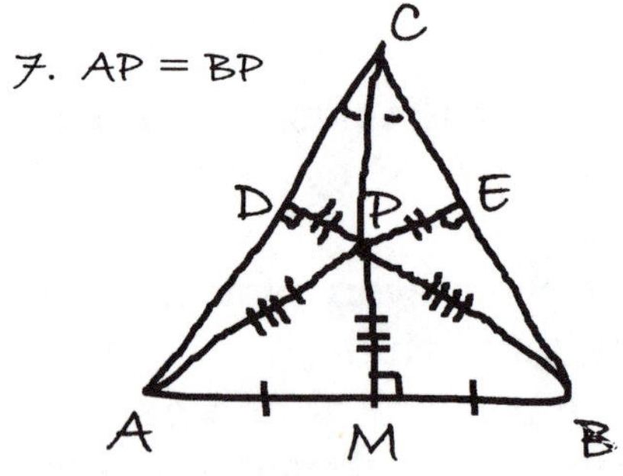

7. Any point on the ⊥ bisector is equidistant from the endpoints of the segment.

8. $\overline{PM} \cong \overline{PM}$

8. Theorem: Every segment is congruent to itself.

9. △APD ≅ △BPE
△APM ≅ △BPM

9. Hypotenuse-Leg Theorem

10. $\angle 1 \cong \angle 2$ and $\angle 3 \cong \angle 4$	10. Def of $\cong$ △s

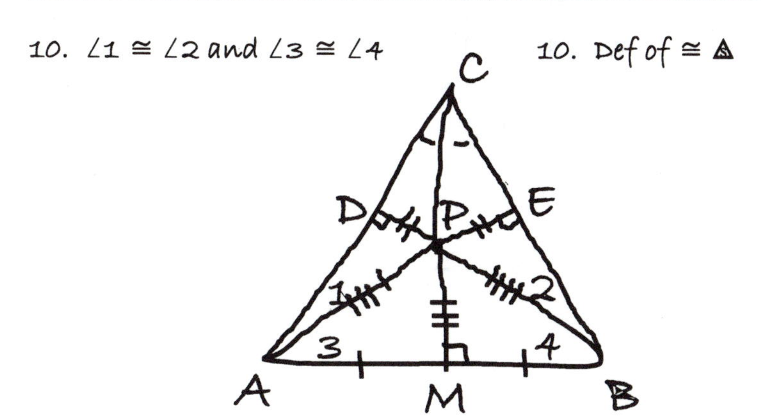

11. $m\angle 1 = m\angle 2$ and $m\angle 3 = m\angle 4$	11. Def of $\cong$ angles.
12. $m\angle 1 + m\angle 3 = m\angle CAB$ $m\angle 2 + m\angle 4 = m\angle CBA$	12. Angle Addition Postulate
13. $m\angle CAB = m\angle CBA$	13. Algebra
14. $\angle CAB \cong \angle CBA$	14. Def of $\cong$ angles.
15. $\triangle ABC$ is isosceles.	15. Theorem: If the base angles are congruent, then the triangle is isosceles. Q.E.D.

Joe puts the chalk down and looks at you. The whole class is looking at you. He has just proven that every triangle is isosceles. And he hasn't made the mistake of saying that C, P and M are collinear.

Every statement in Joe's proof has a legitimate reason and the last line is what he was trying to prove. This is the definition of a correct proof.

A student in the class asks you, "Where did Joe go wrong?"

You look at Joe. "How did you get this proof, Joe?"

"Professor Fred showed it to me. You wanna see another one?"

"Sure," you answer, thinking that your credibility is completely shot anyway. You hand him a yardstick so that his lines will be a little straighter.

Joe erases the board and announces that he's going to show that an obtuse angle (an angle greater than 90°) is congruent to a right angle.

Start with a rectangle.

Draw $\overline{BE}$ so that BE = BC, and m∠CBE is about 10°.

Draw $\overline{DE}$.

Construct the ⊥ bisectors of $\overline{DC}$ and $\overline{DE}$.

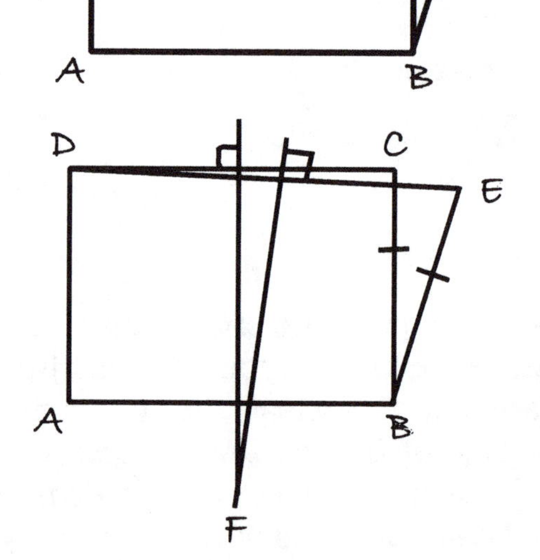

Note that the ⊥ bisectors must intersect. If they didn't, they would be parallel. If they were parallel, then $\overline{DC}$ and $\overline{DE}$ would also have to be parallel, because they are perpendicular to parallel lines. But $\overline{DC}$ and $\overline{DE}$ can't be parallel since they intersect at D.

Note that the ⊥ bisector of $\overline{DC}$ is also the ⊥ bisector of $\overline{AB}$ since the ⊥ bisector of one side of a rectangle is the ⊥ bisector of the other side. It wasn't one of our theorems, but it wouldn't be that hard to prove.

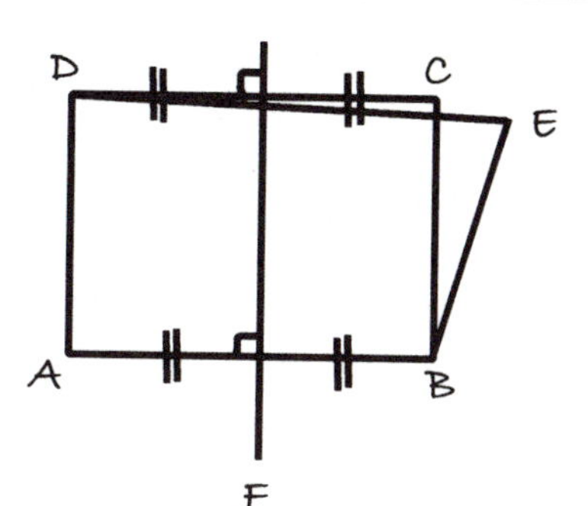

Draw some segments. (Two points determine a line.)

Since F is on the ⊥ bisectors of $\overline{DE}$ and $\overline{AB}$, AF = BF and DF = EF.

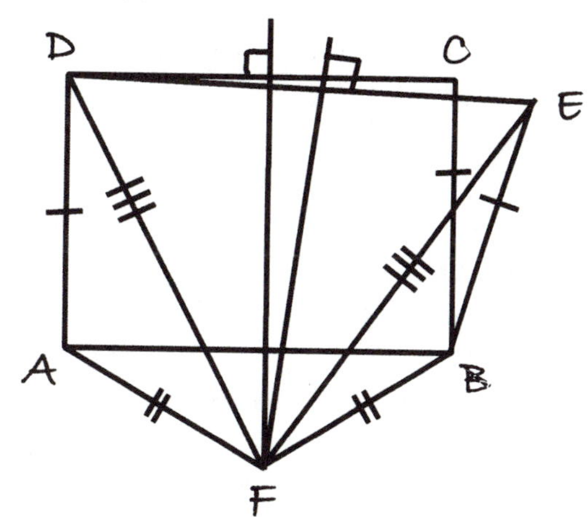

△DAF ≅ △EBF by SSS.
∠DAF ≅ ∠EBF by def of ≅ ⚠ (line *)

∠FAB ≅ ∠FBA since they are base angles of an isosceles △. (line **)

If you subtract the small angles on line ** from the big ones on line *, you get that the right angle (∠DAB) is congruent to the obtuse angle (∠EBA). Q.E.D.

Some of the students in the class are giggling. This is the funniest geometry they have ever seen and they are enjoying the possibility of seeing their $230,000/year teacher squirm a little.

With all the dignity you can muster, you begin, "Some of you have never heard of what happened back in 1899 when a 19-year-old kid named Robert L. Moore faced this situation."

"Did he find the errors in Joe's proofs?" one of your students asks.

"The error isn't so much in Joe as it is in Euclid," you explain. "What Robert L. Moore did was to fix the holes that are in Euclid's system so that you couldn't prove the silly things that Joe just did. This American boy rewrote Euclid's postulates and removed the flaws."

"Well, then, why did we spend all this time on Euclid when we could have been looking at Bob's geometry?" another student asks.

"There is a slight drawback to Robert L. Moore's postulates. The price for perfection is that we don't use diagrams as much as we did with Euclid, and drawing diagrams and making all kinds of marks on them is a fun part of traditional geometry. We lean on the diagram to help us do the proofs. For example, if we take a triangle PQR and draw the angle bisector at P, we look at the diagram and naturally assume that it will hit $\overleftrightarrow{QR}$ somewhere between Q and R.

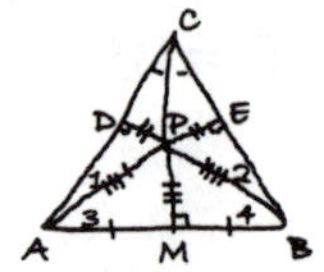

Marking up a diagram is fun

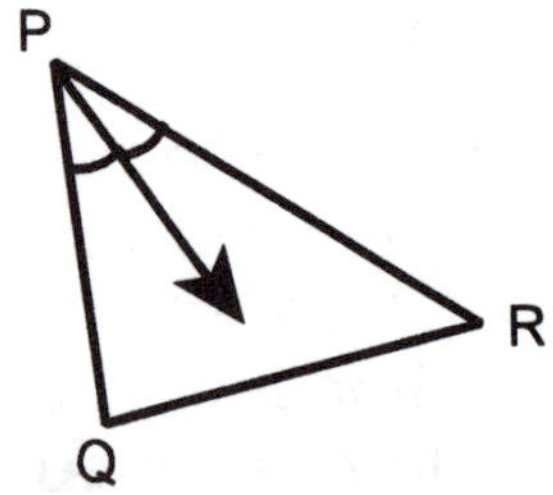

"But we never prove it. The diagrams become hidden postulates, hidden assumptions, which beginning geometry students hardly ever notice. The reason that Joe's proof that *every triangle is isosceles* works is that his diagram is phony. If you take a scalene triangle and accurately draw an angle bisector from the top vertex and accurately draw a perpendicular bisector of the base, they will never meet inside the triangle as they did when Joe drew them. When they meet below the triangle, then his whole proof falls apart."

"And is that also true for his proof that a right angle is congruent to an obtuse angle?" a student asks.

"Sure enough," you say. "If you draw the figure accurately, $\overline{EF}$ will never be to the left of point B."

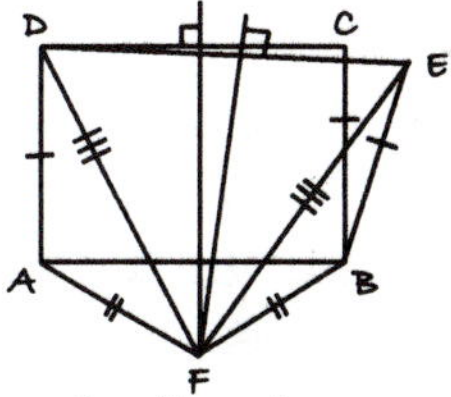

Another phony diagram

Several students at this point will pull out their rulers and T-squares and verify what you've just told them.

Two of your students go into a brief huddle and formulate a Plan of Action. They make up signs and march around the room protesting Euclid.

Nuke the Euke!

Give us more of Moore!

This is usually the time when the Dean of Instruction walks into your room to find out how well you are teaching. One of the students hands him a protest sign to carry. It reads, "We want pure geometry!!!" Later he will call you into his office and ask if you have been teaching "impure" geometry. He doesn't want any parents complaining to him.

"Okay," you announce to your class after the Dean leaves. "I'll give you a glimpse of the pure stuff—the stuff that no other beginning geometry class ever sees."

"A very brief three-sentence history: At the same time (1899) that Moore created his new postulates, David Hilbert created another set of geometry postulates which also got rid of the flaws in Euclid's geometry. The two guys worked separately, but they came up with systems that were essentially equivalent. Later, Tarski came up with another set of postulates which is what I'm going to show you here." [end of history lesson]

GEOMETRY THEORY WITHOUT THE FLAWS

To build a mathematical theory, the first step is the undefined terms. The points will be called a, b, c, d, e, f, g, h, i, j, k, l, m, n, o, p, q, r, s, t, u, v, w, x, y and z. In this theory we use lower-case letters for points, in contrast to Euclidean geometry where we used upper-case letters like P and Q.

We use two upper-case letters: B and D.

B is always followed by three points. For example, B(abc) or B(akx). What it stands for is the idea of betweenness. When you see B(abc), think of a segment whose endpoints are a and c. Point b is somewhere on that segment.

a b c

You can draw the diagram on scratch paper to help imagine B(abc), but there are *no diagrams* in the proofs themselves.

D is always followed by four points. For example, D(abcd). What D(abcd) stands for is the idea that the distance from a to b is the same as the distance from c to d. When you see D(abcd), you might draw on scratch paper:

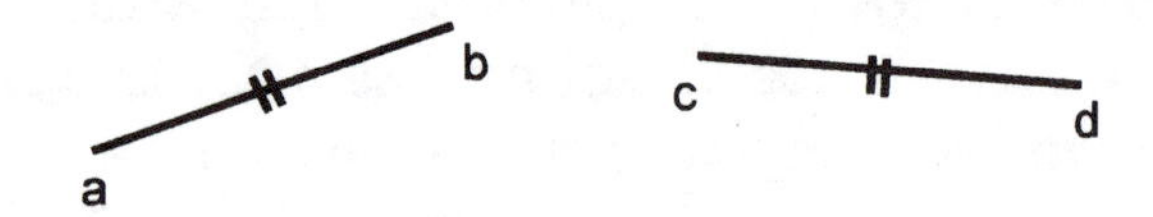

The only other thing that we need to get started are the postulates. Once we have these, we can start "growing" the theory by making definitions and proving theorems.

THE POSTULATES FOR THE FLAWLESS GEOMETRY

A1: If we know that D(abcc), then a = b.

Some notes on A1:
♪#1: "A1" stands for the first axiom. *Axiom* is another word for postulate.
♪#2: A1 says that if the distance from point a to point b is the same as the distance from c to itself, then the distance from a to b must be zero. Point a and point b must be the same point.
♪#2: There are no numbers in any of the postulates in this geometry theory. Even if you never had any arithmetic or algebra, you could study this pure geometry. That wasn't true in the traditional geometry which we studied in this book. The Ruler Postulate, the Angle Measurement Postulate and the use of "Algebra" as a reason in proofs all required a knowledge of the real numbers and of algebra.
♪#3: This axiom is sometimes written: A1: D(abcc) ➜ a = b.
♪#4: If we knew D(ddef), we couldn't use axiom A1 because we need the third and fourth letters to be identical. Similarly, we couldn't use A1 on D(ghhi) or on D(jklj).
♪#4: In a proof, this is how A1 might be used:

Statement	*Reason*
1. D(rtnn)	1. [for some reason]
2. r = t	2. A1

A2: D(abba)

The distance from point a to point b is the same as the distance from b to a.

In a proof you might see A2 used:

Statement	*Reason*
1. D(fyyf)	1. A2

A3: If we know that D(abpq) and we know that D(abrs), then D(pqrs).

The long version: If the distance from a to b is the same as the distance from p to q, and if the distance from a to b is the same as the distance from r to s, then the distance from p to q is the same as the distance from r to s.

The short version: D(abpq) & D(abrs) → D(pqrs)

Your Turn to Play

1. Fill in: D(apht) says that the distance from ___ to ___ is the same as the distance from ___ to ___.

2. Locate d on the figure so that B(adg) will be true.

3. How many undefined terms are there?

4. What axiom would you associate with, "It's as far from my house to yours as it is from your house to mine"?

5. How would you express the diagram in this geometry?

6. Draw a diagram in which both B(ghi) and B(hgi) are true.

7. Fill in the missing reasons. (Your answers will be either A1, A2 or A3.)

Statement	Reason
1. D(affa)	1. ?
2. D(afpr)	2. [Let's suppose this line is Given.]
3. D(fapr)	3. ?
4. D(faaf)	4. ?
5. D(praf)	5. ?
6. D(praa)	6. [Let's suppose this line is Given.]
7. p = r	7. ?
8. D(xxxx)	8. ?

COMPLETE SOLUTIONS

1. From a to p is the same as the distance from h to t.

2. There are several ways to locate d, since d can be any point on the segment from a to g. You could place it between a and g or you could put it on top of either a or g.

3. The undefined terms are B, D, and a, b, c, d . . . x, y, z. There are 28 undefined terms.

4. A2 which states D(abba).

5. This could be either B(tac) or B(cat).

6. If B(ghi) and B(hgi) are both true, the diagram might look like

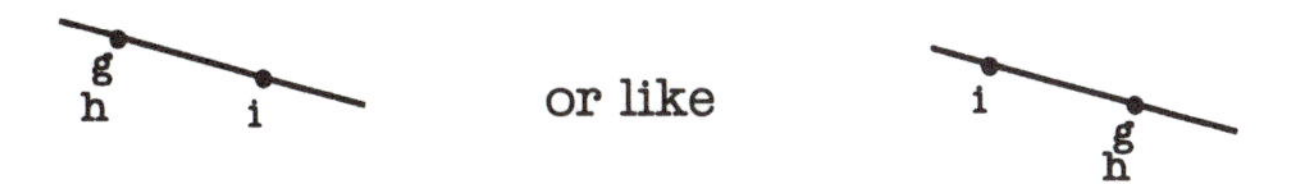

7.

Statement	*Reason*
1. D(affa)	1. A2
2. D(afpr)	2. [Let's suppose this line is Given.]
3. D(fapr)	3. A3
4. D(faaf)	4. A2
5. D(praf)	5. A3 (from lines 3 and 4)
6. D(praa)	6. [Let's suppose this line is Given.]
7. p = r	7. A1
8. D(xxxx)	8. A2

Axiom 3, D(abpq) & D(abrs) → D(pqrs), reminds me of a bit of algebra. In algebra we know that if x = y and y = z, then it must be true that x = z. In algebra this is called the transitive property of equality. One name for A3 is the Transitivity Property for Equidistance.

With one more axiom we can begin proving some theorems. A4 is called the Axiom of Segment Construction.

A4: For any points p, a, b and c, you can find a point x such that B(pax) and D(axbc).

You start with two points, p and x,

and two points b and c.

This axiom states that you can find a point x on the ray from p through a. That point x will be as far from a as point b is from point c.

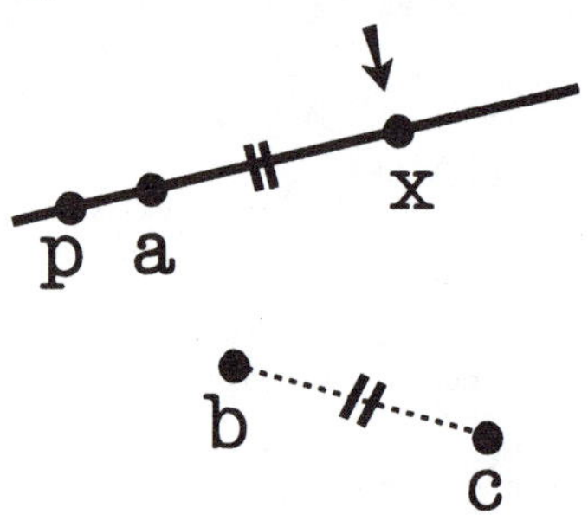

Sometimes A4 is written using a symbol from logic. We have already seen several logic symbols: & ("and"), ∨ ("or"), and → ("implies").

The backwards E, "∃" stands for "there exists." Using ∃, we can abbreviate *There exists an x such that B(pax) and D(axbc)* as ∃x [B(pax) & D(axbc)].

Rules of the Game when using an axiom containing ∃x in a proof: The letter that you introduce using ∃ must be new to the proof. You can't use a letter that has already been used previously in the proof.

Statement	Reason
1. D(affa)	1. A2
2. B(mag) & D(agbc)	2. A4

Okay. g wasn't used before.

Statement	Reason
1. D(affa)	1. A2
2. B(maf) & D(afbc)	2. A4

This isn't okay. f used in line 1

Other Worlds

Chapter 13½ Flawless Geometry

You may be getting the feeling why this flawless geometry is not taught to beginning geometry students. Do you remember the old days when you were given a diagram like this

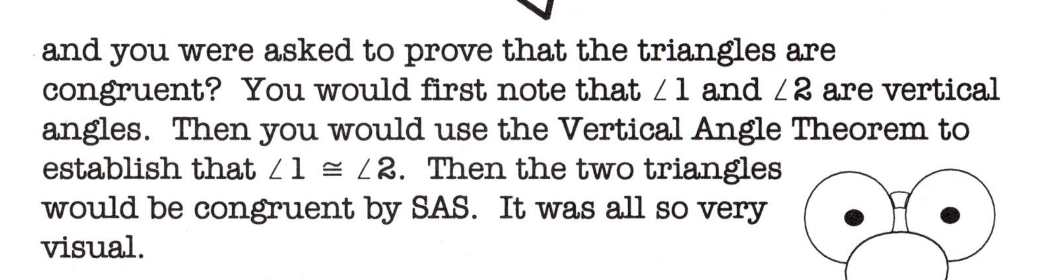

and you were asked to prove that the triangles are congruent? You would first note that ∠1 and ∠2 are vertical angles. Then you would use the Vertical Angle Theorem to establish that ∠1 ≅ ∠2. Then the two triangles would be congruent by SAS. It was all so very visual.

Very visual

In contrast, A4: ∃x [B(pax) & D(axbc)] has more of an algebraic feeling to it. It's the price we pay for a flawless geometry.

Time for our first theorem!

<u>Theorem 1.1</u> D(aabb)

Here are the axioms we have.

<u>A1</u>: D(abcc) → a = b
<u>A2</u>: D(abba)
<u>A3</u>: D(abpq) & D(abrs) → D(pqrs)
<u>A4</u>: ∃x [B(pax) & D(axbc)]

All of the reasons that we used in proofs in the traditional geometry of the previous chapters are still okay to use:

① Given
② Postulate (which we also call "axiom" in this chapter)
③ Definition (we haven't had any yet)
④ Previously proven theorem
⑤ Algebra

⑥ Beginning of an indirect proof

⑦ Contradiction in steps ___ and ___ and therefore the assumption in step ___ is false.

⑧ Cases (which we used only once in the book when we proved the Inscribed Angle Theorem in chapter ten)

In this flawless geometry, there isn't much use for Algebra as a reason since there are no numbers. The only tiny piece of algebra that we will use is the Substitution Property of Equality. That's a fancy name for an easy idea. If you know that two things are equal, then you can substitute one for the other.

For example, if you know that a = x and you know that B(xyz), then by Algebra you can say that B(ayz).

With all of that said, how do we prove Theorem 1.1 D(aabb)? It is not obvious. Robert L. Moore (and Hilbert and Tarski) who created this modern geometry were all pretty clever mathematicians.

The proof of Theorem 1.1 D(aabb)

Statement	*Reason*
1. B(bax) & D(axbb)	1. A4
2. a = x	2. A1
3. D(aabb)	3. Algebra (Substitution Property of Equality)

This is a good example of where the fun part of mathematics is (and where it isn't). It's not especially fun (at least for me) to read a proof. In reading a proof you have to check that every statement has a proper reason and that the last line of the proof is the thing that you wanted to prove. (These are the two criteria for having a valid proof.)

But the fun part is in the puzzle-solving—in finding the proof in the first place. Don't tell anyone, but what all mathematicians really do all day long is play games. They are challenged with, "Can you figure this out?" or "Can you prove this?" They are not some strange mutation of human beings who enjoy adding up long columns of numbers or balancing their checkbooks.

When you tell a mathematician that the second theorem in this modern geometry is D(abab), the most common response might be, "Don't tell me the proof. Let me try and find it." If you give them the proof before they ask for it, you spoil their fun.

Theorem 1.2: D(abab)

Being an engineer is not the same as being a mathematician. Both are fine professions, but they feel quite different from each other. The engineer, of course, uses the mathematics that mathematicians have created over the last two or three thousand years, but doesn't create new math. The creations of the engineer are *things*. And as long as we have corporeal bodies, we can delight in beautiful buildings, graceful freeways, and well-designed sewage treatment plants.

The labors of an engineer are often much more computational than that of a mathematician. They have *answers* that they have to figure out: How many cubic yards of concrete are needed? or What will be the lateral movement of the bridge under gale-force winds?

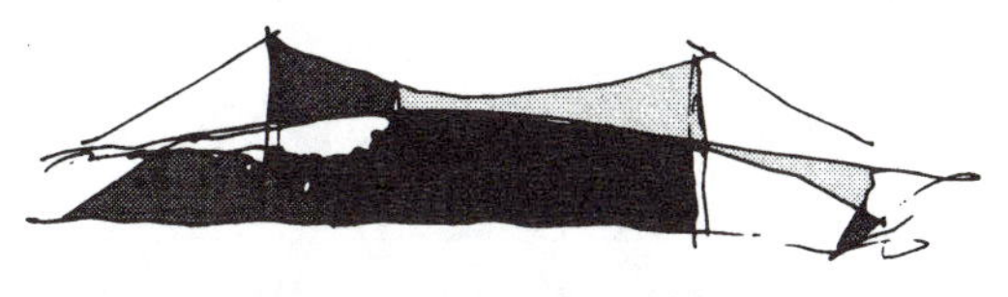

Mathematicians and engineers (and others) sit elbow-to-elbow in calculus for the first two years of college, and then they part company. As juniors, seniors and graduate students, the engineers have textbooks which tend to be filled with numbers and functions from trig and calculus. That's what engineers get in the "divorce."

In contrast, the textbooks for a mathematics major have a lot fewer numbers. It seems that in some upper-division (junior and senior) math books the only numbers are the page numbers. Mathematicians build universes that you can't feel with your hands or see with binoculars. When Aunt Maggie comes to town, the engineer can show her the plastic injection mold factory that he designed and give her a genuine plastic key ring produced by that factory. Aunt Maggie will be so proud.

When Aunt Maggie visits her mathematical nephew who is a mathematician, she will get no trinkets to take home. "So what are you doing?" she'll ask.

You tell her you are trying to prove Theorem 1.2 which states that D(abab).

"So what's that mean?" she asks.

"D(abab) means that the distance from a to b is equal to the distance from a to b," you answer.

"And they pay you to do that?"

"Sure."

"But what can I bring home to show your Uncle Brook? Your engineering cousin gave me this beautiful pink key ring."

You write out the proof of D(abab) and hand it to her.

Statement	*Reason*
1. D(baab)	1. A2
2. D(baab)	2. A2
3. D(abab)	3. A3

She takes it and stuffs it into her purse.

"Isn't it beautiful?" you ask.

"Yeah. Sure. Just like poetry."

Aunt Maggie is closer to the truth than she imagines. The beautiful things of mathematics are those theorems that express the deep order of our world. Just like poetry.

Here's an example of a theorem that expresses the deep order of our world. We need three preliminary steps, each marked with a "✷."

✷ Start with the number e (which is first introduced in trigonometry). It is the number you get when

you take $(1 + 1/n)^n$ and make n infinitely large. e is approximately equal to 2.71828182845904523536028747135266
249775724709369995957496696762772407663035 354759457138
2178525166427427466391932003059921817413596629043572900
3342952605956307381323286279434907632338298807531952510
1901157383418793070215408914993488416750924476146066808
2264800168477411853742345442437107539077744992069551702
7618386062613313845830007520449338265602976067371132007
0932870912744374704723069697720931014169283681902551510
8657463772111252389784425056953696770785449969967946864
4549059879316368892300987931277361782154249992295763514
8220826989519366803318252886939849646510582093923982948
8793320362509443117301238197068416140397019837679320683
28237.

✷ Next, stir in the number π which we learned about in this book. It is the ratio of the circumference of a circle to its diameter. π is approximately equal to 3.14159265
3589793238462643383279502884197169399375105820974944592
3078164062862089986280348253421170679821480865132823066
4709384460955058223172535940812848111745028410270193852
1105559644622948954930381964428810975665933446128475648
2337867831652712019091456485669234603486104543266482133
9360726024914127372458700660631558817488152092096282925
4091715364367892590360011330530548820466521384146951941
5116094330572703657595919530921861173819326117931051185
4807446237996274956735188575272489122793818301194912983
3673362440656643086021394946395224737190702179860943702
7705392171762931767523846748184676694051320005681271452
6356082778577134275778960917363717872146844090122495343
0146549585371050792279689258923542019956112129021960864
0344181598136297747713099605187072113499999983729780499
5105973173281609631859502445945534690830264252230825334
4685035261931188171010003137838752886587533208381420617
1776691473035982534904287554687311595628638823537875937
5195778185778053217122680661300192787661119590921642019
8938095257201065485863278865936153381827968230301952035
3018529689957736225994138912497217752834791315155748572

4245415069595082953311686172785588907509838175463746493
9319255060400927701671139009848824012858361603563707660
1047101819429555961989467678374494482553797747268471040
4753464620804668425906949129331367702898915210475216205
6966024058038150193511253382430035587640247496473263914
1992726042699227967823547816360093417216412199245863150
3028618297455570674983850549458858692699569092721079750
9302955321165344987202755960236480665499119881834797753
5663698074265425278625518184175746728909777727938000816
4706001614524919217321721477235014144197356854816136115
73525552133475741849468438523323907394143334547762416862
5189835694855620992192221842725502542568876717904946016
5346680498862723279178608578438382796797668145410095388
3786360950680064225125205117392984896084128488626945604
2419652850222106611863067442786220391949450471237137869
6095636437191728746776465757396241389086583264599581339
04780275901.

✻ Finally, in our example of a theorem that shows the deep order of the universe, consider the "impossible" square root of –1. In advanced algebra we will call $\sqrt{-1}$ by the name "i."

Now, the theorem itself. In later mathematics we will take e, π and i and prove that: $e^{i\pi}$ equals –1. That's a long way from one of your first theorems in elementary school: 1 + 1 = 2. For many mathematicians, $e^{i\pi} = -1$ is their favorite theorem.

I agree with John L. Casti when he wrote about, ". . . how much closer the mathematician's mind is to that of the poet or artist than to that of the engineer."*

Here's the third theorem in modern geometry.

<u>Theorem 1.3</u>: If D(abcd), then D(cdab).

Using the symbols of logic, this can be written: D(abcd) → D(cdab).

✶ *Mathematical Mountaintops*, p. 1

The first step of the three-step proof will be D(abcd) with "Given" as the reason. You are invited to find the proof before I give it below.

In the meantime, while you're working on the proof, let's consider another aspect of what a mathematician creates. Like good art, a good theorem often has an element of surprise in it. Something new. An "aha" experience.

There were 982 people that attended Fred's sixth birthday party. (*LOF: Trig*, chapter six) At that time Fred might have commented to Betty, who had help plan the party, that he knew that there were at least two people at the party who had exactly the same number of friends at the party.

"How do you know that?" Betty might have asked.

"It's a math theorem," Fred would have answered. "Every mathematician worth his salt can show that: *In any gathering of people, there will always be two people in the crowd that have exactly the same number of friends present.*"*

The very worst way to teach mathematics is to take all the depth and surprise out of the subject and teach it as if it were a cookbook. Do you remember back to your elementary

---

✶ The proof: Given n people at a party. Assume that no two people have the same number of friends at the party. (Beginning of an indirect proof.) The maximum number of friends any person at a party of n people could have would be $n - 1$. That would be the case if he were friends with everyone. The minimum number of friends any person at a party could have would be zero.

If, according to our assumption, no two people had the same number of friends at the party, then one person must have zero friends there, another person must have one friend there, another person must have two friends there, . . . , and finally, one person must have $n - 1$ friends there. Then no two people would have the same number of friends at the party. But the party-goer with $n - 1$ friends is friends with everyone else at the party. Everyone, including the one with zero friends. Contradiction, and hence the beginning assumption is false. Q.E.D.

school days when your teacher taught you how to change a fraction into a decimal? She might have said something like:

"Listen, people. Here is how you change a fraction into a decimal. You start with something like $\frac{7}{8}$ and then you divide the top by the bottom:

```
   0.875
8)7.000
 -6 4
    60
  - 56
     40
   - 40
```

"End of lesson. Here are 40 problems to do tonight.

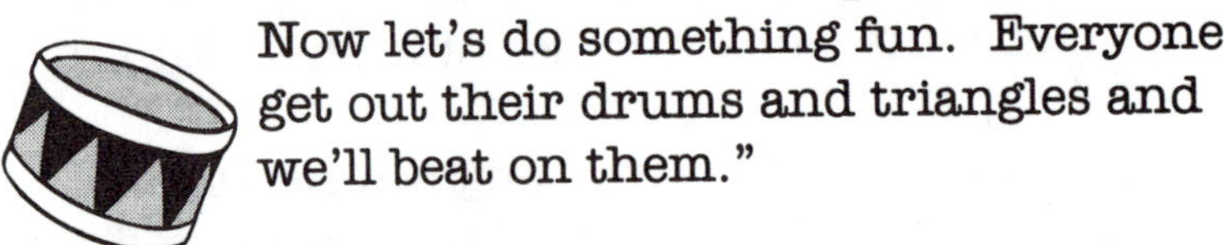

Now let's do something fun. Everyone get out their drums and triangles and we'll beat on them."

Math is b-o-r-i-n-g when it's taught that way. Taught as puzzles to be solved math is much more fun. Here's the solution to the "puzzle" of the proof of

Theorem 1.3: D(abcd) → D(cdab)

Statement	*Reason*
1. D(abcd)	1. Given
2. D(abab)	2. Thm 1.2
3. D(cdab)	3. A3

Theorem 1.4: D(abcd) & D(cdef) → D(abef)

If you were having pizza with a friend, you might trace out the proof on a paper napkin:

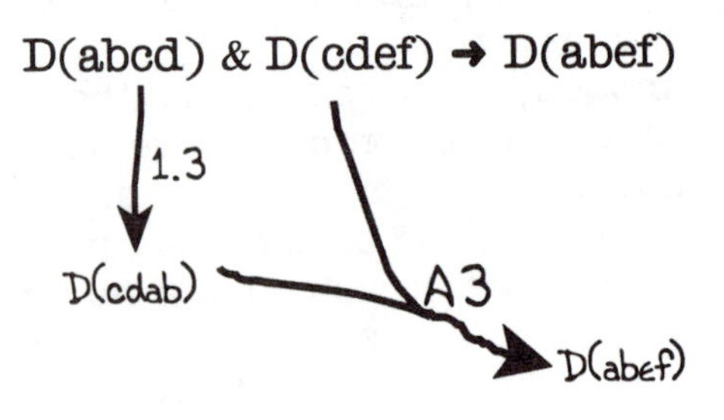

★<u>Theorem 1.5</u>: D(abcd) → D(bacd)*
<u>Theorem 1.6</u>: D(abcd) → D(abdc)

One of the wonderful things about math proofs is that there is often more than one way to do them. They are much less cut-and-dried than solving $3x - 7 = 32$. Almost everyone enjoys the opportunity to be creative.

Here are two different proofs of Theorem 1.6:

Statement	*Reason*
1. D(abcd)	1. Given
2. D(cdab)	2. Thm 1.3
3. D(dcab)	3. Thm 1.5
4. D(abdc)	4. Thm 1.3

Statement	*Reason*
1. D(abcd)	1. Given
2. D(cddc)	2. A2
3. D(abdc)	3. Thm 1.4

THEOREMS ABOUT B(abc)

<u>Theorem 2.1</u>: B(abb)

Statement	*Reason*
1. B(abx) & D(bxbb)	1. A4
2. b = x	2. A1
3. B(abb)	3. Algebra

Right now, you can't prove Theorem 2.2 which states B(abc) → B(cba). Axioms 1–4 are not "strong" enough to give you that result. We need to introduce two more axioms. (There are a total of eleven axioms. You will have them all before the end of this chapter.)

★ Theorems marked with a "★" are ones for which I'm not supplying a proof. In a classroom situation your instructor may wish to use them for either homework or quizzes.

<u>A6</u>: B(aba) → a = b [We'll get to A5 later.]
<u>A7</u>: (Pasch's Axiom)
B(apc) & B(bqc) → ∃x[B(pxb) & B(qxa)]

Just reading Pasch's axiom will probably not make much sense. Drawing a picture will help.

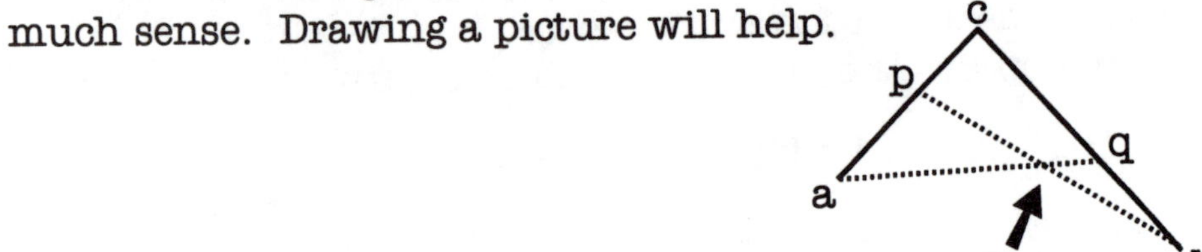

You can't always tell how hard the proof of a theorem will be just by looking. Some hard-looking theorems turn out to be easy to prove and vice versa.

The proof of Theorem 2.2 B(abc) → B(cba) is harder than any of the other proofs we've done so far.

Statement	*Reason*
1. B(abc)	1. Given
2. B(bcc)	2. Thm 2.1
3. B(bxb) & B(cxa)	3. A7 [This may take a little looking to verify that Pasch's Axiom was used correctly. The average time it takes a beginning modern geometry student to verify this step is one minute and 12 seconds.]
4. b = x	4. A6
5. B(cba)	5. Algebra [from steps 3, 4]

★<u>Theorem 2.3</u>: B(aab)
★<u>Theorem 2.4</u>: B(abc) & B(bac) → a = b [Use A7 and A6]
★<u>Theorem 2.5</u>: B(abd) & B(bcd) → B(abc)
★<u>Theorem 2.6</u>: B(abc) & B(acd) → B(bcd) [Use 2.5 and 2.2]

This has given you a little taste of what the flawless modern geometry is like. The proofs have a mechanical feel to them: click, click, click. I've sometimes wondered whether a clever computer programer could write a program that could read a proof and check to see if all the steps were correct.

If the reason supplied was "A2" [which is D(abba)] then the program would have to check if the statement was in the form: D(abba).

D(cvvc)	A2	okay
D(ymmy)	A2	okay
D(rrrr)	A2	okay
D(wxxy)	A2	BEEP! error!

On the other hand, if the reason supplied was "A3" [which is D(abpq) & D(abrs) ➔ D(pqrs)], then the program would have to first check that the statement was in the form D(pqrs), and then it would have to look on the previous lines to find D(abpq) and D(abrs).

It would be a complicated program to write, but once written, it would make the checking of modern geometry proofs very easy.

Once written, it wouldn't be very hard to alter that program so that *it could supply the reasons for each step* in a proof. There are only a finite number of possible reasons for each step in the proof: Given; an Axiom; a Definition; a previously proven Theorem; Algebra; Beginning of an indirect proof; Contradiction; and Cases. All the program would have to do is go through that list one-by-one and see which of the possible reasons matched the statement.

Modern geometry has definitions just as the traditional geometry does. For example, we define three points to be collinear [on the same line] by:

<u>Definition</u>: L(abc) iff B(abc) $\vee$ B(bca) $\vee$ B(cab)

Recall that "$\vee$" means "or."

We define m as the midpoint by:
Definition: M(amb) iff B(amb) & D(ammb)

To deal with triangles we need the fifth axiom:
A5: [The Five Segment Axiom]
a ≠ b &
B(abc) & B(efg) &
D(abef) & D(bcfg) & D(adeh) & D(bdfh)
all imply that D(cdgh)
A diagram would help at this point:

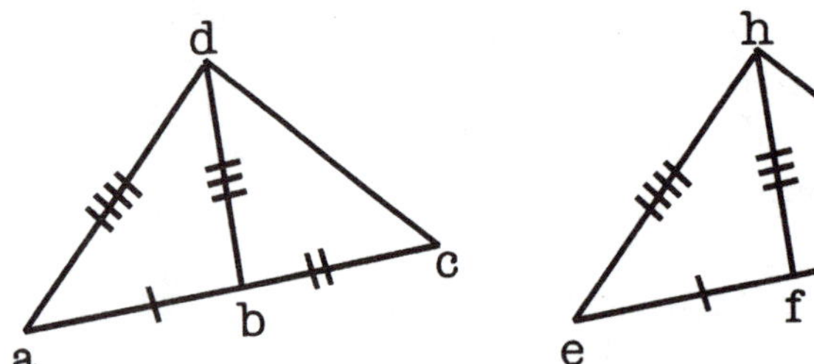

The good old SAS, SSS and ASA postulates begin to look really simple now.

To ensure that not everything in this flawless geometry is all on one line, we need the Lower Dimension Axiom:
A8: $\exists a \exists b \exists c \neg L(abc)$
Recall that "¬" means "not."

To keep everything coplanar, we need the Upper Dimension Axiom. It's really clever how Moore, Hilbert, and Tarski each figured out how to keep everything in two dimensions. They used the ideas from what we know about perpendicular bisectors of segments.
A9: p ≠ q & D(apaq) & D(bpbq) & D(cpcq) → L(abc)

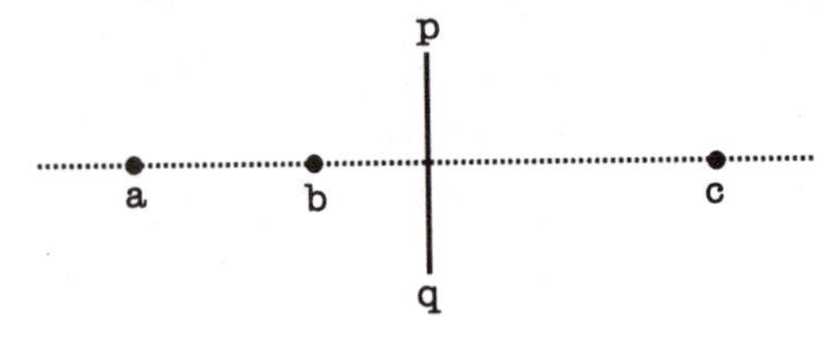

The tenth axiom says that if we have what looks like an "X" [two intersecting segments], then we can wrap a triangle around it.

A10: B(xut) & B(yuz) & x ≠ u →
∃v ∃w (B(xyv) & B(xzw) & B(vtw))

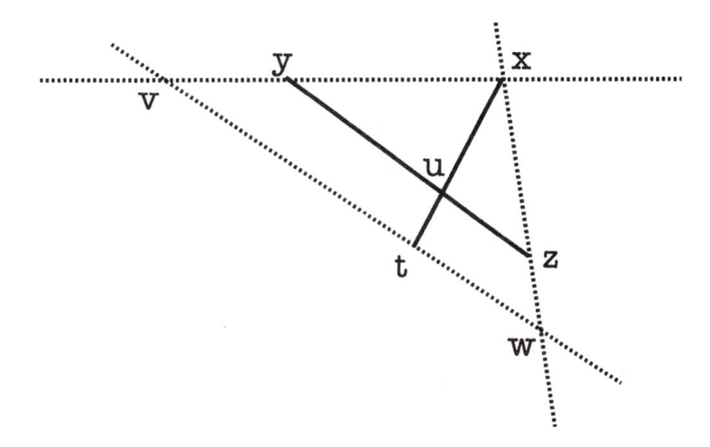

In the traditional Euclidean geometry, we just drew a diagram and found the points v and w. Of course, by "leaning on the diagram" we could arrive at flawed proofs of untrue statements as Joe did earlier in this chapter.

The eleventh axiom is the last axiom for this flawless geometry. It is called the Axiom of Continuity. It states, in essence, that there are no holes in a line, that every line is filled up with points.

You are given a line with two groups of points on it. One group of points we could call set P (the Penguins) and the other group we could call set R (the Rabbits).*

* A set is any collection of objects.

Now suppose that the Penguins and the Rabbits don't intermingle. The Penguins stay with the Penguins, and the Rabbits stay with the Rabbits. If this is the case, then the Axiom of Continuity says that you can find a point on the line between sets P and R.

Suppose point p is any penguin, and suppose that r is any rabbit.

Question: How do you express the condition that the penguins and the rabbits stay in two distinct groupings using either of the two relationships B(- - -) or D(- - - -)?

Answer: You say that there exists a point a on the line so that B(apr) for every p and every r.

Question: How do you express the conclusion that there is a point b between the sets P and R?

Answer: That's easy. You say that there exists a point b so that B(pbr) for every p and every r.

<u>A11</u>: $\exists a\ B(apr) \rightarrow \exists b\ B(pbr)$

If you wanted to toss in lots of symbols, Axiom 11 might read: $\exists a\ \forall p\ \forall r\ [p \in P\ \&\ r \in R \rightarrow B(apr)] \rightarrow$
$\exists b\ \forall p\ \forall r\ [p \in P\ \&\ r \in R \rightarrow B(pbr)]$

"$\forall$" is the logic symbol for "for every" or "for all."

"$\in$" is the set theory symbol for "is an element of" or "is a member of." For example, earth $\in$ the set of all the planets in our solar system.

Other Worlds
Chapter 13½ Flawless Geometry

At this point you have seen the eleven axioms and have seen a half dozen elementary theorems about D(- - - -) and a half dozen theorems about B(- - -). Ahead lie many of the topics that you have already seen in the traditional geometry: triangles, right angles, and parallel lines. The proofs are sometimes much more convoluted than the ones in regular Euclidean geometry.

About half way through the study of the flawless modern geometry is the theorem of the uniqueness of a parallel line through a point not on a given line.

a

"ab ∥ cd" had already been defined, and we arrive at:

Theorem 15.13: ax ∥ cd & ay ∥ cd ➜ L(axy)

The proof is two pages long and one of the four diagrams (used on scratch paper to keep the symbolism straight, but not used in the actual proof) looked like:

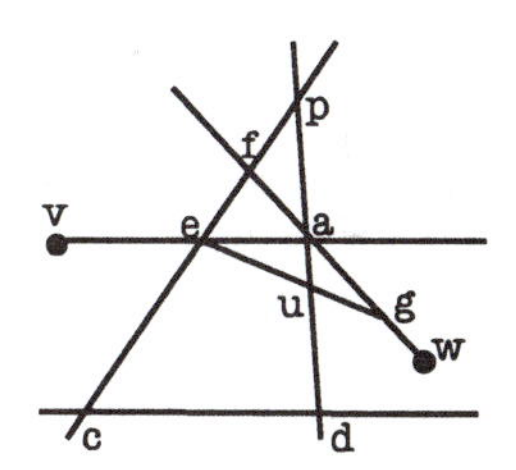

If you are getting the feeling that the price to be paid for this flawless geometry is high, you have learned the most important lesson of this chapter. As a mathematical theory, this is a fairly tough palm tree to climb.

A.R.T.
All Reorganized Together
A Super-condensed Overview of Geometry

Definition 1: H is **between** G and I if and only if GH + HI = GI. (p. 21)

Definition: A set of points is **collinear** iff they lie on the same line. (p. 20 and p. 216)

Definition 2: **Triangle** ABC is defined as non-collinear points A, B and C and the line segments $\overline{AB}$, $\overline{BC}$ and $\overline{CA}$. (p. 21)

Definition 3: The **vertices** of △ABC are A, B and C. They are the "corners" of the triangle. The singular of *vertices* is *vertex*. (p. 22)

Definition: Point M is the **midpoint** of $\overline{AB}$ if and only if AM = MB and A–M–B. (p. 26)

Definition: Two or more lines are **concurrent** if and only if they all share a common point. (p. 23)

A **theorem** is a statement that can be/is being/was proved. (p. 27)

Theorem 1: If P is the midpoint of $\overline{TF}$, then PF = ½ TF. (p. 27)

Postulate 1: One and only one line can be drawn through any two points, or, two points determine a line. Postulates are also called axioms. (p. 33)

A **converse** of an *If ••• then •••* statement interchanges the *If* part with the *then* part. (p. 36)
The converse of *If* P, *then* Q is *If* Q, *then* P.
The converse of P ➔ Q is Q ➔ P.

The **contrapositive** of an *If ••• then •••* statement interchanges the *If* part with the *then* part AND negates both parts. (p. 37)
The contrapositive of P ➔ Q is not-Q ➔ not-P.

A.R.T.
All Reorganized Together

Postulate 2: You can match up every real number with a point on the line.
or
Postulate 2: There is a one-to-one correspondence between the points on a line and the real numbers so that every point matches up with a single real number and every real number matches up with a single point.
or
Postulate 2: There exists a function whose domain is the points on a line and whose codomain is the real numbers such that no two points on the line are mapped to the same real number and every element in the codomain is the image of some point in the domain. (p. 39)

Theorem 2: On any line ℓ, there are five different points on ℓ.
If we have a line ℓ, then there are five different points on ℓ. (p. 41)

A **proposition** is a lightweight or inconsequential theorem. (p. 46)

An if-then statement and its contrapositive must either both be true or both be false. They are said to be **logically equivalent**. (p. 48)

Definition: AB means $|b - a|$. (p. 49)

Definition: **Ray** $\overrightarrow{MP}$ is the set of all points C such that M–C–P or M–P–C. (p. 55)

Definition: $\overrightarrow{AB}$ and $\overrightarrow{AC}$ are **opposite rays** if and only if A is between B and C, or if and only if B–A–C. (p. 55)

Definition: An **angle** is two rays which share a common endpoint. (p. 59)

Postulate 3: (The **Angle Measurement Postulate**) You can match up every angle with a number between 0 and 180, or every angle has a unique measurement which is some number between 0 and 180 degrees inclusive. (p. 59)

Definition: An **acute angle** is an angle whose measure is less than 90°. (p. 60)

A.R.T.
All Reorganized Together

Definition: A **right angle** is an angle whose measure is equal to 90°. (p. 60)

Definition: An **obtuse angle** is an angle whose measure is greater than 90°. (p. 60)

Definition: Two angles, $\angle A$ and $\angle B$, are **congruent** if and only if $m\angle A = m\angle B$. (p. 63)

Definition: $\overline{AB} \cong \overline{CD}$ if and only if $AB = CD$. (p. 64)

There are 60 **minutes** in a degree. In symbols: $60' = 1°$. (p. 65)
There are 60 **seconds** in a minute. In symbols: $60'' = 1'$. (p. 66)

Definition: $\angle A$ and $\angle B$ are **supplementary** if and only if $m\angle A + m\angle B = 180°$. (p. 72)

Definition: $\angle ABC$ and $\angle CBD$ form a **linear pair** if and only if $\overrightarrow{BA}$ and $\overrightarrow{BD}$ are opposite rays. (p. 72)

Definition: Two angles are **vertical angles** if they are formed by intersecting lines and they are not a linear pair,
or
if the sides of two angles form opposite rays, then the angles are called **vertical angles**,
or
if $\angle B$ is formed by taking the opposite rays of $\angle A$, then $\angle A$ and $\angle B$ are **vertical angles**. (p. 72)

Postulate 4: If two angles form a linear pair, then they are supplementary. (p. 73)

Definition: Two angles are **adjacent** if they share a common side and a common vertex. (p. 75)

Theorem 3: Supplements of congruent angles are congruent. (p. 78)

A.R.T.

All Reorganized Together

The only two rules for creating proofs are that every statement have a (proper) reason and the last statement is the thing that you want to prove. (p. 79)

Corollary to Theorem 3: Supplements of the same angle are congruent. (p. 87)

Postulate 5: (Angle Addition Postulate) For any two angles, $\angle AOB$ and $\angle BOC$, if A–B–C, then $m\angle AOB + m\angle BOC = m\angle AOC$. (p. 87)

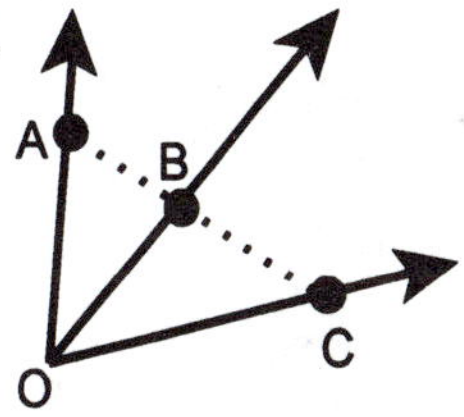

A
B
C
O

Definition: Two lines are **perpendicular** if and only if they form right angles. (p. 88)

Proposition: If two angles form a linear pair and are congruent, then they are right angles. (p. 89)

Definition: $\overrightarrow{OB}$ is the **angle bisector** of $\angle AOC$ if and only if $m\angle AOB = m\angle BOC$. (p. 89)

Definition: A **right triangle** is a triangle with a right angle. The shorter two sides of a right triangle are called the **legs**, and the longest side is called the **hypotenuse**. (p. 92)

Definition: An triangle is an **obtuse triangle** if and only if it contains an obtuse angle. (p. 95)

Definition: An **acute triangle** is a triangle in which all of the angles are acute. (p. 95)

Definition: An **isosceles triangle** is one in which at least two sides are equal in length, or a triangle is **isosceles** if and only if it has at least two congruent sides. (p. 96)

A.R.T.
All Reorganized Together

Definition: A **scalene triangle** is a triangle in which no two sides are congruent. (p. 96)

Definition: ΔABC is **congruent** to ΔDEF if and only if $\angle A \cong \angle D$, $\angle B \cong \angle E$, $\angle C \cong \angle F$, $\overline{AB} \cong \overline{DE}$, $\overline{BC} \cong \overline{EF}$, and $\overline{AC} \cong \overline{DF}$. (p. 99)

Postulate 6: (SSS) If $\overline{AB} \cong \overline{DE}$, $\overline{BC} \cong \overline{EF}$, and $\overline{AC} \cong \overline{DF}$, then $\Delta ABC \cong \Delta DEF$. (p. 101)

Postulate 7: (SAS) If $\overline{AB} \cong \overline{DE}$, $\overline{AC} \cong \overline{DF}$, and $\angle A \cong \angle D$, then $\Delta ABC \cong \Delta DEF$. (p. 103)

Postulate 8: (ASA) If $\angle A \cong \angle D$, $\angle C \cong \angle F$, and $\overline{AC} \cong \overline{DF}$, then $\Delta ABC \cong \Delta DEF$. (p. 104)

The Isosceles Triangle Theorem: The base angles of an isosceles triangle are congruent. (p. 108)

The converse of the Isosceles Triangle Theorem: If the base angles of a triangle are congruent, then the triangle must be isosceles. (p. 108)

Definition: A triangle is **equilateral** if and only if all three sides are congruent. (p. 110)

Definition: A triangle is **equiangular** if and only if all three of its angles are congruent. (p. 110)

Definition: A **median** is the segment in a triangle whose endpoints are a vertex of the triangle and the midpoint of the opposite side. (p. 111)

Definition: Two lines are **parallel** iff they don't intersect. (p. 117)

Postulate 9: (The Parallel Postulate) If you have a line ℓ and a point P not on ℓ, then there is at most one line through P that is parallel to ℓ. (p. 125)

Definition: An **exterior angle** of ΔABC is $\angle BCD$ where A–C–D. (p. 132)

A.R.T.

All Reorganized Together

Theorem: (The Exterior Angle Theorem) The measure of an exterior angle of a triangle is greater than the measure of either remote interior angle. (p. 132)

Theorem: (AI → P) If two lines are cut by a transversal and the alternate interior angles are congruent, then the lines are parallel. (p. 135)

Theorem: If you have a line ℓ and a point P not on ℓ, then there is at least one line through P that is parallel to ℓ. (p. 137)

Theorem: (CAP) If two lines cut by a transversal form congruent corresponding angles, then the lines are parallel. (p. 138)

Theorem: (P → AI) If two parallel lines are cut by a transversal, then the alternate interior angles must be congruent. (p. 139)

Theorem: (The Transitive Property of Congruent Angles) If $\angle 2 \cong \angle 1$ and $\angle 1 \cong \angle 3$, then $\angle 2 \cong \angle 3$. (p. 140)

Theorem: The sum of the angles of any triangle add to 180°. (p. 144)

Theorem: No triangle can contain two right angles. (p. 152)

Theorem: If P is not on ℓ, then there is at most one perpendicular from P to ℓ. (p. 152)

Definition: Two angles are **complementary** iff their measures add to 90°. (p. 153)

Theorem: The measures of each of the angles of an equilateral triangle are equal to 60°. (p. 154)

Theorem: If one pair of alternate interior angles is congruent, then the other pair must also be congruent. (p. 154)

Theorem: (AA → AAA) If two angles of one triangle are congruent, respectively, to two angles of a second triangle, then the third angles of each triangle must be congruent. (p. 155)

A.R.T.
All Reorganized Together

<u>Theorem</u>: (AAS) In triangles ABC and DEF, if $\angle A \cong \angle D$, $\angle B \cong \angle E$, and $\overline{BC} \cong \overline{EF}$, then the triangles are congruent. (p. 155)

<u>Theorem</u>: If two lines are perpendicular to the same line, they are parallel. (p. 160)

<u>Theorem</u>: If a line is perpendicular to one of two parallel lines, then it is perpendicular to the other. (p. 161)

<u>Proposition</u>: (Erect a perpendicular) From every point on a line, there is a perpendicular to that line through that point. (p. 162)

<u>Lemma</u>: If two lines intersect and form a linear pair of congruent angles, then the lines are perpendicular. (p. 164)

<u>Theorem</u>: (Drop a perpendicular) Given a point P not on ℓ, there is at least one perpendicular from P to ℓ. (p. 162)

<u>Theorem</u>: (Hypotenuse-Leg Theorem) Two right triangles are congruent if their hypotenuses and one pair of corresponding sides are congruent. (p. 167)

<u>Lemma</u>: If two lines intersect and form one right angle, then all four angles are right angles and they are all congruent to each other. (p. 171)

<u>Definition</u>: Line ℓ is the perpendicular bisector of $\overline{AB}$ iff it is perpendicular to $\overline{AB}$ and it bisects $\overline{AB}$ (passes through the midpoint of $\overline{AB}$). (p. 173)

<u>Theorem</u>: (Cupid's Bow Theorem) Any point on the perpendicular bisector of a segment is equidistant from the endpoints of the segment. (p. 173)

<u>Theorem</u>: (Converse of Cupid's Bow Theorem) If any point is equidistant from the endpoints of a segment, then it must lie on the perpendicular bisector. (p. 174)

A.R.T.
All Reorganized Together

Theorem: (The Perpendicular Bisector Theorem which is a combination of Cupid's Bow and its converse) A point is on the perpendicular bisector of a segment iff it is equidistant from the endpoints of the segment. (p. 175)

Definition: The **foot** of the perpendicular from P to ℓ is the point where the perpendicular from P to ℓ intersects ℓ. (p. 176)

Definition: The **distance from P to ℓ** is the distance from P to the foot of the perpendicular from P to ℓ. (p. 176)

Theorem: Any point that is equidistant from the sides of an angle is on the angle bisector. (p. 178)

Proposition: In an isosceles triangle, the segment from the vertex which is perpendicular to the opposite side is also a median. (p. 179)

Proposition: If two points, P and Q, are each equidistant from the endpoints of $\overline{AB}$, then $\overleftrightarrow{PQ}$ is the perpendicular bisector of $\overline{AB}$. (p. 180)

Proposition: If $\angle A$ and $\angle B$ are right angles, then $\angle A \cong \angle B$. (p. 180)

Proposition: In ΔABC, if $\overline{CQ}$ is a median and if $\overrightarrow{CQ}$ is also the angle bisector of $\angle ACB$, then ΔABC is isosceles. (p. 180)

Proposition: In an isosceles triangle, the angle bisector is perpendicular to the opposite side. (p. 181)

$P \vee Q$. This is called the "non-exclusive or." (p. 185)

"➔" is the symbol for implies. (p. 36 and p. 152)

The Honors Problem of the Century: Prove that in any triangle, if the angle bisectors of two angles are congruent, then the triangle is isosceles. (p. 190)

Theorem: The opposite sides of a parallelogram are congruent. (p. 195)

A.R.T.

All Reorganized Together

Theorem: ▱ABCD implies that $\triangle ABC \cong \triangle CDA$, namely that a diagonal of a parallelogram divides it into two congruent triangles. (p. 195)

Theorem: The opposite angles of a parallelogram are congruent. (p. 195)

Theorem: ( )If the opposite sides of a quadrilateral are congruent, then it is a ▱. (p. 198)

Theorem: ( )If one pair of opposite sides of a quadrilateral are both parallel and congruent, then it is a ▱. (p. 198)

Theorem: ( )If both pairs of opposite angles of a quadrilateral are congruent, then it is a ▱. (p. 198)

Definition: A **midsegment** is a segment joining the midpoints of two sides of a triangle. (p. 204)

Midsegment Theorem: In any triangle, a midsegment is parallel to the third side and is equal to half of the length of the third side. (p. 205)

Converse of the Midsegment Theorem: In any triangle, a line which passes through the midpoint of one side and is parallel to the third side, must pass through the midpoint of the second side. (p. 205)

Transitive Property of Congruent Segments Theorem: If $\overline{PQ} \cong \overline{RS}$ and if $\overline{RS} \cong \overline{TU}$, then $\overline{PQ} \cong \overline{RS}$. (p. 207)

Theorem: (Hecks Kitchen Theorem a.k.a. Spaghetti Theorem): If three parallel lines intercept congruent segments on one transversal, they will intercept congruent segments on any transversal. (p. 215)

Theorem (Testing the Guarantee Theorem): If three lines intercept congruent segments on two transversals and two of the lines are parallel, then all three lines are parallel. (p. 216)

A.R.T.
All Reorganized Together

Proposition: The diagonals of a parallelogram bisect each other. (p. 222)

Proposition: If the diagonals of a quadrilateral bisect each other, then the quadrilateral is a parallelogram. (p. 222)

Proposition: Connect the midpoints of the sides of any quadrilateral and you get a parallelogram. (p. 224)

Proposition: The base angles of an isosceles trapezoid are congruent. (p. 225)

Proposition: The diagonals of an isosceles trapezoid are congruent. (p. 225)

Proposition: In any quadrilateral, the segments joining the midpoints of the opposite sides bisect each other. (p. 227)

Proposition: If the diagonals of a quadrilateral are perpendicular to each other and bisect each other, then it is a rhombus. (p. 228)

$A = mh$ the area formula for triangles, trapezoids, rectangles, rhombuses, and squares where m is the length of the midsegment and h is the height. (p. 243)

Definition: A **polygon** is a triangle, quadrilateral, pentagon, hexagon, etc. or, more formally, a polygon with n sides is obtained by first taking n points, $A_1, A_2, A_3, \ldots, A_{n-1}, A_n$ that are all distinct (different from each other). Then a polygon is defined as the union of the segments $\overline{A_1A_2}$, $\overline{A_2A_3}$, $\overline{A_3A_4}$, . . . , $\overline{A_{n-1}A_n}$, and $\overline{A_nA_1}$ provided that none of the segments intersect except at their endpoints. (p. 246)

The **vertices** of the polygon are $A_1, A_2 \ldots A_{n-1}, A_n$.

Postulate 10: (Area Postulate) Every polygon has a number attached to it which we will call its area. These areas will be positive numbers. (p. 247)

A.R.T.
All Reorganized Together

Postulate 11: If two triangles are congruent, then their areas are equal. (In symbols: If $\Delta ABC \cong \Delta DEF$, then $a\Delta ABC = a\Delta DEF$. "$a\Delta$" is the area associated with the triangle. We did the same thing with angles: $m\angle A$ is the angle measurement associated with $\angle A$.) (p. 247)

Postulate 12: (Area Addition Postulate) The area of a polygon is the sum of the areas of the (non-overlapping) triangles inside of it, and it doesn't matter which way you cut up the polygon into triangles. (p. 247)

Postulate 13: The area of a square is the length of a side squared. $A = s^2$. (p. 258)

Theorem: The area of a rectangle of width w and length l is given by the formula $A = lw$. (p. 262)

The Pythagorean Theorem: In any right triangle, $a^2 + b^2 = c^2$. (p. 267)

Heron's formula: The area of a triangle whose sides are a, b and c is equal to $A = \sqrt{s(s-a)(s-b)(s-c)}$ where $s = \frac{1}{2}(a + b + c)$. (p. 271)
The length s is called the **semiperimeter**.

The **triangle inequality:** In any triangle whose sides are a, b and c, the lengths of any two sides must always be greater than the length of the third side: $a + b > c$. (p. 273)

Converse of the Pythagorean Theorem: In any triangle, if $a^2 + b^2 = c^2$, then it must be a right triangle. (p. 274)

Definition: ΔABC is **similar** to ΔDEF iff
$\angle A \cong \angle D$, $\angle B \cong \angle E$, $\angle C \cong \angle F$, and $\frac{AB}{DE} = \frac{BC}{EF} = \frac{CA}{FD}$ (p. 293)

Postulate 14: ($AA \rightarrow \sim \Delta$s) ΔABC is similar to ΔDEF if $\angle A \cong \angle D$ and $\angle B \cong \angle E$. (p. 294)

Definition: Proportions are two ratios set equal to each other: $\frac{a}{b} = \frac{c}{d}$ (p. 296)

A.R.T.
All **R**eorganized **T**ogether

Theorem: (The Generalization of the Converse of the Midsegment Theorem or the GCM Theorem) A line parallel to one side of a triangle intercepts proportional segments on the other two sides, assuming, of course, that it hits them. (p. 299)

Theorem: (The Converse of the Generalization of the Converse of the Midsegment Theorem or, more simply, Generalization of the Midsegment Theorem) If a line intercepts proportional segments on two sides of a triangle then it is parallel to the third side.

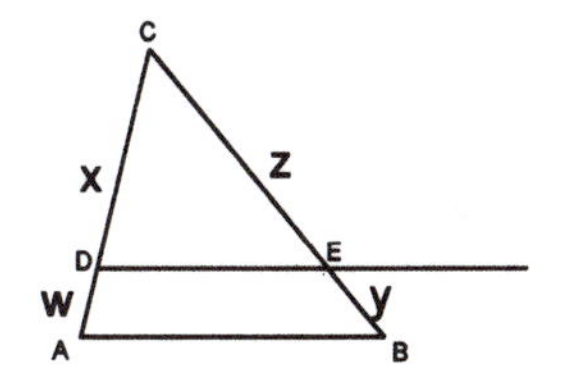

If $\frac{x}{w} = \frac{z}{y}$ then $\overrightarrow{DE} \parallel \overline{AB}$. (p. 299)

Lemma: (for the SAS Similarity Theorem): If the first triangle is similar to the second, and the second is congruent to the third, then the first must be similar to the third. (p. 304)

Theorem: (SAS Similarity) Two triangles are similar if one pair of corresponding angles are congruent and the including sides are proportional. (p. 303)

Theorem: (SSS Similarity) Two triangles are similar if their respective sides are proportional. (p. 304)

Definition: An **altitude** of a triangle is a segment from the vertex drawn perpendicular to the opposite side. (p. 305)

Theorem: In similar triangles, the altitudes are proportional to the corresponding sides. (p. 306)

Theorem: (The Angle Bisector Theorem) In any triangle, an angle bisector divides the opposite side in the same ratio as the other two sides. (p. 309)

Theorem: In similar triangles the areas are proportional to the square of the ratios of corresponding sides. (p. 310)

Proposition: (Transitive Property of Similar Triangles) If $\Delta ABC \sim \Delta DEF$ and if $\Delta DEF \sim \Delta GHI$, then $\Delta ABC \sim \Delta GHI$. (p. 311)

A.R.T.
All Reorganized Together

<u>Theorem:</u> Three parallel lines intercept proportional segments on any two transversals. (p. 311)

<u>Converse of the Angle Bisector Theorem</u>: If a segment from the vertex of a triangle to the opposite side divides the side in the same ratio as the other two sides, then it is an angle bisector. (p. 312)

<u>Proposition</u>: In similar triangles, corresponding angle bisectors are proportional to corresponding sides. (p. 314)

<u>Proposition</u>: In similar triangles, corresponding medians are proportional to corresponding sides. (p. 315)

(The Transitive Property of ➔) If you know that P ➔ Q and if you know that Q ➔ R, then you know that P ➔ R. ((P ➔ Q) & (Q ➔ R)) ➔ (P ➔ R) (p. 319)

<u>Theorem</u>: In a right triangle, the altitude to the hypotenuse divides the triangle into three similar triangles. (p. 326)

<u>Definitions</u>: Given a positive number r and a point O, a **circle** is the set of all points at a distance *r* from O. The distance *r* is called the **radius**. The point O is called the **center of the circle**. (p. 350)

<u>Definitions</u>: A **chord** of a circle is a segment whose endpoints are on the circle. A **diameter** is a chord which passes through the center of the circle. A **secant** is a line which contains a chord. A **tangent** is a line which intersects a circle at only one point. (p. 350)

<u>Definition</u>: A point P is in **the interior of ∠A** if it lies on a segment whose endpoints are on the sides of ∠A. (p. 351)

A.R.T.
All Reorganized Together

Definition: Two circles are **concentric** if they have the same center. (p. 351)

Theorem: If a radius is perpendicular to a chord, it bisects it. (p. 352)

Theorem: A tangent is perpendicular to a radius drawn to the point of tangency. (p. 352)

Definition: An angle is a **central angle** of circle O, if its vertex is at O. (p. 355)

Definition: A **minor arc** is all the points of a circle that are on the interior of a central angle. A **major arc** is all the points of a circle that are not on a minor arc. (p. 356)

Definition: A **half plane whose edge is ℓ** is the set consisting of some point P, not on ℓ, together with all other points Q such that $\overline{PQ}$ does not intersect ℓ. (p. 356)

Definition: **On one side of a line ℓ** is defined to be a half plane whose edge is ℓ. (p. 356)

Definition: A **semicircle** is the set of all points of a circle that are on one side of a line that contains a diameter. (p. 356)

Definition: The **measure of minor arc** $\overset{\frown}{AB}$ in circle O (written $m\overset{\frown}{AB}$) is equal to $m\angle AOB$. (p. 357)

Definition: The measure of a semicircle as 180° and the **measure of a major arc** as 360° minus the measure of its corresponding minor arc. (p. 357)

Definition: An $\angle BAC$ is an **inscribed angle** of a circle iff A, B and C are on the circle. (p. 357)

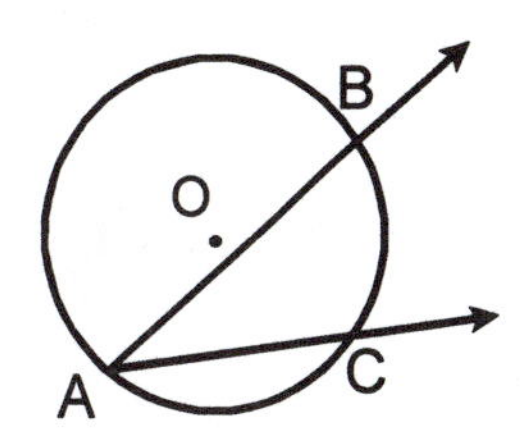

Theorem: (Inscribed Angle Theorem) The measure of an inscribed angle is half of the measure of the intercepted arc. (p. 358)

A.R.T.
All Reorganized Together

Theorem: (The Intersecting Chords Theorem) If any two chords of a circle intersect, then ab = cd. (p. 373)

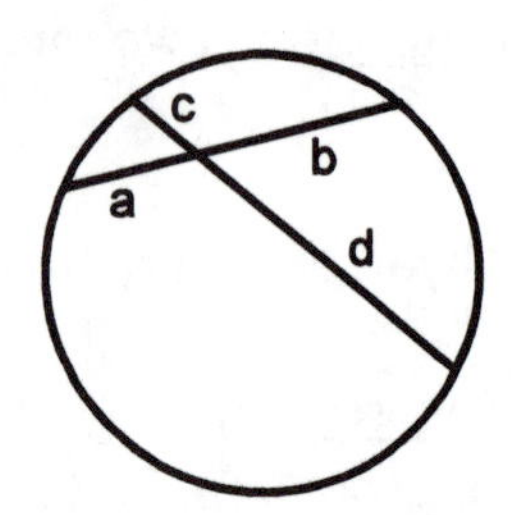

Generalized Intersecting Chords Theorem: If two chords, or their extensions, intersect, then the product of the distances from the endpoints of one of the chords to the point of intersection is equal to the product of the distances from the other chord to the point of intersection. (p. 374)

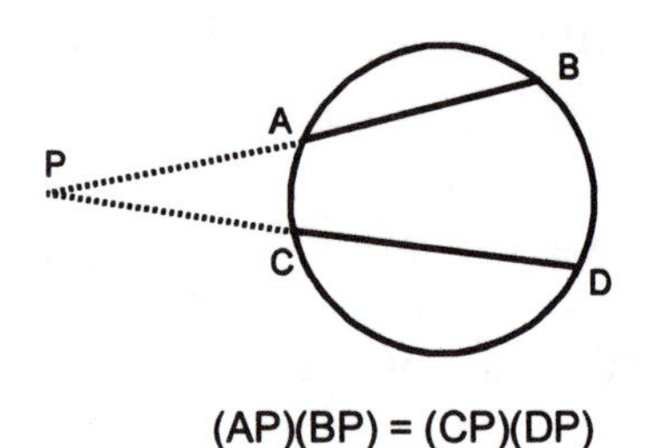

(AP)(BP) = (CP)(DP)

Proposition: If a radius bisects a chord, it is perpendicular to that chord. (p. 382)

Proposition: The perpendicular bisector of a chord passes through the center of the circle. (p. 383)

Proposition: No circle contains three collinear (all-on-the-same-line) points. (p. 383)

Proposition: Two distinct circles can intersect in at most two different points. (p. 383)

Proposition: Three noncollinear points can lie on at most one circle. (p. 383)

Theorem: If a line ℓ is perpendicular to a radius $\overline{OA}$ at A, then ℓ is tangent to the circle. (p. 384)

Theorem: Two chords (in the same circle) are congruent if the measures of the arcs associated with those chords are equal. (p. 385)

A.R.T.
All Reorganized Together

Theorem: If two chords (in the same circle) are congruent, then the measures of the arcs associated with those chords are equal. (p. 385)

Proposition: The two tangents drawn from a point outside a circle are congruent. (p. 386)

Lemma: Midpoints of congruent segments divide them into congruent segments. (p. 387)

Proposition: If two chords are congruent, they are equidistant from the center of the circle. (p. 387)

Proposition: If two chords are equidistant from the center of the circle, they are congruent. (p. 387)

Construction #1: Copy a segment. (p. 392)

Construction #2: Bisect a segment. (Also known as Find the Midpoint.) Also good for constructing the perpendicular bisector of a segment. (p. 394)

Construction #3: Copy an angle. (p. 395)

Construction #4: Construct a triangle given the three sides. (p. 395)

Construction #5R: Construct an equilateral triangle. (Also construct a 60° angle.) (p. 395)

Constructions #6: Construct a parallel line to a given line through a given point not on the line. (p. 395)

Construction #7R: Bisect an angle. (p. 397)

Construction #8: Construct a triangle given two angles and the included side. (p. 397)

Construction #9: Construct a triangle given two sides and the included angle. (p. 397)

A.R.T.
All Reorganized Together

Construction #10: Construct an isosceles triangle given the vertex angle (that's the angle that is not a base angle) and a segment equal to the sum of the two congruent sides. (p. 397)

Construction #11: Construct an isosceles triangle given the vertex angle and the length of the base. (p. 397)

Construction #12: Construct a square. (p. 401)

Construction #13: Construct a golden rectangle. (p. 401)

Construction #14: Erect a perpendicular to ℓ at a given point P on ℓ. (p. 403)

Construction #15: Trisect a segment. (p. 403)

Construction #16: Divide a given segment into 273 equal parts. (p. 403)

Construction #17: Construct a triangle given sides a, b, and the altitude h to side a. (p. 403)

Construction #18: Given three noncollinear points, construct the circle which passes through them and locate the center of the circle. (p. 403)

Construction #19R: Erect a perpendicular to ℓ at a given point P on ℓ. (p. 404)

Construction #20R: Bisect $\overline{AB}$ if $r > AB/2$ where r is the radius of the rusty compass. (p. 404)

Construction #21R: Bisect $\overline{AB}$ if $r < AB/2$. (p. 404)

Construction #22R: Drop a perpendicular from P to ℓ when $r >$ the distance from P to ℓ. (p. 404)

Construction #23R: Drop a perpendicular form P to ℓ when $r <$ the distance from P to ℓ. (p. 404)

A.R.T.
All Reorganized Together

<u>Construction #24R</u>: Construct a line parallel to ℓ through given point P. (p. 404)

<u>Construction #25R</u>: Construct the complement to a given angle. (Two angles are complementary if their measures add to 90°.) (p. 404)

<u>Construction #26R</u>: Locate the center of a circle. (p. 404)

<u>Construction #27</u>: Inscribe a square in a given circle. (p. 406)

<u>Construction #28</u>: Draw a tangent to a circle at a given point on the circle. (p. 406)

<u>Construction #29</u>: Draw a tangent to a circle from a point P which lies outside the circle. (p. 406)

<u>Construction #30</u>: Given a circle and a point P outside the circle, draw a circle with center at P which is tangent to the given circle. (p. 406)

<u>Construction #31</u>: Inscribe a regular hexagon (6 congruent sides) in a given circle. (p. 406)

<u>Construction #32</u>: Inscribe a regular octagon (8 congruent sides) in a given circle. (p. 406)

<u>Construction #33</u>: Given two circles of different radii, draw a tangent to both circles. (p. 406)

<u>Construction #34</u>: Given a triangle, inscribe a circle in it. (p. 408)

<u>Definition</u>: The center of the inscribed circle is called the **incenter** of the triangle. (p. 408)

A.R.T.
All Reorganized Together

Construction #35: Construct an isosceles triangle given the base of the triangle and the radius of the inscribed circle. (p. 408)

Construction #36: Given a triangle, circumscribe a circle around it. (p. 409)

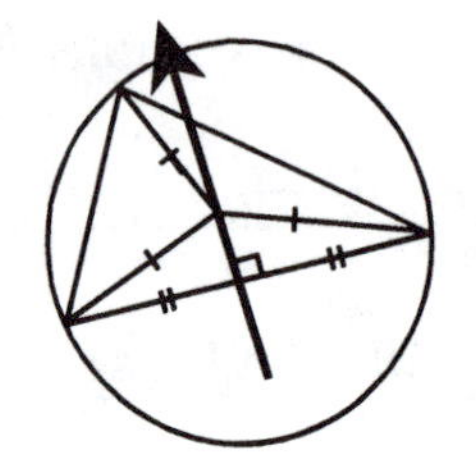

Definition: The center of the circumscribed circle is called the **circumcenter** of the triangle. (p. 409)

Construction #37: Construct an isosceles triangle given the base and the radius of the circumscribed circle. (p. 409)

Theorem: The angle bisectors of a triangle are concurrent. (p. 409)

Theorem: The perpendicular bisectors of the sides of a triangle are concurrent. (p. 409)

Theorem: The altitudes of any triangle are concurrent. (p. 409)

Theorem: The medians of any triangle are also concurrent. (p. 409)

Construction #38: Construct a right triangle given the hypotenuse and one leg. (p. 412)

Construction #39: Construct a right triangle given the median and the altitude to the hypotenuse. (p. 412)

Construction #40: Construct the geometric mean to two given segments. (p. 412)

Construction #41: Construct an isosceles triangle given the altitude to the base and the perimeter. (p. 412)

A.R.T.
All Reorganized Together

The rules for **collapsible compass constructions** are: ① You can take any two points and use one of them for the center and the other for the radius; and ② the compass collapses the second that you pick it up off the paper. That means that you can't take a distance AB and make a circle at point O with radius AB. (p. 412)

Collapsible Compass Construction #42CC: Copy a segment. (p. 412)

Construction #43: Construct a 15° angle. (p. 414)

Construction #44: Construct a 75° angle. (p. 414)

Construction #45: Construct a 24-gon (a regular polygon with 24 congruent sides). (p. 414)

Construction #46: Construct a regular heptadecagon (17-gon). (p. 414)

Theorem: If two lines intersect, then they lie in the same plane. (p. 428)

Postulate 15: Three noncollinear points determine a plane, or given any three noncollinear points, there is exactly one plane which contains them. (p. 429)

Definition: Two lines are **skew** iff they are neither parallel nor intersecting. (p. 429)

Postulate 16: (The Flat Plane Postulate) If two points lie in a plane, then the line which contains them also lies in the plane. (p. 429)

Definition: A **line is perpendicular to a plane** iff it is perpendicular to every line in the plane that intersects it. (p. 432)

Definition: The **distance from a point to a plane** is the length of the perpendicular from the point to the plane. (p. 432)

Proposition: Two lines perpendicular to the same plane are parallel. (p. 432)

A.R.T.
All **R**eorganized **T**ogether

Definition: **Two planes are parallel** iff they don't intersect. (p. 432)

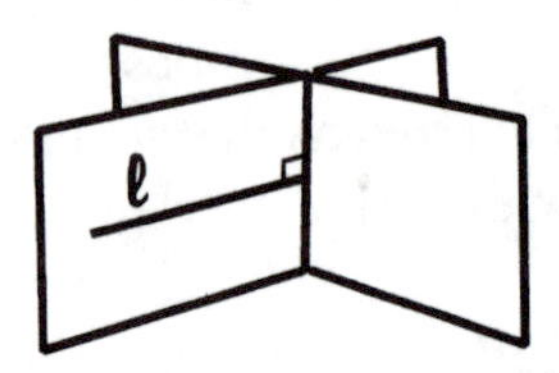

Definition: **Two planes are perpendicular** iff one plane contains a line that is perpendicular to the other plane. (p. 432)

Euler's Theorem: For any polyhedron, F + V – 2 = E where F = number of faces; V = number of vertices; and E = number of edges. (p. 437)

The volume of a cylinder of radius r and height h is $\pi r^2 h$. (p. 440)

The volume for any prism is the same as the volume for a cylinder: base times height. (p. 441)

Polyhedrons are solids with flat faces. (An informal definition.) (p. 434)

Definition: A **prism** is a polyhedron in which two of the faces (called the **bases of the prism**) are congruent polygons lying in parallel planes and in which all of the vertices of the polyhedron are vertices of the two congruent polygons. (p. 441)

Definition: A **pyramid** is a polyhedron in which all but one of the vertices are coplanar. (p. 444)

The volume of a pyramid in a prism is 1/3 × base × height. (p. 444)

A sphere always occupies exactly two-thirds of the volume of the cylinder which contains it. $V_{sphere} = (4/3)\pi r^3$ (p. 449)

The surface area of a sphere is exactly equal to the two-thirds of the cylinder which circumscribes it. Surface $\text{Area}_{sphere} = 4\pi r^2$. That means that the surface area of a sphere is exactly equal to four times the area of a circle which passes through the center of the sphere. (p. 450)

Definition: **Concurrent planes** are planes that all share a common line. (p. 451)

A.R.T.
All Reorganized Together

The distance between (x_1, y_1) and (x_2, y_2) is $d = \sqrt{(x_2 - x_1)^2 + (y_2 - y_1)^2}$. (p. 477)

The midpoint between (x_1, y_1) and (x_2, y_2) is $((x_1 + x_2)/2, (y_1 + y_2)/2)$. (p. 476)

The slopes of perpendicular lines are negative reciprocals of each other. (Reciprocal means turn upside down. The reciprocal of 3/4 is 4/3.) (p. 476)

The equation of a circle of radius r, centered at (a, b) is $(x - a)^2 + (y - b)^2 = r^2$. (p. 481)

Definition: **Locus of points** is the set of points satisfying a given condition. (p. 483)

Index